György Keglevich (Ed.)

Organophosphorus Chemistry

Also of Interest

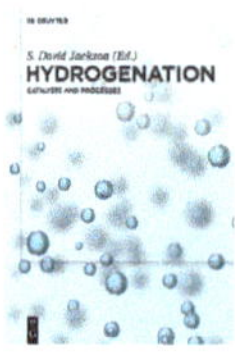

Hydrogenation.
Catalysts and Processes
Jackson (Ed.), 2018
ISBN 978-3-11-054373-5, e-ISBN 978-3-11-054521-0

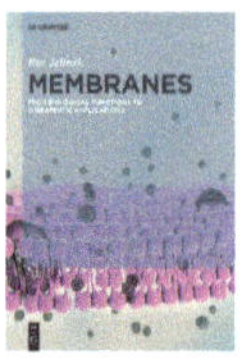

Membranes.
From Biological Function to Therapeutic Applications
Jelinek, 2018
ISBN 978-3-11-045368-3, e-ISBN 978-3-11-045369-0

New-Generation Bioinorganic Complexes.
Jastrzab, Tylkowski (Eds.), 2016
ISBN 978-3-11-034880-4, e-ISBN 978-3-11-034890-3

Modern X-Ray Analysis on Single Crystals.
A Practical Guide
Luger, 2014
ISBN 978-3-11-030823-5, e-ISBN 978-3-11-030828-0

Physical Sciences Reviews.
e-ISSN 2365-659X

Organophosphorus Chemistry

Novel Developments

Edited by
György Keglevich

DE GRUYTER

Editor
Prof. Dr. György Keglevich
Professor of Chemistry
Budapest University of Technology and Economics
Dept of Organic Chemistry and Technology
PO Box 91
Budapest 1521
Hungary

ISBN 978-3-11-053453-5
e-ISBN (PDF) 978-3-11-053583-9
e-ISBN (EPUB) 978-3-11-053457-3

Library of Congress Control Number: 2018935092

Bibliographic information published by the Deutsche Nationalbibliothek
The Deutsche Nationalbibliothek lists this publication in the Deutsche Nationalbibliografie; detailed bibliographic data are available on the Internet at http://dnb.dnb.de.

Cover image: György Keglevich and coworkers
Typesetting: Integra Software Services Pvt. Ltd.
Printing and binding: CPI books GmbH, Leck
♾ Printed on acid-free paper
Printed in Germany

www.degruyter.com

Preface

These days organophosphorus chemistry is one of the most dynamically developing fields of organic and organometallic chemistry. Organophosphorus compounds may be useful starting materials, intermediates and reagents (see the Michaelis–Arbuzov reaction, the Pudovik reaction, the Kabachnik–Fields condensation, the Wittig- and the Wadsworth–Emmons reactions, the Mitsunobu reaction, the Appel reaction, and the Hirao reaction) in synthetic organic chemistry. Phosphonium salts are applied as ionic liquids, tenzides or phase transfer catalysts. Achiral and chiral P-ligands are the components of transition metal (Pt, Pd and Rh, *etc.*) complexes that may serve as homogeneous catalysts. Further exciting fields are the discipline of optically active phosphines and phosphine oxides, as well as related derivatives, P-heterocyclic chemistry, and biologically active substrates used as plant protecting agents or pharmaceuticals.

This book wishes to give a snap-shot on the recent developments and trends in organophosphorus chemistry. Chapters 1 and 2 survey the typical and novel synthetic methods for the preparation of phosphine oxides and phosphinates along with phosphonates, respectively. Chapter 3 summarizes the present knowledge on biologically active organophosphorus compounds that is of special importance due to changes in the regulations in the use of plant protecting agents by the EU and US. In Chapter 4, optical resolutions that are the most suitable methods for the preparation of optically active phosphine oxides are discussed. Chapters 5 and 6 give insights into the synthesis of α-hydroxyphosphonates and α-aminophosphonates, respectively, as biologically active substrates. The utilization of a versatile reagent, cyclic tri(propylphosphonate) (the T3P® reagent) in the synthesis of phosphinic and phosphonic derivatives is demonstrated in Chapter 7, while Chapter 8 gives an overview on a fashionable P–C coupling reaction called the Hirao reaction with special stress on the green chemical accomplishments. The deoxygenation of phosphine oxides (Chapter 9) is important for two reasons: 1.) to be able to recycle the phosphine oxide by-products, 2.) to provide P(III) ligands. The properties and synthesis of α-hydroxymethylenebisphosphonic acid (dronic acid) derivatives used against drug diseases are surveyed in Chapter 10. Chapter 11 reveals an important aspect of environmentally-friendly chemistry, as presents organophosphorus transformations that may be performed at ambient conditions. Chapter 12 comprises the recent results of P-heterocyclic chemistry introducing new heterocyclic systems. Microwave-assisted syntheses related to the organophosphorus field are included in Chapters 2, 5, 6, 8 and 9, while the ultrasound-assisted syntheses were accommodated in Chapter 13. Chapter 14 summarizes the indeed exciting discipline of aromaticity and antiaromaticity within

https://doi.org/10.1515/9783110535839-201

4- and 5-membered P-heterocycles. Last but not least, Chapter 15 guides the reader into the “empire” of special P-heterocycles, crown ethers containing phosphorus in the macroring.

György Keglevich – Editor
Department of Organic Chemistry and Technology
Budapest University of Technology and Economics
1521 Budapest
Hungary

Contents

List of Contributors —— IX

György Keglevich
1 **Typical approaches for the preparation of phosphine oxides; a review on the synthetic methods applied in the last five years —— 1**

Nóra Zsuzsa Kiss and György Keglevich
2 **Methods for the preparation of phosphinates and phosphonates with a focus on recent advances —— 35**

Ádám Tajti and György Keglevich
3 **The importance of organophosphorus compounds as biologically active agents —— 53**

Péter Bagi and Réka Herbay
4 **Resolution of phosphine oxides —— 66**

Zita Rádai, Nóra Zsuzsa Kiss and György Keglevich
5 **Synthesis of α-hydroxyphosphonates, an important class of bioactive compounds —— 91**

Erika Bálint, Anna Tripolszky and Ádám Tajti
6 **Synthesis of α-aminophosphonates by the Kabachnik–Fields reaction and by the Pudovik reaction —— 108**

Réka Henyecz, Mátyás Milen, Károly Kánai and György Keglevich
7 **The use of the T3P® reagent in the synthesis of phosphinic and phosphonic derivatives —— 148**

Réka Henyecz and György Keglevich
8 **P–C couplings by the Hirao reaction —— 158**

Tamara Kovács and György Keglevich
9 **Deoxygenation of phosphine oxides —— 179**

Dávid Illés Nagy, Alajos Grün and György Keglevich
10 **Dronic acid derivatives – An important group of phosphorus-containing drugs —— 199**

Goutam Brahmachari
11 **P-Chemistry at ambient conditions: A recent update — 214**

Péter Ábrányi-Balogh
12 **Recent developments in the synthesis of new P-heterocycles — 232**

Bubun Banerjee, Ádám Tajti and György Keglevich
13 **Ultrasound-assisted synthesis of organophosphorus compounds — 248**

Zoltán Mucsi, Imre G. Csizmadia, Nóra Zs. Kiss and György Keglevich
14 **Computational study of the aromaticity and antiaromaticity of cyclic organophosphorus compounds containing a single phosphorus atom — 264**

Hajnalka Szabó-Szentjóbi, Tamás Szabó, Tünde Tóth and Péter Huszthy
15 **Crown ethers containing phosphorus in the macroring — 284**

Index — 309

List of Contributors

Péter Ábrányi-Balogh
Institute of Organic Chemistry
Hungarian Academy of Sciences
Magyar tudósok körútja 2.
1117 Budapest
Hungary
abpeter@gmail.com

Péter Bagi
Department of Organic Chemistry and Technology
Budapest University of Technology and Economics
Műegyetem rkp. 3.
1111 Budapest
Hungary
pbagi@mail.bme.hu

Erika Bálint
Department of Organic Chemistry and Technology
Budapest University of Technology and Economics
Műegyetem rkp. 3.
1111 Budapest
Hungary
ebalint@mail.bme.hu

Bubun Banerjee
Department of Chemistry
Indus International University
Bathu, Una
Himachal Pradesh 174301
India
banerjeebubun@gmail.com

Goutam Brahmachari
Laboratory of Natural Products & Organic Synthesis
Department of Chemistry
Visva-Bharati (A Central University)
Santiniketan-731235
West Bengal
India
goutam.brahmachari@visva-bharati.ac.in

Imre G. Csizmadia
University of Szeged, Hungary
and University of Toronto, Canada

Alajos Grün
Department of Organic Chemistry and Technology
Budapest University of Technology and Economics
Műegyetem rkp. 3.
1111 Budapest
Hungary
agrun@mail.bme.hu

Réka Henyecz
Department of Organic Chemistry and Technology
Budapest University of Technology and Economics
Műegyetem rkp. 3.
1111 Budapest
Hungary
reka422@gmail.com

Réka Herbay
Department of Organic Chemistry and Technology
Budapest University of Technology and Economics
Műegyetem rkp. 3.
1111 Budapest
Hungary

Péter Huszthy
Department of Organic Chemistry and Technology
Budapest University of Technology and Economics
Műegyetem rkp. 3.
1111 Budapest
Hungary
huszthy@mail.bme.hu

György Keglevich
Department of Organic Chemistry and Technology
Budapest University of Technology and Economics
Műegyetem rkp. 3.
1111 Budapest
Hungary
gkeglevich@mail.bme.hu

Károly Kánai
Department of Organic Chemistry and Technology
Budapest University of Technology and Economics
Műegyetem rkp. 3.
1111 Budapest
Hungary

https://doi.org/10.1515/9783110535839-202

Nóra Zs. Kiss
Department of Organic Chemistry and Technology
Budapest University of Technology and Economics
Műegyetem rkp. 3.
1111 Budapest
Hungary
zsnkiss@mail.bme.hu

Tamara Kovács
Department of Organic Chemistry and Technology
Budapest University of Technology and Economics
Műegyetem rkp. 3.
1111 Budapest
Hungary
kovacstami88@gmail.com

Mátyás Milen
Department of Organic Chemistry and Technology
Budapest University of Technology and Economics
Műegyetem rkp. 3.
1111 Budapest
Hungary

Zoltán Mucsi
Femtonics Kft.
Tűzoltó u. 59
1094 Budapest
Hungary
zoltanmucsi@gmail.com

Dávid Illés Nagy
Department of Organic Chemistry and Technology
Budapest University of Technology and Economics
Műegyetem rkp. 3.
1111 Budapest
Hungary
duwa24@gmail.com

Zita Rádai
Department of Organic Chemistry and Technology
Budapest University of Technology and Economics
Műegyetem rkp. 3.
1111 Budapest
Hungary
radai.zita@mail.bme.hu

Tamás Szabó
Department of Organic Chemistry and Technology
Budapest University of Technology and Economics,
Műegyetem rkp. 3.
1111 Budapest
Hungary

Hajnalka Szabó-Szentjóbi
Department of Organic Chemistry and Technology
Budapest University of Technology and Economics
Műegyetem rkp. 3.
1111 Budapest
Hungary
hajni.szentjobi@gmail.com

Ádám Tajti
Department of Organic Chemistry and Technology
Budapest University of Technology and Economics
Műegyetem rkp. 3.
1111 Budapest
Hungary
tajti.adam@mail.bme.hu

Tünde Tóth
Department of Organic Chemistry and Technology
Budapest University of Technology and Economics
Műegyetem rkp. 3.
1111 Budapest
Hungary

Anna Tripolszky
Department of Organic Chemistry and Technology
Budapest University of Technology and Economics
Műegyetem rkp. 3.
1111 Budapest
Hungary

György Keglevich

1 Typical approaches for the preparation of phosphine oxides; a review on the synthetic methods applied in the last five years

Abstract: Phosphine oxides are an important class of organophosphorus compounds. Their synthesis involves the reaction of >P(O)Cl derivatives with organometallic species. A similar substitution may also be performed in the sphere of P(III) compounds to be followed by oxidation. The Arbuzov reaction and the formation of the P–C bond by addition and coupling reactions are also often applied. Last, but not least there are special protocols for the preparation of phosphine oxides. The chapter begins with a discussion of the most interesting methods.

Keywords: synthesis, phosphine oxides, substitution at P, Arbuzov reaction, coupling reaction.

1.1 Metathetic Reactions

This approach relates to the preparation of α,α'-diarylacenaphtho[*c*]phosphole oxides from 1,8-bis(trimethylsilylethynyl)naphthalene in four steps. In the first step, the bis-acetylene derivative was reacted with the Sato-Urabe low-valent titanium reagent to give a heterocyclic-titanium intermediate that was then treated with $PhPCl_2$ to afford a phosphine oxide after oxidation. Then, the TMS groups were exchanged to bromine atoms, and the resulting dibromophospholes were converted to the bisaryl derivatives by Stille coupling (Figure 1.1) [1].

An efficient one-pot method was developed for the synthesis of polycyclic phospholanes. In the first step, norbornane-annulated aluminacyclopentanes were prepared by catalytic cycloalumination of norbornenes. In the second step, the alumina-heterocycles were converted to the corresponding phospholanes by reaction with $PhPCl_2$. Oxidation afforded a series of cyclic phosphine oxides (Figure 1.2) [2].

1.2 Rearrangement Reactions

Tris(2-hydroxyphenyl)phosphine oxide (THPPO) was obtained from triphenyl phosphate (TPP) by treatment with lithium diisopropylamide (LDA) at *ca.* –85 °C. The reaction takes place via metallation of the phenyl ring in the ortho position followed by migration of phosphorus from oxygen to carbon (Figure 1.3) [3].

https://doi.org/10.1515/9783110535839-001

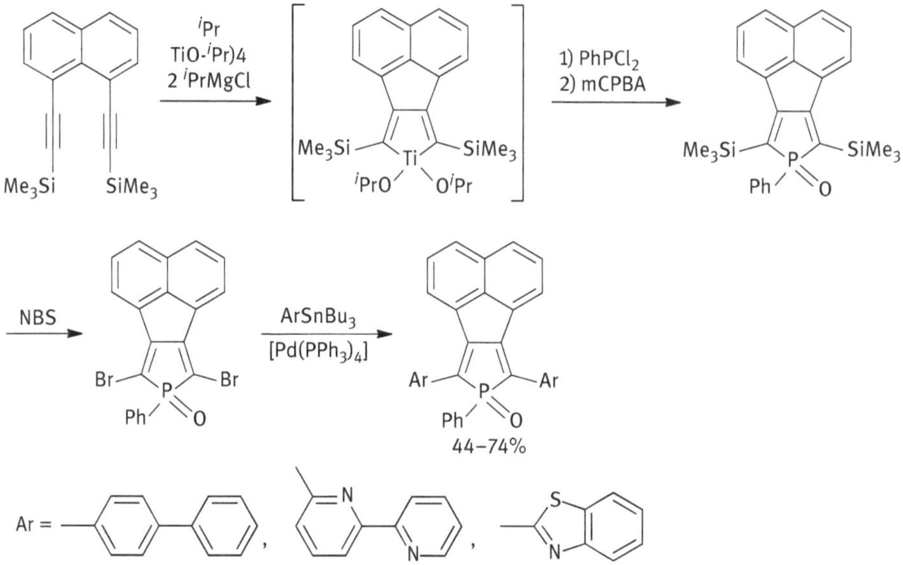

Figure 1.1: Metathetic synthesis of a phosphole oxide by Ti P exchange.

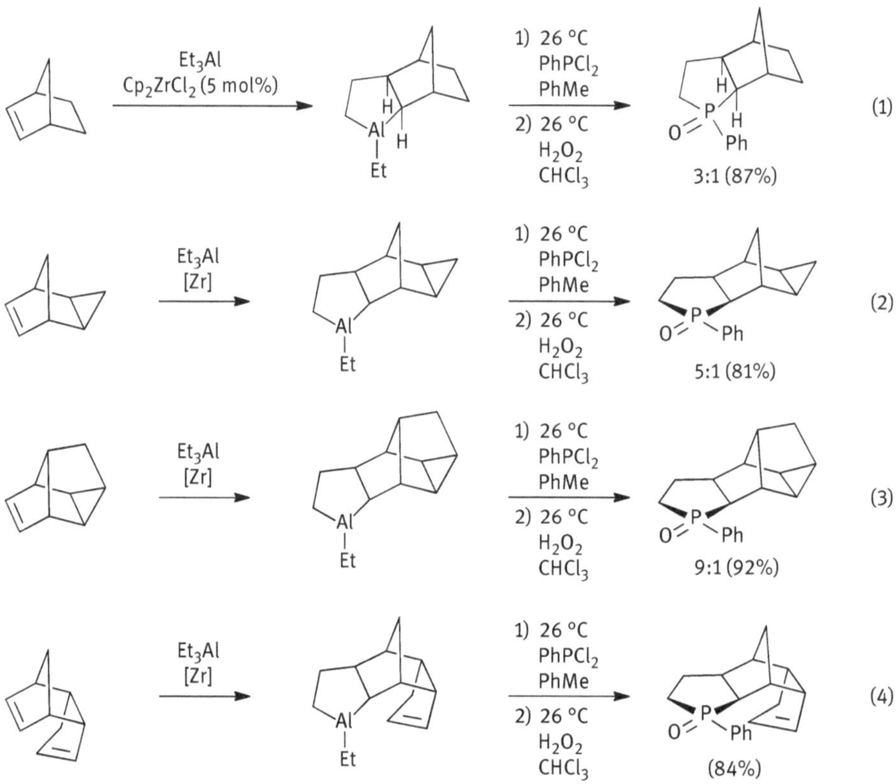

Figure 1.2: Metathetic synthesis of phosphole oxides by Al P exchange.

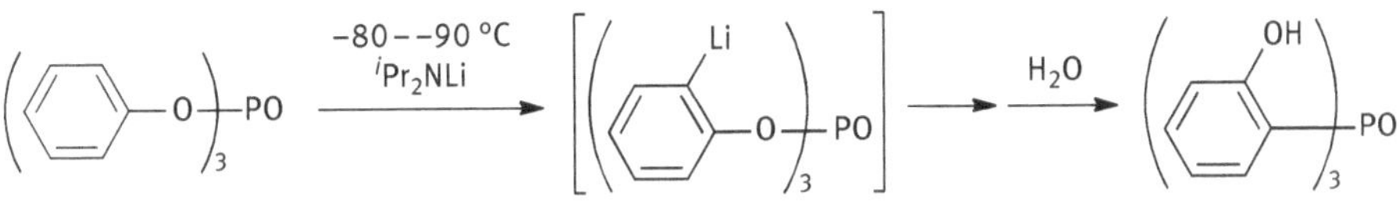

Figure 1.3: A LDA-induced rearrangement of TPP to THPPO.

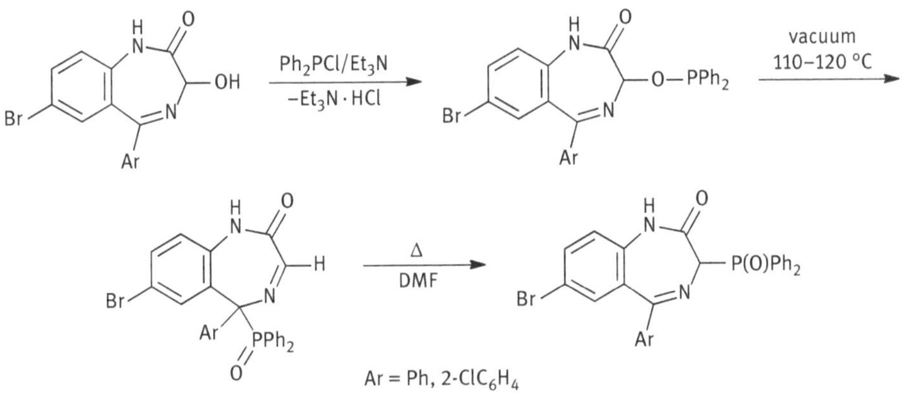

Figure 1.4: A phosphorotopic $Ph_2P–O–C^3 \rightarrow Ph_2P(O)–C^5$ migration.

Y. Gololobov and coworkers reported a new type of phosphorotropic migration in the $C^3–N=C^5$ triad, which was accompanied by the isomerization of $Ph_2P–O–C^3$ into $Ph_2P(O)–C^5$. This sequence together with the preparation of the starting phosphinous ester are shown in Figure 1.4. On heating in DMF, the diphenylphosphinoyl group rearranges back to position 3 [4].

A reaction cascade leading to pyrrolo[1,2-*a*]quinolines with phosphine oxide function was developed starting by the phosphinylation of propargylic alcohols, followed

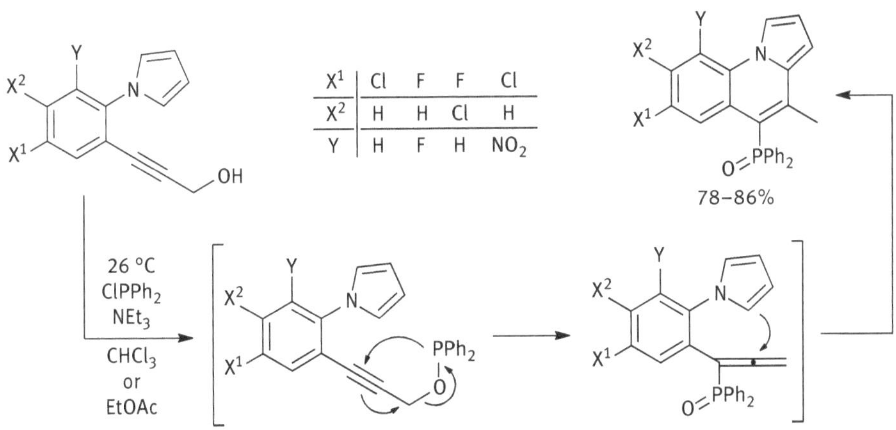

Figure 1.5: A sigmatropic propargyl → allene rearrangement.

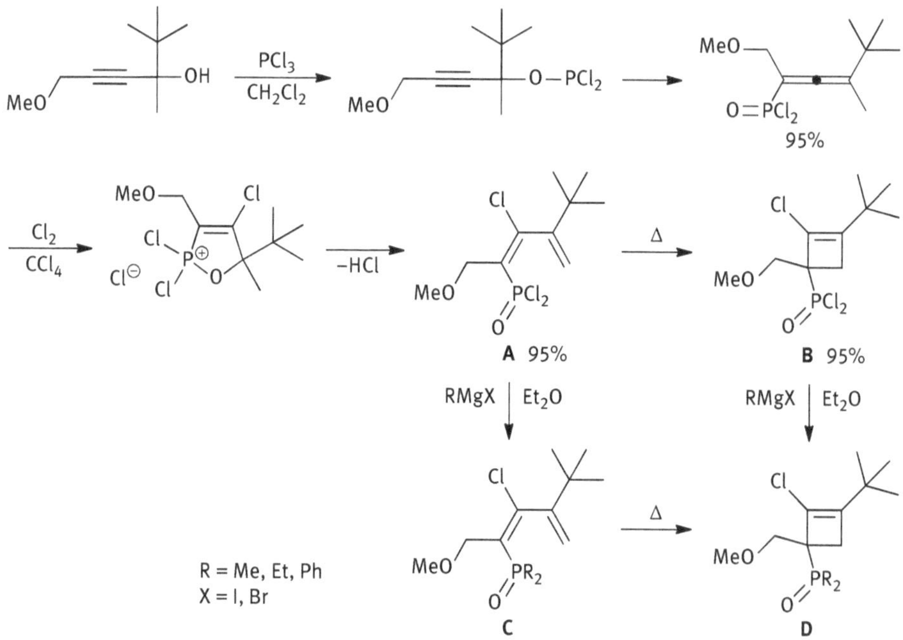

Figure 1.6: A propargyl → allene rearrangement followed by subsequent modifications.

by a [2,3]-sigmatropic rearrangement to give allenes that undergo stabilization by an intramolecular cyclization (Figure 1.5) [5].

The ester formed from the reaction of the adduct of methyl propargyl ether with pinacolone and phosphorus trichloride was stabilized by a spontaneous rearrangement to the corresponding allene derivative. Chlorination of the allene led to an unstable chlorophosphonium salt that gave a dichlorophosphono-1,3-butadiene (**A**) undergoing cyclization to cyclobutene **B** on heating. Dialkyl- or diaryl-cyclobutenylphosphine oxides (**D**) were obtained by thermal electrocyclization of intermediate **C** formed from dichlorophosphonobutadiene **A** in Grignard reaction. It was also possible to convert species **B** to target product **D** (Figure 1.6) [6].

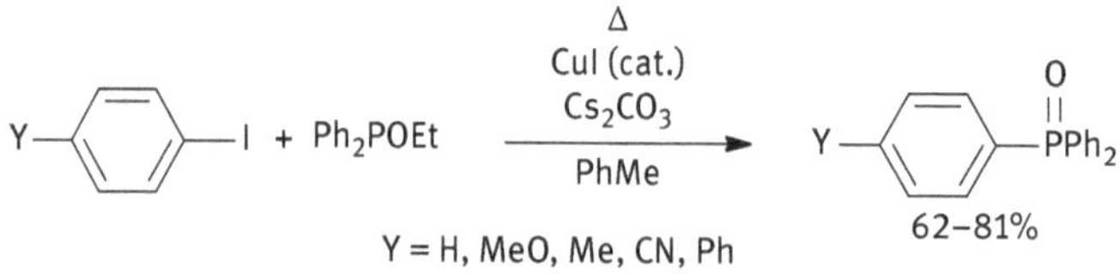

Figure 1.7: Cu-catalyzed Arbuzov reaction of aryl iodides with Ph_2POEt.

1.3 Arbuzov Reactions

A Cu(I)-catalyzed version of the Michaelis–Arbuzov reaction of aryl iodides and the ethyl ester of diphenylphosphinous acid was developed to replace toxic transition metal (Ni and Pd) salts used earlier (Figure 1.7) [7].

Ortho-trimethylsilyl-aryl triflates were used as starting materials in Arbuzov reactions to prepare aryl-diphenylphosphine oxides (Figure 1.8) [8].

In the multistep total synthesis of Herboxidiene methyl ester, one step involved an Arbuzov reaction of an allyl bromide derivative with Ph_2POEt (Figure 1.9) [9].

The reaction of 1,1-dibromo-1-alkenes with Ph_2POEt in the presence of Cs_2CO_3 in boiling toluene furnished alkynylphosphine oxides (Figure 1.10). The first step is the elimination of HBr, followed by Arbuzov reaction of the aryl-bromoacetylene formed [10].

A 2,5-bis(phosphinoylmethyl)thiophene was prepared by the double Arbuzov reaction of 2,5-bis(bromomethyl)thiophene and Ph_2POEt (Figure 1.11) [11].

Dibenzothiophene was converted to the corresponding bis(diphenylphosphinoyl)-dibenzothiophene sulfone in three steps, and to the bis(diphenylphosphinoylmethyl) derivative in five steps. The key steps were the introduction of the

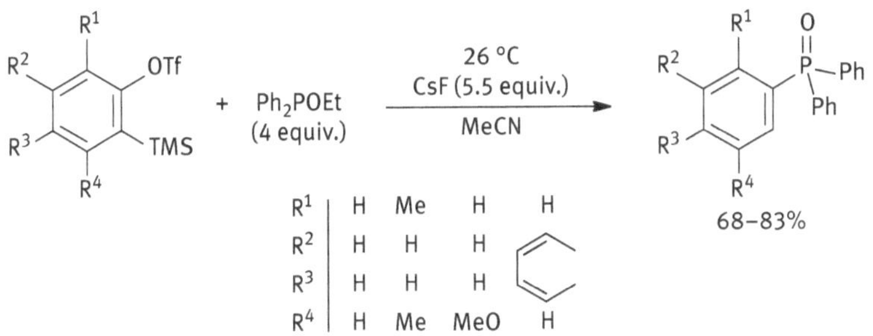

Figure 1.8: Arbuzov reaction of aryl triflates accompanied by de-trimethylsilylation.

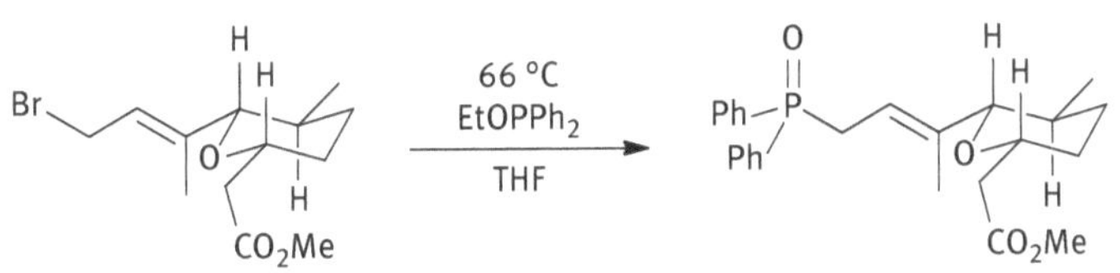

Figure 1.9: Arbuzov reaction of an allyl bromide.

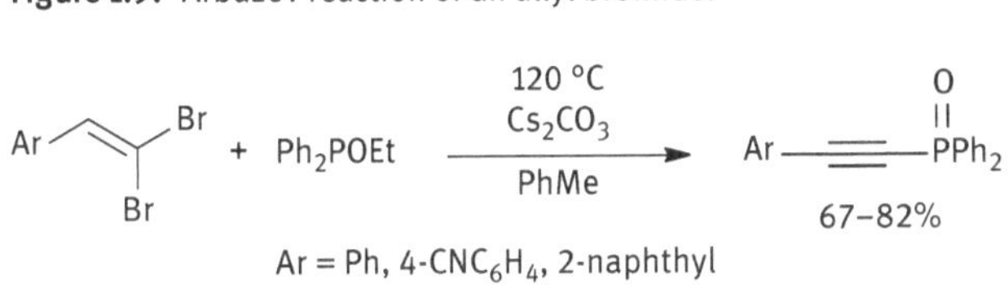

Figure 1.10: Arbuzov reaction of an in situ formed bromoacetylene derivative.

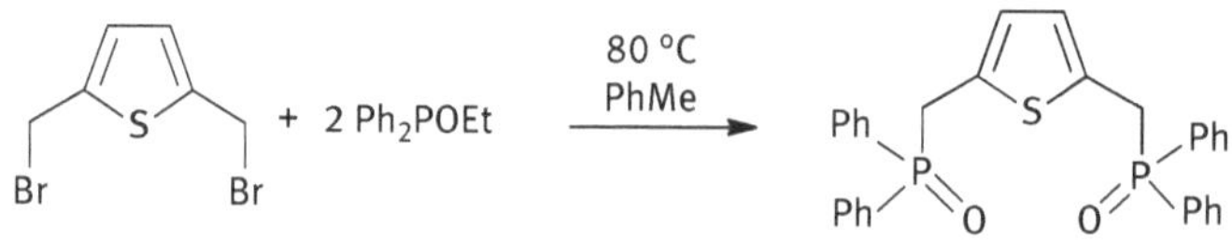

Figure 1.11: Double Arbuzov reaction of 2,5-bis(bromomethyl)thiophene.

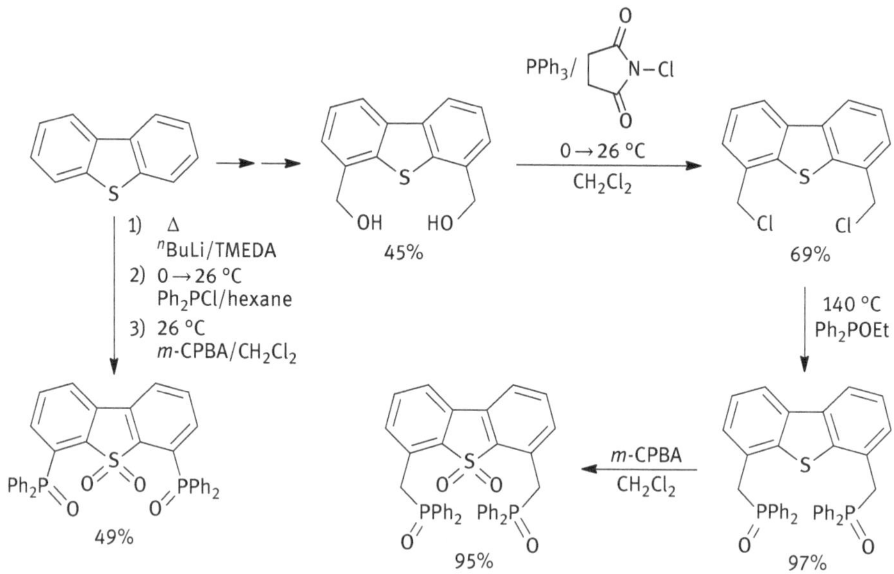

Figure 1.12: Dibenzothiophene functionalizations involving a double Arbuzov reaction.

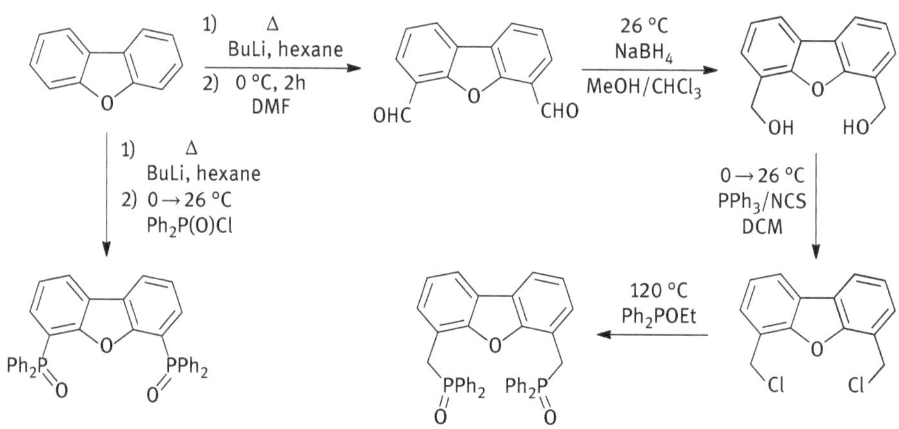

Figure 1.13: Dibenzofuron functionalizations involving a double Arbuzov reaction.

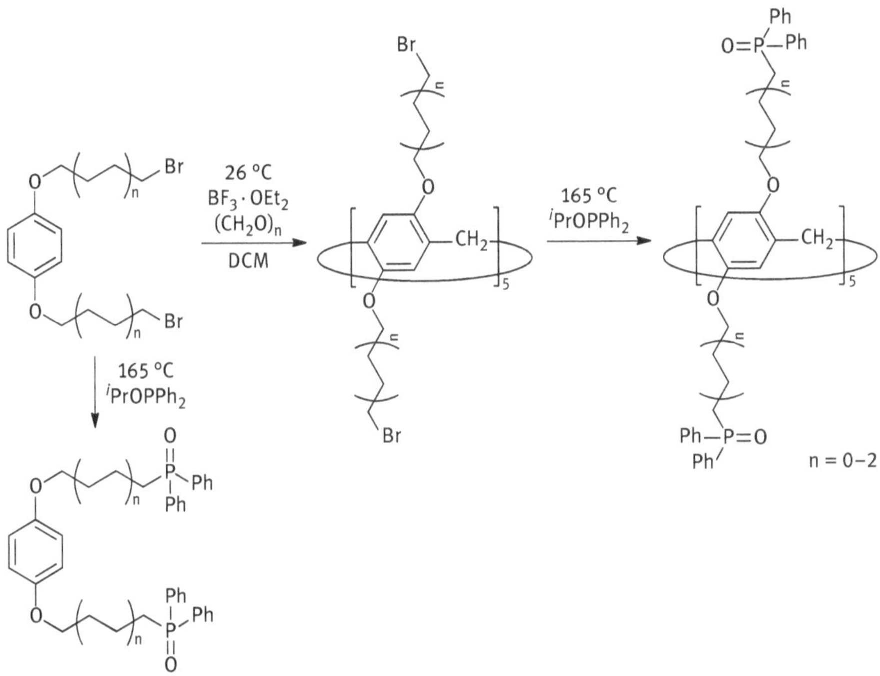

Figure 1.14: Double Arbuzov reaction of 1,4-bis(bromoalkoxy)benzene derivatives.

P-functions into the heterocyclic intermediates by the Arbuzov reaction or by phosphinoylation (Figure 1.12) [12].

A four-step synthesis of 4,6-bis(diphenylphosphinoylmethyl)dibenzofuran including formylation, reduction, substitution, and Arbuzov reaction, together with the two-step preparation of 4,6-bis(diphenylphosphinoyl)dibenzofuran are also described (Figure 1.13) [13].

Pillar [5]arene-based phosphine oxides were synthesized by the reaction of 1,4-bis(bromoalkyoxy)benzenes and paraformaldehyde in the presence of boron trifluoride diethyl etherate followed by an Arbuzov-type bis-functionalization. The starting material was also modified by the Arbuzov reaction (Figure 1.14) [14].

1.4 Hirao Reactions

The Hirao reaction involves primarily the P–C coupling of aryl halides and dialkyl phosphites in the presence of $Pd(PPh_3)_4$ catalyst to afford aryl phosphonates. In order to summarize the classical and novel aspects of this important chemical transformation, the Hirao reaction was reviewed [15, 16]. A variety of secondary phosphine oxides were used in a series of P–C coupling reactions involving a wide range of aryl iodides

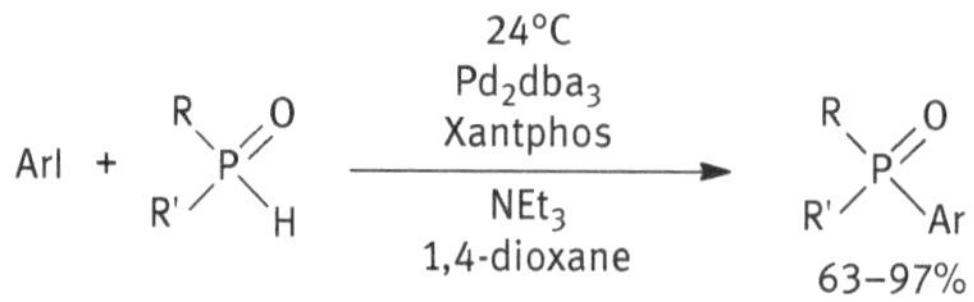

1)

R	Me	Bn	^{t}Bu	CF_3CH_2	Me	Bn	Np	$MesCH_2$	Ph
R'	Me	Cy	Me	Cy	Ph	Ph	Ph	Cy	Ph
Ar	Ph	Ph	Ph	Ph	Ph	Ph	Ph	Ph	Ph

2) R = Bn, R' = Cy

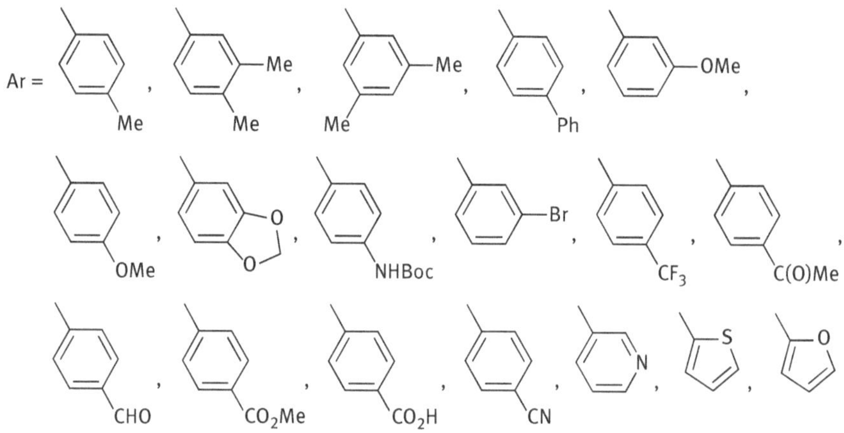

Figure 1.15: The Hirao reaction of iodoarenes and secondary phosphine oxides.

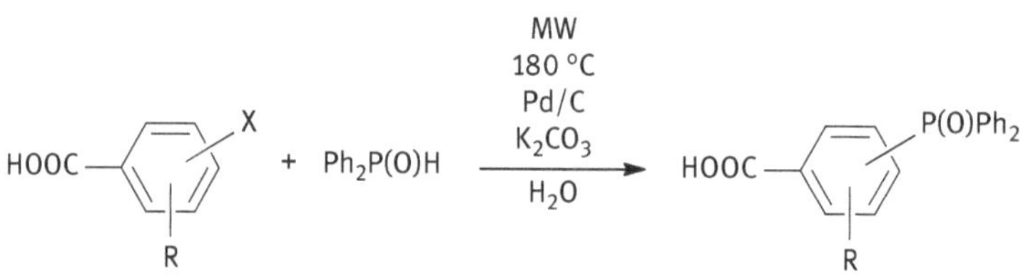

R = H, 4-MeO, 4-Me, 3-NH_2
X = 4-Br, 4-I, 4-Cl, 3-Br, 3-I

Figure 1.16: The Hirao reaction of halogenobenzoic acids and $Ph_2P(O)H$.

in the presence of tris(dibenzylideneacetone)dipalladium(Pd_2dba_3) and Xantphos to furnish the secondary phosphine oxides in good yields. (Figure 1.15) [17].

The coupling of substituted halogenobenzoic acids with diphenylphosphine oxide was achieved by microwave irradiation using water as the reaction medium (Figure 1.16) [18].

Arylboronic acids may also be coupled with diphenylphosphine oxide using $Pd(OAc)_2$/dppb as the catalyst (Figure 1.17) [19].

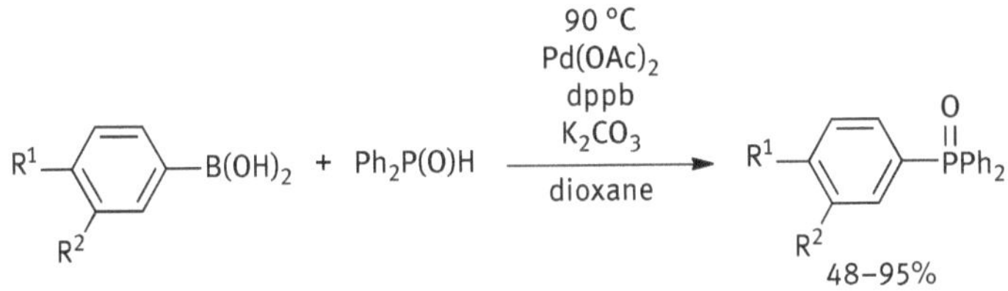

R^1 = H, MeO, MeS, Ph_2N, F_3CO, MeC(O)NH, Me, $HOCH_2$, CH_2=CH, F, Cl, MeOC(O), NC, NO_2, Ph, R^2 = H
R^1 = H, R^2 = CF_3, NO_2

Figure 1.17: The Hirao reaction of arylboronic acids and $Ph_2P(O)H$.

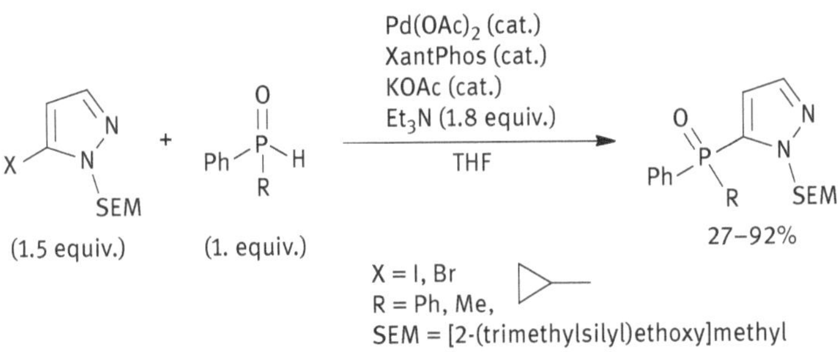

Figure 1.18: The phosphinoylation of halogenopyrazoles by the Hirao reaction.

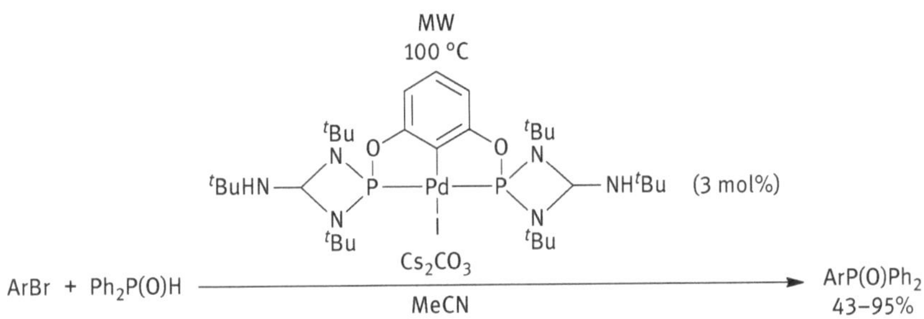

Ar = Ph, 3-MeOPh, 4-MeOPh, 2-MePh, 4-MePh, 4-CNPh, 2-($HOCH_2$)Ph, 2-naphthyl

Figure 1.19: Utilization of a special Pd-catalyst in the Hirao reaction.

Pyrazoles were phosphinoylated under the usual conditions using $Pd(OAc)_2$ together with Xantphos as the P-ligand (Figure 1.18) [20].

The Pd complex of a cyclodiphosphazane-based pincer ligand was successfully applied as a new catalyst in the P–C coupling of aryl bromides with diphenylphosphine oxide (Figure 1.19) [21].

The above examples comprised the use of different Pd-catalysts.

The arylation of diphenylphosphine oxide with aryl halogenides was carried out in the presence of $NiCl_2 \cdot 6H_2O$/Zn, applying 2,2'-bipyridine as the ligand in water under mild conditions (Figure 1.20) [22].

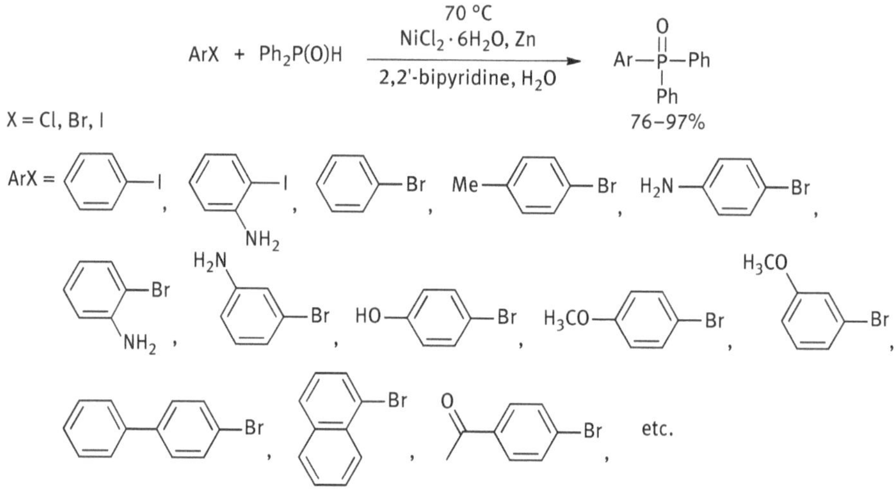

Figure 1.20: Ni-catalyzed P–C couplings involving $NiCl_2$ and 2,2'-bipyridine as the P-ligand.

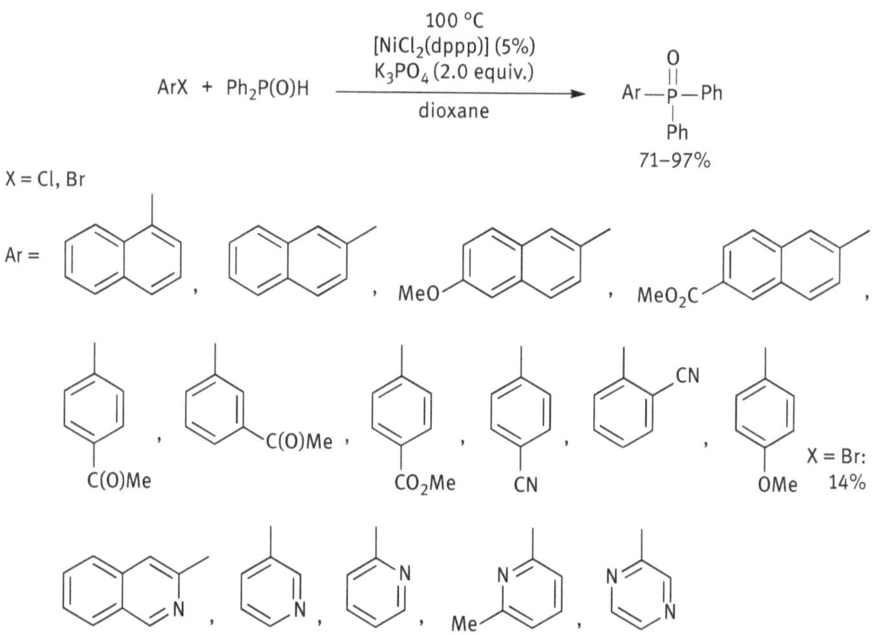

Figure 1.21: Ni-catalyzed P–C couplings involving $NiCl_2$(dppm) as the catalyst.

A series of aryl-diphenylphosphine oxides were prepared by the [$NiCl_2$(dppp)]-catalyzed coupling of aryl halides with diphenylphosphine oxide (Figure 1.21) [23].

P–C bond formation between aryl iodides and secondary phosphine oxides was possible at room temperature by combining Ni catalysis and visible light-induced photoredox catalysis (Figure 1.22) [24].

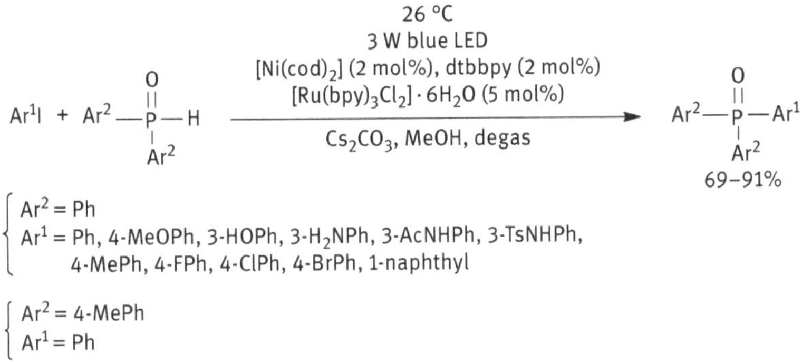

Figure 1.22: A P–C coupling promoted by a combined Ni- and photoredox catalysis.

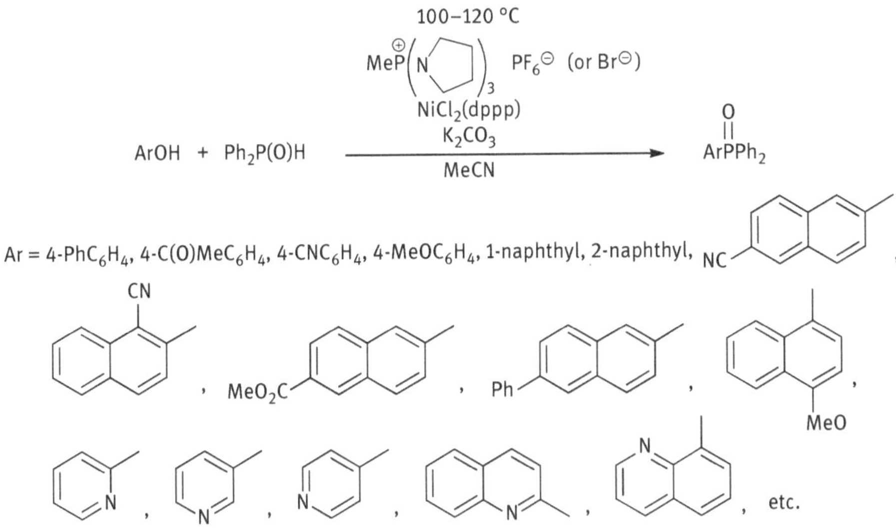

Figure 1.23: Ni-catalyzed P–C coupling of phenols and $Ph_2P(O)H$.

Phenol derivatives could also be coupled with diphenylphosphine oxide using $NiCl_2$(dppp) as the catalyst, and a special phosphonium salt as the activating agent (Figure 1.23) [25].

P–H/C–CN cross coupling reactions of substituted benzonitriles and secondary phosphine oxides were also described utilizing a Ni catalyst (Figure 1.24) [26].

Arylboronic acids and secondary phosphine oxides may also be coupled using a Ni catalyst to afford tertiary phosphine oxides (Figure 1.25) [27].

After showing Pd- and Ni-catalyzed Hirao reactions, let us regard a few cases for Cu catalysis.

The coupling of aryl iodides and secondary phosphine oxides is also possible in the presence of CuI as the catalyst and (*S*)-α-phenylethylamine as the ligand in the

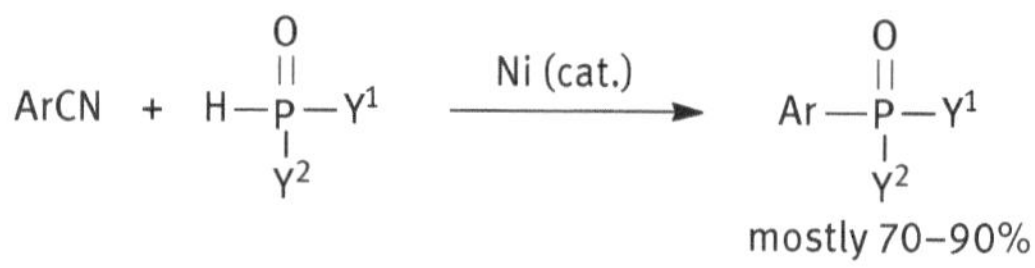

Ar = Ph, 4-MeOPh, 4-Me_2NPh, 4-MePh, 3-MePh, 2-MePh, 4-PhPh, 4-CF_3Ph, 1-naphthyl, 2-naphthyl, 2-(7-MeOnaphthyl),

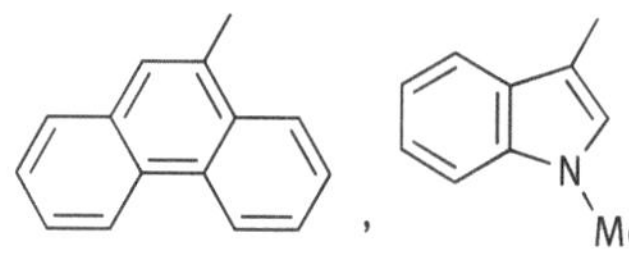

$Y^1 = Y^2 =$ Ph

Ar = Ph

Y^1	4-MePh	Ph	Ph	^{n}Bu
Y^2	4-MePh	^{n}Bu	^{t}Bu	^{n}Bu

Figure 1.24: Ni-catalyzed P–C coupling of arylnitriles and secondary phosphine oxides.

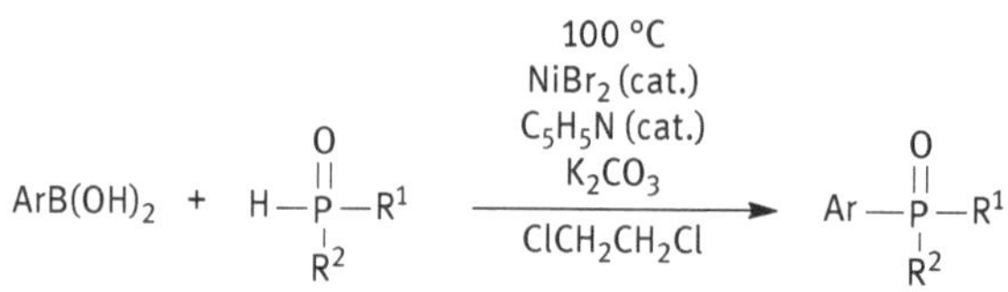

R^1	Ph	4-MePh	4-MeOPh	4-Me_2NPh	4-FPh	Ph	Pent
R^2	Ph	4-MePh	4-MeOPh	4-Me_2NPh	4-FPh	Et	Pent

Ar = 4-MeOPh, 4-MePh, 2-MePh, Ph, 4-ClPh, 4-FPh, 3-CF_3Ph, 4-PhPh, 4-C(O)Me, 4-CO_2Me, 2-naphthyl

Figure 1.25: Ni-catalyzed P–C coupling of arylboronic acids and secondary phosphine oxides.

presence of K_2CO_3 in boiling toluene (Figure 1.26-1) [28]. Under similar conditions, 2-bromo-iodobenzene underwent a double phosphinoylation (Figure 1.26-2) [28].

A copper-catalyzed P-ligand-free Hirao reaction of *ortho*-bromophenols and secondary phosphine oxides was also elaborated (Figure 1.27) [29].

The Gilheany group has developed novel *P*-stereogenic BINAP derivatives. In the first step, *R*-BINOL was bis(trifluoromethanesulfonylated). Then, a diphenylphosphine oxide moiety was introduced (Figure 1.28) [30]. Establishment of another P-function was also made possible.

The bistriflate of BINOL could also be monophosphinoylated by reaction with phenyl-alkylphosphine oxides (Figure 1.29) [31].

The precursor 7'-butoxy-7-(diphenylphosphino)-8,8'-biquinolyl was synthesized from 7,7'-dihydroxy-8,8'-biquinolyl in three steps via Mitsunobu monoetherification, triflation, and phosphinoylation (Figure 1.30). The optically active species was utilized as catalyst in the enantioselective Suzuki–Miyaura synthesis of biaryl derivatives [32].

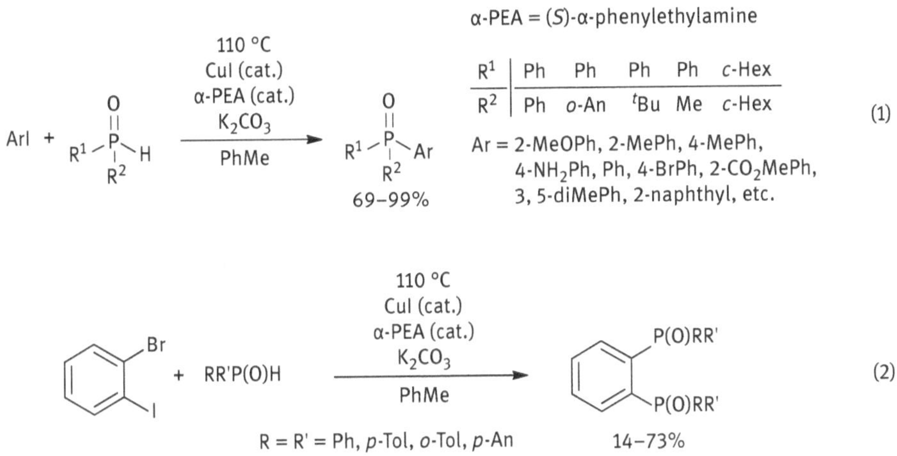

Figure 1.26: Cu-catalyzed Hirao reactions of aryl halogenides.

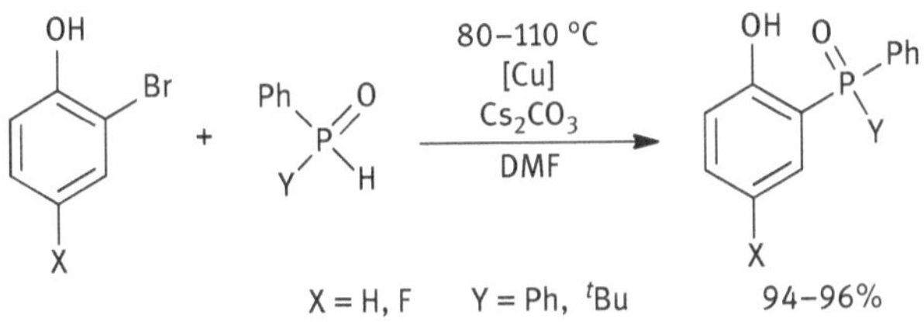

Figure 1.27: Cu-catalyzed Hirao reactions of *ortho*-bromophenols.

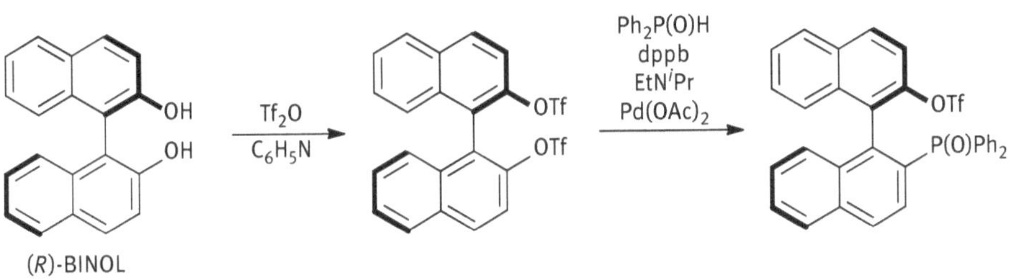

Figure 1.28: Modification of BINOL derivatives by the Hirao reaction.

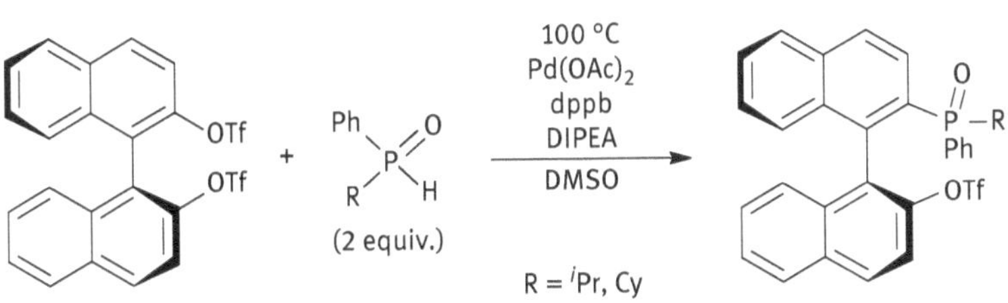

Figure 1.29: The phosphinoylation of the bistriflate of BINOL by the Hirao reaction.

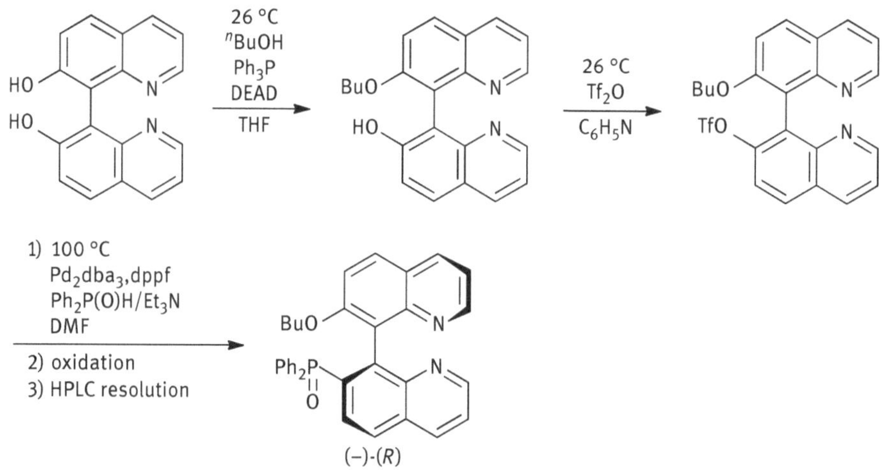

Figure 1.30: Utilization of the Hirao reaction in the functionalization of a biquinolyl derivative.

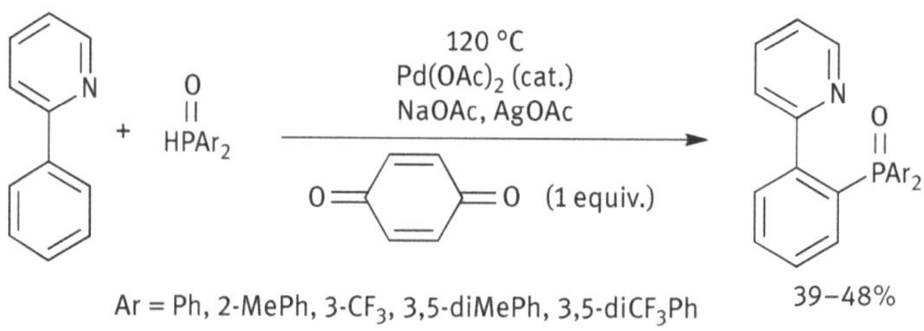

Figure 1.31: Example for an oxidative P–C coupling.

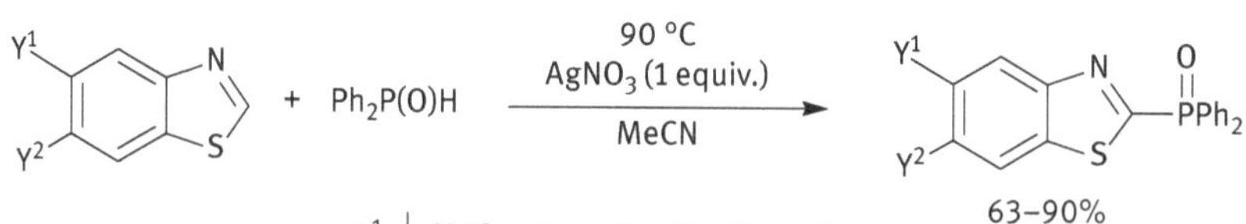

Y^1	MeO	Me	F	H	Br	H
Y^2	H	H	H	Cl	H	CO_2Me

Figure 1.32: Oxidative P–C coupling involving a heterocyclic moiety.

After the usual P–C coupling reactions, let us see a few oxidative versions.

2-Phenylpyridine took part in an oxidative P–C coupling reaction with diarylphosphine oxides using 1 equivalent of 1,4-benzoquinone as the oxidant and $Pd(OAc)_2$ as the catalyst at 120 °C. A heterocycle-directed *ortho*-palladation is responsible for the regioselectivity (Figure 1.31) [33].

Another example for an oxidative P–C coupling is the silver-mediated phosphinoylation of benzothiazoles (Figure 1.32) [34].

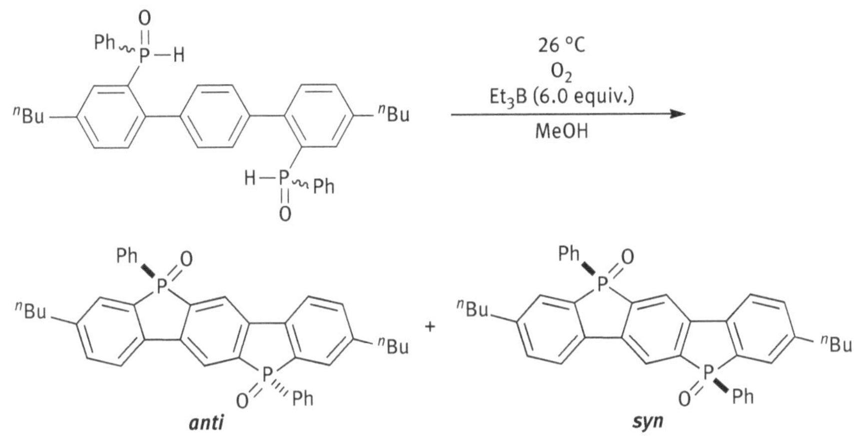

Figure 1.33: A double intramolecular oxidative P–C coupling.

The double oxidative intramolecular cyclization of bis(phenylphosphinyl)-*p*-terphenyl afforded the diastereomers of “twin” bis(dibenzophosphole oxides) (Figure 1.33) [35].

1.5 Phospha-Michael Addition

The phospha-Michael reactions involve the addition of a >P(O)H species, in most cases, to an electron-poor double bond. This approach may be used well for the preparation of phosphine oxides.

An asymmetric Michael addition of diarylphosphine oxides to α,β-unsaturated esters was performed in the presence of chiral magnesium(II) binaphtholates (Figure 1.34) [36].

The uncatalyzed addition of bis(pentafluorophenyl)phosphine oxide to acrylic and cinnamic acid amides afforded (2-carbamoylethyl)-bis(pentafluorophenyl)phosphine oxides (Figure 1.35) [37, 38].

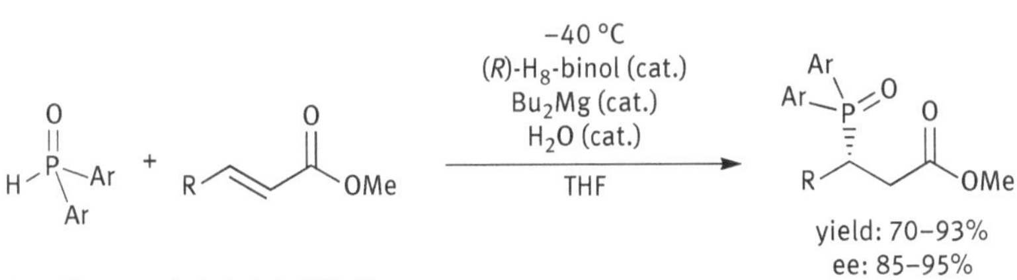

Ar = Ph, 2-naphthyl, 3,5-diMePh
R = 4-MeOC$_6$H$_4$, 4-MeC$_6$H$_4$, Ph, 4-ClC$_6$H$_4$, 2-furyl, 3-pyridyl, PhCH=CH, *c*-C$_6$H$_{11}$

Figure 1.34: An enantioselective phospha-Michael addition.

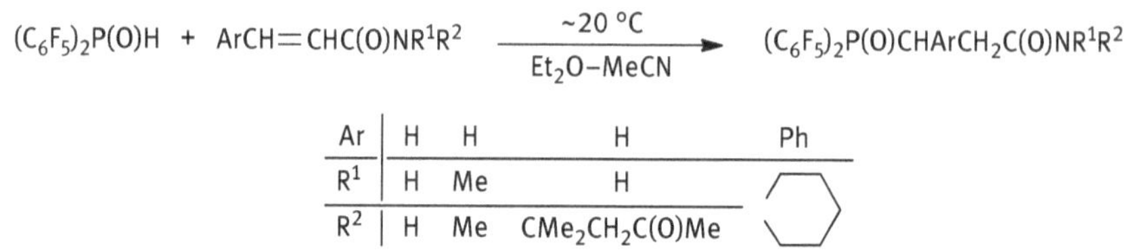

Figure 1.35: An uncatalyzed phospha-Michael addition.

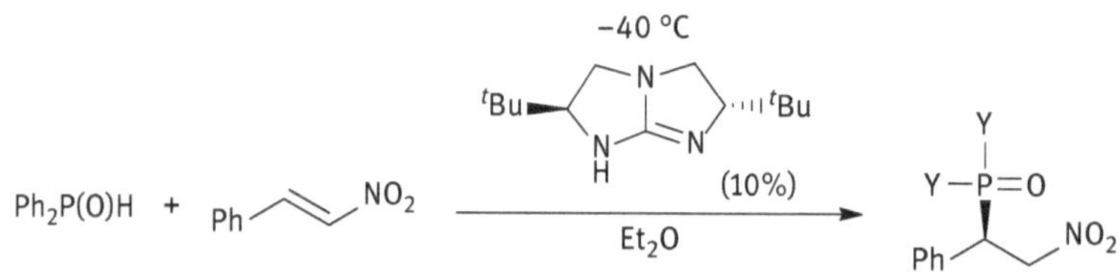

Figure 1.36: An asymmetric phospha-Michael addition.

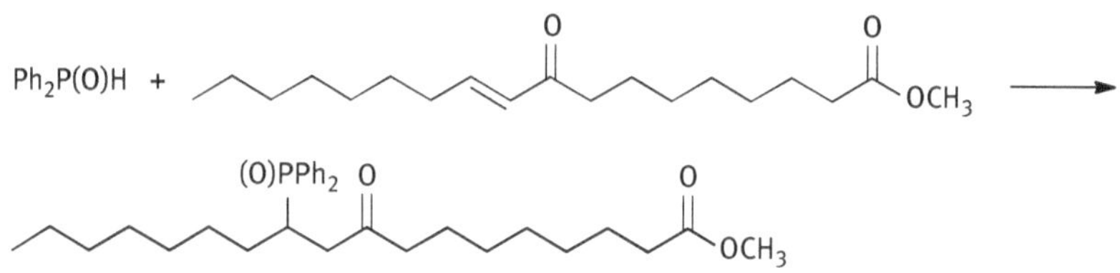

Figure 1.37: Phospha-Michael reaction of an unsaturated ketoester.

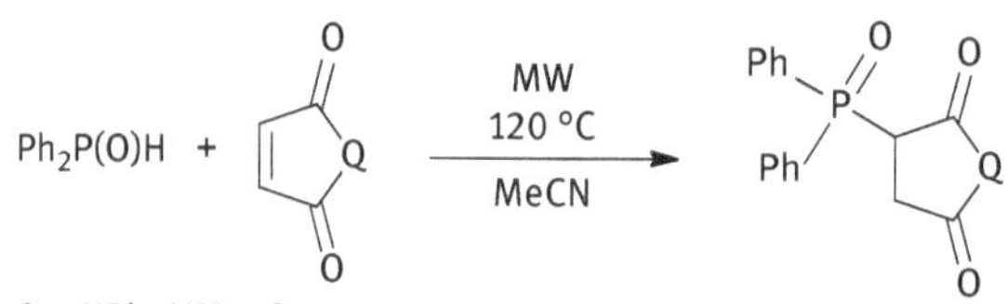

Q = NPh, NMe, O

Figure 1.38: MW-assisted phospha-Michael addition to maleic acid derivatives.

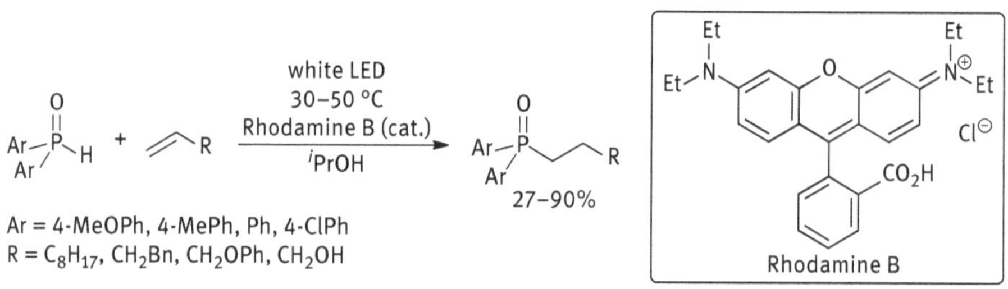

Figure 1.39: Photo-catalyzed phospha-Michael addition.

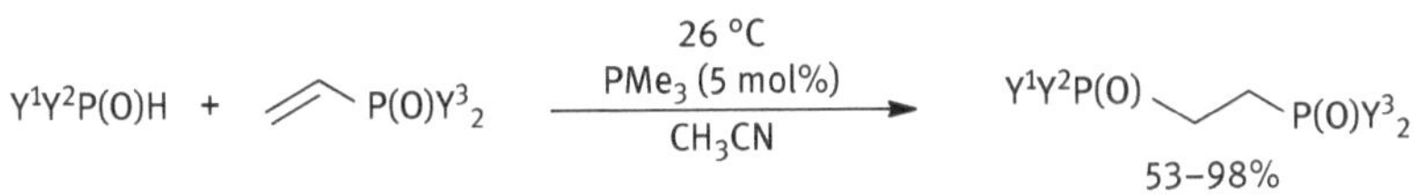

Y^1	MeO	^{i}PrO	Ph	Ph	Ph	^{n}Bu	^{t}Bu
Y^2	MeO	Ph	Ph	Ph	Ph	^{n}Bu	Ph
Y^3	Ph	Ph	MeO	PhO	Ph	MeO	MeO

Figure 1.40: Phosphine-catalyzed phospha-Michael addition.

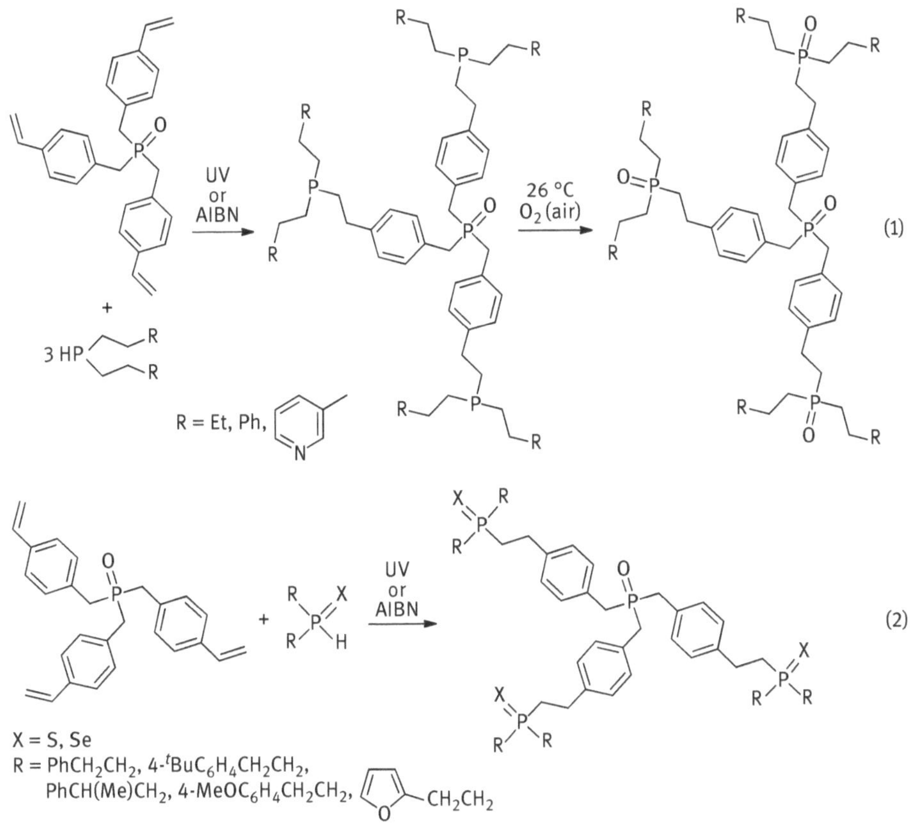

Figure 1.41: Multiple phospha-Michael additions.

The asymmetric addition of secondary phosphine oxides to nitrostyrenes was also investigated (Figure 1.36) [39].

The mechanism and origin of enantioselectivity has been evaluated for a bicyclic guanidine-catalyzed phospha-Michael reaction between diphenylphosphine oxide and β-nitrostyrene by DFT calculations.

In another example, an enone derivative of methyl oleate was reacted with diphenylphosphine oxide (Figure 1.37) [40]. This reaction was then extended to analogous triglyceride derivatives.

The author of this chapter, together with coworkers, investigated the microwave-assisted addition of >P(O)H species, including diphenylphosphine oxide, to maleic acid derivatives. There was no need to use any catalyst (Figure 1.38) [41].

Unactivated alkenes were hydrophosphinylated by reaction with secondary phosphine oxides under visible light photocatalysis (Figure 1.39) [42].

The trimethylphosphine-catalyzed addition of secondary phosphine oxides to vinylphosphine oxides led to 1,2-bisphosphinoylethanes. It is noteworthy that all the three components of the reactions are phosphorus compounds (Figure 1.40) [43].

Trofimov et al. carried out the addition of dialkylphosphines and dialkylphosphine chalcogenides to tris(4-vinylbenzyl)phosphine oxide under radical conditions. The trisphosphine functions were converted to the stable P-oxides (Figure 1.41) [44].

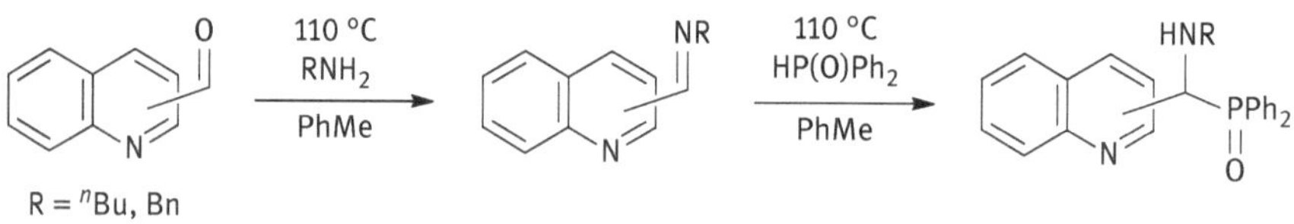

Figure 1.42: Pudovik reaction of quinolinyl-based imines.

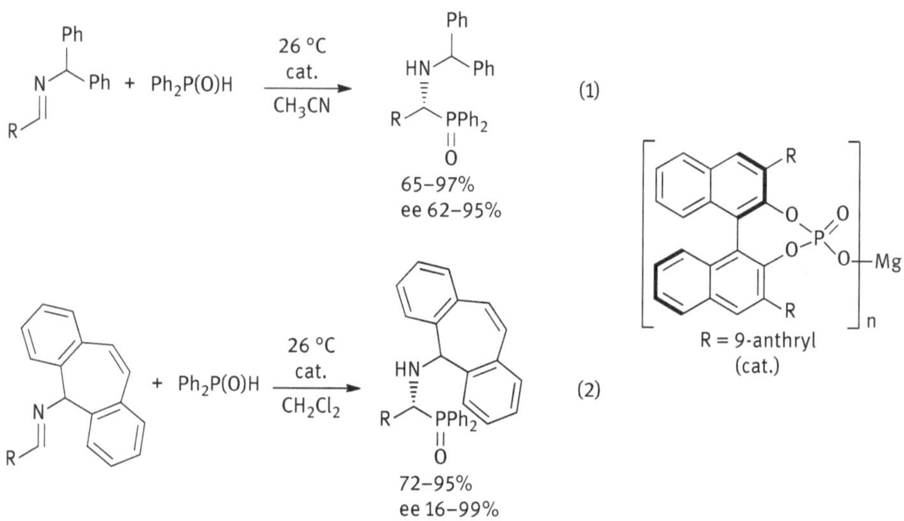

R = Ph, 4-MeOC$_6$H$_4$, 4-FC$_6$H$_4$, 4-BrC$_6$H$_4$, 4-NO$_2$-C$_6$H$_4$, 3-MeC$_6$H$_4$, 2-MeC$_6$H$_4$, ^{c}Hex, ^{n}Pr, ^{i}Pr, ^{n}Bu, etc.

Figure 1.43: Asymmetric Pudovik reaction of imines.

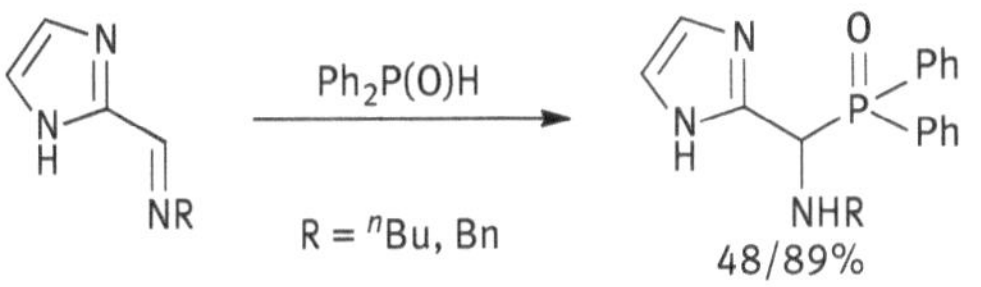

Figure 1.44: Pudovik reaction of imidazol-based imines.

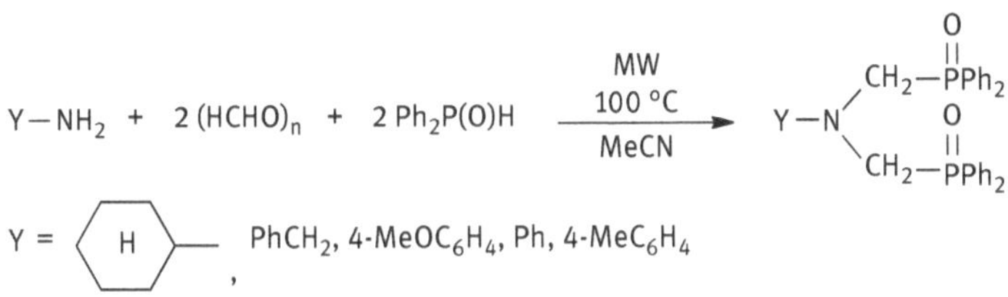

Figure 1.45: A Kabachnik–Fields reaction.

1.6 Pudovik Reaction

The transformations comprising the addition of >P(O)H species to C=O or C=N unsaturations are called the Pudovik reaction. A few examples are known for the addition of secondary phosphine oxides to imines to provide α-aminophosphine oxide derivatives.

New 4- and 2-quinolinyl-aminomethylphosphine oxides were synthesized by the addition of diphenylphosphine oxide to quinoline-derived Schiff bases (Figure 1.42) [45].

Phosphinylation of imines in the presence of a chiral magnesium BINOL phosphate allowed the synthesis of enantioenriched α-aminophosphine oxides. The conversion of two imine derivatives to two series of products is shown in Figure 1.43-1 and 1.43-2 [46].

The addition of diphenylphosphine oxide to imines obtained from imidazole-2-carboxaldehyde and primary amines led to heterocyclic α-aminophosphine oxides (Figure 1.44) [47].

Beside the Pudovik reaction, the Kabachnik–Fields condensation may also provide α-aminophosphine oxides.

The double Kabachnik–Fields (phospha-Mannich) reaction of primary amines, two equivalents of paraformaldehyde, and the same amount of diphenylphosphine oxide afforded bis(diphenylphosphinoylmethyl)amines (Figure 1.45) [48, 49] that may be precursors of bidentate P-ligands after double deoxygenation.

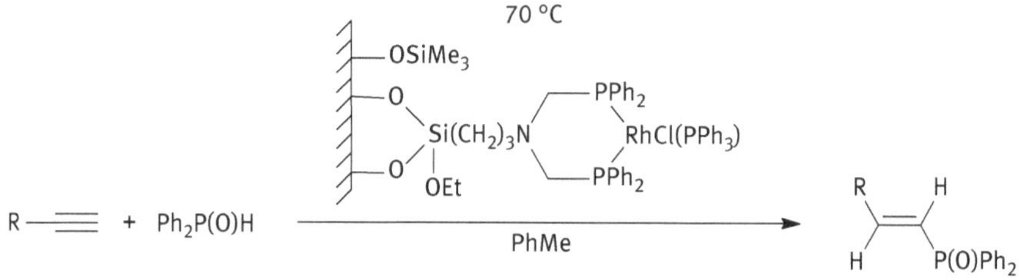

R = ^{n}Hex, ^{n}Bu, ^{t}Bu, Ph, $HOCH_2CH_2$, $MeOCH_2CH_2$, $^tBuCO_2CH_2CH_2$, $ClCH_2CH_2CH_2$, $NCCH_2CH_2CH_2$, 4-MeC_6H_4, $^nBu_2NCH_2$, Me_3SiCH_2, Me_3Si, 2-thienyl, 1-cyclohexenyl

Figure 1.46: Addition of $Ph_2P(O)H$ to acetylene derivatives.

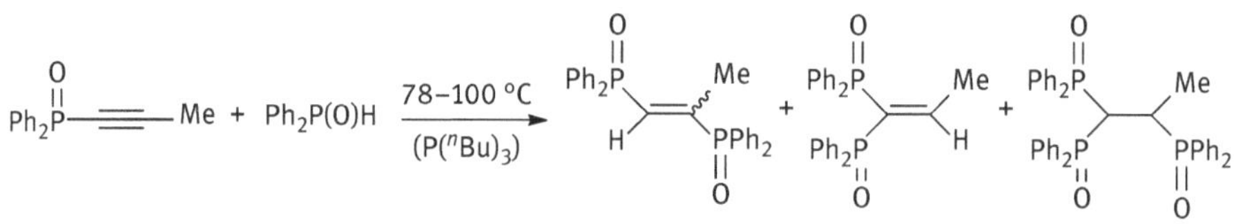

Figure 1.47: Addition of $Ph_2P(O)H$ to a phosphinoylacetylene.

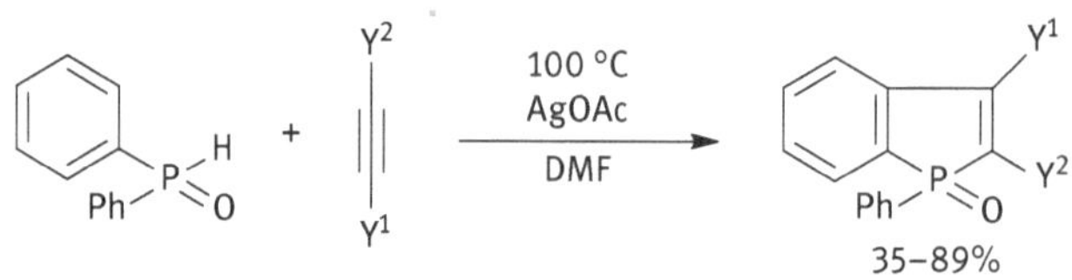

Y^1	4-MeOPh	4-MePh	4-NMe_2Ph	4-Cl	4-Br	^{n}Pr	Ph	Ph	Ph	Ph	Ph	Ph
Y^2	4-MeOPh	4-MePh	4-NMe_2Ph	4-Cl	4-Br	^{n}Pr	Me	Bu	TMS	CO_2Et	Ac	$P(O)(OEt)_2$

Figure 1.48: The addition of $Ph_2P(O)H$ to acetylenes combined by cyclization.

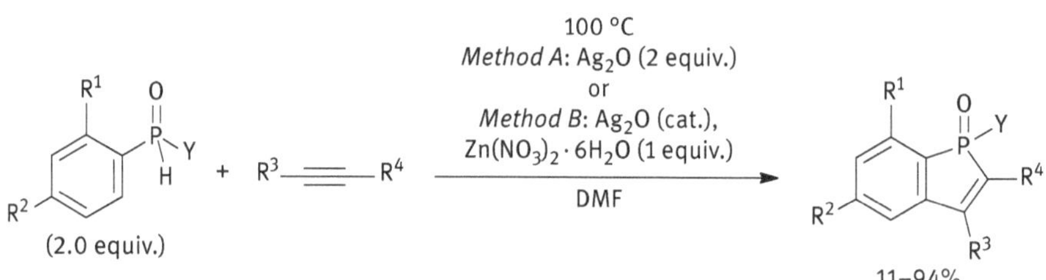

R^1	H	H	H	H	H	H	Me
R^2	H	H	H	F	Cl	MeO	H
Y	Ph	^{t}Bu	Cyclohex	4-FPh	4-ClPh	4-MeOPh	2-MePh

R^3	Ph	4-FPh	Ph	Ph	Ph	Ph	4-FPh	^{n}Pr
R^4	Ph	4-FPh	^{n}Pr	Me	CO_2Et	CN	Me	^{n}Pr

Figure 1.49: Benzophosphole oxides by combined addition and cyclization.

Cherkasov et al. also studied the bis(Kabachnik–Fields) reaction [50] and recent results of the Kabachnik–Fields reaction including the preparation of α-aminophosphine oxides have been reviewed [51].

1.7 Addition to Acetylenic Derivatives

A highly regio- and stereoselective hydrophosphinylation of acetylenes in the presence of an immobilized rhodium-phosphine catalyst may be a useful method for the preparation of vinyl phosphine oxides (Figure 1.46) [52].

The addition of $Ph_2P(X)H$ to a diphenyl-alkynylphosphine oxide theoretically gives a mixture of four products, two different monoadducts (one of them formed as a mixture of isomers) and the double adduct. Without any additive, one of the monoadducts was formed, while in the presence of tributylphosphine, the other monoadduct together with the bisadduct were the products (Figure 1.47) [53].

A series of benzophosphole oxides was synthesized from diphenylphosphine oxide and disubstituted acetylene derivatives utilizing a silver-mediated dehydrogenative annulation involving a P–C and a C–C bond formation (Figure 1.48) [54].

Similar derivatives were prepared by another research group investigating a more diverse reaction model (Figure 1.49) [55].

This approach was utilized further to prepare additional benzophosphole oxides [56].

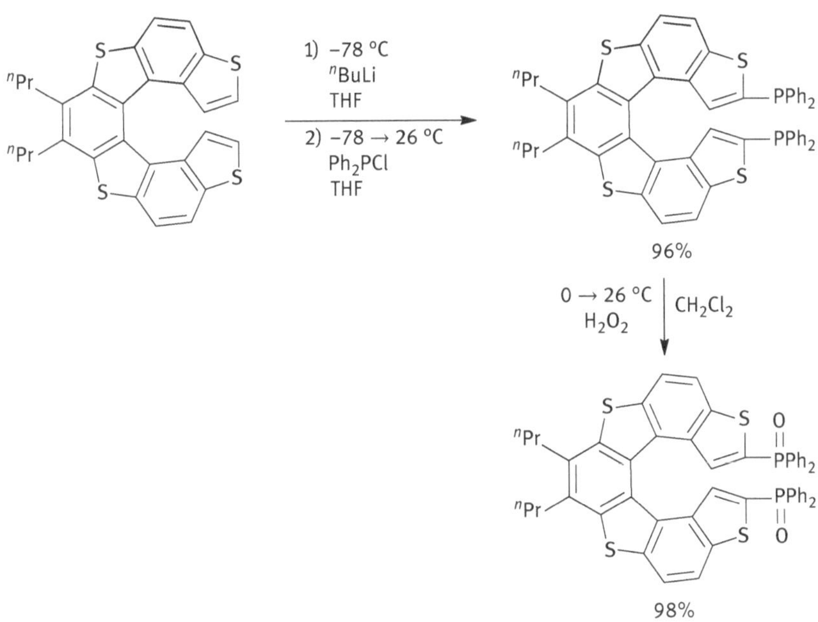

Figure 1.50: Bisphosphinoylation utilizing the lithiation method.

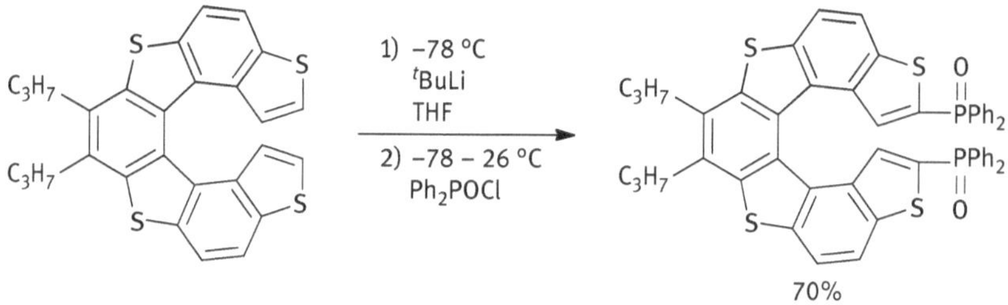

Figure 1.51: A direct bisphosphinoylation via lithiation.

1.8 Lithiation Reactions

The lithiation of either arenes/heteroarenes or bromoarenes/bromo-heteroarenes gives useful intermediates that may be phosphinoylated by reaction with a >PCl derivative following oxidation. It is also possible to apply directly a >P(O)Cl reagent.

Phosphine and phosphine oxide functions were introduced into a tetrathiaheterohelicene via the lithiation technique (Figure 1.50) [57].

Figure 1.51 shows the preparation of a bis(diphenylphosphinoyl)tetrathiahelicene via a double metallation followed by a double phosphinoylation [58].

A cyclic diimidazolophosphine was prepared by the double deprotonation of 1,2-di(*N*-imidazolyl)benzene followed by reaction with dichloro-*tert*-butylphosphine. The dicationic diimidazoliophosphine was obtained by double quaternization with MeOTf. The oxidation of phosphines bearing proximal positive charges is not easy, but could be performed by *m*-CPBA in acetonitrile to afford the corresponding phosphine oxide. As an alternative, the diimidazolophosphine was oxidized directly, and

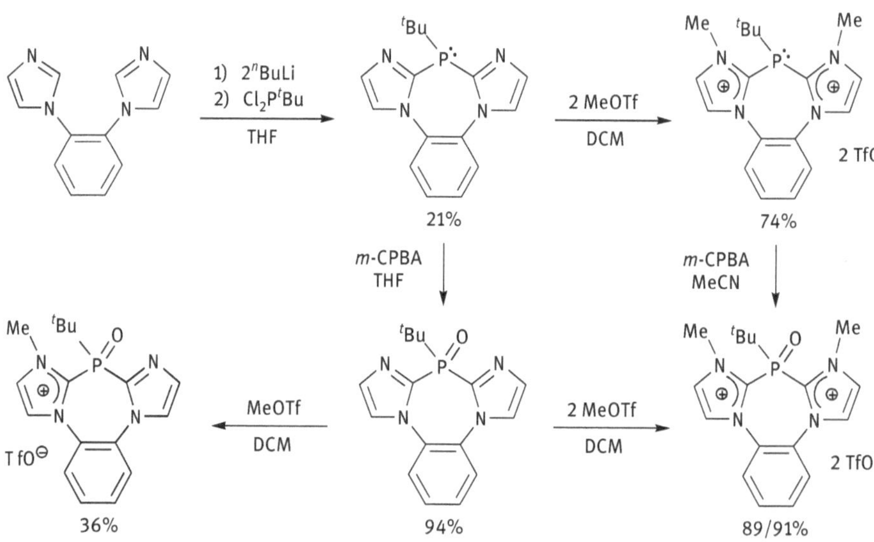

Figure 1.52: A phosphinoylation via lithiation leading to P-heterocycles after other modifications.

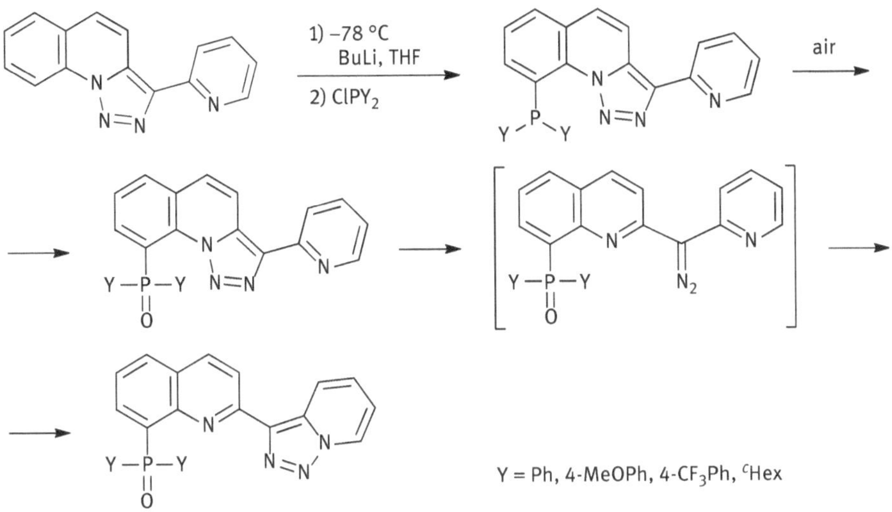

Figure 1.53: A ring-chain isomerization induced by P-functionalization.

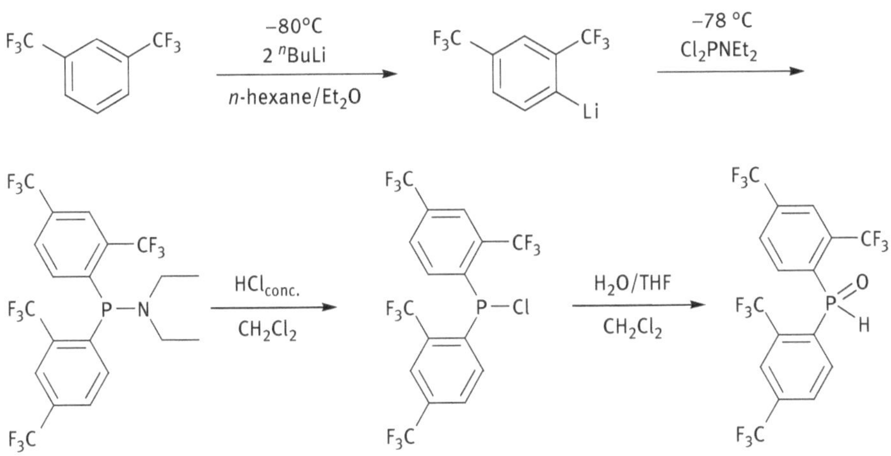

Figure 1.54: Synthesis of a fluorinated secondary phosphine oxide via lithiation.

the phosphine oxide so obtained was quaternized with 1 or 2 equivalents of MeOTf. The monoquaternization was not selective (Figure 1.52) [59].

3-(2-Pyridyl)-[1,2,3]triazolo[1,5-*a*]quinolylphosphines afforded, after P-functionalization, phosphine oxides that were transformed via a ring-chain isomerization to more stable phosphine oxides (Figure 1.53) [60].

Novel perfluoroalkylaryl and perfluoroaryl secondary phosphine oxides were synthesized in order to study the >P(O)H →P–OH tautomeric equilibrium. First, the bis[2,4-bis(trifluoromethyl)phenyl]phosphine oxide was prepared via the lithiation of the starting aromatics followed by reaction with Cl_2PNEt_2. The change in the P-function was achieved by reaction with cc. HCl, followed hydrolysis (Figure 1.54) [61].

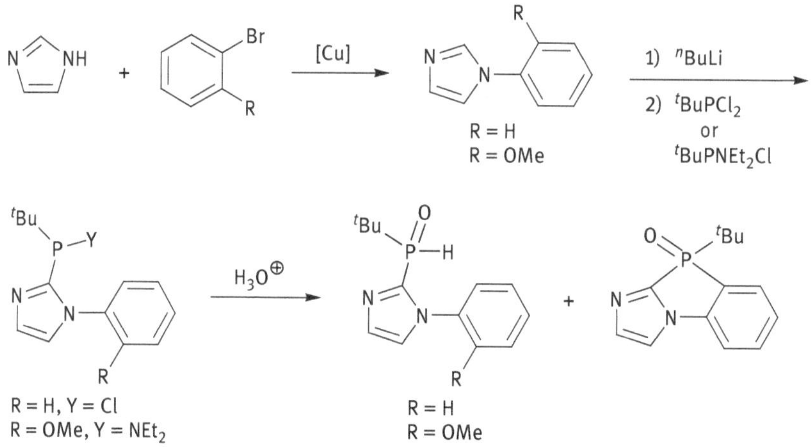

Figure 1.55: P-functionalization of imidazoles via lithiation.

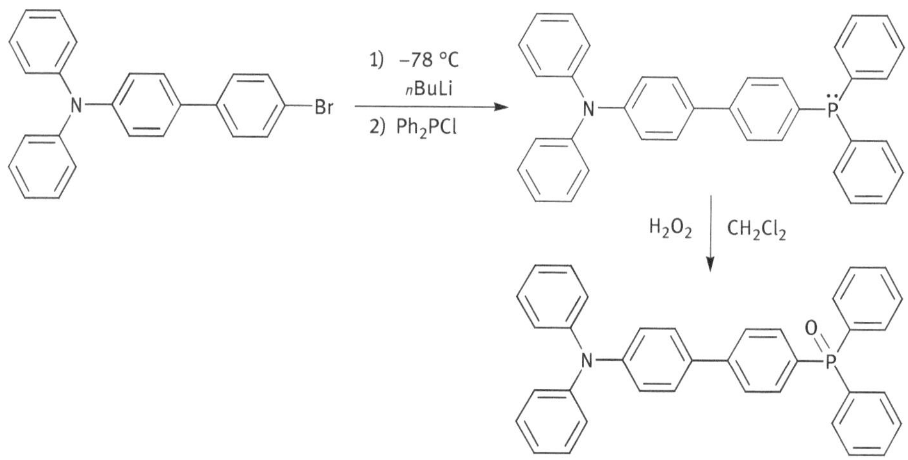

Figure 1.56: Constructing polyaryl compounds by P-functionalization.

Two air-stable secondary phosphine oxides, namely 2-(*tert*-butylhydrophosphoryl)-1-aryl-1*H*-imidazoles were synthesized by lithiathion followed by P-functionalization. The reaction was accompanied by a side-reaction giving rise to a tricyclic phosphine oxide (Figure 1.55) [62].

A 4-diphenylphosphinoyl-4'-diphenylamino-biphenyl was synthesized via metallation and phosphinoylation of the corresponding aryl bromide (Figure 1.56). The end product is a high-performance deep-blue fluorescent emitter and green phosphorescent host [63].

Anthracene-9,10-diylbis(diphenylphosphine oxide) was also made available by this approach (Figure 1.57) [64].

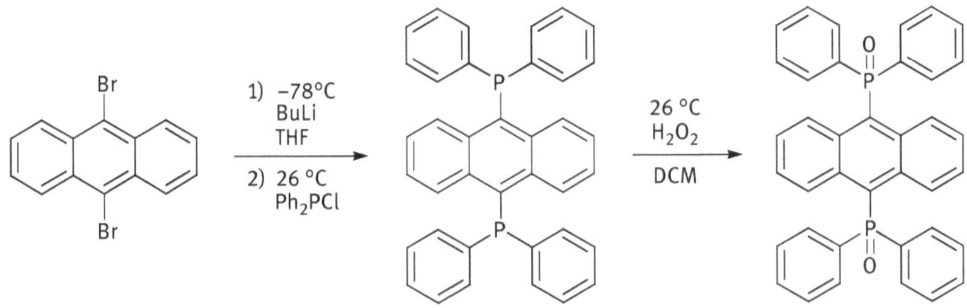

Figure 1.57: Bisphosphinoylation of anthracene following bislithiation.

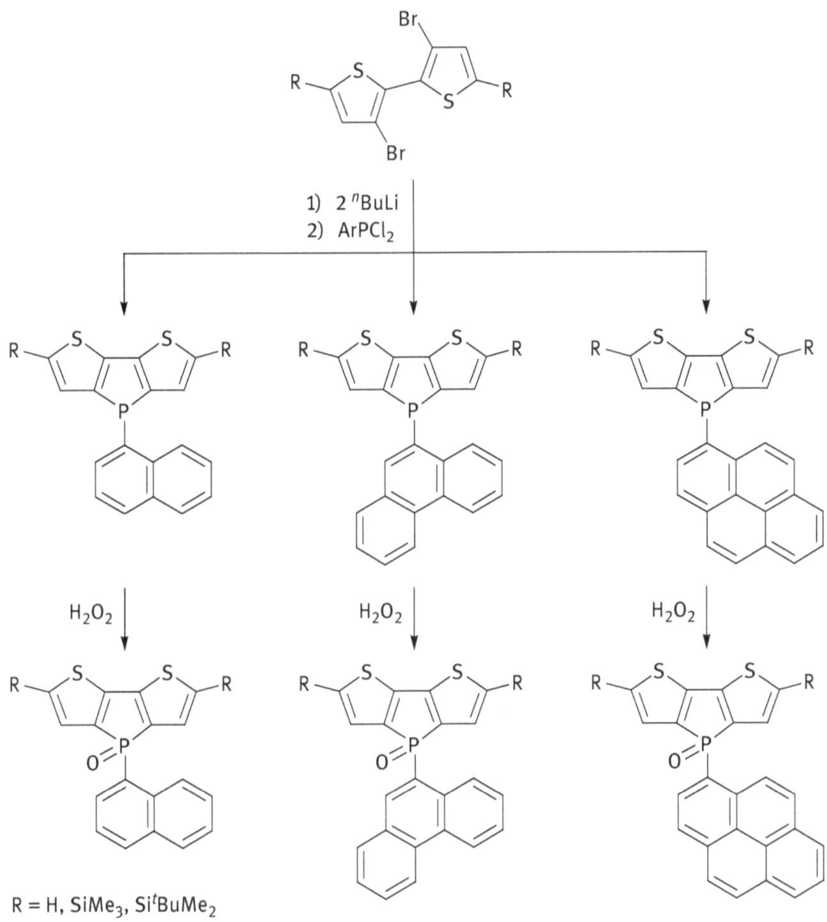

Figure 1.58: Bislithiation followed by cyclization using $ArPCl_2$.

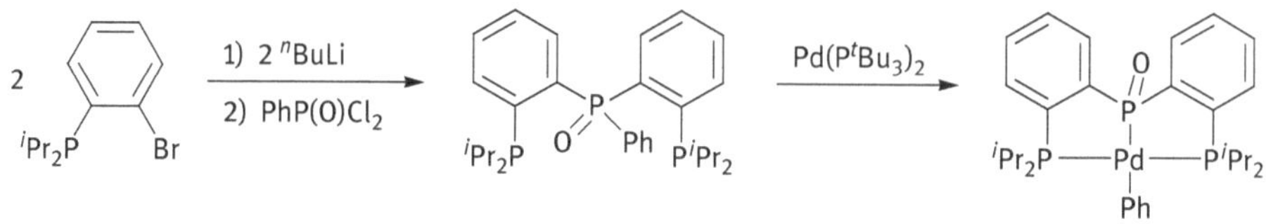

Figure 1.59: Synthesis of a bis(phosphinophenyl)-phenylphosphine oxide as a pincer ligand.

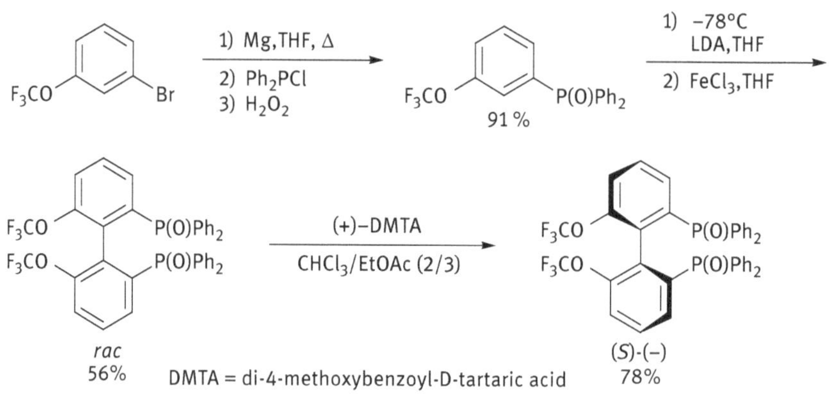

Figure 1.60: Synthesis of a bis(phosphinoyl)biphenyl by an organometallic approach.

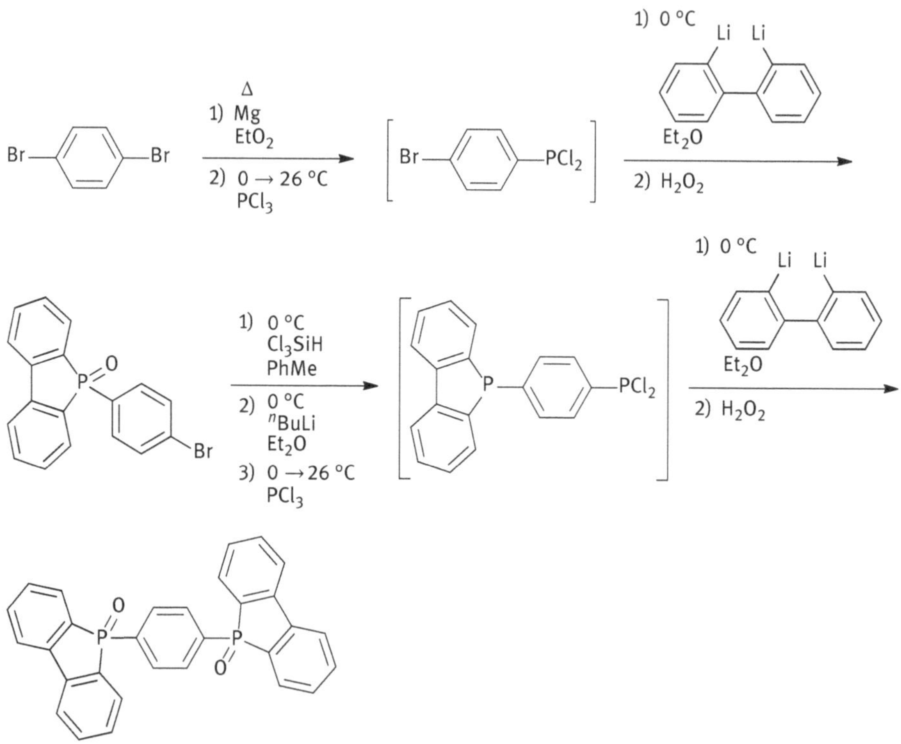

Figure 1.61: Synthesis of a bis(dibenzophosphole oxide) by an organometallic approach.

Bichromophoric dithieno[3,2-*b*:2',3'-*d*]phospholes with polyaromatic hydrocarbon substituents were synthesized by the lithiation of dibromobithiophenes followed by the reaction of the intermediates so formed with arylphosphonous dichlorides. The phosphines obtained were oxidized to the corresponding P-oxides (Figure 1.58) [65].

Another new triarylphosphine oxide was utilized in the preparation of a palladium pincer complex (Figure 1.59) [66].

The precursor of a new atropisomeric bis(phosphine oxide) derivative was synthesized from 1-bromo-3-(trifluoromethoxy)benzene via P-functionalization, coupling, and optical resolution (Figure 1.60) [67]. The series of reactions involved the formation of a Grignard reagent, a phosphinylation and lithiation.

A new bis(dibenzophosphole oxide) derivative was synthesized by the Yamaguchi group utilizing the Grignard reaction and lithiathions (Figure 1.61) [68].

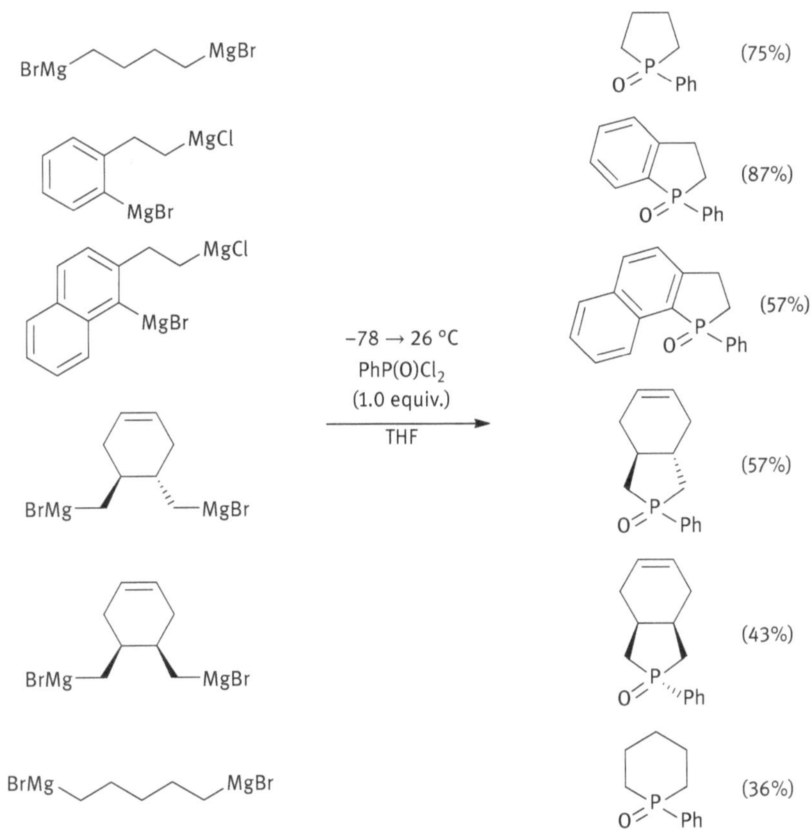

Figure 1.62: Bis(Grignard reagents) in the synthesis of P-heterocycles.

1.9 Grignard Reactions

Grignard reagents may also be the partners of >P(O)Cl or –P(O)Cl_2 intermediates to afford a variety of phosphine oxides. Moreover, >P(O)OR or –P(O)$(OR)_2$ derivatives may also be suitable starting materials.

A series of condensed phospholane oxides were prepared by the cyclization of 1,4-bis-Grignard reagents with phenylphosphonic dichloride. The use of the

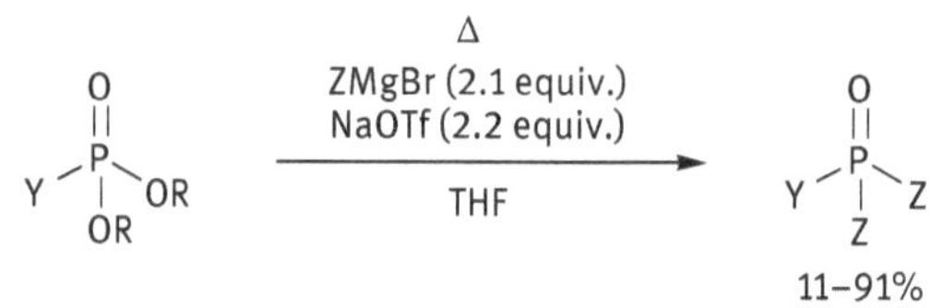

Y = ^{n}Hex, Me, Ph, 4-MeOC_6H_4, 3-MeOC_6H_4, 2-MeOC_6H_4, 2-MeC_6H_4, 4-FC_6H_4, 3-FC_6H_4, 2-FC_6H_4, 2-ClC_6H_4, 2-BrC_6H_4, 2-PhC_6H_4
R = Me, Et, ^{i}Pr
Z = Me, Et, ^{i}Pr, ^{n}Bu, ^{t}Bu, Cy, Ph

Figure 1.63: Preparation of tertiary phosphine oxides from phosphonates.

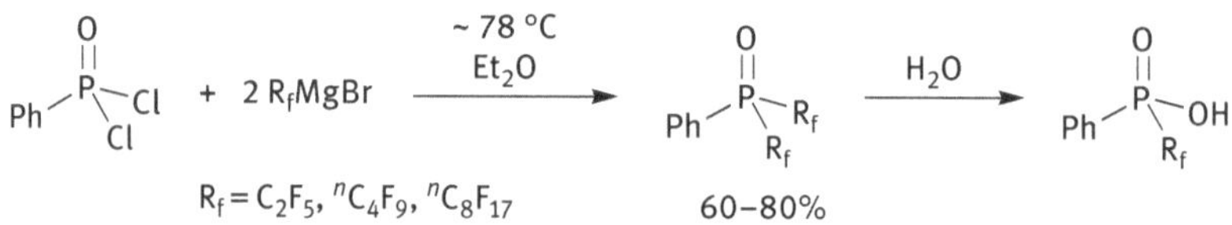

Figure 1.64: Synthesis of tertiary phosphine oxides from phenylphosphonic dichloride.

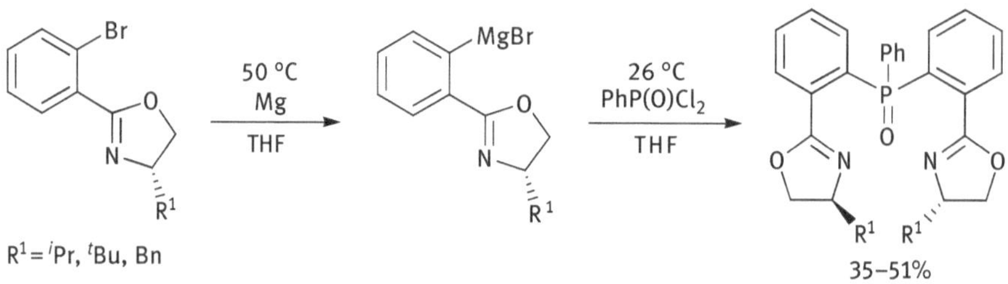

Figure 1.65: Synthesis of a triarylphosphine oxide by the Grignard approach.

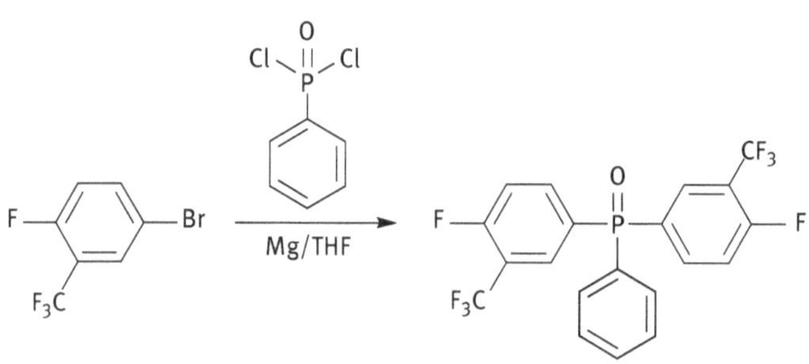

Figure 1.66: Synthesis of a fluoro-containing triarylphosphine oxide by the Grignard approach.

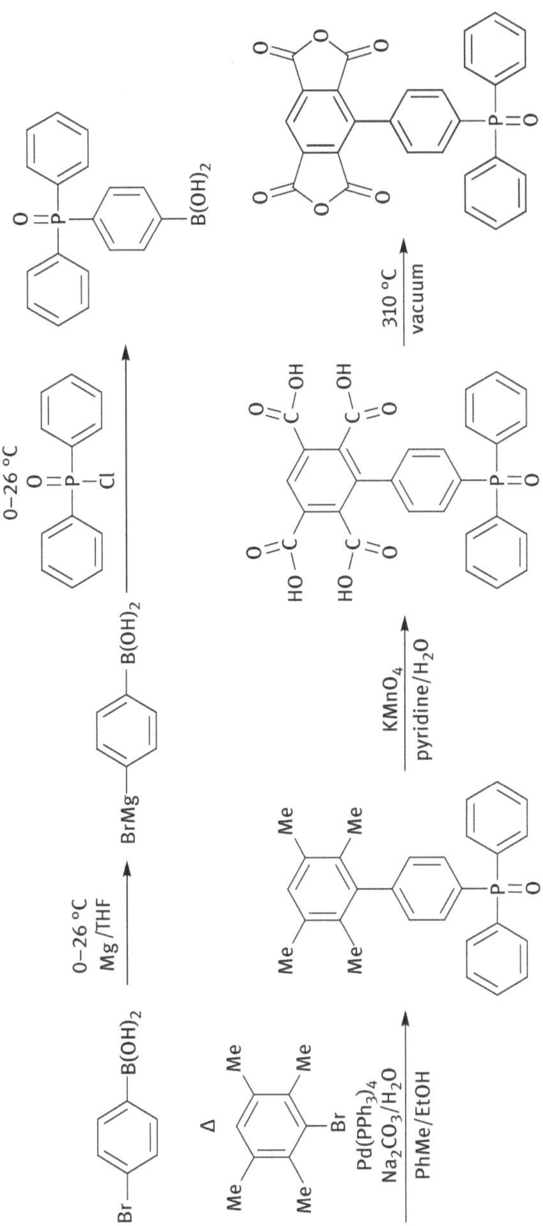

Figure 1.67: Synthesis of a boron-functionalized triarylphosphine oxide and its further utilization.

1,5-dibromopentane-based bis(Grignard reagent) led to a hexahydrophosphinine oxide (Figure 1.62) [69].

Dialkyl phosphonates were converted to tertiary phosphine oxides by reaction with two equivalents of alkyl or aryl Grignard reagents in the presence of sodium trifluoromethanesulfonate as an activator (Figure 1.63) [70].

The interaction of phenylphosphonic dichloride with perfluoroalkyl Grignard reagents furnished the corresponding phenyldi(perfluoroalkyl)phosphine oxides. Surprisingly it was observed that one of the electron-poor P–C bonds could be hydrolyzed to give the corresponding phenyl-perfluoroalkylphosphine oxide (Figure 1.64) [71].

2-Oxazolyl-bromobenzenes were converted to the corresponding Grignard reagents that were then reacted with a half equivalent of phenylphosphonic

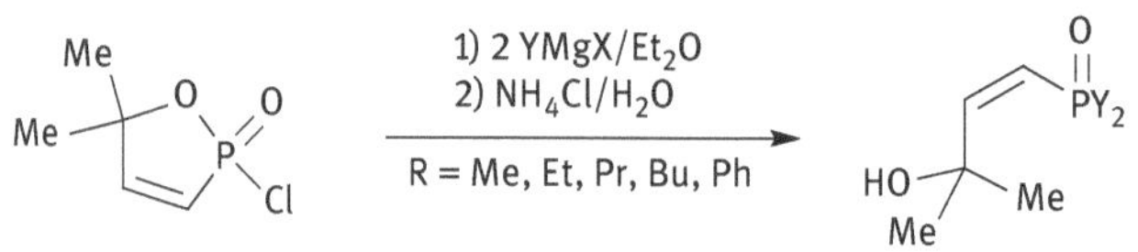

Figure 1.68: Grignard reagent-promoted ring opening of a chloro-oxaphospholene oxide.

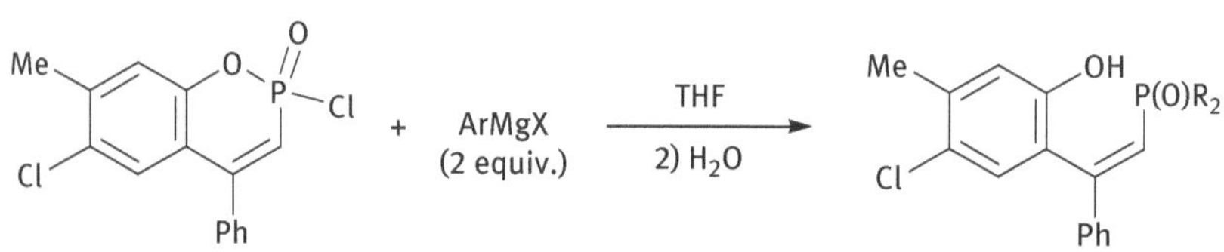

Ar = 2-MeOC$_6$H$_4$, 3,4-(MeO)$_2$C$_6$H$_3$, 3,4,5-(MeO)$_3$C$_6$H$_2$

Figure 1.69: Grignard reagent-promoted ring opening of a chloro-oxaphosphinine oxide.

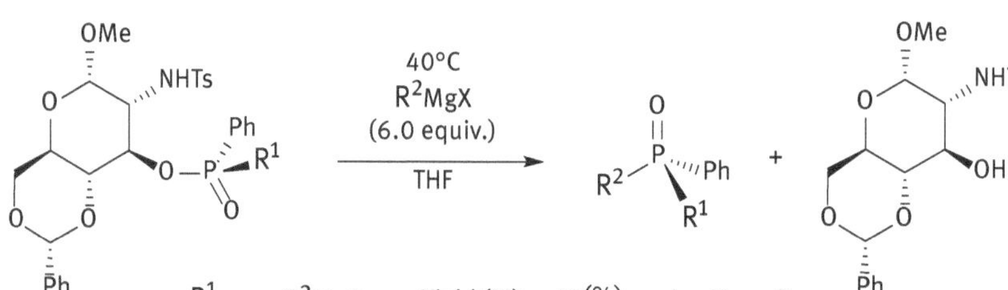

R^1	R^{2}MgX	Yield (%)	er (%)	abs. P-config.
o-An	MeMgCl	75	90	S_p
o-An	EtMgBr	85	92	S_p
Me	EtMgBr	99	94	S_p
Me	*o*-AnMgBr	84	98	R_p
Me	^{i}PrMgCl	34	90	S_p

Figure 1.70: Utilization of the Grignard reaction in an asymmetric synthesis of tertiary phosphine oxides.

dichloride. The products may be regarded as phosphine oxide-linked bis(oxazolines) (Figure 1.65) [72].

In a similar fashion, bis(4-fluoro-3-trifluoromethylphenyl-)phenylphosphine oxide was prepared (Figure 1.66), and reacted with bisphenol derivatives to furnish poly(arylene ether phosphine oxides) [73].

Additional triarylphosphine oxides were also prepared from $ArP(O)Cl_2$ and arylmagnesium bromide, and the resulting species were involved in polycondensation. The resulting poly(arylene ether phosphine oxide)-type plastics are of interest as novel materials [74].

A novel dianhydride monomer with a phosphine oxide function was prepared via the Suzuki coupling reaction of 4-(diphenylphosphinoyl)phenyl boronic acid synthesized in two steps. The methyl groups in the aromatic ring of the Suzuki product were then oxidized to carboxyl functions, and dehydrated to the corresponding bis(anhydride) (Figure 1.67). This monomer was then used in polycondensation with bis(3-aminophenyl)phenylphosphine oxide [75].

In reaction with two equivalents of a Grignard reagent, a chloro-oxaphospholene oxide was transformed to 4-hydroxy-4-methyl-butenylphosphine oxides (Figure 1.68) [76].

A chloro-oxaphosphinine oxide was converted to aryl-phenylvinylphosphine oxides in a similar fashion by reaction with two equivalents of Grignard reagents (Figure 1.69) [77].

A D-glucosamine-based phosphinate was converted to chiral tertiary phosphine oxides by reaction with a Grignard reagent (Figure 1.70). High ee values could be attained [78].

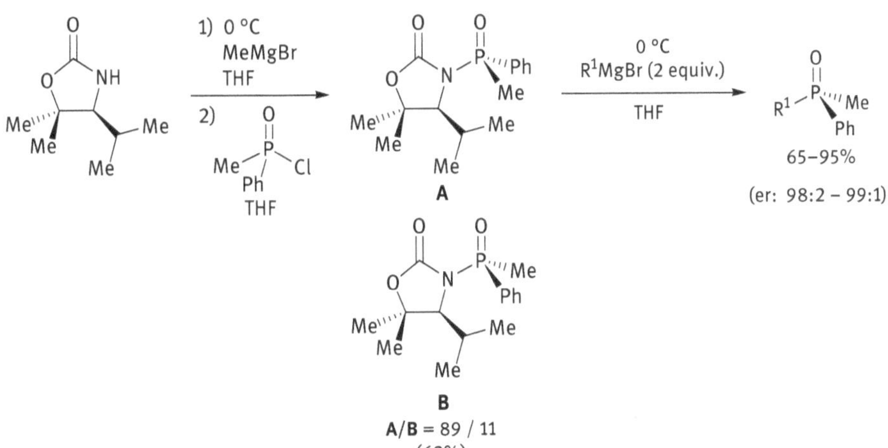

R^1 = 4-$^tBuC_6H_4$, 4-$Me_2NC_6H_4$, 4-FC_6H_4, 2-$MeOC_6H_4$, 2-MeC_6H_4, 3,5-diMeC_6H_3, Et, C_9H_{19}

Figure 1.71: Utilization of the Grignard reaction in another asymmetric synthesis of tertiary phosphine oxides.

Another chiral auxiliary-based strategy was also developed for the synthesis of P-chiral phosphine oxides. According to this, P-chiral oxazolidines were formed in a stereoselective manner that then underwent displacements with Grignard reagents to afford the desired phosphine oxides in high ee values (Figure 1.71) [79].

This chapter attempted to give an overview of the recently used approaches for the synthesis of phosphine oxides. One could see that the classical methods remained important, but new protocols were also elaborated.

Acknowledgment: This work was supported by the National Research, Development and Innovation Office (K119202).

References

[1] Y. Matano, A. Saito, Y. Suzuki, T. Miyajima, S. Akiyama, S. Otsubo, E. Nakamoto, S. Aramaki, H. Imahori, *Chem. Asian J.*, 2012, **7**, 2305.
[2] V. A. D'yakonov, A. L. Makhamatkhanova, L. K. Dilmukhametova, R. A. Agliullina, T. V. Tyumkina, U. M. Dzhemilev, *Organometallics*, 2015, **34**, 221.
[3] I. Yu. Kudryavtsev, T. V. Baulina, V. N. Khrustalev, P. V. Petrovskii†, M. P. Pasechnik, E. E. Nifant'ev, *Doklady Chem.*, 2013, **448**, 55.
[4] Y. G. Gololobov, I. Yu. Krasnova, S. V. Barabanov, V. N. Khrustalev, S. A. Andronati, V. I. Pavlovsky, *Tetrahedron Lett.*, 2014, **55**, 4879.
[5] M. Baumann, I. R. Baxendale, *J. Org. Chem.*, 2015, **80**, 10806.
[6] A. S. Bogachenkov, M. M. Efremova, B. I. Ionin, *Tetrahedron Lett.*, 2012, **53**, 2100.
[7] J. Ballester, J. Gatignol, G. Schmidt, C. Alayrac, A.-C. Gaumont, M. Taillefer, *ChemCatChem.*, 2014, **6**, 1549.
[8] R. A. Dhokale, S. B. Mhaske, *Org. Lett.*, 2013, **15**, 2218.
[9] R. Premraj, M. D. McLeod, G. W. Simpson, M. G. Banwell, *Heterocycles*, 2012, **85**, 2949.
[10] Y. Wang, J. Gan, L. Liu, H. Yuan, Y. Gao, Y. Liu, Y. Zhao, *J. Org. Chem.*, 2014, **79**, 3678.
[11] J.-M. Camus, S. Pailloux, C. E. Shirima, R. T. Paine, *Main Group Chem.*, 2011, **10**, 25.
[12] D. Rosario-Amorin, S. Ouizem, D. A. Dickie, R. T. Paine, R. E. Cramer, B. P. Hay, J. Podair, L. H. Delmau, *Inorg. Chem.*, 2014, **53**, 5698.
[13] D. Rosario-Amorin, E. N. Duesler, R. T. Paine, B. P. Hay, L. H. Delmau, S. D. Reilly, A. J. Gaunt, B. L. Scott, *Inorg. Chem.*, 2012, **51**, 6667.
[14] Y. Fang, L. Wu, J. Liao, L. Chen, Y. Yang, N. Liu, L. He, S. Zou, W. Feng, L. Yuan, *RSC Adv.*, 2013, **3**,12376.
[15] E. Jablonkai, G. Keglevich, *Curr. Org. Synth.*, 2014, **11**, 429.
[16] E. Jablonkai, G. Keglevich, *Org. Prep. Proc. Int.*, 2014, **46**, 281.
[17] A. J. Bloomfield, S. B. Herzon, *Org. Lett.*, 2012, **14**, 4370.
[18] S. M. Rummelt, M. Ranocchiari, J. A. van Bokhoven, *Org. Lett.*, 2012, **14**, 2188.
[19] T. Fu, H. Qiao, Z. Peng, G. Hu, X. Wu, Y. Gao, Y. Zhao, *Org. Biomol. Chem.*, 2014, **12**, 2895.
[20] G. Tran, D. G. Pardo, T. Tsuchiya, S. Hillebrand, J.-P. Vors, J. Cossy, *Org. Lett.*, 2013, **15**, 5550.

[21] G. S. Ananthnag, J. T. Mague, M. S. Balakrishna, *Dalton Trans.*, 2015, **44**, 3785.
[22] X. Zhang, H. Liu, X. Hu, G. Tang, J. Zhu, Y. Zhao, *Org. Lett.*, 2011, **13**, 3478.
[23] Y.-L. Zhao, G.-J. Wu, Y. Li, L.-X. Gao, F.-S. Han, *Chem. Eur. J.*, 2012, **18**, 9622.
[24] J. Xuan, T.-T. Zeng, J.-R. Chen, L.-Q. Lu, W.-J. Xiao, *Chem. A Eur. J.*, 2015, **21**, 4962.
[25] Y.-L. Zhao, G.-J. Wu, F.-S. Han, *Chem. Commun.*, 2012, **48**, 5868.
[26] J.-S. Zhang, T. Chen, J. Yang, L.-B. Han, *Chem. Commun.*, 2015, **51**, 7540.
[27] G. Hu, W. Chen, T. Fu, Z. Peng, H. Qiao, Y. Gao, Y. Zhao, *Org. Lett.*, 2013, **15**, 5362.
[28] M. Stankevič, A. Włodarczyk, *Tetrahedron*, 2013, **69**, 73.
[29] B. Xiong, M. Li, Y. Liu, Y. Zhou, C. Zhao, M. Goto, S.-F. Yin, L.-B. Han, *Adv. Synth. Catal.*, 2014, **356**, 781.
[30] E. Rafter, J. Muldoon, H. Müller-Bunz, D. G. Gilheany, *Tetrahedron: Asymmetry*, 2011, **22**, 1680.
[31] M.-C. Duclos, Y. Singjunla, C. Petit, A. Favre-Reguillon, E. Jeanneau, F. Popowycz, E. Metay, M. Lemaire, *Tetrahedron Lett.*, 2012, **53**, 5984.
[32] Z. Wu, C. Wang, L. N. Zakharov, P. R. Blakemore, *Synthesis*, 2014, **46**, 678.
[33] C.-G. Feng, M. Ye, K.-J. Xiao, S. Li, J.-Q. Yu, *J. Am. Chem. Soc.*, 2013, **135**, 9322.
[34] H.-J. Zhang, W. Lin, Z. Wu, W. Ruan, T.-B. Wen, *Chem. Commun.*, 2015, **51**, 3450.
[35] S. Furukawa, S. Haga, J. Kobayashi, T. Kawashima, *Org. Lett.*, 2014, **16**, 3228.
[36] M. Hatano, T. Horibe, K. Ishihara, *Angew. Chem. Int. Ed.*, 2013, **52**, 4549.
[37] V. K. Brel, E. I. Goryunov, I. B. Goryunova, N. G. Frolova, E. A. Latokhina, A. G. Buyanovskaya, L. V. Gumileva, E. E. Nifant'ev, *Doklady Chem.*, 2015, **462**, 152.
[38] E. I. Goryunov, I. B. Goryunova, K. A. Lysenko, N. G. Frolova, E. A. Latokhina, V. K. Brel, *Russ. J. Gen. Chem.*, 2015, **85**, 2112.
[39] B. Cho, C.-H. Tan, M. W. Wong, *Org. Biomol. Chem.*, 2011, **9**, 4550.
[40] M. Moreno, G. Lligadas, J. C. Ronda, M. Galia, V. Caldiz, *J. Polym. Sci. Pol. Chem.*, 2012, **50**, 3206.
[41] E. Bálint, J. Takács, L. Drahos, G. Keglevich, *Heteroat. Chem.*, 2012, **23**, 235.
[42] W.-J. Yoo, S. Kobayashi, *Green Chem.*, 2013, **15**, 1844.
[43] Y. Saga, D. Han, S.-i. Kawaguchi, A. Ogawa, L.-B. Han, *Tetrahedron Lett.*, 2015, **56**, 5303.
[44] N. K. Gusarova, V. A. Kuimov, S. F. Malysheva, N. A. Belogorlova, A. I. Albanov, B. A. Trofimov, *Tetrahedron*, 2012, **68**, 9218.
[45] J. Michalska, B. Boduszek, T. K. Olszewski, *Heteroat. Chem.*, 2011, **22**, 617.
[46] G. K. Ingle, Y. Liang, M. G. Mormino, G. Li, F. R. Fronczek, J. C. Antilla, *Org. Lett.*, 2011, **13**, 2054.
[47] B. Boduszek, T. K. Olszewski, W. Goldeman, K. Grzegolec, P. Blazejewska, *Tetrahedron*, 2012, **68**, 1223.
[48] E. Bálint, E. Fazekas, G. Pintér, Á. Szöllősy, T. Holczbauer, M. Czugler, L. Drahos, T. Körtvélyesi, G. Keglevich, *Curr. Org. Chem.*, 2012, **16**, 547.
[49] E. Bálint, E. Fazekas, P. Pongrácz, L. Kollár, L. Drahos, T. Holczbauer, M. Czugler, G. Keglevich, *J. Organomet. Chem.*, 2012, **717**, 75.
[50] R. A. Cherkasov, A. R. Garifzyanov, S. A. Koshkin, N. V. Davletshina, *Russ. J. Gen. Chem.*, 2012, **82**, 1453.
[51] G. Keglevich, E. Bálint, *Molecules*, 2012, **17**, 12821.
[52] Y. Huang, W. Hao, G. Ding, M.-Z. Cai, *J. Organomet. Chem.*, 2012, **715**, 141.
[53] V. Srinivas, E. Balaraman, K. V. Sajna, K. C. Kumara Swamy, *Eur. J. Org. Chem.*, 2011, 4222.
[54] Y. Unoh, K. Hirano, T. Satoh, M. Miura, *Angew. Chem. Int. Ed.*, 2013, **52**, 12975.
[55] Y.-R. Chen, W.-L. Duan, *J. Am. Chem. Soc.*, 2013, **135**, 16754.
[56] W. Ma, L. Ackermann, *Synthesis*, 2014, **46**, 2297.
[57] M. Monteforte, S. Cauteruccio, S. Maiorana, T. Benincori, A. Forni, L. Raimondi, C. Graiff, A. Tiripicchio, G. R. Stephenson, E. Licandro, *Eur. J. Org. Chem.*, 2011, 5649.
[58] S. Cauteruccio, D. Dova, M. Benaglia, A. Genoni, M. Orlandi, E. Licandro, *Eur. J. Org. Chem.*, 2014, 2694.

[59] C. Maaliki, Y. Canac, C. Lepetit, C. Duhayon, R. Chauvin, *RSC Adv.*, 2013, **3**, 20391.
[60] R. Ballesteros-Garrido, B. Abarca, R. Ballesteros, F. Colobert, F. R. Leroux, E. García-España, *RSC Adv.*, 2015, **5**, 29809.
[61] B. Kurscheid, W. Wiebe, B. Neumann, H.-G. Stammler, B. Hoge, *Eur. J. Inorg. Chem.*, 2011, 5523.
[62] D.-F. Hu, C.-M. Weng, F.-E Hong, *Organometallics*, 2011, **30**, 1139.
[63] X.-K. Liu, C.-J. Zheng, M.-F. Lo, J. Xiao, C.-S. Lee, M.-K. Fung, X.-H. Zhang, *Chem. Commun.*, 2014, **50**, 2027.
[64] Y. Zhao, L. Duan, X. Zhang, D. Zhang, J. Qiao, G. Dong, L. Wang, Y. Qiu, *RSC Adv.*, 2013, **3**, 21453.
[65] C. J. Chua, Y. Ren, T. Baumgartner, *Organometallics*, 2012, **31**, 2425.
[66] E. J. Derrah, S. Ladeira, G. Bouhadir, K. Miqueu, D. Bourissou, *Chem. Commun.*, 2011, **47**, 8611.
[67] D.-Y. Zhang, C.-B. Yu, M.-C. Wang, K. Gao, Y.-G. Zhou, *Tetrahedron Lett.*, 2012, **53**, 2556.
[68] A. Fukazawa, M. Kiguchi, S. Tange, Y. Ichihashi, Q. Zhao, T. Takahashi, T. Konishi, K. Murakoshi, Y. Tsuji, A. Staykov, K. Yoshizawa, S. Yamaguchi, *Chem. Lett.*, 2011, **40**, 174.
[69] A. M. Gregson, S. M. Wales, S. J. Bailey, A. C. Willis, P. A. Keller, *J. Org. Chem.*, 2015, **80**, 9774.
[70] A. J. Kendall, C. A. Salazar, P. F. Martino, D. R. Tyler, *Organometallics*, 2014, **33**, 6171.
[71] A. I. Hosein, A. J. M. Caffyn, *Dalton Trans.*, 2012, **41**, 13504.
[72] Y. Jin, D.-M. Du, *Tetrahedron*, 2012, **68**, 3633.
[73] A. Ghosh, D. Bera, D.-Y. Wang, H. Komber, A. K. Mohanty, S. Banerjee, B. Voit, *Macromol. Mater. Eng.*, 2012, **297**, 145.
[74] H. Satpathi, A. Ghosh, H. Komber, S. Banerjee, B. Voit, *Eur. Polym. J.*, 2011, **47**, 196.
[75] Y.-U. Bae, T.-H. Yoon, *J. Appl. Polym. Sci.*, 2012, **123**, 3298.
[76] D. A. Tatarinov, V. K. Brel', V. F. Mironov, *Russ. J. Org. Chem.*, 2015, **51**, 1245.
[77] R. R. Zairov, D. A. Tatarinov, N. A. Shamsutdinova, A. R. Mustafina, I. Kh. Rizvanov, V. V. Syakaev, V. F. Mironov, A. I. Konovalov, *Russ. J. Org. Chem.*, 2014, **50**, 547.
[78] A. D'Onofrio, L. Copey, L. Jean-Gérard, C. Goux-Henry, G. Pilet, B. Andrioletti, E. Framery, *Org. Biomol. Chem.*, 2015, **13**, 9029.
[79] H. Adams, R. C. Collins, S. Jones, C. J. A. Warner, *Org. Lett.*, 2011, **13**, 6576.

Nóra Zsuzsa Kiss and György Keglevich

2 Methods for the preparation of phosphinates and phosphonates with a focus on recent advances

Abstract: Phosphinic and phosphonic acid esters (phosphinates and phosphonates) are useful starting materials and intermediates in synthetic organic chemistry. Their traditional synthesis involves the reaction of the corresponding acid chlorides with alcohols in the presence of a base. This chapter aims at summarizing the alternative possibilities including the "greener" approaches for the synthesis of phosphinates and phosphonates, starting preferably from the corresponding acids. One of the possibilities discussed involves the conversion of acid to a more active intermediate that may then react with the nucleophile efficiently. Other green methods involve the microwave (MW)-assisted direct esterification, or the alkylating esterification. Besides these, other methods are also discussed, with the Arbuzov and the Hirao reactions being the most important. The majority of the methods discussed are suitable, with slight modifications, for both the preparation of phosphinates and phosphonates.

Keywords: phosphinates, phosphonates, esterification.

2.1 Esterification of Phosphinic Chlorides

The traditional, although not environmentally friendly, yet widely used method for the preparation of phosphinates (**2**) involves the reaction of phosphinic chlorides (**1**) with an alcohol or phenol as the nucleophile (Figure 2.1) [1–6]. To remove the hydrochloric acid formed, mostly a tertiary amine [7] was used in an apolar solvent. Alcoholates or phenolates may also be applied [8, 9].

Cyclic phosphinic chlorides (such as 1-chloro-3-phospholene oxide **3**) may also serve as starting materials in the reaction with alcohols or phenols to afford the corresponding alkyl- or aryl phosphinates (**4**) (Figure 2.2) [10–12].

Analogously, phosphonic acid dichlorides (**5**) may be converted to the corresponding phosphonates (**7**) by reaction with two equivalents of the alcohol [13], or to phosphonic acid monoesters (**8**) by reaction with one equivalent of the alcohol, followed by hydrolysis of the phosphonic ester-chloride (**6**) intermediate (Figure 2.3) [14].

https://doi.org/10.1515/9783110535839-002

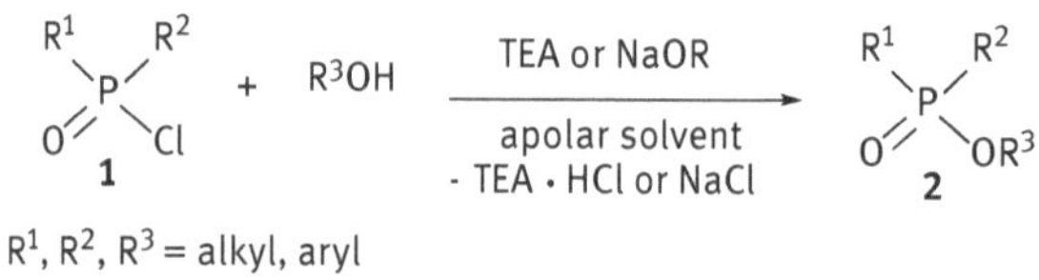

Figure 2.1: Traditional method for the synthesis of phosphinates.

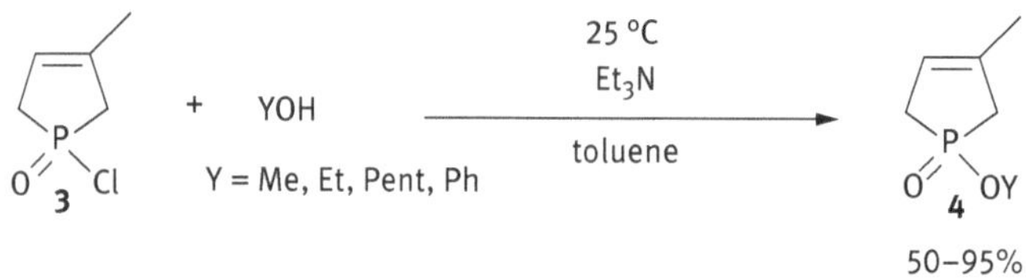

Figure 2.2: The synthesis of alkoxy- and aryloxy-3-phospholene oxides by traditional routes.

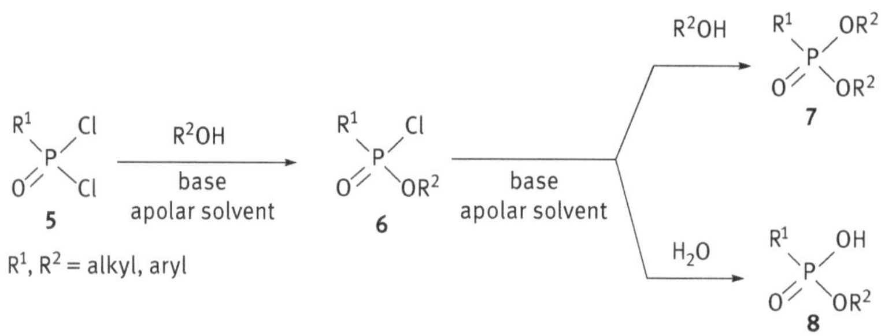

Figure 2.3: Simplest synthesis of phosphonic mono- and diesters.

2.2 Esterification of Phosphinic Acids Utilizing Activating Agents

Transforming phosphinic acids into an intermediate with a better leaving group by an activating/condensating agent makes the nucleophilic substitution on the phosphorus atom feasible. Usually, a large excess of the alcohol is required, and the yields are dependent on the components being coupled. One such example is the activation of the phosphinic acid by reaction with dicyclohexylcarbodiimide (DCC) [15]. As a refinement, the esterification of *H*-phosphinic acids (**9**) was reported in the presence of DCC and 10% of *N,N*-dimethylaminopyridine by applying a slight excess of the alcohol in THF. The phosphinates (**10**) were obtained in good yields (Figure 2.4) [16–18].

Phosphonic acid monoesters can also be synthesized from the corresponding phosphonic acids using DCC as the activating agent in the presence of trimethylamine or pyridine as the base [19–21]. Alkyl- and aryl phosphinates were successfully synthesized by applying *N,N'*-carbonyl diimidazole [22] or uranium-based salts, such as *N,N,N',N'*-tetramethyl-*O*-(benzotriazol-1-yl) [23] as the coupling agents.

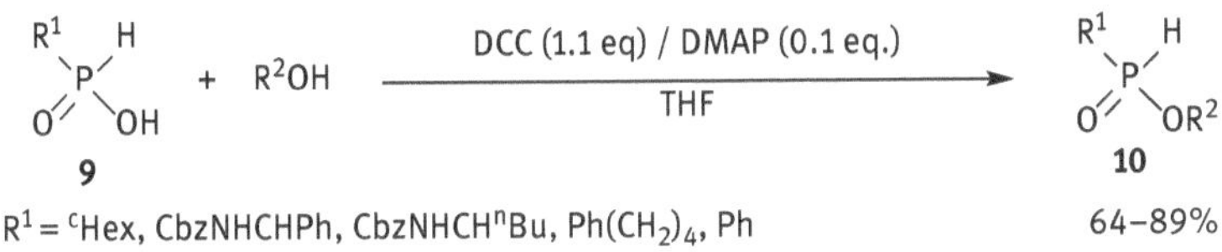

R^1 = ^{c}Hex, CbzNHCHPh, CbzNHCH^{n}Bu, $Ph(CH_2)_4$, Ph
R^2 = Et, ^{i}Pr, Bu, ^{t}Bu, etc.

Figure 2.4: Esterification of *H*-phosphinic acids using DCC as the activating agent.

Trichloroacetonitrile is another activating agent that can be applied in the preparation of phosphate monoesters [24]. A MW-assisted variation was also described for the preparation of biologically interesting phosphate diesters (**12**) containing alkyl, benzyl, and prenyl substituents using pyridine as the solvent (Figure 2.5) [25].

A better option for an activating agent is propylphosphonic anhydride (T3P®) applied widely in synthetic chemistry [26] (see also Chapter 7). Cyclic phosphinic acids (**13**) could be esterified efficiently with a series of alcohols using 1.1 equivalents of the T3P® reagent at room temperature (Figure 2.6) [27].

The role of the T3P® reagent is to form a reactive mixed anhydride (**15**) from the phosphinic acid (Figure 2.7), which may then react with the alcohol at milder conditions [27].

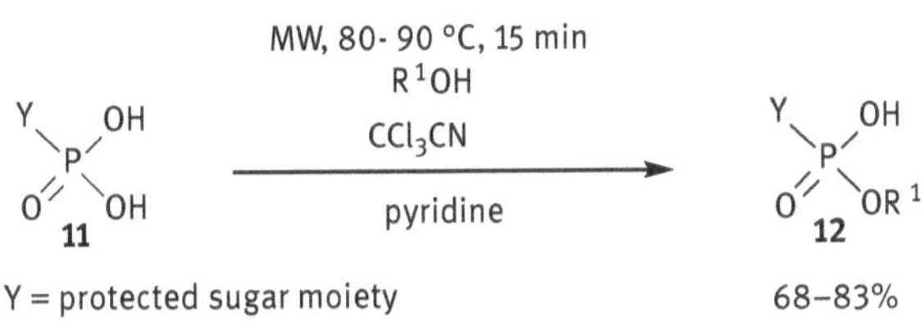

Y = protected sugar moiety
R^1 = Dodecil, Bn, farnesyl, solanesyl

Figure 2.5: Trichloroacetonitrile-mediated esterification of phosphoric acid monoesters.

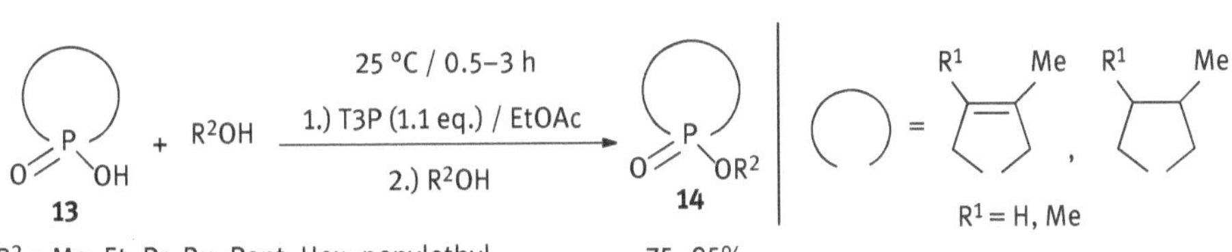

R^2 = Me, Et, Pr, Bu, Pent, Hex, penylethyl,
2-(1-naphthyl)ethyl, menthyl

Figure 2.6: Esterification of cyclic phosphinic acids in the presence of T3P® as the activating agent.

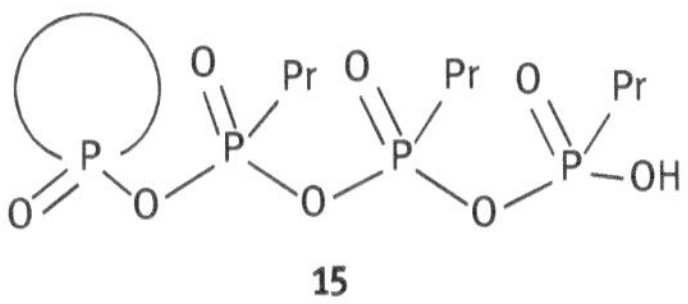

Figure 2.7: The reactive intermediate formed from the phosphinic acid and the T3P® reagent.

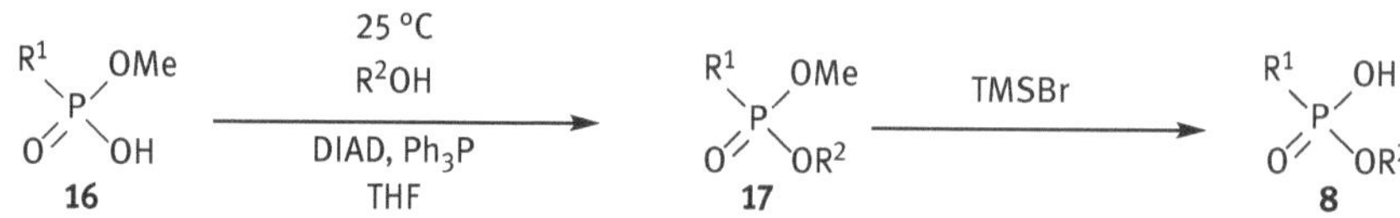

R^1 = ^{t}Bu, Bn, [*N*-CBZamino]methyl, (R,S)-[*N*-CBZvalinyl] 64-94%
R^2 = Et, ^{i}Pr
CBZ: Carbobenzyloxy

Figure 2.8: Mitsunobu reaction in the synthesis of phosphonates.

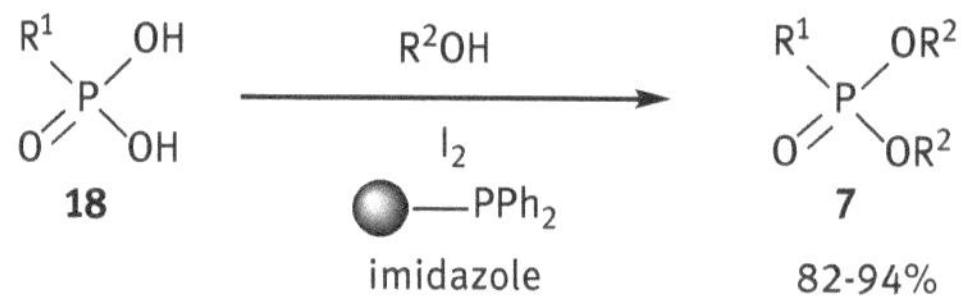

R^1 = Me Et, Pr
R^2 = Me, Pent, Dec

Figure 2.9: Esterification of alkylphosphonic acids using polymer-bound triphenylphosphine and iodine as the activating agent.

With regard to the question if the two other units of the T3P® reagent may also be utilized, it was found that in the case of more reactive phosphinic acids, such as *H*-phenylphosphinic acid, it is enough to use 0.66 equivalents of the T3P® reagent. Moreover, when applying a higher temperature of 85°C under MW conditions, already 0.44 equivalents of the coupling reagent was found to be enough [28].

The Mitsunobu reaction utilizing the redox system of triphenylphosphine and a dialkyl azodicarboxylate is a mild and efficient method to condense a phosphonic acid with an alcohol providing a convenient route to mono- and diphosphonates [29]. Starting from a phosphonic acid monoester (**16**), mixed diesters (**17**) were obtained. In the example shown in Figure 2.8, the methyl ester function was cleaved by TMSBr to afford eventually another monoester (**8**).

A mild and efficient method for the esterification of alkylphosphonic acids (**11**) was developed using primary alcohols, iodine, imidazole and polymer-bound triphenylphosphine (Figure 2.9). In this way, the triphenylphosphine oxide formed may be removed by simple filtration [30].

2.3 Special Activation of P-acids by Orthoesters, Orthosilicates, or Trialkyl Phosphites

Reactants, such as orthoesters (n=0), orthoformates/orthoacetates (n=1), and ketals or acetals (n=2) may undergo reaction with hypophosphorous acid (Figure 2.10) [31].

The reaction with orthoesters took place at room temperature, while the reaction with ketals and acetals required higher temperatures to result in the formation of

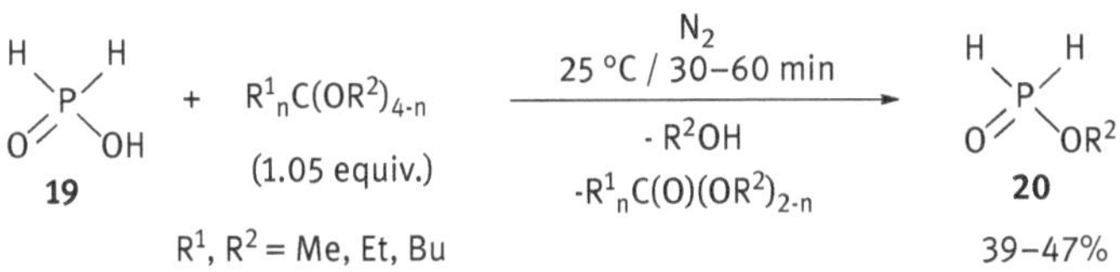

Figure 2.10: Esterification of hypophosphorous acid by orthoesters and related derivatives.

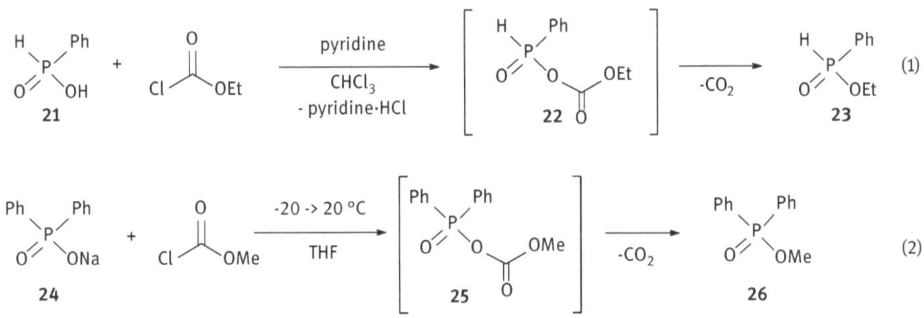

Figure 2.11: Esterification of phosphinic acids by alkyl chloroformates.

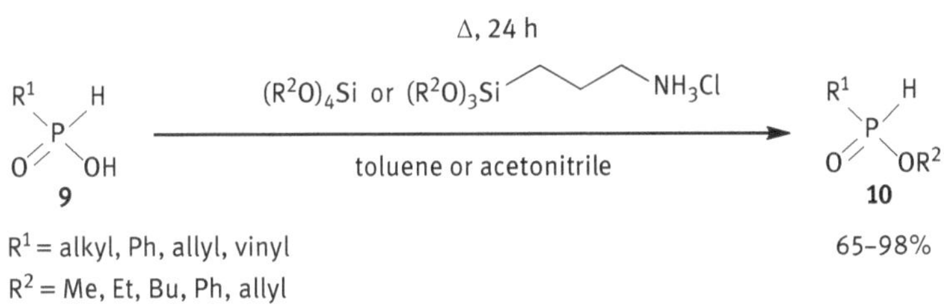

Figure 2.12: Esterification of *H*-phosphinic acids by orthosilicates or aminoalkyl-alkoxysilanes.

hypophosphorous acid esters (**20**) [31, 32]. The moderate yields can be explained by the decomposition of the esters. Yields could be improved applying trialkyl orthocarboxylates, such as $MeC(OMe)_3$ in ionic liquids (ILs) (e.g. [bmim][PF_6], [bmim][BF_4], or [bmim][Cl]) [33].

Alkyl chloroformates were also used in the preparation of *H*-phosphinates (**23**). See the esterification of phenyl-*H*-phosphinic acid (**21**) with ethyl chloroformate (Figure 2.11-1) [34], or the reaction of the sodium salt of diphenylphosphinic acid (**24**) with methyl chloroformate accompanied by the elimination of CO_2 (Figure 2.11-2) [35].

The use of allyl- or benzyl chloroformate in dichlorlomethane afforded the corresponding esters in better yields [36].

H-phosphinic acids (**9**) may react with orthosilicates or aminoalkyl-alkoxysilanes to furnish *H*-phosphinates (**10**) in good yields (Figure 2.12) [37, 38].

In a similar way, trialkoxysilanes [$(RO)_3SiR'$] or dialkoxysilanes could also be applied, showing a somewhat decreased reactivity. Trialkyl phosphites may also be effective esterifying agents, although this method is of limited use [39, 40].

2.4 MW-assisted Direct Esterification of Phosphinic and Phosphonic Acids

Phosphinic and phosphonic acids fail to undergo direct esterification by alcohols on conventional heating. Their direct esterification has only been reported in a few cases applying special catalysts, extreme reaction conditions, and long (up to 22 days) reaction times, providing the esters in moderate yields [41–45]. According to a novel observation, the direct esterifications may take place under MW conditions. A series of phosphinic acids (**27**) underwent esterification with normal alcohols at 160–230°C on MW irradiation. The direct esterification of the reactive phenyl-*H*-phosphinic acid (**27A**) required a temperature of 160–190°C, and afforded the corresponding phosphinates (**28A**) in good (73–90%) yields (Figure 2.13) [46]. The similar esterification of cyclic phosphinic acids, such as 1-hydroxy-3-phospholene oxides, 1-hydroxyphospholane oxides, and a 1-hydroxy-1,2,3,4,5,6-hexahydrophosphinine oxide (**27B-D**, respectively), at 180–235°C gave the alkyl phosphinates (**28B-D**) in acceptable to excellent yields (54–95%) (Figure 2.13) [47–49]. The esterifications were more efficient with non-volatile and sterically non-hindered alcohols.

The MW-assisted direct esterification was also extended to the synthesis of phenylphosphonic-diesters [50]. However, in these cases, the yields were low.

The only drawback of the method outlined above is the relatively high reaction temperature required. To overcome this shortcoming, ionic liquids (ILs) were tested as additives. It was found, that a catalytic amount (10%) of [bmim][PF_6] significantly enhanced

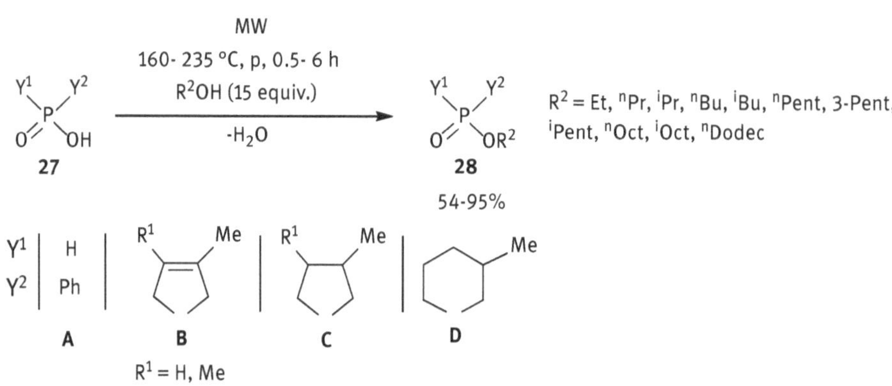

Figure 2.13: MW-assisted direct esterification of phosphinic acids.

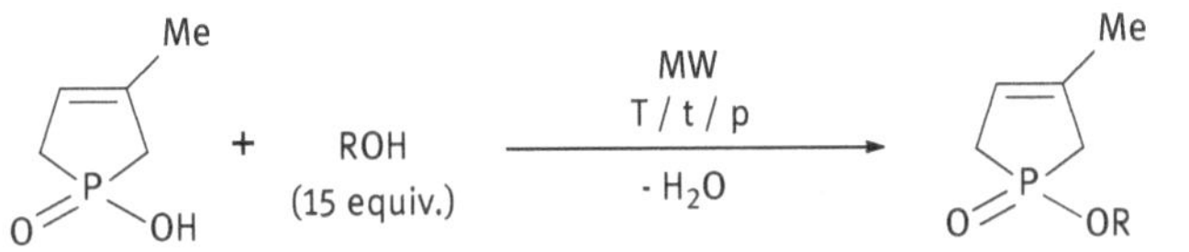

27B ($R^1 = H$) **28B** ($R^1 = H$)

Figure 2.14: Direct esterification of phosphinic acids in the presence of a catalytic amount of an IL.

Table 2.1: Direct esterification of 1-hydroxy-3-methyl-3-phospholene 1-oxide (**27B**) in the absence or presence of 10% of [bmim][PF_6] as the catalyst.

Entry	R	IL (%)	T (°C)	t (h)	Yield (%)
1	Me	–	160	4	32
2	Me	10	160	3	38
3	Et	–	160	4	30
4	Et	10	160	3	60
5	^{i}Pr	–	180	4	22
6	^{i}Pr	10	180	3	60
7	^{n}Bu	–	200	2	58
8	^{n}Bu	10	180	0.5	83
9	^{n}Pent	–	220	2.5	82
10	^{n}Pent	10	180	0.5	94
11	^{i}Pent	–	235	3	76
12	^{i}Pent	10	180	0.5	95
13	^{n}Oct	–	220	2	71
14	^{n}Oct	10	180	0.33	85
15	^{n}Dodecyl	–	230	2	95
16	^{n}Dodecyl	10	180	0.33	94

the esterification reactions: quantitative conversions were attained already at lower temperatures in shorter reaction times. One example is shown in Figure 2.14, Table 2.1.

2.5 Alkylating Esterification of Phosphinic Acids

Another possibility for the synthesis of phosphinates is the alkylating esterification of phosphinic acids by alkyl halides [1, 2, 51]. The acid (**29**) may be transformed into its sodium or potassium salt (**30**) that is reacted with an alkyl halide in the second step (Figure 2.15).

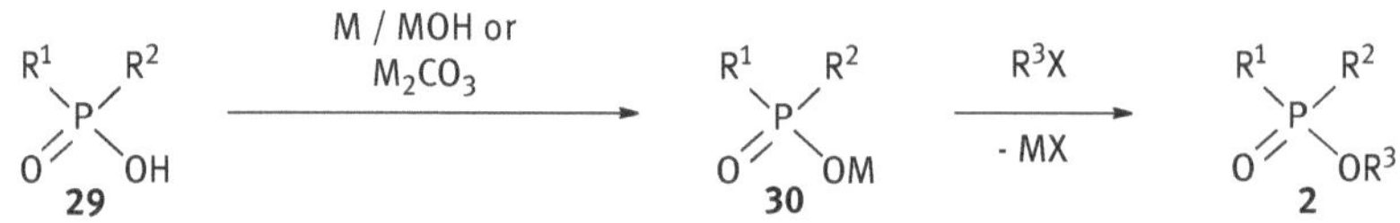

Figure 2.15: Alkylating esterification of phosphinic acids.

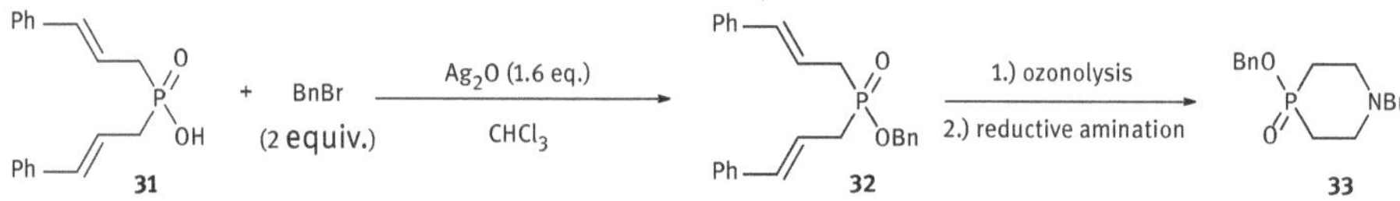

Figure 2.16: A special protocol utilizing an alkylation esterification and leading eventually to a [1,4]-azaphosphinane oxide.

The alkylating esterification of bis(cinammyl)phosphinic acid (**31**) was carried out in the presence of Ag_2O. The intermediate obtained (**32**) was then converted into an [1,4]azaphosphinane 4-oxide (**33**) via ozonolysis, followed by reductive amination (Figure 2.16) [52].

The salt of an acid may also be the starting material that can be transformed to the corresponding phosphinates by reaction with alkyl halides in the presence of crown ethers as the phase transfer (PT) catalyst [53]. This is exemplified by the alkylation of the potassium salt of a hydroxy-dibenzooxaphosphorine oxide (**34**) with alkyl halides in acetonitrile (Figure 2.17) [54].

The alkylating esterification of phenyl-*H*-phosphinic acid (**21**) with alkyl halides may also take place in a homogeneous medium using triethylamine as the base under MW conditions (Figure 2.18) [55].

PT catalysis may be a more general tool to make such transformations more efficient and environmentally-friendly. The alkylating esterification of cyclic phosphinic acids (**27B-D**) was carried out under PT catalytic and solvent-free conditions under MW irradiation using the alkyl halides in the presence of potassium carbonate as the base, and triethylbenzylammonium chloride (TEBAC) as the catalyst (Figure 2.19) [56, 57]. In case of using akyl halides with increased reactivity (e.g. R^2X = BnBr), there was no need for the catalyst. However, when alkyl halides of normal reactivity were chosen as reaction partners, the application of 5% of TEBAC increased significantly the yields of the phosphinates (**28B-D**). *O*-alkylation of the saturated 1-hydroxyphospholane oxides (**27C**) and 1-hydroxy-3-methyl-1,2,3,4,5,6-hexahydrophosphinine oxide (**27D**) resulted in a mixture of isomers. Thermally unstable cyclic phosphinic acids, such as 1-hydroxy-1,2-dihydrophosphinine 1-oxide (**27E**) and a

3-hydroxy-3-phosphabicyclo[3.1.0]hexane 3-oxide (**27F**), could be transformed to the corresponding phosphinates (**28E** and **F**) by reaction with alkyl halides in the presence of K_2CO_3 in boiling acetone (Figure 2.19) [58].

A one-pot procedure was described for the synthesis of phenylphosphonic monoesters (**38**) via the monoalkylation of the phosphonic acids (**37**) (Figure 2.20) [59].

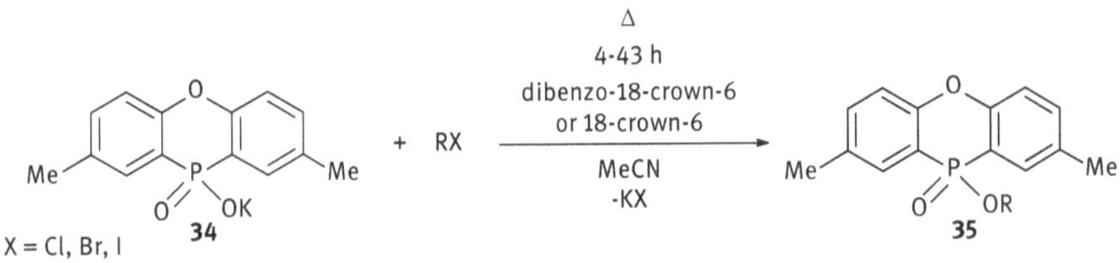

X = Cl, Br, I

R = Me, NC-CH$_2$, H$_2$C=CH-CH$_2$, MeCOOCH$_2$, Bn, *N*-phthalimidomethyl 47–99%

Figure 2.17: *O*-alkylation of a phosphinic acid salt under phase transfer catalytic conditions.

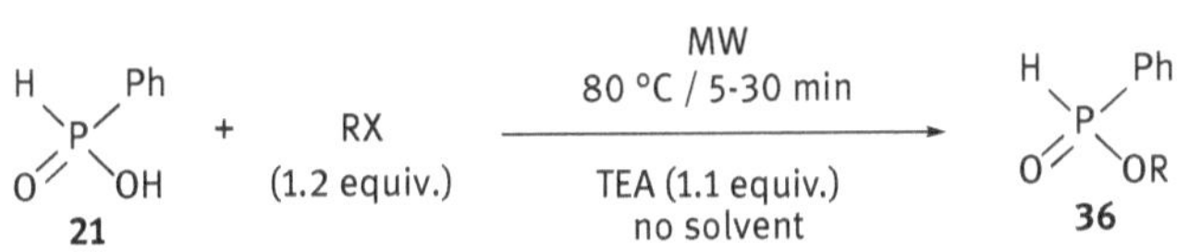

RX = EtI, ^{n}PrBr, ^{i}PrBr, ^{n}BuBr, ^{i}BuBr, ^{i}PentBr 76–96%

Figure 2.18: Alkylating esterification of phenyl-*H*-phosphinic acid in a homogeneous medium.

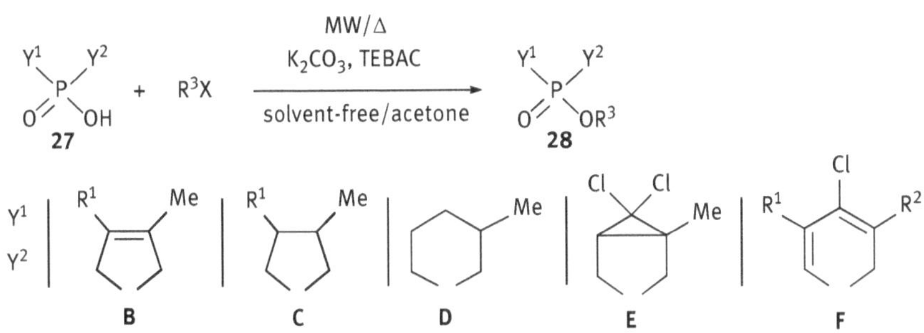

R^1, R^2 = H, Me

R^{3}X = EtI, ^{n}PrBr, ^{i}PrBr, ^{n}BuBr, ^{i}BuBr, ^{i}PentBr, BnBr

Figure 2.19: PT-catalyzed *O*-alkylation of cyclic phosphinic acids.

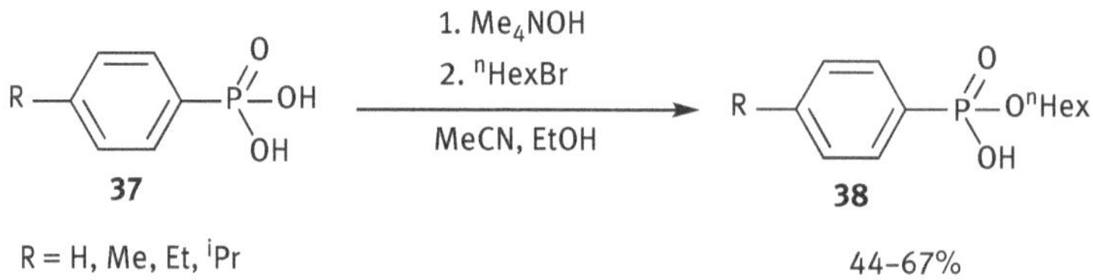

Figure 2.20: The monoesterification of arylphosphonic acids.

The addition of phosphinic acids to terminal alkynes in the presence of a transition metal catalyst may also lead to phosphinates [60], as well as the reaction of phosphinic acids with diazoalkanes [61–65]. However, these methods are of lower importance.

2.6 Michaelis–Arbuzov Reaction of Phosphonous Diesters and Trialkyl Phosphites

The Michaelis–Arbuzov reaction is an elegant way to form phosphinates ($R^2YP(O)OR^1$, **2**), or phosphonates ($R^2P(O)(OR^1)_2$, **7**) via phosphonium salts (**40**) obtained by the reaction of alkyl/arylphosphonous diesters ($(R^1O)_2PY$, Y = R, Ar), or trialkyl phosphites ($(R^1O)_3P$) with alkyl halides (Figure 2.21) [66, 67].

The intramolecular Arbuzov reaction of 4-chloro-butylphosphonous diester (**41**) afforded 1-butoxyphospholane 1-oxide **42** (Figure 2.22) [68].

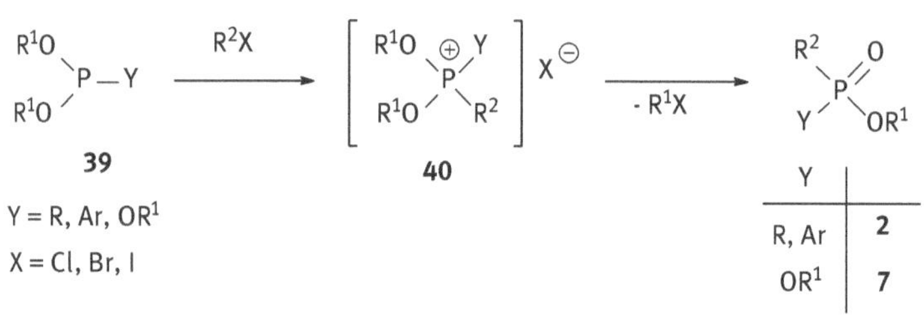

Figure 2.21: The synthesis of phosphinates or phosphonates by the Michaelis–Arbuzov reaction.

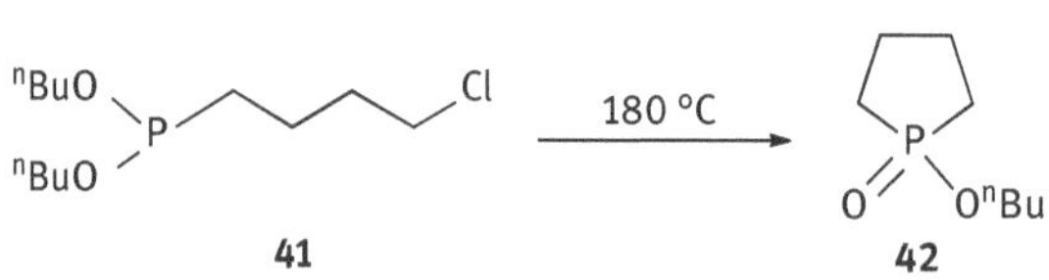

Figure 2.22: Cyclization by an intramolecular Michaelis–Arbuzov reaction.

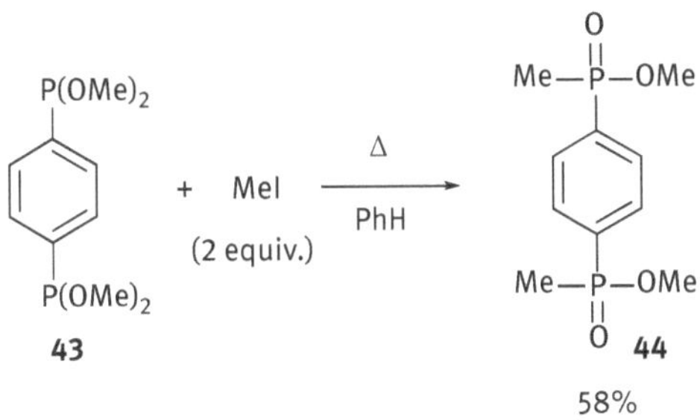

Figure 2.23: A double Michaelis–Arbuzov reaction.

Aryl halides with decreased reactivity may be applied in the presence of Ni salts at higher temperatures [69, 70], while vinyl halides may react in the presence of Cu(I) bromide [71]. However, the reaction of alkynyl halides does not require the use of any catalyst [72]. The reaction of diphosphonites (**43**) with two equivalents of alkyl halides gave bisphosphonates (**44**) (Figure 2.23) [73].

Further examples on interesting Arbuzov reactions are provided in Chapter 1.

2.7 Synthesis of P-esters by the Hirao Reaction

The Hirao reaction is a widely used method to form a P–C bond by the coupling of an aryl or vinyl halide with a >P(O)H reagent, such as a dialkyl phosphite (**45**) in the presence of tetrakis(triphenylphosphine)palladium ($Pd(PPh_3)_4$) (Figure 2.24) [74] (see also Chapter 8).

Instead of $Pd(PPh_3)_4$, a Pd precursor/ligand system may also be used, when the active Pd(0) complex is formed in situ in the reaction mixture [75–77]. Different catalysts have also been applied for incorporating other transition metals, such as Ni salts [78] or CuI, the latter together with proline [79]. However, the most beneficial approach involves the use of $Pd(OAc)_2$ catalyst precursor utilizing the excess of the trivalent tautomer form of the P-reagent under MW-assisted conditions. This protocol represents a greener approach [80–82].

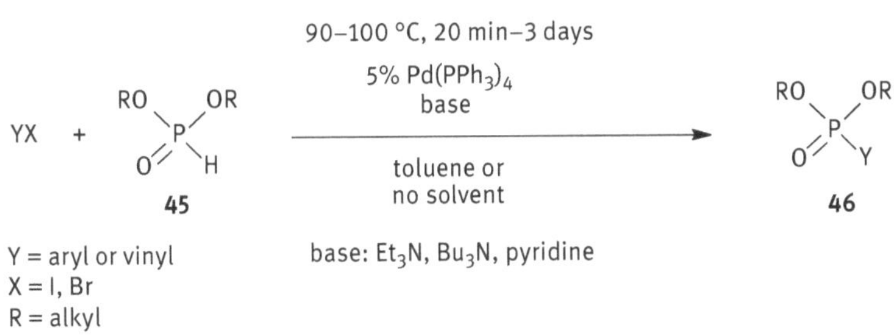

Figure 2.24: General scheme for the P–C coupling by the Hirao reaction.

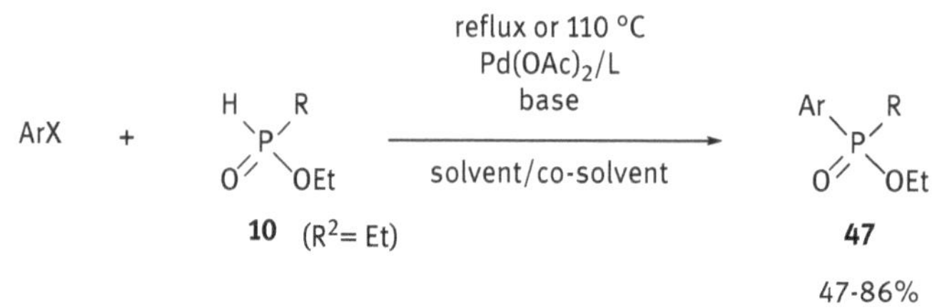

ArX = PhI, PhBr, PhCl, PhOTf, 2-Cl-pyridine, 2-Cl-pyrimidine, 2-Cl-pyrazine, *etc.*

R = octyl, Cy, $(CH_2)_4Ph$

L = dppf, xantphos, polystyrene-supported Pd-NiXantphos

Figure 2.25: Synthesis of phosphinates by the Hirao reaction.

Phosphinates may also be synthesized by the Hirao reaction using ethyl alkyl-*H*-phosphinates (**10**, R^2=Et) in reaction with aryl halides to afford aryl alkylphosphinates (**47**) (Figure 2.25) [83].

2.8 Cyclic Phosphinates by the Alcoholysis of Phosphonium Salts

Cyclic phosphinates may be synthesized by the alcoholysis of the phosphonium salts (**48** or **49**) obtained in the McCormack cycloaddition of 1,3-dienes and phosphorus trihalides (Figure 2.26) [84]. Alcoholysis of the adducts led to 3- (**28B**) or 2-phospholene (**50**) derivatives [85, 86]. The solvolytic accomplishment applying the alcohol in an excess in the presence of sodium carbonate resulted in increased (up to 57–68%)

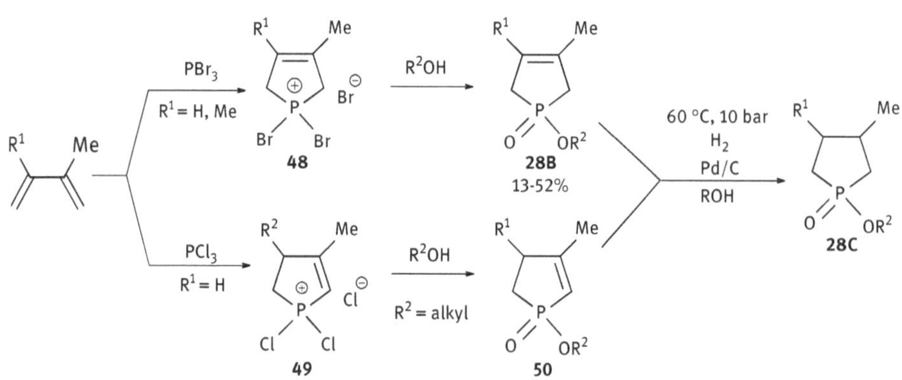

Figure 2.26: Synthesis of cyclic phosphinates via phosphonium salts.

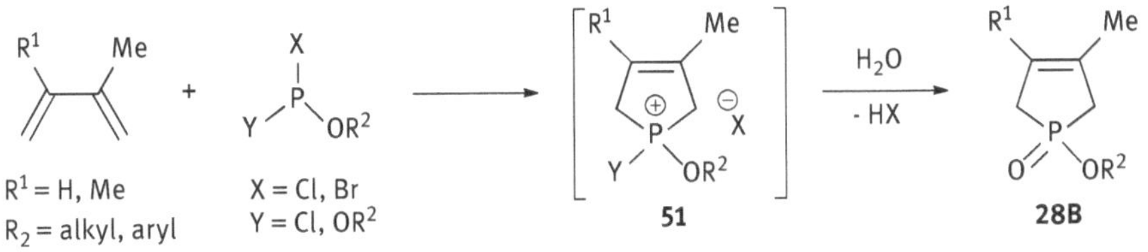

Figure 2.27: Another approach to obtain 1-alkoxy-3-phospholene oxides.

yields [87, 88]. Saturated cyclic phosphinates (1-alkoxyphospholane oxides **28C**) were obtained by a subsequent catalytic hydrogenation [89].

Other P-reactants, such as $ROPX_2$, $(RO)_2PX$, or $ArOPX_2$ may also be used in the McCormack cycloaddition, which should be followed by a hydrolysis step (Figure 2.27) [90–92].

2.9 Miscellaneous Methods for the Esterification of P-acids

2.9.1 The Atherton–Todd Reaction

The reaction of secondary phosphine oxides (**52**) with alcohols in the presence of carbon tetrachloride and triethylamine (called the Atherton–Todd reaction) gives phosphinates (**2**), while starting from *H*-phosphinates (**10**), phosphonates (**7**) may be obtained (Figure 2.28) [93].

A somewhat greener approach involves the chloroform-based Atherton–Todd reactions of alcohols with secondary phosphine oxides to provide phosphinates under milder reaction conditions [94]. Starting from chiral >P(O)H compounds, a stereospecific coupling with alcohols gives rise to a variety of optically active phosphorus acid derivatives [93]. Heterocyclic phosphinic acids may also be used as starting materials, resulting in cyclic phosphinates [95].

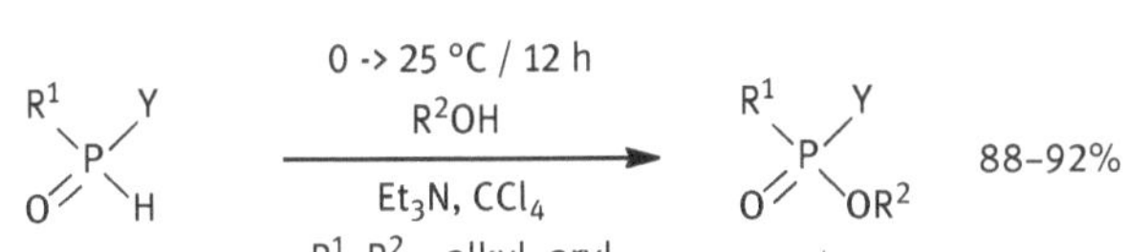

Y	
R1	**52**
OR2	**10**

Y	
R1	**2**
OR2	**7**

Figure 2.28: Synthesis of phosphinates and phosphonates by the Atherton–Todd reaction.

2.9.2 The Copper-catalyzed Esterification of P(O)OH Compounds with Phenols

A novel copper-catalyzed method involving the esterification of phosphinic and phosphonic acids with phenols was described to furnish aryl phosphinates (**53**) and phosphonates (**54**) (Figure 2.29) [96].

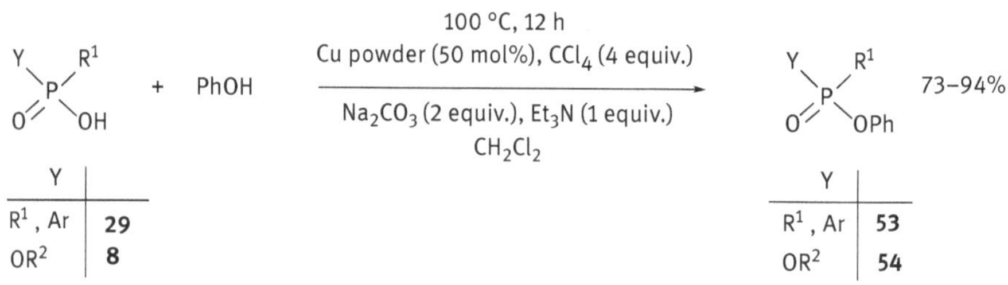

Figure 2.29: Cu-catalyzed esterification of P-acids.

2.9.3 The Oxidation of P(III) Esters

Phosphinates (**2**) and phosphonates (**7**) may be obtained by the oxidation of the corresponding trivalent species (**55**) by mild oxidizing agents (Figure 2.30) [97–99].

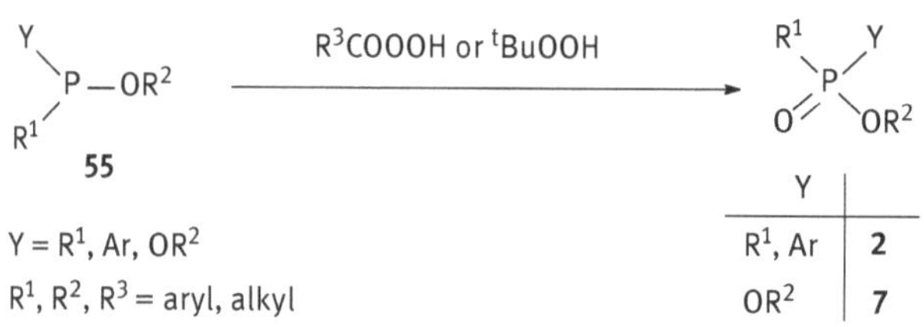

Figure 2.30: Preparation of phosphinates and phosphonates by the oxidation of trivalent P-esters.

2.9.4 Fragmentation-related Phosphinylation of Strained P-heterocycles

H-Phosphinates (**58**) may be synthesized via the UV light-mediated fragmentation of suitable bridged *P*-heterocycle precursors (e.g. 7-phosphanorbornene derivative **56**) in the presence of an alcohol that is phosphinylated by the low-coordinated fragment ejected (Figure 2.31) [100–103].

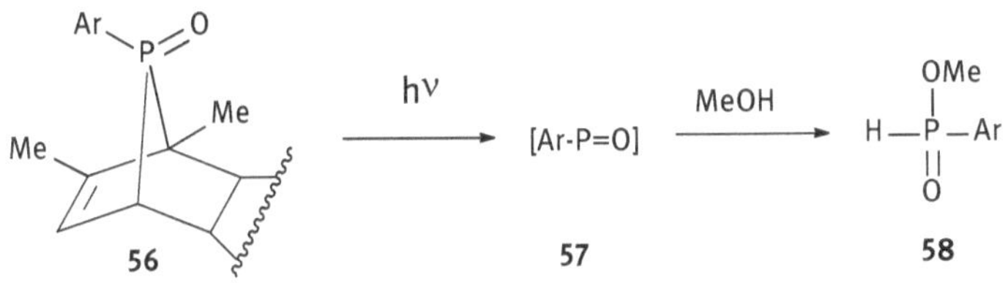

Figure 2.31: Synthesis of *H*-phosphinates by fragmentation-related phosphinylation.

2.10 Summary

Different approaches for the synthesis of phosphinic and phosphonic esters have been summarized. Besides the widespread method of converting phosphinic or phosphonic chlorides to the corresponding esters, it is also possible to derivatize the phosphinic acids by activating them with the use of condensing agents, or to convert the acids to esters by alkylation, or by direct esterification under MW-assisted conditions. Additional methods involve the Arbuzov reaction of phosphonous diesters or trialkyl phosphites, the Hirao reaction, and the alcoholysis of phosphonium salts.

Acknowledgments: This work was supported by the National Research, Development and Innovation Office (K119202) and by the New National Excellence Program of the Ministry of Human Capacities (ÚNKP-17-4-I-BME-133).

References

[1] L. D. Quin, *A Guide to Organophosphorus Chemistry*; Wiley: New York, 2000.
[2] G. M. Kosolapoff, L. Maier, *Organic Phosphorus Compounds*, Vol. 6. Wiley: New York, 1973.
[3] D. Barton, W. D. Ollis, *Comprehensive Organic Chemistry, Phosphorus Compounds*, Vol. 2. Pergamon: Oxford, 1979.
[4] T. Minami, T. Okauchi, In: *Comprehensive Organic Functional Group Transformations II*; Katritzky, A. R., Taylor, R. J. K., Eds.; Elsevier: Oxford, 2005.
[5] G. Marosi, A. Toldy, G. Parlagh, Z. Nagy, K. Ludányi, P. Anna, G. Keglevich, *Heteroatom Chem.*, 2002, **13**, 126.
[6] S. Begeç, S. Alataş, A. Kılıç, *Phosphorus, Sulfur, Silicon*, 2011, **186**, 1531.
[7] A. Michaelis, W. LaCoste, *Ber. Deut. Chem. Ges.*, 1885, **18**, 2109.
[8] G. M. Kosolapoff, R. M. Watson, *J. Am. Chem. Soc.*, 1951, **73**, 4101.
[9] K. Sasse, *Methoden der Organischen Chemie*; Georg Thieme Verlag: Stuttgart, 1963; Vol. 12, Part 1.

[10] A. O. Vizel, L. I. Shchukina, T. A. Zyablikova, V. K. Krupnov, *J. Gen. Chem. USSR (Engl. Transl.)*, 1989, **59**, 121.
[11] L.-D. Quin, X.-P. Wu, N. D. Sadanani, I. Lukeŝ, A. S. Ionkin, R. O. Day, *J. Org. Chem.*, 1994, **59**, 120.
[12] U. Hasserodt, K. Hunger, F. Korte, *Tetrahedron*, 1963, **19**, 1563.
[13] R. Engel, *Handbook of Organophosphorus Chemistry*; Marcel Dekker Inc.: New York, 1992.
[14] R. Rabinowitz, *J. Am. Chem. Soc.*, 1980, **82**, 4564.
[15] W. F. Gilmore, H. A. McBride, *J. Pharm. Sci.*, 1974, **63**, 965.
[16] D. S. Karanewsky, M. C. Badia, *Tetrahedron Lett.*, 1986, **27**, 1751.
[17] D.S. Karanewsky, M. C. Badia, D. W. Cushman, J. M. DeForrest, T. Dejneka, M. J. Loots, M. G. Perri, E. W. Petrillo, J. R. Powell, *J. Med. Chem.*, 1988, **31**, 204.
[18] S. Chen, J. K. Coward, *J. Org. Chem.*, 1998, **63**, 502.
[19] F. W. Gimore, H. A. McBride, *Pharm. Sci.*, 1974, **63**, 965.
[20] J. Houben, K. H. Büchel, H. G. Padeken, *Methoden der Organischen Chemie (Houben-Weyl), Organische Phosphorverbindungen*, Vol. 2. G. Thieme: Stuttgart, 1982.
[21] U. Pradere, E. C. Garnier-Amblard, S. J. Coats, F. Amblard, R. F. Schinaz, *Chem. Rev.*, 2014, **114**, 9154.
[22] B. Xiong, C. Hu, H. Li, C. Zhou, P. Zhang, Y. Liu and K. Tang, *Tetrahedron Lett.*, 2017, **58**, 2482.
[23] B. Xiong, Q. Cheng, C. Hu, P. Zhang, Y. Liu, K. Tang, *Chem. Select*, 2017, **2**, 6891.
[24] C. Wasieleweki, M. Hoffmann, E. Witkowska, J. Rachon, *Rocz. Chem.*, 1976, **50**, 1613.
[25] H.-W. Shih, K.-T. Chenb, W.-C. Cheng, *Tetrahedron Lett.*, 2012, **53**, 243.
[26] Basavaprabhu; T. M. Vishwanatha, N. R. Panguluri, V. V. Sureshbabu, *Synthesis*, 2013, **45**, 1569.
[27] E. Jablonkai, M. Milen, L. Drahos, G. Keglevich, *Tetrahedron Lett.*, 2013, **54**, 5873.
[28] E. Jablonkai, R. Henyecz, M. Milen, J. Kóti, G. Keglevich, *Tetrahedron* 2014, **70**, 8280.
[29] D. A. Campbell, *J. Org. Chem.*, 1992, **57**, 6331.
[30] A. K. Purohit, D. Pardasani, V. Tak, A. Kumara, R. Jain, D. K. Dubey, *Tetrahedron Lett.*, 2012, **53**, 3795.
[31] S. J. Fitch, *J. Am. Chem. Soc.*, 1964, **86**, 61.
[32] A. W. Schwabacher, A. D. Stefanescu, *Tetrahedron Lett.*, 1996, **37**, 425.
[33] T. Yoshino, S. Imori, H. Togo, *Tetrahedron*, 2006, **62**, 1309.
[34] D. G. Hewitt, *Aust. J. Chem.*, 1979, **32**, 463.
[35] M. Aresta, M. Dibenedetto, *Chem. Eur. J.*, 2002, **8**, 685.
[36] A. Kamyar, H. Yu, *Tetrahedron Lett.*, 2003, **44**, 781.
[37] Y. R. Dumond, R. L. Baker, J.-L. Montchamp, *Org. Lett.*, 2000, **2**, 3341.
[38] S. Deprèle, J.-L. Montchamp, *J. Organomet. Chem.*, 2002, **643–644**, 154.
[39] I. V. Berezovskaya, L. A. Eliseeva, E. V. Kuznetsov, E. Mukhametzyanova, *Ser. Khim.*, 1968, 1369.
[40] S. Vassiliou, E. Tzouma, *J. Org. Chem.*, 2013, **78**, 10069.
[41] M. J. Gallagher, M. G. Ranasinghe, I. D. Jenkins, *Phosphorus, Sulfur, Silicon*, 1996, **115**, 255.
[42] A. Mais, J. J. McBride, *Preparing esters of phosphinic acids*. US3092650, 1963.
[43] E. Cherbuliez, G. Weber, J. Rabinowitz, *Helv. Chim. Acta*, 1963, **46**, 2464.
[44] E. E. Nifant'ev, L. P. Levitan, *Zh. Obshch. Khim.*, 1965, **35**, 758.
[45] H. Denham, A. R. Stiles, *Allyl-type phosphinates and polymers of the same*. US2659714 A, 1953.
[46] N. Z. Kiss, K. Ludányi, L. Drahos, G. Keglevich, *Synth. Commun.*, 2009, **39**, 2392.
[47] G. Keglevich, N. Z. Kiss, Z. Mucsi, T. Körtvélyesi, *Org. Biomol. Chem.*, 2012, **10**, 2011.
[48] Z. Mucsi, N. Z. Kiss, G. Keglevich, *RSC Adv.*, 2014, **4**, 11948.
[49] N. Z. Kiss, É. Böttger, L. Drahos, G. Keglevich, *Heteroat. Chem.*, 2013, **24**, 283.
[50] N. Z. Kiss, Z. Mucsi, É. Böttger, L. Drahos, G. Keglevich, *Curr. Org. Synth.*, 2014, **11**, 767.
[51] K. A. Andrianov, I. K. Kuznetsova, *Bull. Acad. Sci., USSR*, 1961, **10**, 1670.
[52] L. Coudray, K. Bravo-Altamirano, J.-L. Montchamp, *Org. Lett.*, 2008, **10**, 1123.
[53] M. Skwarczynski, P. Kafarski, *Synth. Commun.*, 1995, **25**, 3565.

[54] Z. E. Golubski, *Synthesis*, 1980, **1980**, 632.
[55] G. Keglevich, E. Jablonkai, L. B. Balázs, *RSC Adv.*, 2014, **4**, 22808.
[56] G. Keglevich, E. Bálint, N. Z. Kiss, E. Jablonkai, L. Hegedűs, A. Grün, I. Greiner, *Curr. Org. Chem.*, 2011, **15**, 1802.
[57] E. Bálint, E. Jablonkai, M. Bálint, G. Keglevich, *Heteroat. Chem.*, 2010, **21**, 211.
[55] E. Jablonkai, E. Bálint, G. T. Balogh, L. Drahos, G. Keglevich, *Phosphorus, Sulfur, Silicon*, 2012, **187**, 357.
[59] C. Yuan, H. Feng, *Synthesis* 1990, **1990**, 140.
[60] R. Hua, M. Tanaka, *Chem. Lett.*, 1998, **27**, 431.
[61] N. Tokutake, J. Hiratake, M. Katoh, T. Irie, H. Kato, J. Oda, *Bioorg. Med. Chem.*, 1998, **6**, 1935.
[62] M. T. Martin, T. S. Angeles, R. Sugasawara, N. I. Aman, A. D. Napper, M. J. Darsley, R. I. Sanchez, P. Booth, R. C. Titmas, *J. Am. Chem. Soc.*, 1994, **116**, 6508.
[63] T. L. Emmick, R. L. Letsinger, *J. Am. Chem. Soc.*, 1968, **90**, 3459.
[64] F. F. Blicke, S. Raines, *J. Org. Chem.*, 1964, **29**, 204.
[65] L. D. Quin, M. R. Dysart, *J. Org. Chem.*, 1962, **27**, 1012.
[66] D. Barton, W. D. Ollis, *Comprehensive Organic Chemistry, Phosphorus Compounds*, Vol. 2. Pergamon: Oxford, 1979.
[67] A. K. Bhattacharya, G. Thyagarajan, *Chem. Rev.*, 1981, **81**, 415.
[68] B. Helferich, E. Aufderhaar, *Liebigs Ann. Chem.*, 1962, **658**, 100.
[69] P. Tavs, *Chem. Ber.*, 1970, **103**, 2428.
[70] G. Keglevich, A. Grün, A. Bölcskei, L. Drahos, M. Kraszni, G. T. Balogh, *Heteroat. Chem.*, 2012, **23**, 574.
[71] G. Axelrad, S. Laosooksathit, R. J. Engel, *Org. Chem.*, 1981, **46**, 5200.
[72] E. P. Kyba, S. P. Rines, P. W. Owens, S.-S. P. Chou, *Tetrahedron Lett.*, **1981**, **22**, 1875.
[73] R. A. Baldwin, C. O. Wilson, R. I. Wagner, *J. Org. Chem.*, 1967, **32**, 2172.
[74] E. Jablonkai, G. Keglevich, *Curr. Org. Synth.*, 2014, **11**, 429.
[75] M. Kalek, J. Stawinski, *Organometallics*, 2007, **26**, 5840.
[76] M. Kalek, M. Jezowska, J. Stawinski, *Adv. Synth. Catal.*, 2009, **351**, 3207–3216.
[77] M. Lilley, B. Mambwe, R. F. W. Jackson, R. Muimo, *Chem. Commun.*, 2014, **50**, 9343.
[78] X. Zhang, H. Liu, X. Hu, G. Tang, J. Zhu, Y. Zhao, *Org. Lett.*, 2011, **13**, 3478.
[79] C. Huang, X. Tang, H. Fu, Y. Jiang, Y. Zhao, *J. Org. Chem.*, 2006, **71**, 5020.
[80] E. Jablonkai, G. Keglevich, *Advances in Engineering Research*, Vol. 10, Nova Science Publishers Inc.: New York, 2015.
[81] E. Jablonkai, G. Keglevich, *Tetrahedron*, 2013, **54**, 4185.
[82] G. Keglevich, E. Jablonkai, L. B. Balázs, *RSC Adv.*, 2014, **4**, 22808.
[83] E. L. Deal, C. Petit, J.-L. Montchamp, *Org. Lett.*, 2011, **13**, 3270.
[84] L. D. Quin, *The Heterocyclic Chemistry of Phosphorus: System based on the Phosphorus–Carbon Bond*; Wiley: New York, 1981.
[85] U. Hasserodt, K. Hunger, F. Korte, *Tetrahedron*, 1963, **19**, 1563.
[86] K. Hunger, U. Hasserodt, F. Korte, *Tetrahedron*, 1964, **20**, 1593.
[87] G. Keglevich, L. Tőke, A. Kovács, G. Tóth, K. Újszászy, *Heteroat. Chem.*, 1993, **4**, 61.
[88] G. Keglevich, J. Brlik, F. Janke, L. Tőke, *Heteroat. Chem.*, 1990, **1**, 419.
[89] E. Bálint, N. Z. Kiss, E. Jablonkai, L. Hegedűs, A. Grün, I. Greiner, G. Keglevich, *Curr. Org. Chem.*, 2011, **15**, 1802.
[90] N. K. Bliznyuk, Z. N. Kvasha, A. F. Kolomiets, *Zh. Obshch. Khim.*, 1967, **37**, 1811.
[91] N. A. Razumova, L. Z. Evtikhov, L. I. Zubtsova, A. A. Petrov, *Zh. Obshch. Khim.*, 1968, **38**, 2342.
[92] B. A. Arbuzov, L. A. Shapshinskaya, *Bull. Acad. Sci., USSR, Div. Chem. Sci.*, 1962, 57.
[93] G. Wang, R. Shen, Q. Xu, M. Goto, Y. Zhao, L.-B. Han, *J. Org. Chem.*, 2010, **75**, 3890.
[94] S. Li, T. Chen, Y. Saga, L.-B. Han, **RSC Adv.**, 2015, **5**, 71544.
[95] S. P. Ivonin, A. O. Pushechnikov, A. A. Tolmachev, *Heteroat. Chem.*, 2000, **11**, 107.

[96] B. Xiong, K. Zeng, S. Zhang, Y. Zhou, C.-T. Au, S.-F. Yin, *Tetrahedron*, 2015, **71**, 9293.
[97] D. Barton, W. D. Ollis, *Comprehensive Organic Chemistry*; Vol. 2. Pergamon: Oxford, 1979.
[98] D. M. Brown, P. R. Hammond, *J. Chem. Soc.*, 1960, 4229.
[99] R. F. Stockel, *Chem. Comm.*, 1968, 1594.
[100] H. Tomioka, Y. Hirano, Y. Izawa, *Tetrahedron Lett.*, 1974, **15**, 4477.
[101] H. Tomioka, S. Miura, Y. Izawa, *Tetrahedron Lett.*, 1983, **24**, 3353.
[102] G. Keglevich, R. Kovács, L. Drahos, *Phosphorus, Sulfur, Silicon*, 2011, **186**, 2172.
[103] L. D. Quin, J. Szewczyk, *Multiple Bonds and Low Coordination in Phosphorus Chemistry*; Thieme Verlag: Stuttgart, 1990.

Ádám Tajti and György Keglevich

3 The importance of organophosphorus compounds as biologically active agents

Abstract: Phosphorus is one of the most important elements in living organisms [1, 2]. While, on the one hand, phosphorus compounds can sustain life (endogen compounds), cure or prevent illnesses (drugs), on the other hand they are capable of limiting the growth, or destroying undesired plants, insects and fungi (pesticides) or even humans (nerve agents). In this chapter, a brief summary of the evolution of bioactive organophosphorus chemistry, as well as the most important types of biologically active compounds, will be provided *via* demonstrative examples.

Keywords: bioactivity, endogen organophosphorus compounds, drugs, pesticides, nerve agents.

3.1 From Alchemy to up to date Organophosphorus Chemistry

The story of phosphorus began in the seventeenth century in the laboratory of the German alchemist Henning Brandt [3]. Like other alchemists of the time, he searched for the "Philosopher's Stone", a substance that was believed to turn basic metals to elemental gold. The fact that phosphorus could be isolated from urine clearly indicated the biocompatibility of phosphorous compounds. In the eighteenth century, Hensing detected phosphorus in human brain [4]; at the same time, Scheele discovered that it is also an important ingredient in the bones and teeth of animals [5]. During the 1800s, plant science studies recognized phosphorus to be essential for plant nutrition. The first patents on phosphate-containing fertilizers were also obtained at that time [6]. Lecithin, the first identified organophosphorus compound, was isolated from brain fat in 1811 [7]; however it was characterized as a P-containing lipid only in 1846 [8]. Another scientific milestone was the isolation of impure DNA (called "nuclein") from human tissues by Miescher in 1868 [9]. The exact formula was not known at that time, but the presence of phosphorus in human tissues was confirmed. One of the most important molecules involved in energy transfer, adenosine triphosphate (ATP), was isolated from muscle extracts in 1929 by Fiske and Subbarow [10], and the significance of its high-energy bonds was reported by Lipmann in 1941 [11]. The total synthesis of ATP was reported in 1949 by Todd [12]. In 1944, studies conducted by Avery proved that the DNA is the carrier of genetic information [13]. A decade later (in 1953), the structure of DNA and RNA was established by Crick and Watson [14].

https://doi.org/10.1515/9783110535839-003

Table 3.1: Milestones in the evolution of bioactive phosphorus compounds.

Event	Researcher	Year
Invention of phosphorus	Henning Brandt	1669
Discovery of phosphorus in human brain	Hensing	1719
Isolation of phosphorus from bones and teeth	Scheele	1770
Isolation of the first organophosphorus compound	Vauquelin	1811
Isolation of impure DNA (nuclein)	Miescher	1868
Isolation of ATP	Fiske and Subbarow	1929
Significance of ATP demonstrated	Lipmann	1941
Experiment demonstrating that DNA carries genetic information	Avery	1944
Total synthesis of ATP	Todd	1949
Structure of DNA and RNA revealed	Crick and Watson	1953

Nowadays, it is an accepted fact that organophosphorus compounds play an important role in living organisms. Almost all biological and biochemical processes, such as energy transfer and storing genetic information involve some kind of P-containing compounds. Milestones in the evolution of bioactive phosphorus compounds are summarized in Table 3.1.

3.2 Organophosphorus Compounds in the Living Body

Thousands of phosphorus compounds are responsible for the proper functioning of the human body [1, 15–17]. However, phosphorus accounts for *ca.* 1% of human weight, which shows that it has more biochemical functions than any other element. The largest amount of phosphorus is present in bones and teeth (*ca.* 85%), 14% located in soft tissues, and the remaining 1% can be found in blood and extravascular fluids.

3.2.1 Phosphorus in Bones and Teeth

The largest amount of phosphorus in the human body is located in the skeleton and the teeth. The main component is hydroxyapatite; however, the synthetic and biological apatites differ from each other in their Ca/P molar ratio, cell dimensions and the composition of natural impurities. Hydroxyapatite in the body may also contain carbonates. As for the teeth, dentine contains *ca.* 72% apatite, the characteristic difference from the mineral form is the OH–F substitution in certain amounts, which can decrease the acid solubility of dentine.

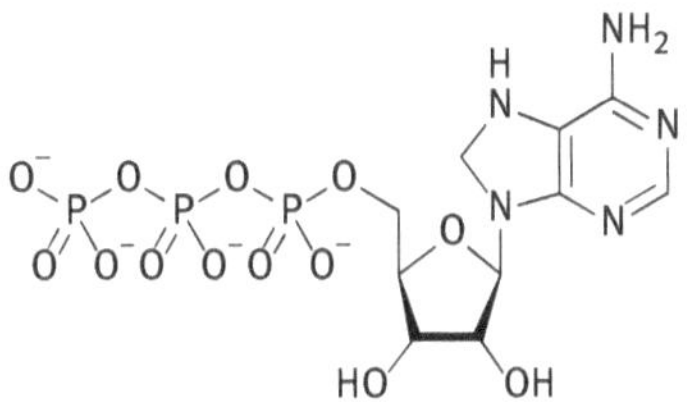

ATP

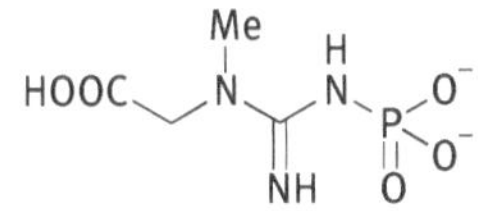

Phosphocreatine

Figure 3.1: ATP and phosphocreatine as endogen phosphorus compounds.

3.2.2 Phosphorus in Brain

In the brain, phosphorus is present in the second-highest concentration after the bones and teeth. Any mental activity involves a high amount of two main substrates, ATP and phosphocreatine (Figure 3.1). Other P-compounds are nicotinamide adenine dinucleotide phosphate, phospholipids, and sugar phosphates.

3.2.3 Phosphorus in Muscles

As muscle cells need a lot of energy to function properly, the main components of P-containing species are phosphocreatine and ATP (Figure 3.1). Besides the two major components, inorganic phosphates are also essential.

3.2.4 Phosphorus in Blood

In blood, besides the bicarbonate system, the HPO_4^{2-} and $H_2PO_4^{-}$ ions act as buffering agents; they are responsible for maintaining a pH of 7.4. Blood contains DNA, ATP, as well as phospholipids and 2,3-diphosphoglycerate (Figure 3.2). The latter compound competes with oxygen for hemoglobin, which explains its importance to the respiratory system.

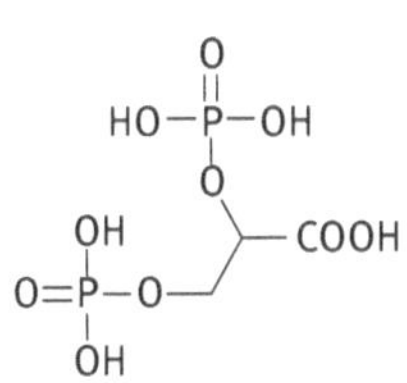

2,3-Diphosphoglycerate

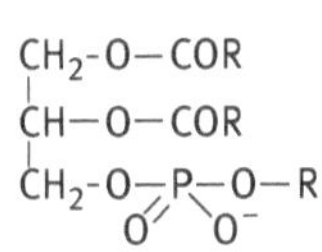

Phospholipids

Figure 3.2: The structure of 2,3-diphosphoglycerate and phospholipids.

3.3 Organophosphorus Drugs

3.3.1 Elemental Phosphorus and Inorganic Phosphates

After Hensing detected phosphorus in the human brain, the use of elemental phosphorus as nerve medicine became common in the beginning of the nineteenth century. As phosphorus is a poison in elemental form, the first treatments using phosphorous often ended with death. Later, the use of inorganic phosphates, such as sodium and aluminum phosphates, as stomach antacids and mixtures of Na_2HPO_4 and KH_2PO_4 as laxatives became a common practice [1].

3.3.2 Phosphate Salts of Organic Drugs

There are a few drugs that are formulated as phosphate salts, due to their better compatibility with physiological pH or better solubility [1, 18]. Several examples of phosphate salts of well-known organic drugs are listed in Table 3.2.

3.3.3 Bisphosphonate Derivatives

Bisphosphonates are the most important class of organophosphorus compounds in the treatment of osteoporosis and the Paget's disease (see also Chapter 10) [19–21]. They show high affinity for the Ca-ions of hydroxyapatite, and therefore can prevent Ca loss from bones. Bisphosphonic acid is isosteric with pyrophosphoric acid, but the P–C bond is hydrolytically stable (Figure 3.3).

The first generation of bisphosphonic derivatives, including etidronic acid, clodronic acid, and tiludronic acid, were used as effective drugs to treat osteoporosis, but caused several serious side effects (Figure 3.4).

After the development of early bisphosphonates, members of the second generation appeared. These derivatives contain an aminoalkyl side chain, as in pamidronic

Table 3.2: Phosphate salts of some well-known organic drugs.

Drug	Function
Codeine phosphate	Analgesic
Amphetamine phosphate	Antidepressant
Histamine phosphate	Anthelmintic
Diisopyramide phosphate	Antiarrhythmic
Piperazine phosphate	Anthelmintic

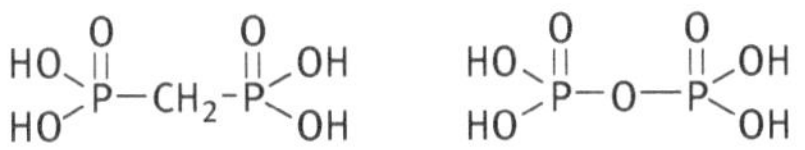

Bisphosphonic acid **Pyrophosphoric acid**

Figure 3.3: Bisphosphonic acid and pyrophosphoric acid.

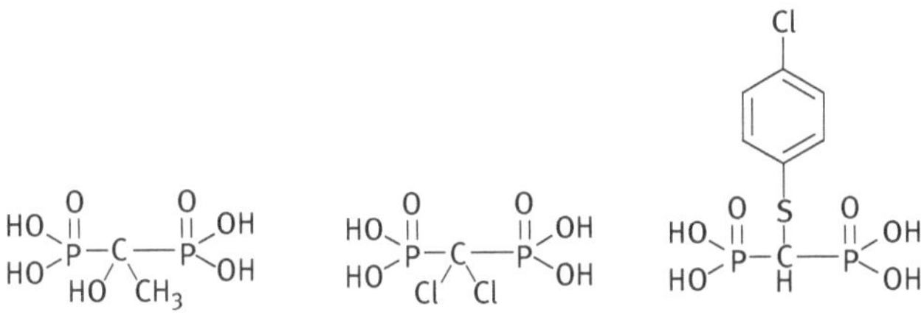

Etidronic acid **Clodronic acid** **Tiludronic acid**

Figure 3.4: A few members of the first generation of bisphosphonates.

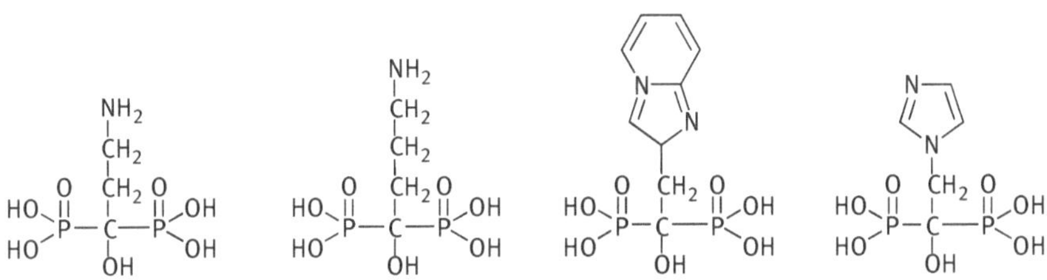

Pamidronic acid **Alendronic acid** **Minodronic acid** **Zolendronic acid**

Figure 3.5: Examples for the second and third generations of bisphosphonic derivatives.

acid and alendronic acid. The second-generation bisphosphonates have increased effectiveness and reduced side effects. The presently applied representatives of the third generation involve an *N*-containing heterocycle in the side chains. Two examples for this group are minodronic acid and zolendronic acid (Figure 3.5).

3.3.4 α-Aminophosphonate Derivatives

α-Aminophosphonate derivatives are considered the structural analogs of α-amino acids (Figure 3.6). Due to this similarity, α-aminophosphonates and related compounds possess real and potential bioactivity [22]. Their derivatives have shown to possess various biological activities, for example they have been used as antibiotics, antiviral or anticancer agents, or enzyme inhibitors (see also Chapter 6). Beside their medicinal applications, α-aminophosphonates can be used as herbicides (such as glyphosate, see in Section 3.4.1), fungicides, and growth regulators in agrochemistry.

α-Aminophosphonic acid **α-Amino acid**

Figure 3.6: The structure of α-aminophosphonic acids and α-amino acids.

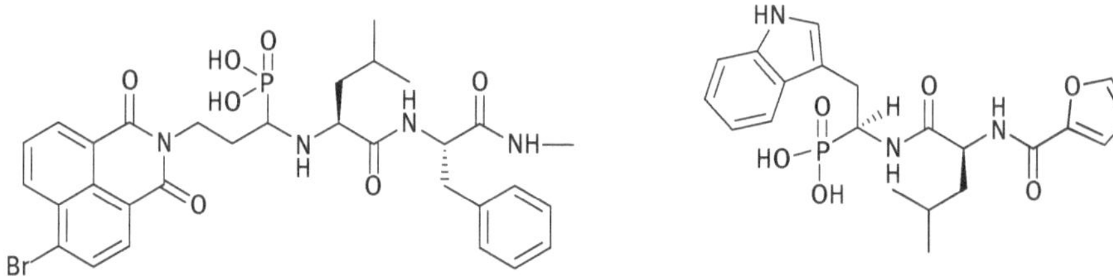

Figure 3.7: α-Aminophosphonates used as MMP inhibitors.

Peptidomimetic α-aminophosphonates can be applied as matrix metalloproteinase (MMP) inhibitors [21]. Derivatives showed activity toward MMP-1, 2, 3, 8, and 9. Examples of active molecules can be seen on Figure 3.7.

3.3.5 Anticancer Organophosphorus Drugs

On the basis of literature data, the use of a wide spectra of organophosphorus compounds as anticancer drugs has been reported [21]. A part of these compounds are double alkylating agents (Figure 3.8), and they react with DNA or RNA, and create

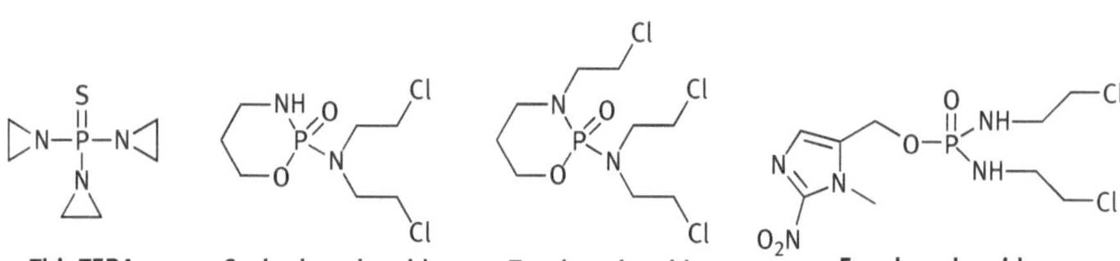

ThioTEPA **Cyclophosphamide** **Trophosphamide** **Evophosphamide**

Figure 3.8: Organophosphorus alkylating agents applied in cancer therapy.

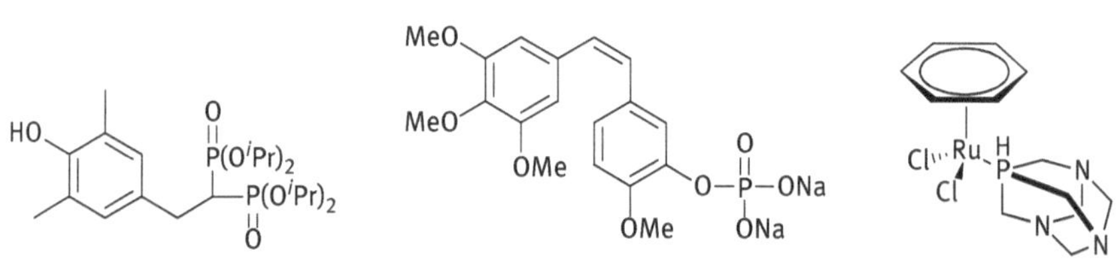

Apomine **Combretastatine-phosphate** **Rapta-B**

Figure 3.9: P-containing anticancer drugs interacting with tumor cells.

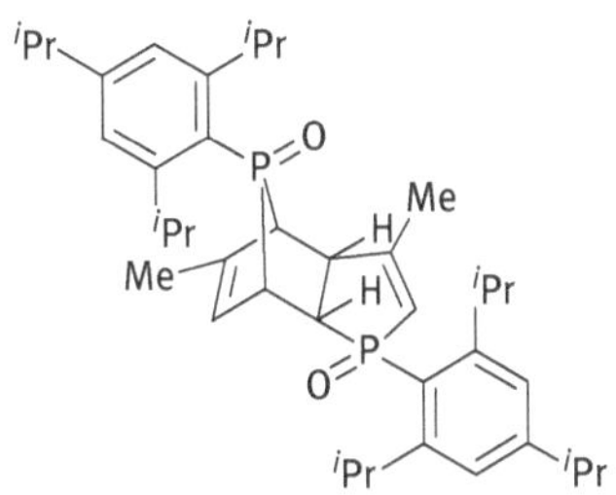

Figure 3.10: A phosphanorbornene derivative as a potential anticancer agent.

a crosslink between the two nucleic acid chains, preventing their duplication. N,N′,N″-triethylenethiophosphoramide, discovered in the 1950s, acts as an alkylating agent, and has a broad spectrum of antitumor activity [23]. Due to its high activity, it is rather toxic. Cyclophosphamide, which acts in a similar manner as ThioTEPA, but has significantly less side effects, is one of the most widely used cytostatic drug [24]. Trophosphamide, a drug similar to cyclophosphamide, has a third N-mustard chain. Evofosfamide is a new-generation alkylating agent, and can be used to treat even hypoxia-activated tumors [21].

Other types of organophosphorus anticancer agents react with tumor cells (Figure 3.9) [21]. Apomine containing a bisphosphonate moiety induces apoptosis in cancer cells. Combretastatin phosphate is an example of drugs destabilizing microtubules. Ruthenium-phosphine complexes, like Rapta-B show antiproliferative activity.

Anticancer activity of *P*-heterocycles was also studied on human cancer cells [25, 26]. A phophanorbornene derivative was shown to have curative effects against leukemia, lung cancer, colon cancer, and melanoma (Figure 3.10). The mode of action has not been clarified.

3.3.6 Antiviral Organophosphorus Drugs

Antiviral organophosphorus compounds can block the reproduction of viruses [1, 21, 27]. Foscarnet, an endogen pyrophosphate analogue works as a DNA polymerase inhibitor. It is used in the treatment of herpes, HIV, and other viruses. Cidofovir, another DNA polymerase inhibitor has been found to be active against HPV,

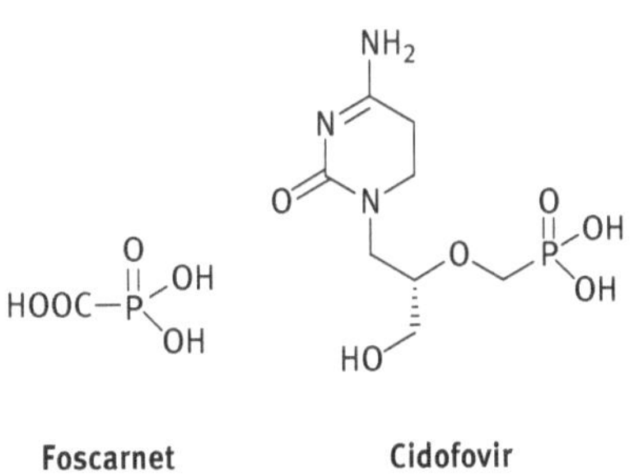

Figure 3.11: Organophosphorus antiviral agents blocking the viral DNA polymerase enzyme.

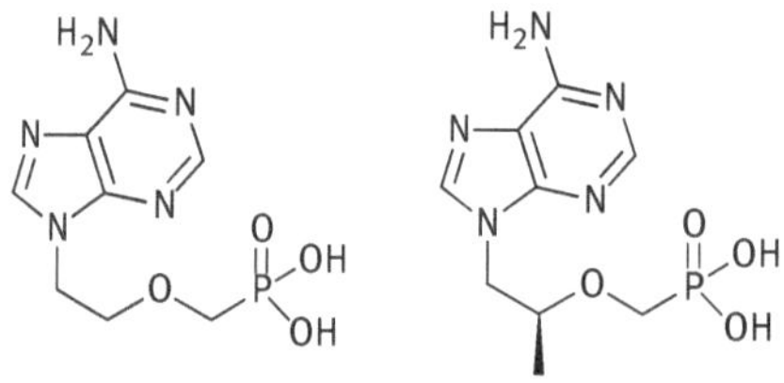

Adefovir Tenofovir

Figure 3.12: Organophosphorus viral reverse transcriptase enzyme inhibitors.

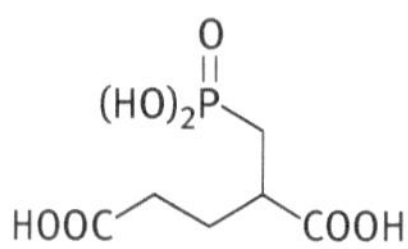

PMPA

Figure 3.13: PMPA as an organophosphorus antistroke drug.

adenoviruses, and also herpes. Foscarnet and cidofovir are only available as an injection (Figure 3.11).

Another way of blocking the reproduction of viruses is to inhibit the viral reverse transcriptase enzyme (Figure 3.12). Adefovir and tenofovir act by this mechanism. Adefovir can be applied in the treatment of herpes and hepatitis-B, while tenofovir has been shown to be active against HIV and hepatitis-B. Both agents may be dosed *per os*.

3.3.7 Antistroke Organophosphorus Drugs

Poor blood flow in brain can cause cell death, a kind of stroke. PMPA is an effective inhibitor of important neuropeptidase enzymes in brain, and has an antistroke activity (Figure 3.13) [28].A research was conducted to find analogous agents [29].

3.4 Organophosphorus Pesticides

Modern agricultural plant protection practices comprise the parallel use of herbicides, insecticides, fungicides, and rodenticides as the main type of pesticides [1, 30–32]. Pesticides need to be very toxic to the target, at the same time, safe for the user. Their effectiveness, low price, and environmentally-friendliness are also of importance. A huge number of organophosphorus pesticides have been used all over the world, from the 1950s until now. It has to be noted that the control and regulatory measures imposed by the US and EU have completely changed the previous practices, and many pesticides have been banned.

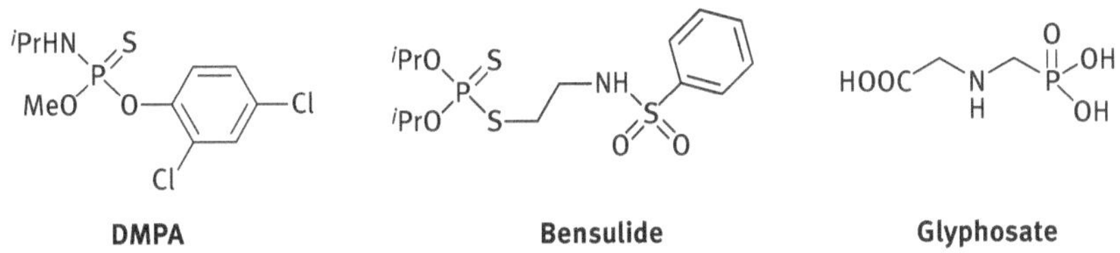

Figure 3.14: A few organophosphorus herbicides.

3.4.1 Herbicides

The two main types of herbicides in use are the pre- and post-emergent herbicides, depending on whether undesired plants are present or not in the target area. DMPA is an early organophosphorus herbicide, introduced in 1958, used as a contact pre-emergent herbicide. Bensulide can be used as a pre- and also a post-emergent herbicide, with a residual action of 4–12 months. The discovery of glyphosate (mentioned in Section 3.3.4) in 1971 was a milestone event in the development of organophosphorus herbicides [33]. Due to its low toxicity for mammals and the effectiveness against target plants, it is used in huge quantities all over the world as a post-emergent herbicide (Figure 3.14).

3.4.2 Insecticides

Organophosphorus insecticides, in most cases, act by inhibiting the cholinesterase enzyme. One of the first insecticides was the TEPP, a type of pyrophosphate, which is very toxic to humans, as it can easily be absorbed through the skin. On the one hand, it is very effective against a wide range of insects, while on the other hand, it is very dangerous for the user. Dichlorvos is used on pets in households, or to protect stored products in warehouses. Avenin is a phosphoric amide with a narrower selectivity. It is much less toxic than organophosphates (Figure 3.15).

Another type of organophosphorus insecticides is the group of thiophosphates (Figure 3.16). Parathion was one of the first insecticides with this structure discovered in the 1940s. Due to its high toxicity to humans, its use is prohibited. The toxic behavior could be somewhat decreased by the change of the ethoxy groups to methoxy units. However, the methyl parathion so obtained is not used either these days, although it was produced and applied widely earlier. Fenitrothion is much less toxic, however, it is

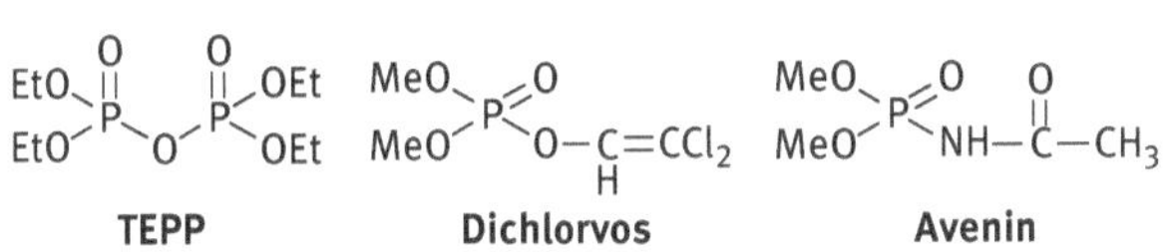

Figure 3.15: Phosphoric acid derivatives used as insecticides.

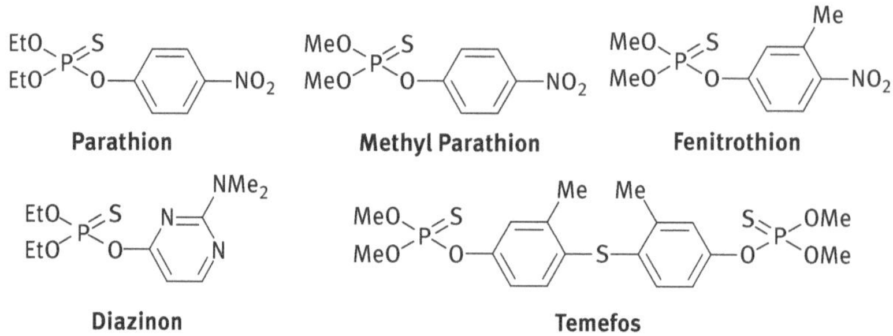

Figure 3.16: Examples for thiophosphate insecticides.

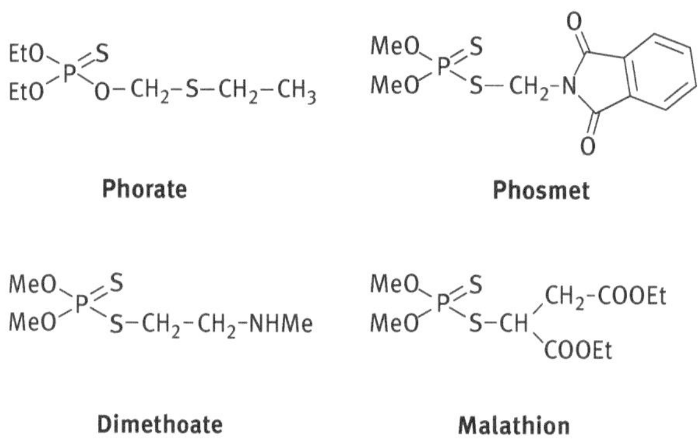

Figure 3.17: Representatives of dithiophosphate insecticides.

of limited use. Diazinon was introduced in 1952. This compound was used against flies, ants, and cockroaches in non-food buildings. Temefos is a much less toxic derivative, and it finds application even now.

Dithiophosphates also show insecticidal activities (Figure 3.17). An early compound is phorate, which is a wide-spectrum contact insecticide with high toxicity to mammals. Phosmet can be used on plants and animals, usually against fruit flies and moths. It is very dangerous to bees, which limits its application. Dimethoate can be applied in households for the control of house flies or cockroaches, due to the low mammal toxicity. Malathion is one of the most commonly used and ever-green organophosphorus insecticides [34]. It is widely employed in agriculture and mosquito eradication.

3.4.3 Fungicides

Beside the well-known inorganic and organometallic fungicides, a few organophosphorus compounds also find application as fungicides (Figure 3.18). Kitazin P is a

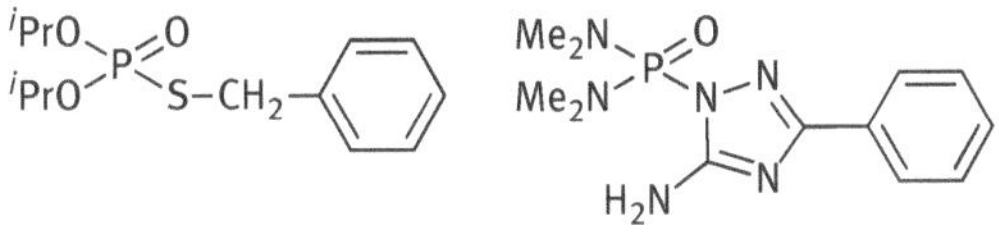

Kitazin P   Wepsyn

Figure 3.18: Organophosphorus compounds with antifungal activities.

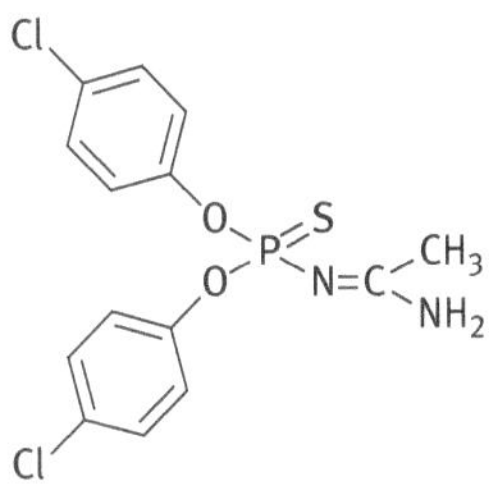

Phosacetim

Figure 3.19: Phosacetim, a cholinesterase inhibitor rodenticide.

systemic fungicide. It inhibits chitin synthase enzyme to limit cell growth in fungal tissues. Another compound with fungicidal activity is Wepsyn, which is widely used in rose and apple cultures.

3.4.4 Rodenticides

A well-established widely used rodenticide is zinc phosphide (Zn_3P_2). Due to the acidic media in the stomach of mice and rats, phosphine gas is liberated from this agent. To protect mammals, usually 20% antimony potassium tartrate is mixed with zinc phosphide, which is an effective human emetic. Phosacetim is an acetylcholinesterase inhibitor used as rodenticide (Figure 3.19).

3.5 Organophosphorus Nerve Agents

Organophosphorus nerve agents are among the most lethal poisons ever known in history (Figure 3.20) [1, 35]. Their development started in World War II, and unfortunately continued after the war ended. Most nerve agents are volatile liquids or gases, and can be absorbed through inhalation as well as by eyes or skin. Their mechanism of action is the irreversible inhibition of the cholinesterase enzyme, causing fatal injuries in the nervous system. Three extremely toxic agents, Sarin, Tabun and Soman, were developed in Germany during World War II. VX was discovered in the 1950s, and together with Soman, is perhaps the most dangerous nerve agent that has been produced and applied in large quantities during "biological wars".

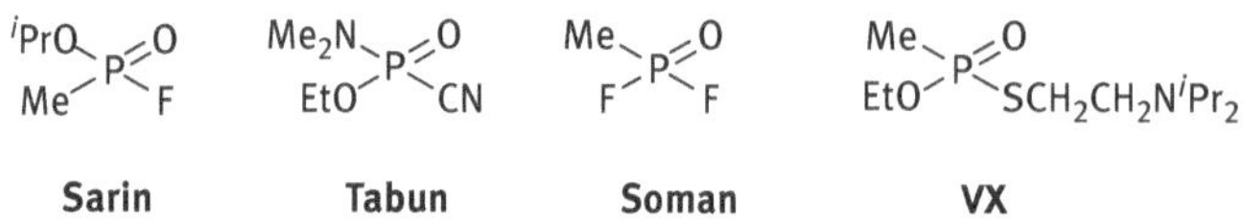

Figure 3.20: Organophosphorus nerve agents.

Table 3.3: LD_{50} values of organophosphorus pesticides and nerve agents.

Compound	LD_{50}	Compound	LD_{50}
DMPA	270	Phorate	2
Bensulide	1,082	Phosmet	230
Glyphosate	5,000	Dimethoate	600
TEPP	1	Malathion	1,200
Dichlorvos	80	Wepsyn	20
Avenin	5,000	Kitazin P	3,500
Parathion	6	Phosacetim	4
Diazinon	10	Tabun	3
Methyl parathion	20	Sarin	0.67
Fenitrothion	250	Soman	0.37
Temephos	2,000	VX	0.19

The LD_{50} values of the pesticides and nerve agents mentioned in Sections 3.5 and 3.6 are listed in Table 3.3.

3.6 Conclusions

Bioactive organophosphorus compounds are involved in thousands of biological and biochemical processes, as they are inevitable in maintaining a healthy body, in the medical treatments of a number of diseases, in fighting against undesired plants, insects or fungi, as well as in biological warfare. The purpose of this summary was to stir up interest in bioactive organophosphorus compounds by highlighting the most important families of the relevant compounds.

Acknowledgments: Ádám Tajti thanks Gedeon Richter Talentum Foundation and Pro Progressio Foundation for financial support.

References

[1] D. E. C. Corbridge, *Phosphorus: Chemistry, Biochemistry and Technology*; CRC Press: Boca Raton, 2013.

[2] E. N. Walsh, E. J. Griffith, R. W. Parry, L. D. Quin, *Phosphorus Chemistry: Developments in American Science*; American Chemical Society: New York, 1992.

[3] K. Ashley, D. Cordell, D. Mavinic, *Chemosphere*, 2011, **84**, 737.

[4] D. B. Tower, J. T. Hensing, *Hensing, 1719: An Account of the First Chemical Examination of the Brain and the Discovery of Phosphorus Therein: Set Against the Background of Europe in the 17th and early 18th Centuries*; Raven Press: New York, 1983.

[5] A. Lenartson, *The Chemical Works of Carl Wilhelm Scheele*, Springer: Switzerland, 2017.

[6] N. K. Fageria, Z. He, V. C. Baligar, *Phosphorus Management in Crop Production*; CRC Press: Boca Raton, 2017.

[7] M. Bockisch, *Fats and Oils Handbook*, AOCS Press: Illinois, 1998.

[8] T. L. Sourkes, *Bull. Hist. Chem.*, 2004, **29**, 9.

[9] R. Dahm, *Hum. Genet.*, 2008, **122**, 565.

[10] C. H. Fiske, Y. Subbarow, *J. Biol. Chem.*, 1929, **81**, 629.

[11] F. F. Nord, C. H. Werkman, *Advances in Enzymology and Related Areas of Molecular Biology: Volume 1*, Interscience Publishers: New York, 1941.

[12] M. A. Shampo, R. A. Kyle, D. P. Steensma, *Mayo Clin. Process*, 2012, **87**, e19.

[13] M. Cobb, *Curr. Biol.*, 2014, **24**, R55.

[14] J. D. Watson, F. H. Crick, *Nature*, 1953, **171**, 737,

[15] H. Seto, T. Kuzuyama, *Nat. Prod. Rep.*, 1999, **16**, 589.

[16] M. G. M. G. Penido, U. S. Alon, *Pediatr. Nephrol.*, 2012, **27**, 2039.

[17] W. J. Marshall, M. Lapsley, A. Day, R. Ayling, *Clinical Biochemistry: Metabolic and Clinical Aspects;* Elsevier: Churchill Livingstone, 2014.

[18] P. H. Stahl, C. G. Wermuth, *Pharmaceutical Salts: Properties, Selection, and Use, Verlag Helvetica Chimica Acta: Zürich*; Wiley: Weinheim, 2002.

[19] H. R. Hudson, N. J. Wardle, S. W. A. Bligh, I. Greiner, A. Grün, G. Keglevich, *Mini Rev. Med. Chem.*, 2012, **12**, 313.

[20] Russell, R. G. G., *Bone*, 2011, **49**, 2.

[21] S. Demkowicz, J. Rachon, M. Dasko, W. Kozak, *RSC Adv.*, 2016, **6**, 7101.

[22] V. P. Kukhar, H. R. Hudson, *Aminophosphonic and Aminophosphinic Acids: Chemistry and Biological Activity*; Wiley: Chichester, 2000.

[23] M. J. van Maanen, C. J. M. Smeets, J. H. Beijnen, *Cancer Treat. Rev.*, 2000, **26**, 257.

[24] A. R. Ahmed, S. M. Hombal, *J. Am. Acad. Dermatol.*, 1984, **11**, 115.

[25] H. R. Hudson, G. Keglevich, *Phosphorus, Sulfur, Silicon*, 2008, **183**, 2256.

[26] G. Keglevich, L. D. Quin, Z. Böcskei, G. M. Keserü, R. Kalgutkar, P. M. Lahti, *J. Organomet. Chem.*, 1997, **532**, 109.

[27] S. S. Long, L. Pickering, C. Prober, *Principles and Practice of Pediatric Infectious Diseases*; Elsevier: Churchill Livingstone, 2003.

[28] R. Schwarcz, *Neuropsychopharmacology: A Tribute to Joseph T. Coyle*; Elsevier: Cambridge; San Diego; Oxford; London, 2016.

[29] K. Haráanyi, G. Domány, I. Greiner, H. Forintos, G. Keglevich, *Heteroatom Chem.*, 2005, **16**, 562.

[30] N. N. Melnikov, *Chemistry of Pesticides*; Springer: New York, 1971.

[31] K. Naumann, *Chemie der Pflanzenschutzund Schädlingsbekämpfungsmittel*; Springer: Berlin; Heidelberg; New York, 1981.

[32] O. Lopez, J. Fernandez-Bolanos, *Green Trends in Insect Control*; RSC Publishing: Cambridge, 2011.

[33] J. E. Franz, US3799758, 1971.

[34] M. R. Bonner, J. Coble, A. Blair, L. E. B. Freeman, J. A. Hoppin, D. P. Sandler, M. C. R. Alavanja, *Am. J. Epidemiol.*, 2007, **166**, 1023.

[35] J. Misik, R. Pavlikova, J. Cabal, K. Kuca, *Drug Chem. Toxicol.*, 2015, **38**, 32.

Péter Bagi and Réka Herbay
4 Resolution of phosphine oxides

Abstract: *P*-Stereogenic compounds are among the most well-known ligands used in enantioselective catalysis. Besides its historical importance, the enantiomeric separation of the *P*-chiral organophosphorus compounds is still the mainly used approach in preparing corresponding optically active derivatives. This chapter summarizes the optical resolution methods developed for *P*-stereogenic phosphine oxides and phosphinates.

Keywords: P-stereogenic, phosphine oxide, phosphinate, resolution, diastereomer.

4.1 Introduction

Since the pioneering work of Knowles and Horner, chiral organophosphorus compounds have been among the most important ligands used in asymmetric catalysis [1, 2]. Although there are many species bearing a chiral backbone, the *P*-stereogenic P(III)-compounds are of significance among the ligands used [3–10]. Moreover, the preparation and application of *P*-chiral organophosphorus compounds are at its renaissance in the past decades as many novel compounds and methods continue to be reported in the literature.

As *P*-stereogenic organophosphorus compounds cannot be found in the natural pool of chirality, asymmetric synthesis and resolution of racemates are the two main sources of the corresponding optically active compounds [11, 12]. There is a wide variety of unsymmetrically substituted organophosphorus compounds, which can be prepared in optically active form: phosphines, complexes of phosphines, phosphine chalcogenides, phosphonium salts, and phosphoranes. From the large pool of synthetic methods, the optical resolution of *P*-stereogenic phosphine oxides and phosphinates was chosen as the main topic of this chapter. Moreover, the resolution of a few axially dissymmetric phosphine oxides is also covered. There are several enantioselective syntheses, which are not part of the present work, and can be found in other reviews [6, 10, 13–20].

4.2 Resolution of Phosphine Oxides via the Formation of Covalent Diastereomers

Many enantioseparations of phosphine oxides have been elaborated via the formation of covalent diastereomers, in which case a covalent bond is formed between the resolving agent and the racemic compound during the resolution. There are several

https://doi.org/10.1515/9783110535839-004

examples when the *P*-stereogenic center is involved in the formation of the covalent bond. On the other hand, there are examples when the resolving agent is introduced to another part of the molecular scaffold.

In the resolution techniques developed for the enantioseparation of phosphine oxides, the formation of covalent diastereomers has a few drawbacks. A specific functional group is required in the racemic compound, which is suitable for the formation of the covalent bond. In a few instances, the scaffold of the resolving agent remains in the target molecule, but most of the methods involve the removal of the resolving agent. However, these steps often require demanding reaction conditions, and the optically active auxiliary cannot be recycled. Despite all these limitations, many methods for resolution via the formation of covalent diastereomers have been elaborated over the decades, and novel methods still appear in the literature.

Campbell and Way used (+)-phenylethylamine for the resolution of a phenyl-dibenzophosphole oxide bearing a carboxylic acid functional group. Amides were formed over the course of the resolution. The diastereomeric species were separated, and acidic hydrolysis afforded the enantiomer of the phenyl-dibenzophosphole oxide derivative (Figure 4.1) [21].

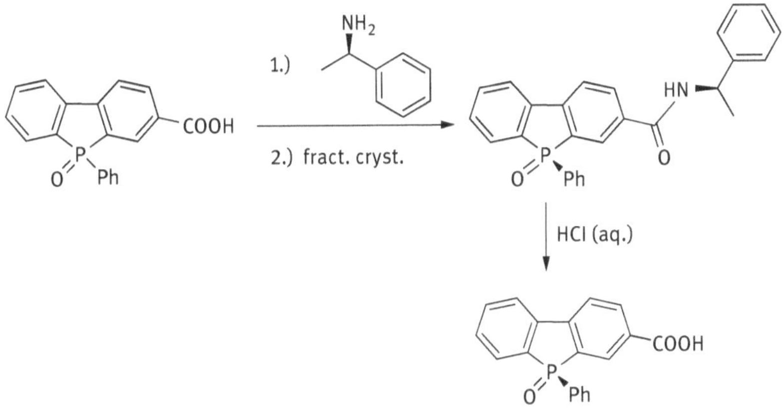

Figure 4.1: Resolution of a phenyl-dibenzophosphole oxide derivative with 1-phenylethylamine.

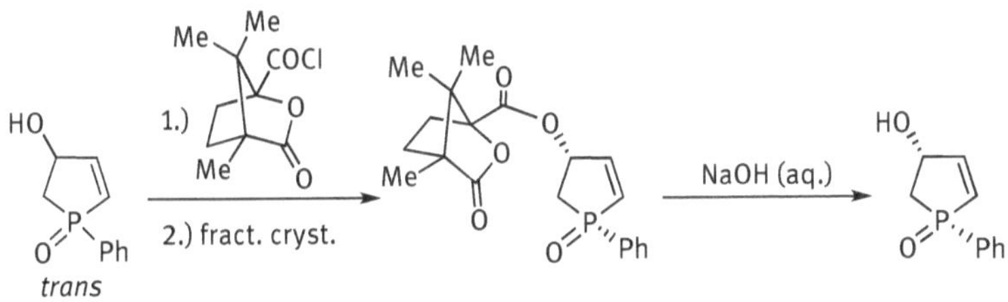

Figure 4.2: Preparation of optically active *trans*-4-hydroxy-1-phenyl-2-phospholene 1-oxide with ω-camphanyl chloride.

Bodalski et al. reacted racemic *trans*-4-hydroxy-1-phenyl-2-phospholene 1-oxide with (–)-ω-camphanyl chloride. The diastereomeric esters formed were separated by fractional crystallization, and subsequent hydrolysis led to optically active *P*-heterocyclic phosphine oxide (Figure 4.2) [22].

During the synthesis of novel dihydrobenzooxaphosphole-based ligands, Tang et al. elaborated the resolution of *P*-stereogenic *t*-butyl-hydroxy-benzodihydrooxaphosphole oxide. The phenolic OH group of the racemic compound was reacted with menthyl chloroformate, and both covalent diastereomers could be prepared in a pure form by crystallization. The hydrolysis of the corresponding carbonate intermediate afforded the enantiopure dihydrobenzooxaphosphole core [23, 24]. Recently, the authors revisited and fine-tuned their original synthesis, and they found that the undesired diastereomer could be racemized in the presence of oxalyl chloride via the formation of a chlorophosphonium salt. Using this epimerization protocol, the overall yield of this process could be improved (Figure 4.3) [25].

In the literature, oxidative resolution methods involving a P(III)–P(V) transformation is of special interest. One variant of this approach relies on a Staudinger reaction between a *P*-chiral phosphine and an enantiomerically pure azide. The iminophosphorane diastereomers were separated by either crystallization or column chromatography, and subsequent stereospecific hydrolysis afforded enantiopure phosphine oxides. In this reaction, a sulfonyl azide having isoborneol scaffold was found to be the best resolving agent, but 10-camphorsulfonyl azide was also used in a few instances. Several enantiopure dialkylaryl-, diarylalkyl-, and triarylphosphine oxides were prepared in this manner (Figure 4.4) [26, 27].

Keay and his research group extended this resolution method based on the formation of covalent diastereomeric iminophosphoranes for the enantioseparation of a few axially dissymmetric biaryl biphosphine oxides (Figure 4.5) [27–30].

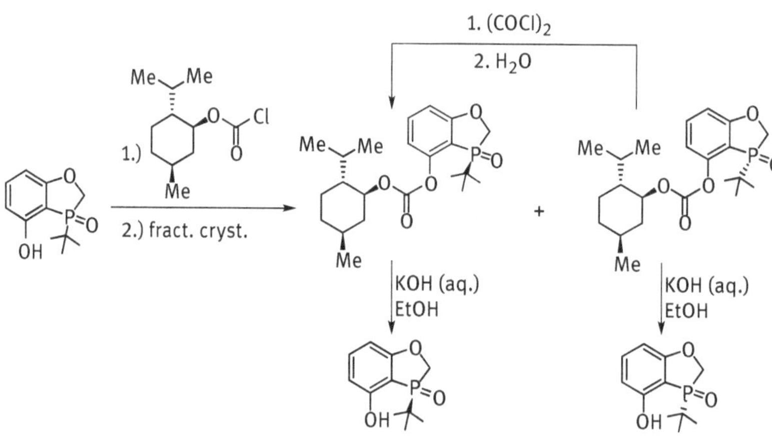

Figure 4.3: Preparation of both enantiomers of *t*-butyl-hydroxy-benzodihydrooxaphosphole oxide.

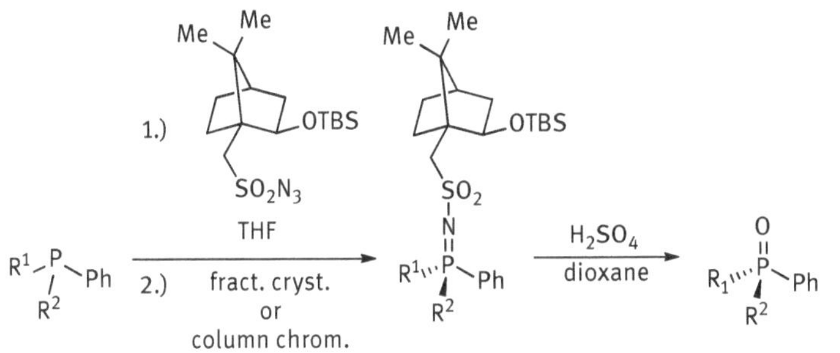

R^1 = Me, 1-Napht

R^2 = ^{i}Pr, ^{c}Pent, ^{c}Hex, 1-Napht, 2-Me-1-Napht, 2-MeO-1-Napht, 2-Napht, 9-Phenantryl, 4-Ph-C_6H_4

Figure 4.4: Preparation of optically active *P*-chiral phosphine oxides via Staudinger reaction.

The enantiomers of a few 3,3′-disubstituted biaryl bis(phosphine oxides) were separated via the formation of covalent diastereomers. Axially dissymmetric diastereomeric esters or ethers were prepared by reacting the phenolic OH group with (*S*)-2-acetoxypropanoyl chloride or with an alcohol derived from Naproxen. The covalent diastereomers were separated by column chromatography followed by the removal of the chiral auxiliary (Figure 4.6) [31–33].

Another variant of oxidative resolution methods involves the formation of diastereomeric phosphonium salts. Pietrusiewicz et al. reacted racemic cyclic phosphines, such as phenyl-2-phospholene oxide or phenylhexahydrocyclopenta[b]phosphole with (–)-menthyl bromoacetate, and the phosphonium salts were separated by fractional crystallization. The optically active phosphine oxide was obtained from the corresponding diastereomer under hydrolytic or Wittig-conditions [34, 35]. This methodology was also suitable for the preparation of a triaryl-phosphine oxide in enantiopure form (Figure 4.7) [36].

The menthyl bromoacetate was also a proper resolving agent for converting the *P*-chiral butyl phenylvinylphosphinite into a diastereomerically pure tertiary phosphine oxide in a Michaelis–Arbuzov reaction [37]. The chiral auxiliary could be removed by

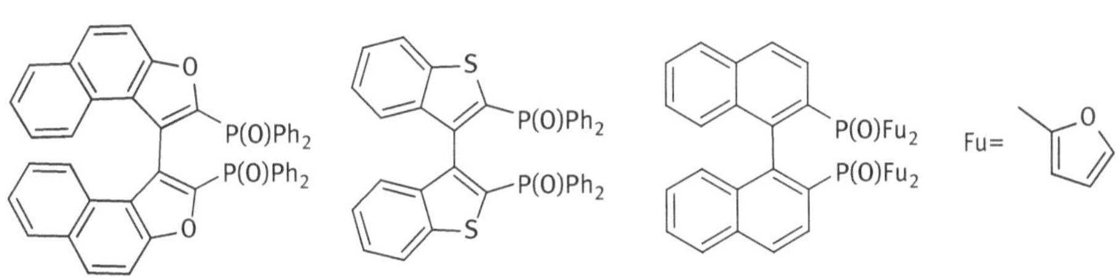

BINAPFu BITIANP TetFuBINAP

Figure 4.5: Axially dissymmetric biaryl biphosphine oxides resolved via Staudinger reaction.

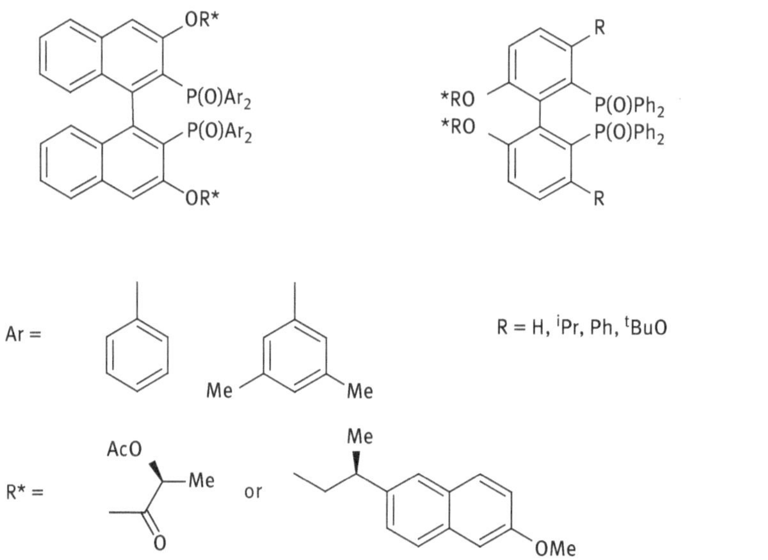

Figure 4.6: Axially dissymmetric biaryl biphosphine oxides resolved via the formation of diastereomeric esters or ethers.

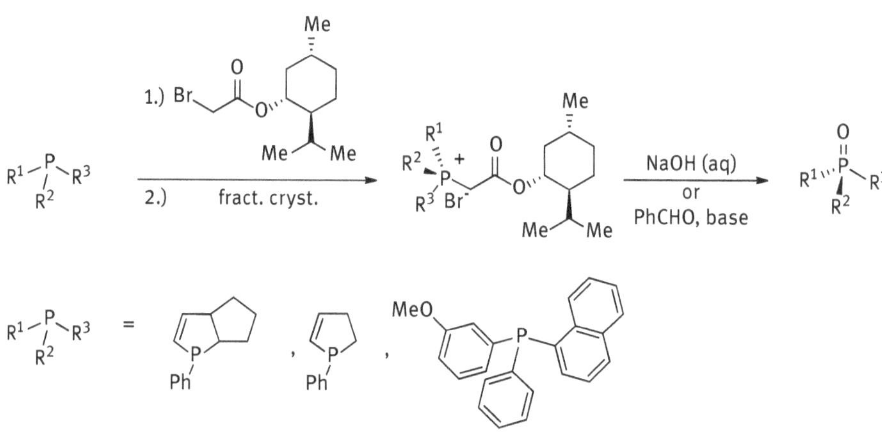

Figure 4.7: Optically active phosphine oxides prepared via the formation of diastereomeric phosphonium salts.

decarboxylation using LiCl, DMSO, and water, and the tertiary vinylphosphine oxide prepared in this way served as a valuable intermediate for the preparation of dialkylarylphosphine oxides (Figure 4.8) [37–39].

Imamoto, Johnson, and their co-worker also used menthyl haloacetate to prepare the diastereomers of tertiary phosphine oxides. According to their method, secondary

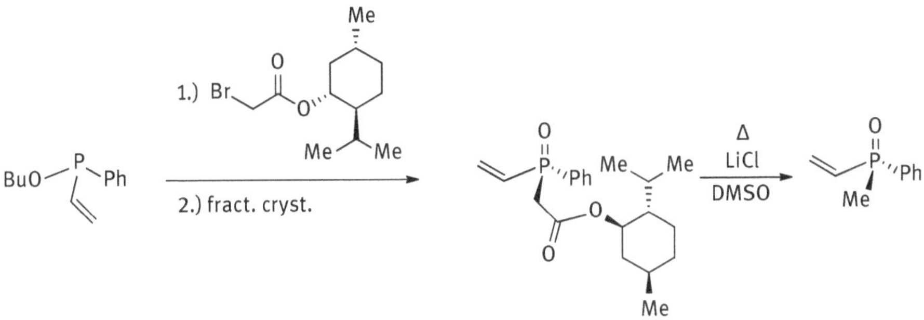

Figure 4.8: Optically active phosphine oxides prepared by a Michaelis–Arbuzov reaction.

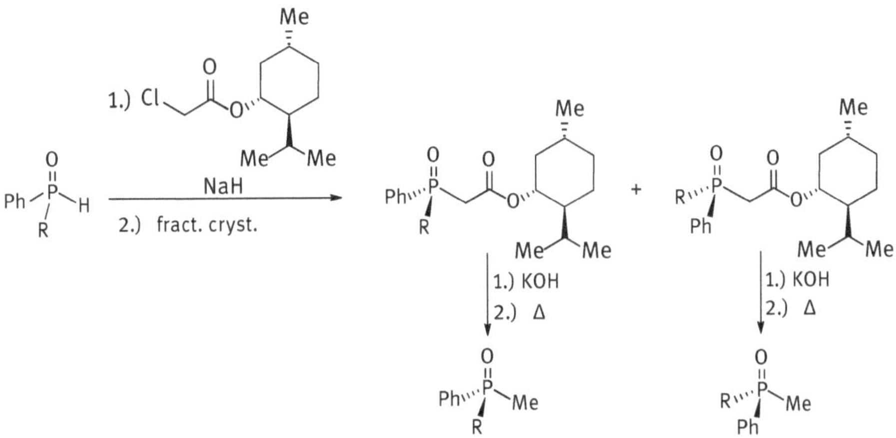

R = ^{t}Bu, 2-CF_3-C_6H_5, 2,4,6-tri^{i}Pr-C_6H_2, 2,4,6-triMe-C_6H_2, 2,6-MeO-C_6H_4, 8-MeO-1-Napht

Figure 4.9: Preparation of optically active tertiary phosphine oxides from *P*-stereogenic secondary phosphine oxides with menthyl chloroacetate.

phosphine oxides were first deprotonated with NaH, and the anions formed were reacted with menthyl chloroacetate. The diastereomeric tertiary phosphine oxides were separated by fractional crystallization, and the menthol auxiliary, as well as the carboxylic group could be removed by hydrolysis and decarboxylation (Figure 4.9) [40].

Recently, Gilheany and his research group elaborated a dynamic resolution method for the preparation of optically active *P*-chiral tertiary phosphine oxides. In the original publication, they described an oxidative resolution method, as racemic phosphines were converted to optically active phosphine oxide [41], but racemic phosphine oxides were also used as starting materials in the subsequent publications [42–45].

During the reaction, racemic phosphines or phosphine oxides were reacted with hexachloroacetone or oxalyl chloride, respectively. It was proved, by NMR, that the enantiomers of the chlorophosphonium salts thus prepared dynamically interconverted into each other, which was a prerequisite of a dynamic resolution [46]. The chlorophosphonium salts were then reacted with chiral alcohols to form the diastereomeric species in unequal amounts. On heating, the diastereomerically enriched alkoxyphosphonium salts were converted to optically active phosphine oxides in an Arbuzov-type collapse.

The research group proved that the enantiomeric excess of the phosphine oxide products was limited by the diastereomeric excess of the alkoxyphosphonium salts, which depended on the chiral auxiliary used. Many chiral alcohols were screened for this reaction, and (–)-menthol was found to be the best auxiliary [42–44]. It was found that *ee* of the phosphine oxides could be improved by exploiting the different thermal decomposition rates of the alkoxyphosphonium salt diastereomers or by using acid-scavengers in the Arbuzov-collapse [44, 45]. Moreover, it turned out that Arbuzov-collapse of the alkoxyphosphonium salts involved retention of the *P*-chiral center, whereas the alkaline hydrolysis of the same species led to inversion. In this manner, both enantiomers of a given phosphine oxide could be prepared from the same diastereomeric intermediate (Figure 4.10) [44].

Considering the history of the preparation and application of *P*-stereogenic compounds, the synthesis of menthyl phoshinates is of special importance among the methods based on the formation of covalent diastereomers. According to Mislow and his co-workers, menthyl phoshinates were prepared by reacting phosphinic chloride with (–)-menthol in the presence of a base. The 1:1 mixture of diastereomers obtained in this reaction was generally separated by crystallization. Menthyl phosphinates were also extensively used in transformations involving a *P*-chiral center. These investigations comprised the stereospecific nucleophilic displace-

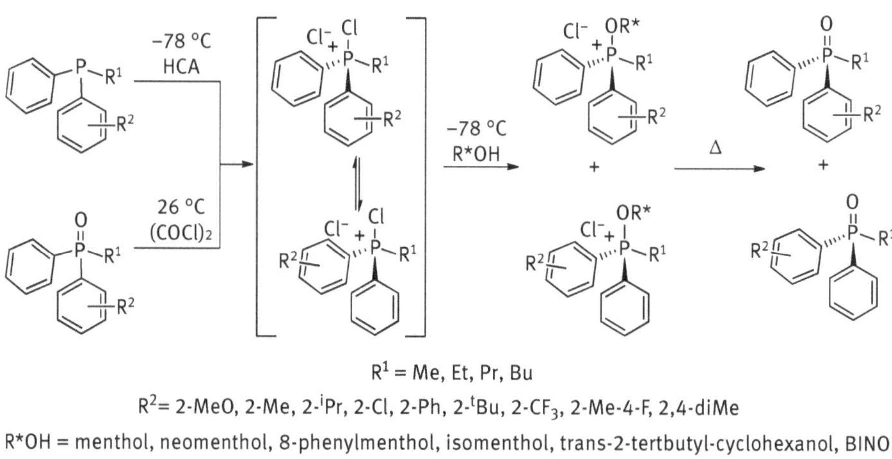

Figure 4.10: Dynamic resolution of phosphines or phosphine oxides via the formation of alkoxyphosphonium salts.

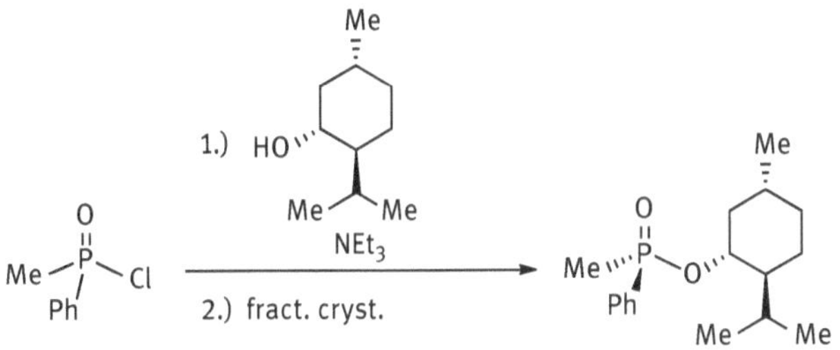

Figure 4.11: Preparation of diastereomerically pure menthyl methylphenylphosphinate.

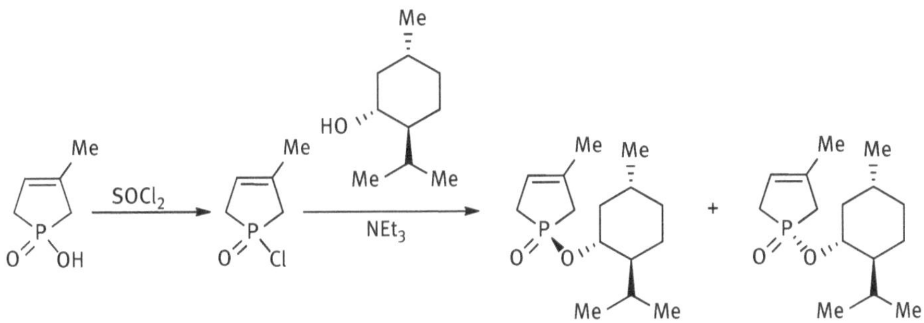

Figure 4.12: Preparation of the diastereomers of 1-[(1*R*,2*S*,5*R*)-menthyloxy]-3-methyl-3-phosphole 1-oxide.

ment of the menthyloxy group by organometallic reagents, which reaction was key for the preparation of optically active tertiary phosphine oxides [47–49]. The fact that Knowles followed this synthetic route during the synthesis of DIPAMP, a cornerstone ligand used in asymmetric catalysis, also underlines the significance of menthyl phosphinates (Figure 4.11) [50, 51].

Keglevich et al. also prepared a menthyl ester of a cyclic phosphinic acid. The 1-[(1*R*,2*S*,5*R*)-menthyloxy]-3-methyl-3-phosphole 1-oxide was synthesized from the corresponding phosphinic chlorides and (–)-menthol. The diastereomers could not be separated under achiral conditions. However, diastereomerically pure compounds were prepared by complexation with TADDOL derivatives (Figure 4.12) [52].

The diastereomeric esters also became important molecules in the sphere of *H*-phosphinates, in which class of compounds the menthyl phenyl-*H*-phosphinate is the most well-known species. The preparation of menthyl phenyl-*H*-phosphinate was first reported by Emmick and Letsinger in 1968 [53], and the first separation of the corresponding diastereomers was published by Mislow and his co-workers [54]. The straightforward synthesis of menthyl phenyl-*H*-phosphinate involved the reaction of phenylphosphonous dichloride and (–)-menthol in the presence of a base, which was followed by hydrolysis [53, 55]. Alternatively, MenthOPCl_2 was first reacted with phenylmagnesium chloride and then water was added to the reaction mixture. This latter approach is more flexible, as different aryl groups can be attached to the

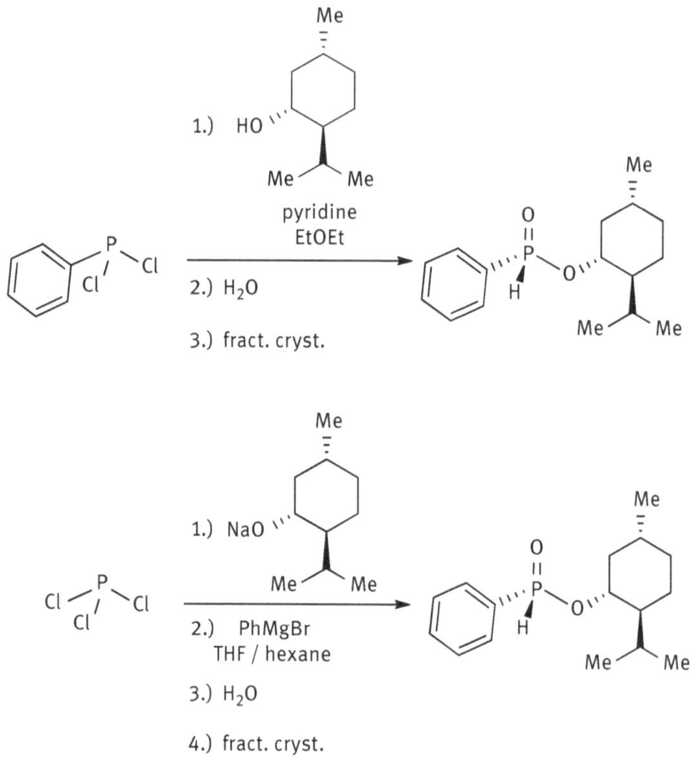

Figure 4.13: Preparation of diastereomerically pure menthyl phenyl-*H*-phosphinate.

P-chiral center [56]. None of these procedures were diastereoselective, so the separation of the diastereomers was important. The purification step was probably the biggest difficulty of this synthetic method, as the repeated fractional crystallizations of diastereomers required cryogenic conditions, and the yields were generally low [57]. Despite all of these drawbacks, diasteromerically pure menthyl phenyl-*H*-phosphinate was prepared on a multi-gram scale. This compound is still the subject of many current papers, as it is the key starting material for the preparation of several *P*-chiral compounds (e.g. secondary phosphine oxides) in stereospecific syntheses (Figure 4.13) [55, 56, 58–62].

In the recent years, Berger and Montchamp have contributed to the topic of *P*-stereogenic menthyl *H*-phosphinate chemistry. According to their multistep synthesis, the cheap and environmentally friendly hypophosphorus acid was the starting material which was reacted with paraformaldehyde and (−)-menthol. The diastereomerically pure menthyl hydroxymethyl-*H*-phosphinate was obtained after one crystallization at 0 °C, but the yield was low (<15%). This simple process does not require either anhydrous or cryogenic recrystallization conditions. These advantages make this approach attractive and the corresponding *H*-phosphinate could be prepared on a multi-gram scale. In their publications, Berger and Montchamp emphasized the synthetic importance of the menthyl hydroxymethyl-*H*-phosphinate by describing several stereoselective transfomations (Figure 4.14) [57, 63].

Considering the covalent diastereomers of phosphinates, (–)-menthol is the most frequently used chiral auxiliary, which fact can be attributed to its availability and price. However, other alcohols were also employed in a few instances. Schmidt and his co-workers prepared and separated the bornyl ester of methylphenylphosphinic acid [64]. Fernandez et al. applied diacetone- and dicyclohexylide-*D*-glucose, (–)-borneol, (+)-isomenthol, and (+)-isopinocampheol for the synthesis of phosphinate esters [65]. From this pool of chiral compounds, the *D*-glucose derivatives were the best auxiliaries, which compounds were also used in other studies allowing the preparation of several *P*-chiral phosphinic esters with a *de* above 90% [66–68]. It is noteworthy that the diastereoselectivity was dependent on the structure of the phosphinic chloride and on the reaction conditions, especially on the nature of the base (Figure 4.15).

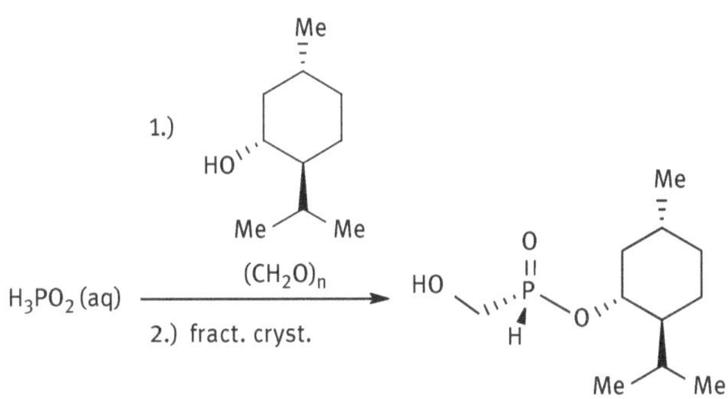

Figure 4.14: Preparation of diastereomerically pure menthyl hydroxymethyl-*H*-phosphinate.

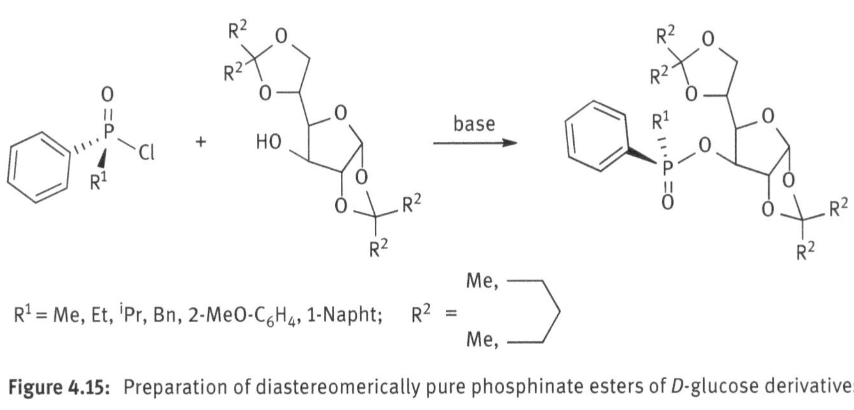

Figure 4.15: Preparation of diastereomerically pure phosphinate esters of *D*-glucose derivatives.

4.3 Resolution of Phosphine Oxides via the Formation of Diastereomeric Salts

The number of resolution methods utilizing the separation of diastereomeric salts are significantly lower compared to the techniques based on the formation of covalent diastereomers. Generally, basic or acidic functional groups of the racemic compound are a prerequisite for the formation of diastereomeric salts, and these functional groups are often utilized for other synthetic transformation of the pure enantiomer. Although there are a few examples within the scope of chiral phosphine oxides, resolution via the formation of diastereomeric salts is the main method for the enantiosepartion of phosphonium salts [11, 12], as well as species bearing P(S)OH, P(Se)OH, or $P(BH_3)OH$ functional groups [69–74].

Ostrogovich and Kerek reported the optical resolution of 1-(*p*-diethylaminophenyl)-3-methyl-2-phospholene 1-oxide with (+)-9-camphorsulfonic acid. The diastereomeric salts were separated by fractional recrystallization from a mixture of benzene and hexane, and the optically active cyclic phosphine oxide was recovered by extraction (Figure 4.16) [75].

Horner and Simons elaborated the resolution of acyclic *P*-stereogenic phosphine oxides bearing amino functional group(s). Despite the fact that the corresponding phosphine oxides incorporated another *C*-chiral center, the separation of the corresponding diastereomers could be effectively accomplished via salt-formation with (+)-(*R*,*R*)-tartaric acid (Figure 4.17) [76].

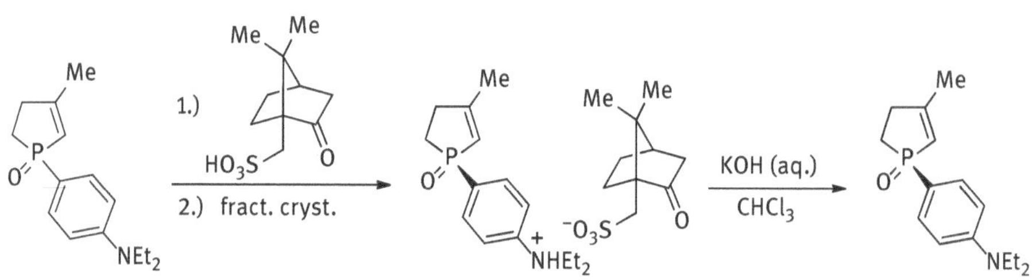

Figure 4.16: Resolution of l-(*p*-diethylaminophenyl)-3-methyl-2-phospholene 1-oxide with (+)-9-camphorsulfonic acid.

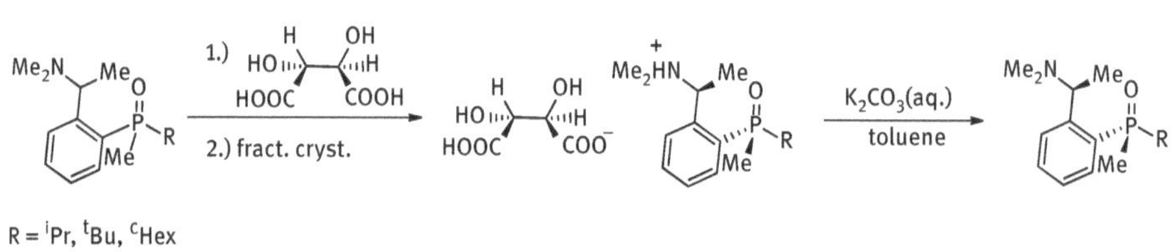

Figure 4.17: Resolution of *P*-chiral phosphine oxides via the formation of diastereomeric salts.

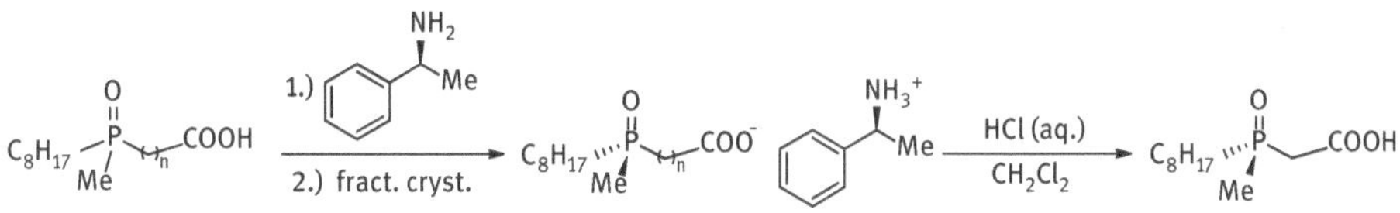

C_8H_{17} = 1,1,3,3-tetramethylbutyl
n = 1,2

Figure 4.18: Resolution of [methyl(1,1,3,3-tetramethylbutyl)phosphinoyl]acetic acid and 3-[methyl(1,1,3,3-tetramethylbutyl)phosphinoyl]propionic acid with (–)-1-phenylethylamine.

(–)-1-Phenylethylamine was used for the optical resolution of [methyl(1,1,3,3-tetramethylbutyl)phosphinoyl]acetic acid and 3-[methyl(1,1,3,3-tetramethylbutyl)phosphinoyl]propionic acid. Several recrystallizations of the corresponding diastereomers were necessary to prepare enantiopure phosphine oxides, which was homocoupled in a Kolbe electrolytic coupling reaction to form the precursors of *P*-stereogenic bidentate ligands (Figure 4.18) [77].

The optical resolution of a phosphonic ester bearing a terminal carboxylic group was elaborated with quinine or cinchonine. As the distereomeric salts were not crystalline, they were separated by column chromatography. The enantiopure phosphonic ester was recovered by extraction (Figure 4.19) [78].

4.4 Resolution via the Formation of Diastereomeric Molecular or Coordination Complexes

Considering all the examples shown in Sections 4.2 and 4.3, functional groups suitable for the formation of covalent diastereomers or diastereomeric salts were necessary for resolution. However, in the absence of these specific functional groups of the racemic compound, the phosphine oxide enantiomers may be prepared via the formation of diastereomeric molecular or coordination complexes, where only non-bonding interactions are responsible for the enantiomeric recognition.

The first resolution of phosphine oxides via the formation of a diastereomeric molecular complex is of particular historical importance, as it was the first proof that unsymmetrically substituted organophosphorus compounds can be separated into their enantiomers. In 1911, Meisenheimer and Lichtenstadt reported the resolution of ethylmethylphenylphosphine oxide using (+)-bromocamphorsulfonic acid as the resolving agent [79]. They also used the same methodology to prepare optically active benzylmethylphenylphosphine oxide with camphorsulfonic acid (Figure 4.20) [80]. It may be assumed that there is an *H*-bond between the sulfonic acid and the P=O function of the phosphine oxide.

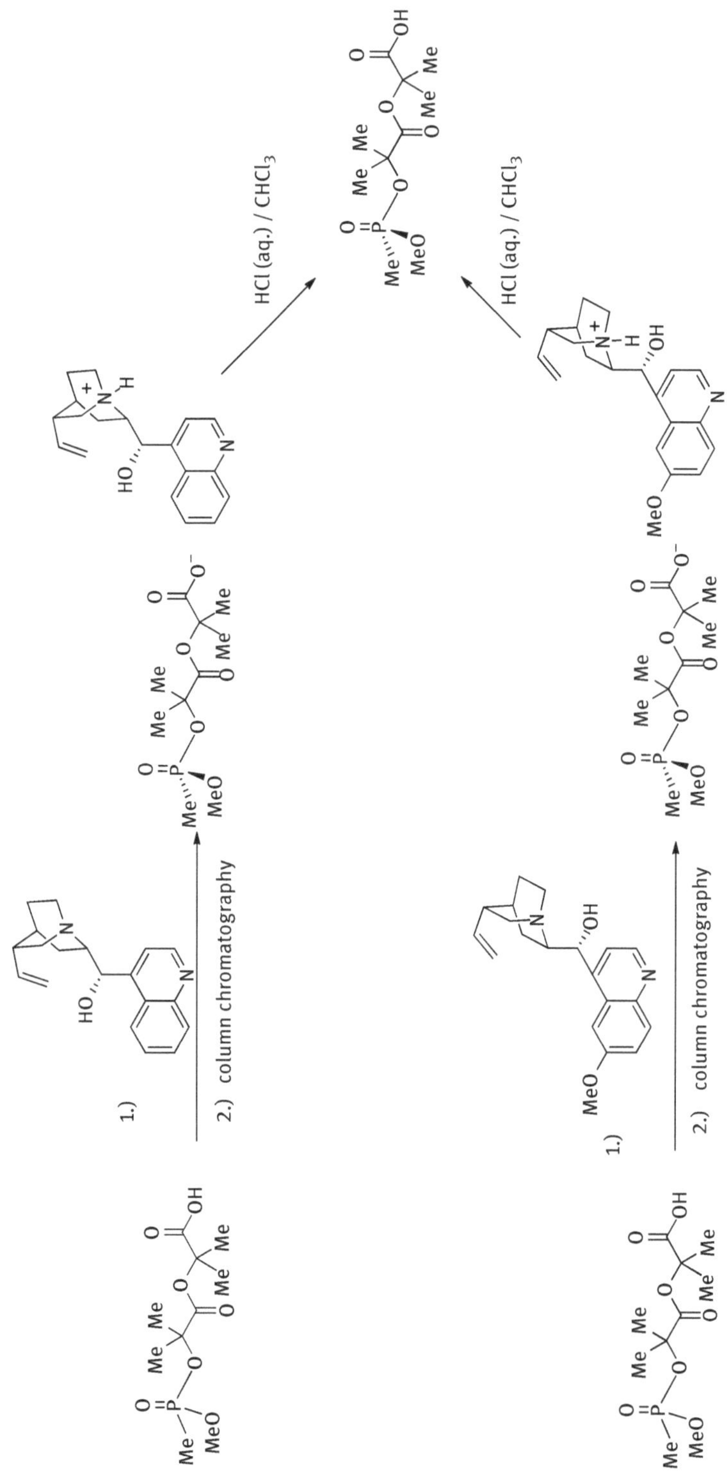

Figure 4.19: Resolution of a phosphonic ester with quinine or cinchonine.

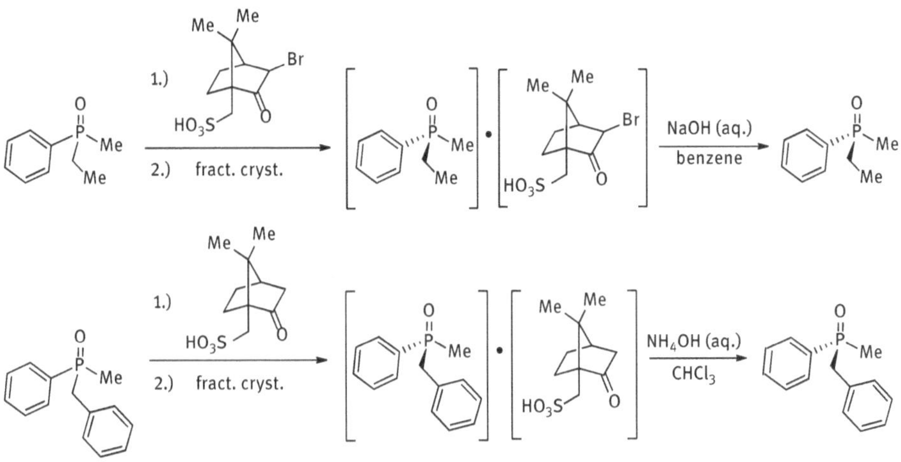

Figure 4.20: Resolution of ethylmethylphenylphosphine oxide and benzylmethylphenylphosphine oxide with camphorsulfonic acid derivatives.

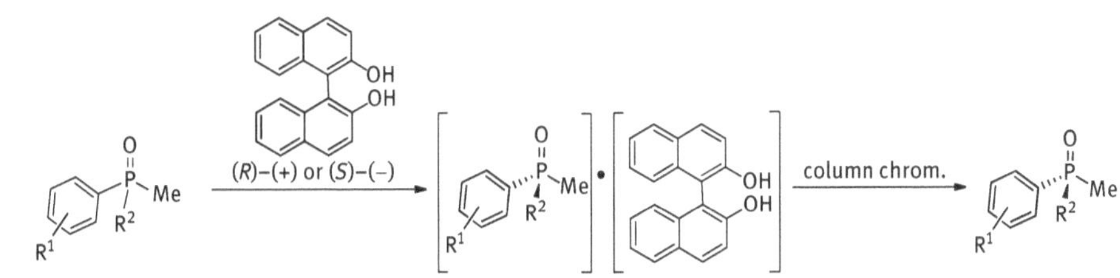

R^1 = H, 2-Me, 3-Me, 4-Me
R^2 = OMe, Et

Figure 4.21: Resolution of P-chiral phosphine oxides and phosphinates with BINOL.

Toda and his co-workers applied 2,2'-dihydroxy-1,1'-binaphthyl (BINOL) for the enantioseparation of phosphine oxides and phosphinates. The corresponding *P*-chiral racemic compound and the BINOL were dissolved in benzene, and the diastereomeric molecular complexes precipitated. The diastereomers were purified by recrystallizations, and the enantiopure phosphinates and phosphine oxides were liberated by column chromatography. Interestingly, the efficiency of this resolution method was sensitive to the substituents on the *P*-stereogenic center. A few methyl arylmethylphosphinates, ethylmethylphenylphosphine oxide, and ethylmethyl(3-methylphenyl) phosphine oxide could be separated into their enantiomers with BINOL. In contrast, the resolution was inefficient or no diastereomeric complexes were formed when arylethylmethylphosphine oxides, methylphenylpropylphosphine oxide, ethyl methylphenylphosphinate, methyl or ethyl phenyl-*H*-phosphinate was the racemic compound. In a few instances, the intermolecular interactions were characterized by X-ray crystallographic methods (Figure 4.21) [81].

Keglevich et al. found that TADDOL derivatives are efficient resolving agents for the enantioseparation of cyclic phosphine oxides. According to our method, the phosphine oxide and the resolving agent were co-crystallized. The diastereomeric molecular complexes were purified by several recrystallizations, and the optically active phosphine oxides were recovered by column chromatography. Interestingly, the solvent used for the crystallizations also influenced the enantiomeric recognition. In many cases, both antipodes of a racemic compound could be prepared with the same resolving agent by changing the solvent. X-ray crystallography confirmed a similar enantiomeric recognition pattern in the diastereomers, as a *H*-bond between the OH group of the given TADDOL derivative and the P=O function of the phosphine oxide was the main interaction in all of the molecular complexes prepared. These resolution methods were originally developed for cyclic phosphine oxides, and five- and six-membered heterocyclic *P*-oxides with various substitution patterns could be resolved with TADDOL derivatives (Figure 4.22) [82–89]. Recently, these resolving

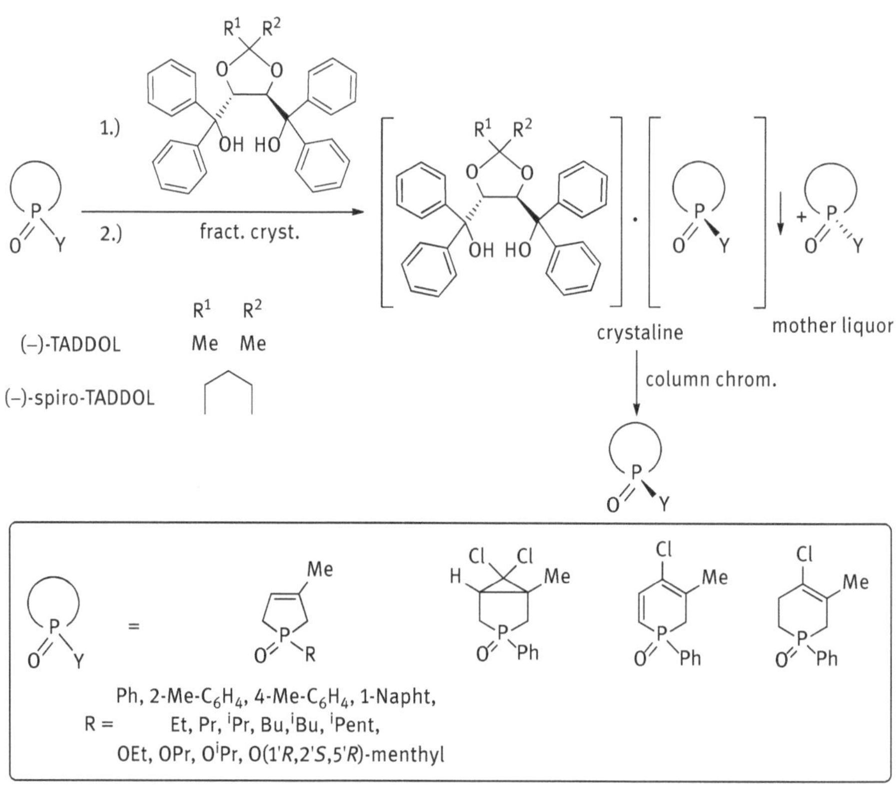

Figure 4.22: Resolution of *P*-stereogenic cyclic phosphine oxides and phosphinates with TADDOL-derivatives.

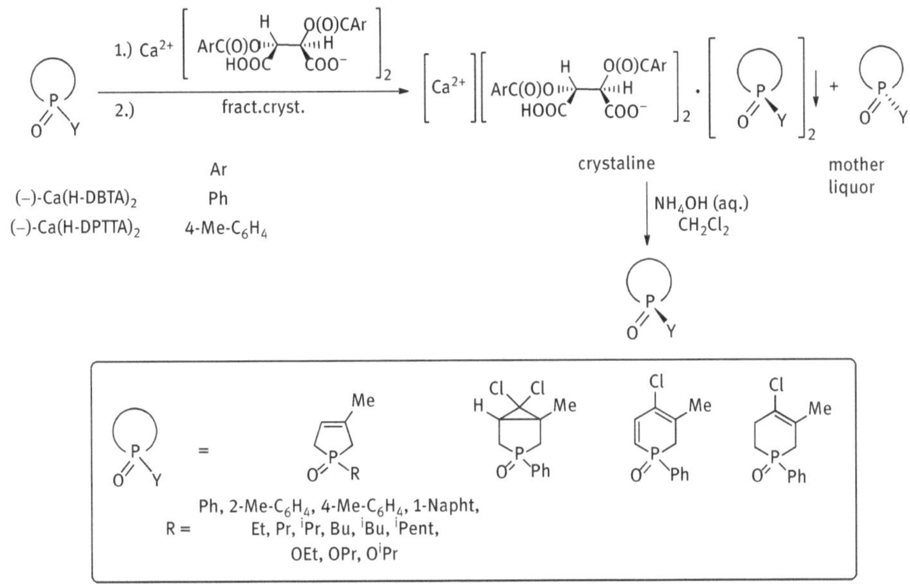

Figure 4.23: Resolution of *P*-stereogenic cyclic phosphine oxides and phosphinates with the Ca^{2+}-salts of tartaric acid derivatives.

agents were also applied for the enantioseparation of acyclic secondary and tertiary phosphine oxides.

The Ca^{2+} salts of the *O,O'*-dibenzoyl- and *O,O'*-di-*p*-toluoyl-(2*R*,3*R*)-tartaric acids were also applicable resolving agents for five- and six-membered heterocyclic phosphine oxides. The diastereomeric coordination complexes precipitated from the solution, were purified by recrystallizations, and the optically active phosphine oxides were liberated by extraction. These resolving agents can also be used for the enantioseparation of a few acyclic phosphine oxides (Figure 4.23) [84–91]. The Ca^{2+} ions determined the mode of binding within the diastereomers, as the coordination of the P=O group to the Ca^{2+} ion of the resolving agent was the main interaction over the H-bonds observed in all the other instances with tartaric acid derivatives.

Besides the optical resolution of tertiary phosphine oxides, a few enantioseparation methods for secondary phosphine oxides were also elaborated. The importance of the secondary phosphine oxides lies in the fact that the *P–H* bond can be reacted in various substitution reactions without the loss of optical activity. Thus, optically active *P*-stereogenic moieties can be incorporated into other molecules. To the best of our knowledge, the *tert*-butylphenylphosphine oxide is the only derivative whose resolution was investigated in detail, despite the relatively large number of *P*-chiral secondary phosphine oxides.

Drabowicz et al. elaborated the resolution of *tert*-butylphenylphosphine oxide via the formation of diastereomeric molecular complexes using (+)-(*R*)-BINOL or

(+)-(*S*)-mandelic acid as the resolving agent. Enantiopure (–)-(*S*)-*tert*-butylphenylphosphine oxide could be obtained with mandelic acid, whereas only partial enantiomeric separation was achieved with BINOL (Figure 4.24) [92].

Pietrusiewicz, Minnaard, and their co-workers screened a few acidic resolving agents, among which the *O,O*'-dibenzoyl-(*R,R*)-tartaric acid was also found to be a suitable resolving agent for the optical resolution of the *tert*-butylphenylphosphine oxide. One equivalent of *O,O*'-dibenzoyl-(*R,R*)-tartaric acid was applied, and the diastereomers were separated by crystallization. Interestingly, one crystallization was sufficient, and both enantiomers of the secondary phosphine oxide were obtained in enantiopure form. Single-crystal X-ray measurement confirmed that a *H*-bond between the P=O and the COOH functions was the main interaction within the molecular coordination complex (Figure 4.25) [93].

Recently, Minnaard et al. optimized further the resolution of *tert*-butylphenylphosphine oxide with *O,O*'-dibenzoyl-(*R,R*)-tartaric acid. They found that the secondary phosphine oxide can be converted into a phosphinyl iodide derivative, which racemizes via a radical process. This racemization protocol was compatible with the resolution conditions, thus a crystallization-induced dynamic resolution could be elaborated. According to this method, the *tert*-butylphenylphosphine oxide was reacted with one equivalent of *O,O*'-dibenzoyl-(*R,R*)-tartaric acid, and the iodine was used in a catalytic amount to promote racemization. In this reaction, nearly all of the racemic starting material was converted to the corresponding single diastereomer,

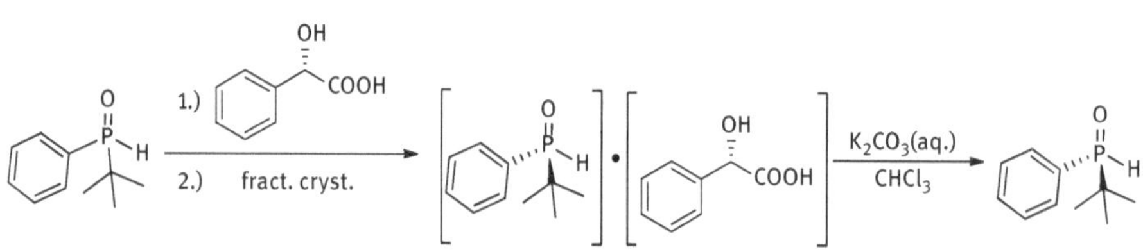

Figure 4.24: Resolution of *tert*-butylphenylphosphine oxide with mandelic acid.

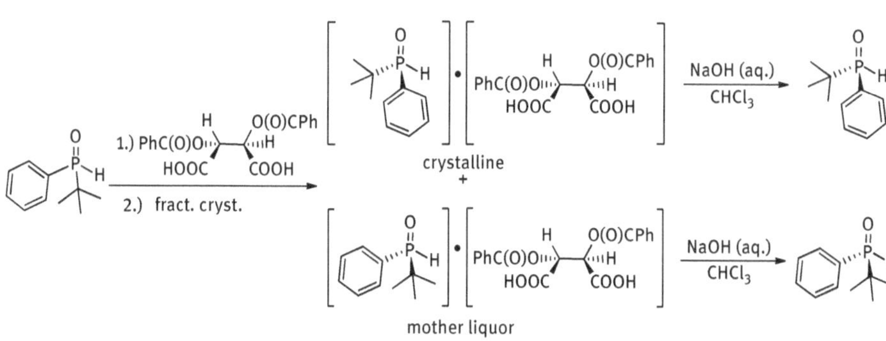

Figure 4.25: Resolution of *tert*-butylphenylphosphine oxide with *O,O*'-dibenzoyl-(*R,R*)-tartaric acid.

from which the enantiopure *tert*-butylphenylphosphine oxide could be liberated by extraction (Figure 4.26) [94].

Pietrusiewicz reported a partial enantioseparation of 1-phenyl-2-phospholene oxide using also *O*,*O*'-dibenzoyl-(*R*,*R*)-tartaric acid as the resolving agent, but the exact conditions were not published [11].

Besides these examples of a *P*-chiral monophosphine oxide, the bis(phosphine oxides) were the main substrate scope of the diastereomeric complex formation with *O*,*O*'-dibenzoyl tartaric acid. In this manner, *P*-chiral bis(phosphine oxides) having different aryl backbones, such as 1,2-bis(isopropylmethyl-phosphino)benzene dioxide [95], 1,8-bis(*tert*-butylphenylphosphino)naphthalene dioxide [96], and 2,7-di-*tert*-butyl-9,9-dimethyl-4,5-bis(methylphenylphosphino)xanthene dioxide [97], as well as 1,1'-di-*tert*-butyl-2,2'-dibenzophosphetanyl dioxide [98] and DuanPhos dioxide [99] were separated into their enantiomers. All of the resolutions were accomplished by co-crystallization of the given racemic bis(phosphine oxide) and the *O*,*O*'-dibenzoyl tartaric acid. The diastereomeric molecular complexes so obtained could be purified by recrystallization, which was followed by extractive liberation. As both enantiomers of *O*,*O*'-dibenzoyl tartaric acid are commercially available, both antipodes of a few bis(phosphine oxides) were obtained (Figure 4.27) [96–99].

The *O*,*O*'-dibenzoyl tartaric acid is also a widely used resolving agent for the enantioseparation of axially chiral bis(phosphine oxides). Since its first appearance [100–103], this method remains practically unchanged. The solutions of the given racemic compound and the resolving agent were combined, and the corresponding diastereomeric complexes appeared on cooling. In many instances, a single crystallization afforded pure diastereomers, but the diastereomeric purity could be enhanced by recrystallizations if it was required. The enantiopure bis(phosphine oxides) could

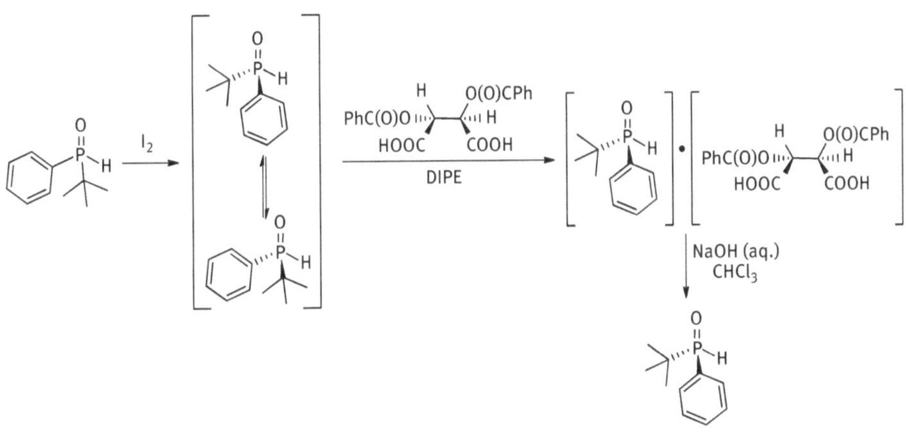

Figure 4.26: Dynamic resolution of *tert*-butylphenylphosphine oxide with *O*,*O*'-dibenzoyl-(*R*,*R*)-tartaric acid.

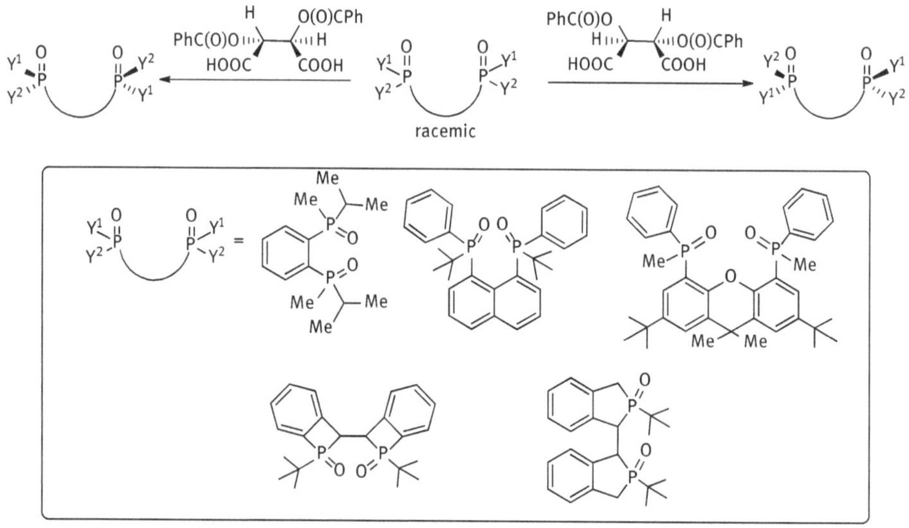

Figure 4.27: Resolution of *P*-chiral bis(phosphine oxides) with *O,O*'-dibenzoyl-tartaric acid.

be liberated by extraction. It turned out that the dibenzoyl tartaric acid is a unique resolving agent for the enantioseparation of axially chiral bis(phosphine oxides), as many compounds having various biaryl structures could be resolved [102, 104–115]. Moreover, the enantiomers of a few *C*-chiral and planar chiral bis(phosphine oxides) were also prepared by this method [100, 101, 103, 116–119]. However, other resolving agents, such as di-*p*-toluoyl-, di-*p*-methoxybenzoyl, or di[(phenylamino)carbonyl]-tartaric acid, or camphorsulfonic acid had to be used in a few instances [102, 109, 110, 112–115, 118]. The X-ray analysis of a few diastereomers confirmed that a *H*-bond is the main interaction between the corresponding bis(phosphine oxides) and the resolving agent (Figure 4.28) [102, 111, 114].

4.5 Resolution by Chromatography

The separation of the enantiomers on a chiral stationary phase is also a technique for preparing optically active organophosphorus compounds, or any kind of chiral molecules. The flexibility of chiral chromatography makes the enantioseparation feasible within a large scope of racemic compounds. Chromatographic method development is relatively straightforward and quick compared to asymmetric synthesis or optical resolution. In our opinion, chiral chromatography is a valuable technique in the early stages of a development of compounds, when time is the most crucial factor for the preparation of an optically active chiral compound. When a given optically active compound is needed on a larger scale, the scalability

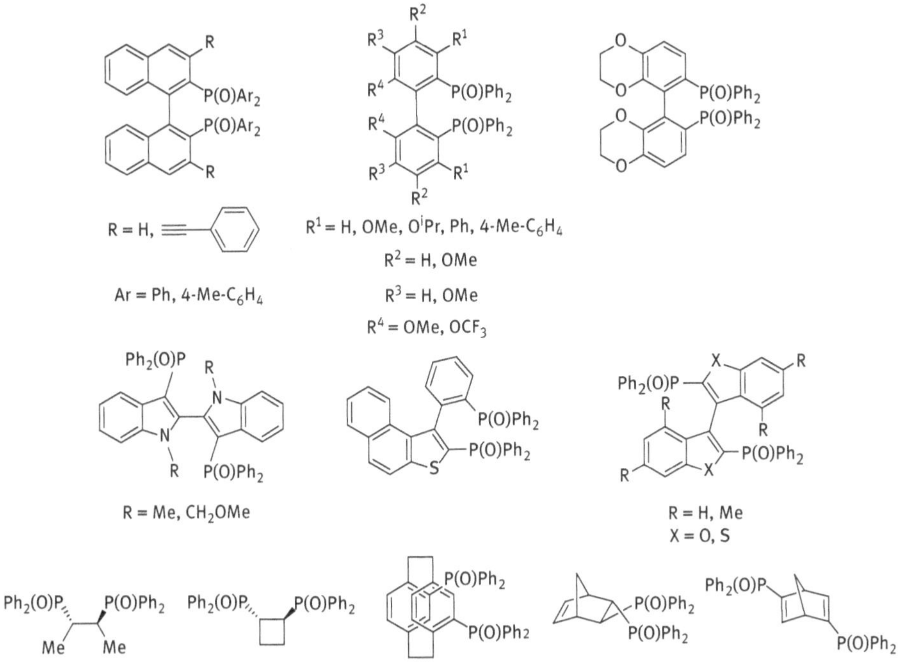

Figure 4.28: Chiral bis(phosphine oxides) resolved with *O*,*O*'-dibenzoyl-tartaric acid.

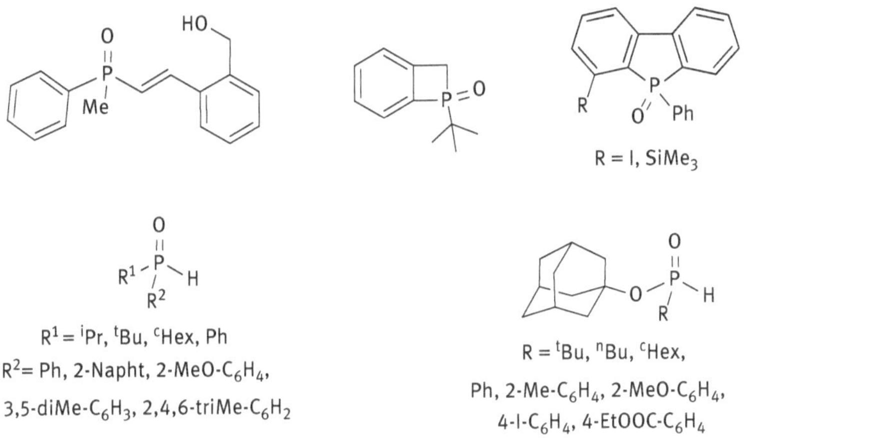

Figure 4.29: P-stereogenic phosphine oxides or *H*-phosphinates resolved by chiral chromatography.

and productivity of this technique may become questionable [120–123]. Recent publications have shown that the cost of such enantiomeric purification may be decreased, especially when a supercritical fluid chromatographic technique is used [124]. This approach makes chiral chromatography viable in the industry, but the cost of equipment is beyond the financial limits of many research groups. Over the decades, the enantiomers of many phosphine oxides were separated by chromatography on chiral columns. The applicability of this method was obvious when different preparative approaches failed to produce the corresponding organophosphorus compounds in enantiopure form. The number of preparative chiral columns is increasing, which may indicate the spreading of this technique. Figure 4.29 shows a few selected examples of racemic phosphine oxides and phosphinates which were separated into their enantiomers on at least a 100 mg scale by chromatography on a chiral stationary phase [98, 125–130].

4.6 Conclusions

In this chapter, resolution procedures based on the formation of diastereomers were summarized. There is a wide variety of enantioselective synthetic methods developed for the preparation of *P*-stereogenic compounds, but the substrate scope of these methods is often limited. Moreover, there might be instances when the asymmetric synthesis of a given chiral compound cannot be accomplished due to chemical reasons or functional group intolerance. In these instances, the resolution of the racemic compounds might be the only possibility to obtain the desired optically active derivative. Besides the phosphine boranes, the *P*-oxides are also stable compounds, which are easy to handle. Phosphine oxides are also important precursors of chiral phosphines, and the stereochemical background of deoxygenation is also well known [131]. In our opinion, the resolution of *P*-stereogenic phosphine oxides is an attractive way for the preparation of the corresponding enantiopure chiral compounds. Over the decades, many enantioseparation methods have been developed for chiral phosphine oxides and phosphinates. However, there is no generally applicable resolution method, although there are a few resolving agents that can be applied in a wider scope of racemic *P*-oxides. The resolution of such *P*-chiral derivatives which have functional groups suitable for substitution are of great importance. Moreover, resolution methods which also involve in situ racemization are also a hot topic. These facts also show that there are still challenges facing this field, and there is a need for developing novel enantioseparation techniques for this group of organophosphorus compounds.

Acknowledgments: This work was supported by the National Research, Development and Innovation Office – NKFIH (Grant No. OTKA PD 116096) and by the ÚNKP-17-4-I new national excellence program of the ministry of human capacities (Grant No. ÚNKP-17-4-I-BME-102). The authors also thank the help of Sarolta Timári.

References

[1] W. S. Knowles, M. J. Sabacky, *Chem. Commun.*, 1968, 1445.
[2] L. Horner, H. Siegel, H. Büthe, *Angew. Chem. Int. Ed.*, 1968, **7**, 942.
[3] R. Noyori, *Asymmetric Catalysis in Organic Synthesis;* John Wiley & Sons: New York, 1994.
[4] A. Börner *Ed.*, *Phosphorus Ligands in Asymmetric Catalysis;* Wiley-VCH: Weinheim, 2008.
[5] I. Ojima *Ed.*, *Catalytic Asymmetric Synthesis*; Wiley: Weinheim, 2010.
[6] K. V. L. Crépy, T. Imamoto, *Adv. Synth. Catal.*, 2003, **345**, 79.
[7] W. Tang, X. Zhang, *Chem. Rev.*, 2003, **103**, 3029.
[8] T. Jerphagnon, J.-L. L. Renaud, C. Bruneau, *Tetrahedron: Asymmetry,* 2004, **15**, 2101.
[9] W. Fu, W. Tang, *ACS Catal.*, 2016, **6**, 4814.
[10] M. Dutartre, J. Bayardon, S. Jugé, *Chem. Soc. Rev.*, 2016, **45**, 5771.
[11] K. M. Pietrusiewicz, M. Zablocka, *Chem. Rev.*, 1994, **94**, 1375.
[12] A. Grabulosa, *P-Stereogenic Ligands in Enantioselective Catalysis*; The Royal Society Of Chemistry: Cambridge, 2010.
[13] M. Ohff, J. Holz, M. Quirmbach, A. Börner, *Synthesis,* 1998, **1998**, 1391.
[14] O. I. Kolodiazhnyi, *Tetrahedron: Asymmetry,* 1998, **9**, 1279.
[15] P.-H. Leung, *Acc. Chem. Res.*, 2004, **37**, 169.
[16] M. J. Johansson, N. Kann, *Mini-Rev. Org. Chem.*, 2004, **1**, 233.
[17] D. S. Glueck, *Synlett,* 2007, 2627.
[18] A. Grabulosa, J. Granell, G. Muller, *Coord. Chem. Rev.*, 2007, **251**, 25.
[19] O. I. Kolodiazhnyi, *Tetrahedron: Asymmetry,* 2012, **23**, 1.
[20] O. I. Kolodiazhnyi, V. P. Kukhar, A. O. Kolodiazhna, *Tetrahedron: Asymmetry,* 2014, **25**, 865.
[21] I. G. M. Campbell, J. K. Way, *J. Chem. Soc.*, 1961, 2133.
[22] R. Bodalski, T. Janecki, Z. Gałdecki, M. Główka, *Phosphorus Sulfur Relat. Elem.*, 1982, **14**, 15.
[23] W. Tang, B. Qu, A. G. Capacci, S. Rodriguez, X. Wei, N. Haddad, B. Narayanan, S. Ma, N. Grinberg, N. K. Yee, D. Krishnamurthy, C. H. Senanayake, *Org. Lett.*, 2010, **12**, 176.
[24] W. Tang, A. G. Capacci, A. White, S. Ma, S. Rodriguez, B. Qu, J. Savoie, N. D. Patel, X. Wei, N. Haddad, N. Grinberg, N. K. Yee, D. Krishnamurthy, C. H. Senanayake, *Org. Lett.*, 2010, **12**, 1104.
[25] G. Li, X. Wang, Y. Zhang, Z. Tan, P. DeCroos, J. C. Lorenz, X. Wei, N. Grinberg, N. K. Yee, C. H. Senanayake, *J. Org. Chem.*, 2017, **82**, 5456.
[26] N. G. Andersen, P. D. Ramsden, D. Che, M. Parvez, B. A. Keay, *Org. Lett.*, 1999, **1**, 2009.
[27] N. G. Andersen, P. D. Ramsden, D. Q. Che, M. Parvez, B. A. Keay, *J. Org. Chem.*, 2001, **66**, 7478.
[28] Neil G. Andersen, M. Parvez, B. A. Keay, *Org. Lett.*, 2000, **2**, 2817.
[29] N. G. Andersen, R. McDonald, B. A. Keay, *Tetrahedron: Asymmetry,* 2001, **12**, 263.
[30] N. G. Andersen, M. Parvez, R. McDonald, B. A. Keay, *Can. J. Chem.*, 2004, **82**, 145.
[31] J. M. Hopkins, S. A. Dalrymple, M. Parvez, B. A. Keay, *Org. Lett.*, 2005, **7**, 3765.
[32] E. Gorobets, B. M. M. Wheatley, J. M. Hopkins, R. McDonald, B. A. Keay, *Tetrahedron Lett.*, 2005, **46**, 3843.
[33] D. Rankic, J. Hopkins, M. Parvez, B. Keay, *Synlett,* 2009, 2513.
[34] K. M. Pietrusiewicz, *Phosphorus Sulfur. Silicon Relat. Elem.*, 1996, **109**, 573.
[35] Z. Pakulski, O. M. Demchuk, J. Frelek, R. Luboradzki, K. M. M. Pietrusiewicz, *Eur. J. Org. Chem.*, 2004, 3913.
[36] K. Dziuba, A. Flis, A. Szmigielska, K. M. Pietrusiewicz, *Tetrahedron: Asymmetry,* 2010, **21**, 1401.
[37] R. Bodalski, E. Rutkowska-Olma, K. Michał Pietrusiewicz, *Tetrahedron,* 1980, **36**, 2353.
[38] K. M. Pietrusiewicz, M. Zablocka, J. Monkiewicz, *J. Org. Chem.*, 1984, **49**, 1522.
[39] K. M. Pietrusiewicz, M. Kuźnikowski, M. Koprowski, *Tetrahedron: Asymmetry,* 1993, **4**, 2143.
[40] T. Imamoto, K. Sato, C. R. Johnson, *Tetrahedron Lett.*, 1985, **26**, 783.

[41] E. Bergin, C. T. O'Connor, S. B. Robinson, E. M. McGarrigle, C. P. O'Mahony, D. G. Gilheany, *J. Am. Chem. Soc.*, 2007, **129**, 9566.
[42] K. V. Rajendran, J. S. Kudavalli, K. S. Dunne, D. G. Gilheany, *Eur. J. Org. Chem.*, 2012, 2720.
[43] K. V. Rajendran, D. G. Gilheany, *Chem. Commun.*, 2012, **48**, 10040.
[44] K. Nikitin, K. V Rajendran, H. Müller-Bunz, D. G. Gilheany, *Angew. Chem. Int. Ed.*, 2014, **53**, 1906.
[45] K. V. Rajendran, K. V. Nikitin, D. G. Gilheany, *J. Am. Chem. Soc.*, 2015, **137**, 9375.
[46] E. V. Jennings, K. Nikitin, Y. Ortin, D. G. Gilheany, *J. Am. Chem. Soc.*, 2014, **136**, 16217.
[47] O. Korpium, K. Mislow, *J. Am. Chem. Soc.*, 1967, **89**, 4784.
[48] O. Korpiun, R. A. Lewis, J. Chickos, K. Mislow, *J. Am. Chem. Soc.*, 1968, **90**, 4842.
[49] R. A. Lewis, K. Mislow, *J. Am. Chem. Soc.*, 1969, **91**, 7009.
[50] B. D. Vineyard, W. S. Knowles, M. J. Sabacky, G. L. Bachman, D. J. Weinkauff, *J. Am. Chem. Soc.*, 1977, **99**, 5946.
[51] W. S. Knowles, *Angew. Chem. Int. Ed.*, 2002, **41**, 1998.
[52] G. Keglevich, J. Dvorszki, V. Ujj, K. Ludányi, *Heteroatom Chem.*, 2010, **21**, 271.
[53] T. L. Emmick, R. L. Letsinger, *J. Am. Chem. Soc.*, 1968, **90**, 3459.
[54] W. B. Farnham, R. K. Murray, K. Mislow, *J. Am. Chem. Soc.*, 1970, **92**, 5809.
[55] W.-M. Wang, L.-J. Liu, C.-Q. Zhao, L.-B. Han, *Eur. J. Org. Chem.*, 2015, 2342.
[56] D. Gatineau, L. Giordano, G. Buono, *J. Am. Chem. Soc.*, 2011, **133**, 10728.
[57] O. Berger, J.-L. Montchamp, *Org. Biomol. Chem.*, 2016, **1**, 7552.
[58] L.-B. Han, C.-Q. Zhao, S. Onozawa, M. Goto, M. Tanaka, *J. Am. Chem. Soc.*, 2002, **124**, 3842.
[59] L. B. Han, C. Q. Zhao, *J. Org. Chem.*, 2005, **70**, 10121.
[60] A. Leyris, J. Bigeault, D. Nuel, L. Giordano, G. Buono, *Tetrahedron Lett.*, 2007, **48**, 5247.
[61] Q. Xu, C.-Q. Zhao, L.-B. Han, *J. Am. Chem. Soc.* 2008, **130**, 12648.
[62] T. Chen, L.-B. Han, *Synlett,* 2015, **26**, 1153.
[63] O. Berger, J.-L. Montchamp, *Angew. Chem. Int. Ed.*, 2013, **52**, 11377.
[64] U. Schmidt, B. Riedl, H. Griesser, C. Fitz, *Synthesis,* 1991, 655.
[65] I. Fernández, N. Khiar, A. Roca, A. Benabra, A. Alcudia, J. Espartero, F. Alcudia, *Tetrahedron Lett.*, 1999, **40**, 2029.
[66] A. Benabra, A. Alcudia, N. Khiar, I. Fernández, F. Alcudia, *Tetrahedron: Asymmetry,* 1996, **7**, 3353.
[67] O. I. Kolodiazhnyi, E. V Grishkun, *Tetrahedron: Asymmetry,* 1996, **7**, 967.
[68] M. Oliana, F. King, P. N. Horton, M. B. Hursthouse, K. K. (Mimi) Hii, *J. Org. Chem.*, 2006, **71**, 2472.
[69] R. K. Haynes, R. N. Freeman, C. R. Mitchell, S. C. Vonwiller, *J. Org. Chem.*, 1994, **59**, 2919.
[70] T. Kimura, T. Murai, *Chem. Commun.*, 2005, 4077.
[71] Y. Kobayashi, J. Maeda, K. Saigo, *Tetrahedron: Asymmetry,* 2006, **17**, 1617.
[72] Y. Kobayashi, J. Maeda, F. Morisawa, K. Saigo, *Tetrahedron: Asymmetry,* 2006, **17**, 967.
[73] M. Stankevič, K. M. Pietrusiewicz, *J. Org. Chem.*, 2007, **72**, 816.
[74] M. Stankevič, K. M. Pietrusiewicz, *Tetrahedron: Asymmetry,* 2007, **18**, 552.
[75] G. Ostrogovich, F. Kerek, *Angew. Chem. Int. Ed.*, 1971, **10**, 498.
[76] L. Horner, G. Simons, *Phosphorus Sulfur Relat. Elem.*, 1984, **19**, 65.
[77] M. Sugiya, H. Nohira, *Bull. Chem. Soc. Jpn.*, 2000, **73**, 705.
[78] P. Bałczewski, A. Szadowiak, T. Białas, W. M. Wieczorek, A. Balińska, *Tetrahedron: Asymmetry,* 2006, **17**, 1209.
[79] J. Meisenheimer, L. Lichtenstadt, *Chem. Ber.*, 1911, **44**, 356.
[80] J. Meisenheimer, J. Casper, M. Höring, W. Lauter, L. Lichtenstadt, W. Samuel, *J. Liebig. Ann. Chem.*, 1926, **449**, 213.
[81] F. Toda, K. Mori, S. Zafra, I. Goldberg, *J. Org. Chem.*, 1988, **53**, 308.

[82] T. Novák, J. Schindler, V. Ujj, M. Czugler, E. Fogassy, G. Keglevich, *Tetrahedron: Asymmetry*, 2006, **17**, 2599.
[83] T. Novák, V. Ujj, J. Schindler, M. Czugler, M. Kubinyi, Z. A. Mayer, E. Fogassy, G. Keglevich, *Tetrahedron: Asymmetry*, 2007, **18**, 2965.
[84] P. Bagi, A. Laki, G. Keglevich, *Heteroat. Chem.*, 2013, **24**, 179.
[85] P. Bagi, M. Kállay, D. Hessz, M. Kubinyi, T. Holczbauer, M. Czugler, E. Fogassy, G. Keglevich, *Tetrahedron: Asymmetry*, 2014, **25**, 318.
[86] P. Bagi, A. Fekete, M. Kállay, D. Hessz, M. Kubinyi, T. Holczbauer, M. Czugler, E. Fogassy, G. Keglevich, *Chirality*, 2014, **26**, 174.
[87] P. Bagi, A. Fekete, M. Kállay, D. Hessz, M. Kubinyi, T. Holczbauer, M. Czugler, E. Fogassy, G. Keglevich, *Heteroat. Chem.*, 2015, **26**, 79.
[88] P. Bagi, K. Juhász, I. Timári, K. E. Kövér, D. Mester, M. Kállay, M. Kubinyi, T. Szilvási, P. Pongrácz, L. Kollár, K. Karaghiosoff, M. Czugler, L. Drahos, E. Fogassy, G. Keglevich, *J. Organomet. Chem.*, 2015, **797**, 140.
[89] P. Bagi, V. Ujj, M. Czugler, E. Fogassy, G. Keglevich, *Dalton Trans.*, 2016, **45**, 1823.
[90] V. Ujj, J. Schindler, T. Novák, M. Czugler, E. Fogassy, G. Keglevich, *Tetrahedron: Asymmetry*, 2008, **19**, 1973.
[91] V. Ujj, P. Bagi, J. Schindler, J. Madarász, E. Fogassy, G. Keglevich, *Chirality*, 2010, **22**, 699.
[92] J. Drabowicz, P. Łyżwa, J. Omelańczuk, K. M. Pietrusiewicz, M. Mikołajczyk, *Tetrahedron: Asymmetry*, 1999, **10**, 2757.
[93] J. Holt, A. M. Maj, E. P. Schudde, K. M. Pietrusiewicz, L. Sieroń, W. Wieczorek, T. Jerphagnon, I. W. C. E. Arends, U. Hanefeld, A. J. Minnaard, *Synthesis*, 2009, 2061.
[94] F. A. Kortmann, M.-C. Chang, E. Otten, E. P. A. Couzijn, M. Lutz, A. J. Minnaard, *Chem. Sci.*, 2014, **5**, 1322.
[95] T. Miura, T. Imamoto, *Tetrahedron Lett.*, 1999, **40**, 4833.
[96] J. Omelańczuk, A. Karaçar, M. Freytag, P. G. Jones, R. Bartsch, M. Mikołajczyk, R. Schmutzler, *Inorg. Chim. Acta*, 2003, **350**, 583.
[97] Y. Hamada, F. Matsuura, M. Oku, K. Hatano, T. Shioiri, *Tetrahedron Lett.*, 1997, **38**, 8961.
[98] T. Imamoto, K. V. . Crépy, K. Katagiri, *Tetrahedron: Asymmetry*, 2004, **15**, 2213.
[99] D. Liu, X. Zhang, *Eur. J. Org. Chem.* 2005, 646.
[100] H. Brunner, W. Pieronczyk, *Angew. Chem. Int. Ed.*, 1979, **18**, 620.
[101] H.-J. Kreuzfeld, C. Döbler, *Z. Chem.*, 1984, **24**, 405.
[102] H. Takaya, K. Mashima, K. Koyano, M. Yagi, H. Kumobayashi, T. Taketomi, S. Akutagawa, R. Noyori, *J. Org. Chem.*, 1986, **51**, 629.
[103] T. Minami, Y. Okada, R. Nomura, S. Hirota, Y. Nagahara, K. Fukuyama, *Chem. Lett.*, 1986, **15**, 613.
[104] H. Takaya, S. Akutagawa, R. Noyori, *Org. Synth.*, 1989, **67**, 20.
[105] R. Schmid, J. Foricher, M. Cereghetti, P. Schönholzer, *Helv. Chim. Acta*, 1991, **74**, 370.
[106] T. Benincori, E. Brenna, F. Sannicolò, L. Trimarco, P. Antognazza, E. Cesarotti, F. Demartin, T. Pilati, *J. Org. Chem.*, 1996, **61**, 6244.
[107] T. Benincori, O. Piccolo, S. Rizzo, F. Sannicolo, *J. Org. Chem.*, 2000, **65**, 8340.
[108] T. Benincori, S. Gladiali, S. Rizzo, F. Sannicolo, *J. Org. Chem.*, 2001, **66**, 5940.
[109] W. Tang, Y. Chi, X. Zhang, *Org. Lett.*, 2002, **4**, 1695.
[110] T. Shibata, H. Tsuruta, H. Danjo, T. Imamoto, *J. Mol. Catal. A Chem.*, 2003, **196**, 117.
[111] S. D. de Paule, S. Jeulin, V. Ratovelomanana-Vidal, J.-P. Genêt, N. Champion, P. Dellis, *Eur. J. Org. Chem.*, 2003, 1931.
[112] E. Gorobets, G.-R. Sun, B. M. M. Wheatley, M. Parvez, B. A. Keay, *Tetrahedron Lett.*, 2004, **45**, 3597.
[113] S. Wu, M. He, X. Zhang, *Tetrahedron: Asymmetry*, 2004, **15**, 2177.
[114] D.-Y. Zhang, C.-B. Yu, M.-C. Wang, K. Gao, Y.-G. Zhou, *Tetrahedron Lett.*, 2012, **53**, 2556.
[115] D. A. Rankic, M. Parvez, B. A. Keay, *Tetrahedron: Asymmetry*, 2012, **23**, 754.

[116] U. Matteoli, V. Beghetto, C. Schiavon, A. Scrivanti, G. Menchi, *Tetrahedron: Asymmetry*, 1997, **8**, 1403.
[117] P. J. Pye, K. Rossen, R. A. Reamer, N. N. Tsou, R. P. Volante, P. J. Reider, *J. Am. Chem. Soc.*, 1997, **119**, 6207.
[118] K. Vandyck, B. Matthys, M. Willen, K. Robeyns, L. Van Meervelt, J. Van Der Eycken, *Org. Lett.*, 2006, **8**, 363.
[119] H. Brunner, M. Muschiol, M. Zabel, *Synthesis*, 2008, 405.
[120] G. B. Cox, Ed. , *Preparative Enantioselective Chromatography*; Blackwell Publishing Ltd: Oxford, 2005.
[121] G. Subramanian, Ed., *Chiral Separation Techniques;* Wiley-VCH: Weinheim, 2006.
[122] L. Miller, C. Orihuela, R. Fronek, D. Honda, O. Dapremont, *J. Chromatogr. A*, 1999, **849**, 309.
[123] S. Andersson, S. G. Allenmark, *J. Biochem. Biophys. Methods*, 2002, **54**, 11.
[124] H. Leek, L. Thunberg, A. C. Jonson, K. Öhlén, M. Klarqvist, *Drug Discovery Today*, 2017, **22**, 133.
[125] A. Brandi, S. Cicchi, F. Gasparrini, F. Maggio, C. Villani, M. Koprowski, K. M. Pietrusiewicz, *Tetrahedron: Asymmetry*, 1995, **6**, 2017.
[126] X. Jiang, A. J. Minnaard, B. Hessen, B. L. Feringa, A. L. L. Duchateau, J. G. O. Andrien, J. A. F. Boogers, J. G. de Vries, *Org. Lett.*, 2003, **5**, 1503.
[127] X. Jiang, M. van den Berg, A. J. Minnaard, B. L. Feringa, J. G. de Vries, *Tetrahedron: Asymmetry*, 2004, **15**, 2223.
[128] J. Bigeault, L. Giordano, G. Buono, *Angew. Chem. Int. Ed.*, 2005, **44**, 4753.
[129] M. Widhalm, L. Brecker, K. Mereiter, *Tetrahedron: Asymmetry*, 2006, **17**, 1355.
[130] D. Gatineau, D. H. Nguyen, D. Hérault, N. Vanthuyne, J. Leclaire, L. Giordano, G. Buono, *J. Org. Chem.*, 2015, **80**, 4132.
[131] D. Herault, D. H. Nguyen, D. Nuel, G. Buono, *Chem. Soc. Rev.*. 2015, **44**, 2508.

Zita Rádai, Nóra Zsuzsa Kiss and György Keglevich

5 Synthesis of α-hydroxyphosphonates, an important class of bioactive compounds

Abstract: The synthesis of α-hydroxyphosphonates represents a hot topic in organophosphorus chemistry due to the potential biological activity of these derivatives.

The so-called Pudovik or Abramov reaction has been performed using a large number of catalysts, including bases and metal complexes. In the last two decades, "green" accomplishments were developed to avoid exotic catalysts, harsh reaction conditions and the excessive use of organic solvents in the reactions, and during the work-up procedures.

Keywords: α-hydroxyphosphonates, Pudovik reaction, green chemistry.

5.1 Bioactivity of α-hydroxyphosphonates

Since 1950, when the reaction between aldehydes and dialkyl phosphites affording α-hydroxyphosphonates was first reported, the method has attracted great interest (Figure 5.1).

Insecticide *O*,*O*-dimethyl (2,2,2-trichloro-1-hydroxyethyl)phosphonate (**1**), commonly known as trichlorfon or metrifonate, is a well-known α-hydroxyphosphonate that was discovered in the 1950s, but is no longer in use due to its toxicity [1]. Compound **1** shown in Figure 5.2 is a prodrug that is transformed to 2,2-dichlorovinyl dimethyl phosphate (DDVP) (**2**) and acts on flies, ticks, fleas and cockroaches by inhibiting acetylcholinesterase. As pyrimidyl carboxylic acids form a well-known family of herbicides, a new family of α-hydroxyphosphonates containing the pyrimidine scaffold was synthesized and tested as herbicides against amaranth pigweed [2]. In the study, compound **3** was shown to possess high activity against *Amaranthus retroflexus*. The most well-known bioactive property of α-hydroxyphosphonates is their antibacterial activity [3–6]. Quinoline-based α-hydroxyphosphonate **4** was found to be an efficient antibiotic agent against Gram positive (*Staphylococai*, *Bacillus megtesium-I*) as well as Gram negative (*E. coli*, *Salmonella typhi* and *Proteus vulgaris*) bacteria (Figure 5.2) [3]. Quinoline-3-carbaldehydes showed moderate activity, however, converting it to the corresponding α-hydroxyphosphonate, the bioactivity increased significantly. Anellating a tetrazole ring to the ring of a **4**-like compound, a new class of antibacterial and antifungal α-hydroxyphosphonates (e.g. **5**) was synthesized [4]. It was found that the substituents in the benzene ring determined the efficiency of the compounds against the examined strains. Phosphonate **5** showed a remarkable activity against Gram positive *Bacillus subtilis* and Gram negative *E. coli* as well. With regard to cyto-

https://doi.org/10.1515/9783110535839-005

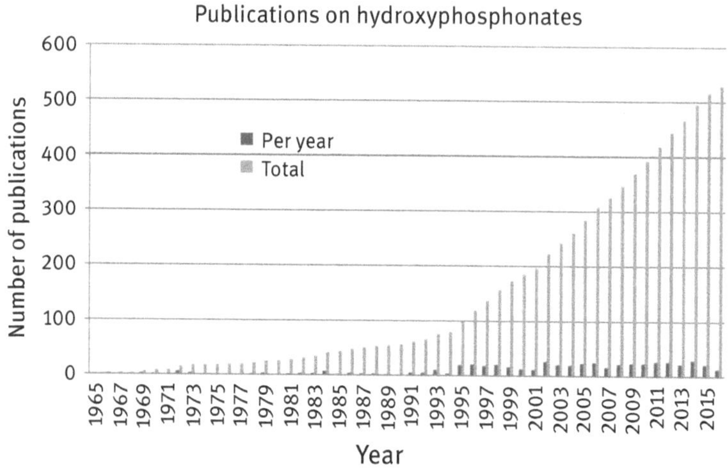

Figure 5.1: Increasing number of publications on α-hydroxyphosphonates.

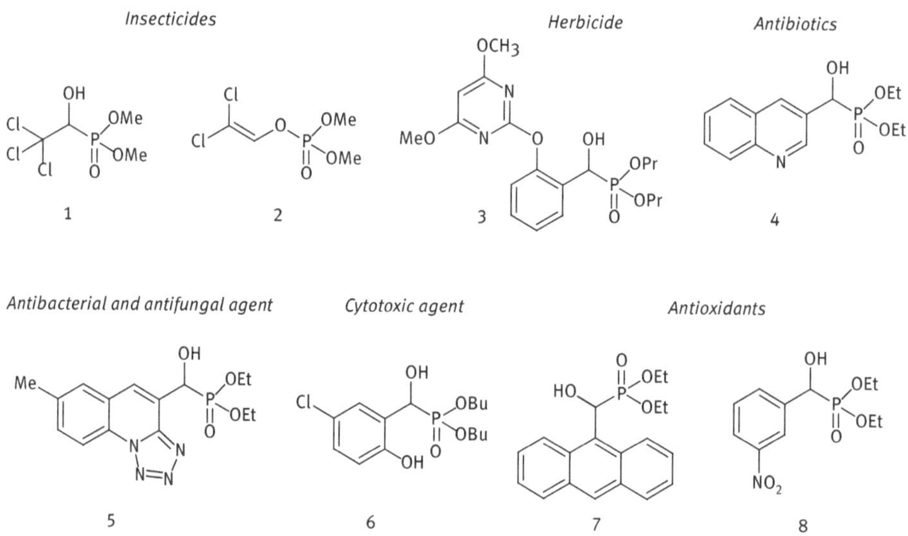

Figure 5.2: Selected examples of bioactive α-hydroxyphosphonates.

toxicity, a family of α-hydroxyphosphonates bearing the hydroxy group in the aromatic ring was reported as potential anticancer agents [7]. The compounds prepared were tested on two human cancer lines (A594 and KB), and a few of the derivatives (including **6**) showed remarkable activity against both cell lines. The positive tests of the application of α-hydroxyphosphonates as antioxidants promoted their versatile use in organic chemistry and biochemistry. Derivatives **7** and **8** showed a free radical scavenging activity nearly as high as ascorbic acid [8, 9].

Tripeptides containing an α-hydroxyphosphonate moiety were tested as human renin inhibitors. By varying the amino acids and synthesizing the corresponding phosphine oxide analogs, the structure–activity relationship of the resulting species was evaluated. A few of the derivatives tested were found to be promising inhibitors of human renin [10].

5.2 Synthesis of α-hydroxyphosphonates from Aldehydes/Ketones and Dialkyl Phosphites

The synthesis of α-hydroxyphosphonates represents an "evergreen" field in organophosphorus chemistry. Two main approaches exist for the preparation of α-hydroxyphosphonic esters (Figure 5.3). One of them involves the nucleophilic addition of dialkyl phosphites to an aldehyde or ketone (method "A"). The alternative route is the condensation of the corresponding oxo compound with trialkyl phosphite (method "B").

Generally, both reactions are carried out in the presence of a catalyst (which is, in the majority of the cases, a base) under mild conditions. The denomination of the reactions is not consistent in the literature, as both of them are referred as the Pudovik, Abramov, or phospha-Aldol reaction.

Traditionally, the addition of dialkyl phosphites to ketones or aldehydes (method "A") was catalyzed by alkali alcoholates, as they can deprotonate the >P(O)H reagent [11, 12]. The reaction may also be performed on heating without the application of any catalyst, but this method required relatively harsh conditions for the completion of reaction [13].

Another variation is when the addition is carried out under solvent-free conditions and on the surface of a solid catalyst. For example, the addition may be performed on the surface of aluminum oxide [14]. By applying potassium fluoride with Al_2O_3, the reaction time of 72 h could be shortened to 30 min [15, 16]. Procedures involving magnesium oxide [17], inorganic phosphates [18], a sodium-modified fluoroapatite [19], MoO_2Cl_2 [20], silica-supported tungstic acid [21], or polyethylene glycol

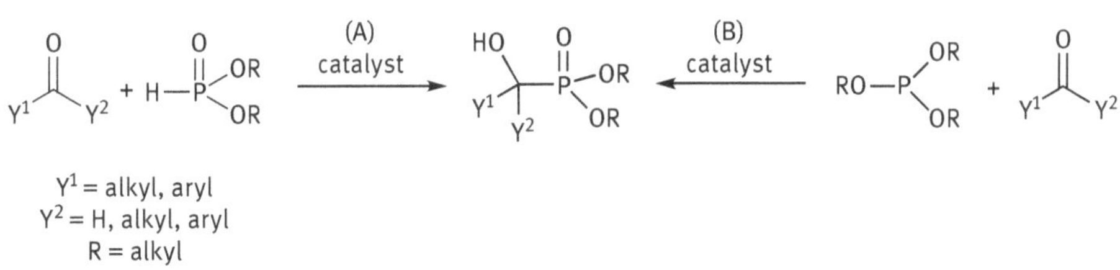

Figure 5.3: Possible approaches for the synthesis of α-hydroxyphosphonates.

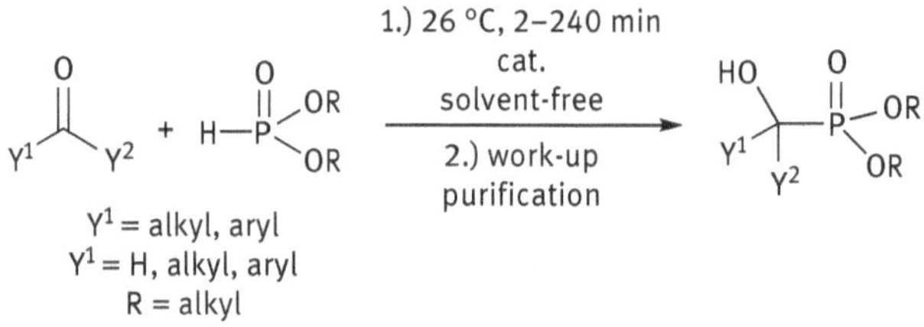

Figure 5.4: Green protocol for the synthesis of α-hydroxyphosphonates.

[22] are also solvent-free, solid-catalyzed methods. The use of paramagnetic magnesium ferrite ($MgFe_2O_4$) nanoparticles [23] or magnetic nanoparticle-supported guanidine [24] as the catalyst has also been reported.

The spread of green chemical principles and tools in organic chemistry could not be disregarded in papers published recently on the synthesis of α-hydroxyphosphonates. More and more studies are focusing on the application of simple and inexpensive catalysts together with mild and solvent-free conditions [25] (Figure 5.4, Table 5.1).

The reaction of aldehydes and dialkyl phosphites has been carried out in the presence of barium hydroxide as the catalyst [26, 27]. Table 1/Entry 1 presents a solvent-free procedure applying 10 mol% of $Ba(OH)_2$ [27]. The work-up comprised extraction, washing with hexane, and purification by column chromatography to afford α-hydroxyphosphonates in yields of 72–96%. The use of potassium phosphate (Table 5.1/Entry 2) or potassium hydrogen sulfate (Table 5.1/Entry 3) as the catalyst provides a simple and inexpensive method [8, 28]. K_3PO_4 proved to be an efficient catalyst even in amounts as low as 5 mol% [28], while a quantity of 20 mol% of $KHSO_4$ was necessary to promote the reaction [8]. In both cases, the work-up procedure involved extraction, followed by column chromatography in the latter case. The $KHSO_4$-catalyzed procedure was carried out in a series of different solvents, however, the highest yields were obtained under solvent-free conditions [8]. An interesting method is the ionic liquid (IL)-catalyzed reaction of aldehydes and diethyl phosphite (Table 5.1/Entry 4) [29]. An addition of 10 mol% choline hydroxide allowed the reaction to take place at 25 °C after 5–10 min. The desired products were obtained by extraction, and purified by recrystallization. In a few cases, the interaction of the starting components was ensured by grinding. One method (Table 5.1/Entry 5) involves the use of piperazine as the catalyst [30], while another one uses sodium carbonate (Table 5.1/Entry 6) [31]. Both procedures were carried out at 25 °C, and reached completion within 10 min. The work-up involved washing with water, extraction by ethyl acetate, and, in the former case, purification by chromatography. Microwave (MW)-assisted variations have also been reported for the synthesis of α-hydroxyphosphonates [32–35]. Sodium carbonate present in an amount of 0.75 equivalents was an efficient catalyst under MW irradiation at 110 °C (Table 5.1/Entry 7) [32]. The crude product was extracted with ethyl acetate, and recrystallized from acetone-pentane solvent mixture to provide the corresponding α-hydroxyphosphonates in yields of 71–88%.

Table 5.1: Details of novel "green" syntheses of α-hydroxyphosphonates from dialkyl phosphites and benzaldehydes (Entries 1–7) or ketones (Entries 8–12) (method "A").

Entry	Catalyst	Amount of the catalyst	Conditions	Work-up procedure	Yield (%)	Ref
1	$Ba(OH)_2$	10 mol%	25 °C, 4–10 min	extr (CH_2Cl_2) washing (hexane) cryst (EtOAc-hexane)	72–96	[27]
2	K_3PO_4	5 mol%	25 °C, 4–8 min	extr (CH_2Cl_2)	90–98*	[28]
3	$KHSO_4$	20 mol%	25 °C, 2–4 h	extr (EtOAc) chrom	82–91	[8]
4	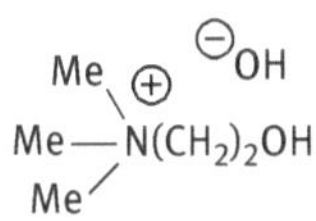	10 mol%	25 °C, 5–10 min	extr (EtOAc) cryst (EtOH)	90–98	[29]
5	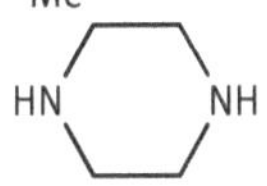	1 equiv.	grinding 25 °C, 2–10 min	washing (H_2O) extr (EtOAc) chrom	78–96	[30]
6	Na_2CO_3	1 equiv.	grinding 25 °C, 10 min	washing (H_2O) extr (EtOAc)	75–83	[31]
7	Na_2CO_3	0.75 equiv.	MW 110 °C, 20 min	extr (EtOAc) cryst (acetone-pentane)	71–88	[32]
8	MgO	2 equiv.	25 °C, 1–6 h	extr (H_2O) chrom (Et_2O-hexane)	70–82	[42]
9	Et_3N	1 equiv.	40 °C, 2 h	cryst (EtOAc)	75–89	[43]
10	Na-modified fluoroapatite	1 g/1.25 mol acetophenone	20–25 °C, 31–60 min stirring with a spatula	washing (CH_2Cl_2) cryst	80–98	[44]
11	BuLi	0.1 mol%	10 °C, 5 min in hexane	quenching (EtOAc) washing (hexane) chrom	40–99	[45]
12	$Ti(O^iPr)_4$	5 mol%	30 °C, 15 min	chrom	81–97	[46]

extr = extraction, chrom = chromatography, cryst = recrystallization, *calculated for the crude product

The reaction of aldehydes and dialkyl phosphites takes place rather easily, however, starting from ketones, the realization of the Pudovik reaction is challenging due to the lower reactivity. In this case, procedures involving expensive and "exotic" compounds (such as rare earth metal complexes [36–40] or $NbCl_5$ [41]) were developed. A few reactions catalyzed by common bases or acids required the reagents in an excess. α-Hydroxyphosphonates bearing a cyano group in the α-alkyl chain were synthesized by applying two equivalents of MgO (Table 5.1 / Entry 8) [42]. Among the series of catalysts (NaOEt, Et_3N, piperazine, K_2CO_3,

$Ba(OH)_2$) tested in the reaction of γ-ketonitriles and diethyl phosphite, only MgO proved to be efficient. The reaction of ketones and dimethyl phosphite was also performed in the presence of one equivalent of triethylamine (Table 5.1/Entry 9) [43]. The crude product was recrystallized from ethyl acetate to afford the products in good to excellent yields (75–89%). The next method utilized sodium-modified fluoroapatite as an efficient catalyst (Table 5.1/Entry 10) [44]. The starting materials were stirred with a spatula. After 31–60 min, the crude product was washed with water and recrystallized. Organometallic compounds, such as butyl lithium (Table 5.1/Entry 11) [45], and titanium tetraisopropoxide [46] (Table 5.1/Entry 12) were also used for the activation of the >P(O)H reagent. According to the BuLi-catalyzed method, the work-up involved quenching with water, and washing with hexane. In both instances, purification of the crude products by column chromatography afforded the desired products.

It can be seen from Table 5.1 that in most of the cases, α-hydroxyphosphonates were synthesized under mild and solvent-free conditions. The "green" aspects of the procedures were overemphasized, however, the excessive amount of organic solvents applied during the complex work-up procedures outlined above was disregarded. The crude product was usually washed with water and/or extracted with an organic solvent (Table 5.1/Entries 1–8), and this was followed by purification steps comprising (re)crystallization (Table 5.1/Entries 1, 4, 7, 9, and 10) or column chromatography (Table 5.1/Entries 3, 5, 8, 11, 12). These additional steps require the excessive use of organic solvents, thus despite the reactions being solvent-free, the methods discussed above [8, 27–32, 42–46] do not meet the requirements of green chemistry.

According to a benign method, the equimolar mixture of benzaldehyde derivative and dialkyl phosphite was stirred at reflux in a minimal amount of acetone in the presence of 10 mol% of triethylamine as the catalyst (Figure 5.5) [47]. Starting from sterically hindered ketones, the application of one equivalent of Et_3N was necessary.

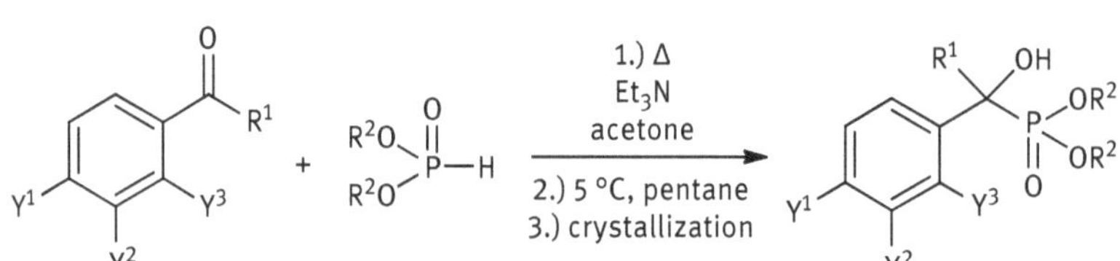

R^1 = H, Me

R^2 = Me, Et

Y^1	H	Cl	NO_2	H	Me	OMe
Y^2	H	H	H	H	H	OMe
Y^3	H	H	H	NO_2	H	H

Figure 5.5: The greenest synthesis of α-hydroxyphosphonates.

After the completion of the reaction (5–10 min in the case of electron-withdrawing substituents in the aromatic ring, and 90–120 min in the case of electron-releasing methyl or methoxy groups), some *n*-pentane was added to the reaction mixture. On cooling, the α-hydroxyphosphonate precipitated, and could be filtered easily. As a compromise, minimal amount (a few mLs) of solvent was used in the synthesis, and further purification was unnecessary, saving the excessive use of organic solvents [47].

5.3 Synthesis of α-hydroxyphosphonates from Aldehydes/Ketones and Trialkyl Phosphites

An alternative route for the synthesis of α-hydroxyphosphonates involves the reaction of the corresponding oxo compound with a trialkyl phosphite (Figure 5.2, method "B"). Due to the reduced atom efficiency as compared to the procedures applying dialkyl phosphite, this method is of less importance.

In contrast to method "A," bases do not belong to the typical catalysts of this approach. Instead, different acids, including hydrogen chloride gas [48], camphorsulfonic acid [49], sulphamic acid [50], oxalic acid [51], tartaric or fumaric acid [52], and pyridine-2,6-dicarboxylic acid [53] are applied usually. Metal salts ($ZnBr_2$ [54], $Bi(NO_3)_3{\cdot}5H_2O$ [55]), organometallic compounds (TMSCl [56–58], and AuPd bimetallic nanoparticles [59]) were also used to promote the addition of trialkyl phosphite to oxo compounds. Lanthanide complexes and polyethylene glycol, previously mentioned as efficient catalysts for the Pudovik reaction using dialkyl phosphites, also promoted the addition of trialkyl phosphites to oxo compounds [60, 61]. According to an interesting method, the reaction was catalyzed by iodine in water [62]. Aromatic aldehydes and triethyl phosphite were stirred at 80 °C for 30–60 min. Washing the crude product with aqueous $Na_2S_2O_3$ and brine solution, followed by extraction and column chromatography, led to α-hydroxyphosphonates in yields of 89–96%. A similar, β-cyclodextrin-catalyzed procedure was carried out in water at 60–70 °C [63]. After 8–12 h, the reaction mixture was extracted, and washed. The crude product was purified by column chromatography to furnish the product in good yields (81–93%).

As in the case of method "A," the reaction of oxo compounds and trialkyl phosphites was also carried out under mild and solvent-free conditions (Table 5.2). Sonication and MW irradiation were used in many cases to promote the reaction [49, 50, 58, 64, 65]. According to a benign method, the ultrasound-assisted reaction of aldehydes or ketones and trimethyl or triethyl phosphite took place without applying any catalyst (Table 5.2/Entry 1) [64]. The work-up included a simple recrystallization of the crude product to afford the α-hydroxyphosphonates in yields of 85–94%. Another approach is a KH_2PO_4-catalyzed synthesis under ultrasonic irradiation (Table 5.2/Entry 2) [65].

Table 5.2: "Green" syntheses of α-hydroxyphosphonates from benzaldehydes or ketones and trialkyl phosphites (method "B").

Entry	Catalyst	Amount of the catalyst	Conditions	Work-up procedure	Yield (%)	Ref
1	–	–	sonication 25 °C 10–35 min	cryst	85–94	64
2	KH_2PO_4	5 mol%	sonication 25 °C, 5–120 min	poured on crushed ice extr (EtOAc)	56–90	65
3	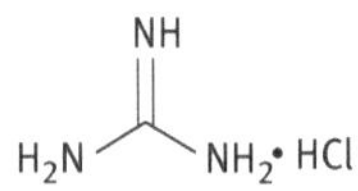	0.5 mol%	50 °C, 2 h in H_2O	quenching (aq. $NaHCO_3$ and brine solution) extr (CH_2Cl_2) chrom	63–92	66
4	Et_3N and $MgCl_2$	3 equiv. and 1 equiv.	50 °C, 2 h	extr (EtOAc) cryst	85–98	67
5	NH_4VO_3	10 mol%	25 °C, 5–20 min	extr ($CHCl_3$) chrom	80–94	68

extr = extraction, chrom = chromatography, cryst = recrystallization

The reaction was complete after 5–120 min at 25 °C. The crude product was poured on ice, and extracted with ethyl acetate. Guanidine hydrochloride was also found to be an efficient catalyst for the Pudovik reaction (Table 5.2/Entry 3) [66]. The work-up procedure was rather complex, as it comprised quenching, extraction, and column chromatography. The next procedure applied three equivalents of $MgCl_2$ and one equivalent of Et_3N to promote the addition (Table 5.2/Entry 4) [67]. The reaction was carried out at 50 °C, and was complete after 2 h. The reaction mixture was extracted with ethyl acetate, and the crude product was further purified by crystallization. Last, but not least, the reaction was also carried out in the presence of ammonium metavanadate as the catalyst (Table 5.2/Entry 5) [68]. Applying 10 mol% of this salt, the reaction took place after 5–20 min at 25 °C. The work-up procedure included extraction and chromatography. It is noteworthy that NH_4VO_3 did not promote the reaction of aldehydes and dialkyl phosphites [68].

5.4 Syntheses of α-hydroxyphosphonates in Reductive and Oxidative Ways

The Pudovik reaction is not the only way to synthesize α-hydroxyphosphonates. α-Ketophosphonates may be converted to the corresponding α-hydroxy derivatives by reaction with organometallic compounds [69, 70]. One of the procedures involved the addition of trialkyl aluminum on the C=O group, followed by hydrolysis to afford the desired products with variable (35–85%) yields (Figure 5.6, method "A") [69]. According to another method, a Grignard reagent (R^2MgBr) was the partner in the addition step (Figure 5.6, method "B") [70].

A unique approach is the introduction of the α-hydroxy group to the corresponding α-alkyl phosphonates by oxidation (Figure 5.7) [71, 72]. Method "A" employed oxygen as the oxidizing agent in the presence of copper(II) chloride and 10 mol% of *N*-hydroxyphthalimide as the catalysts [71]. (The authors of this review note that the real reactant may have been air.) Presumably, the reaction takes place through a radical intermediate that is transformed into an α-hydroperoxyphosphonate. This is followed by treatment with triphenyl phosphine to afford the α-hydroxyphosphonates in yields of 53–73%. Further on, $CuCl_2$, PPh_3, and *N*-hydroxyphthalimide were replaced by a simple ionic liquid, 1-butyl-3-methylimidazolium hydroxide that made possible the efficient and green synthesis of α-hydroxyphosphonates (method "B") [72].

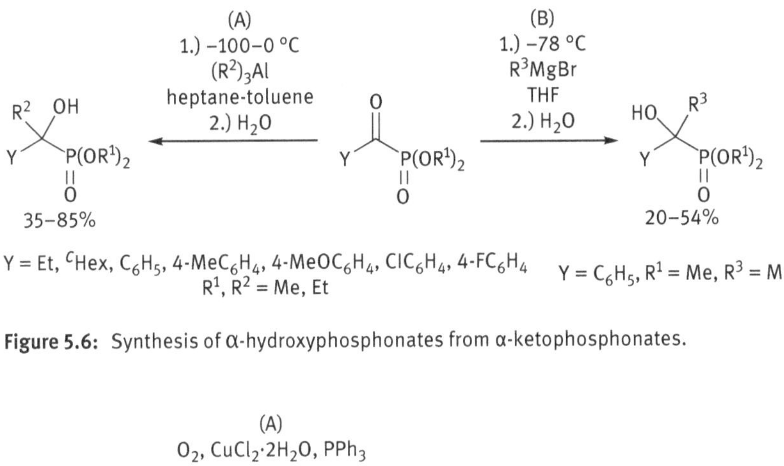

Y = Et, ᶜHex, C_6H_5, 4-MeC_6H_4, 4-$MeOC_6H_4$, ClC_6H_4, 4-FC_6H_4
R^1, R^2 = Me, Et

Y = C_6H_5, R^1 = Me, R^3 = Me, C_6H_5

Figure 5.6: Synthesis of α-hydroxyphosphonates from α-ketophosphonates.

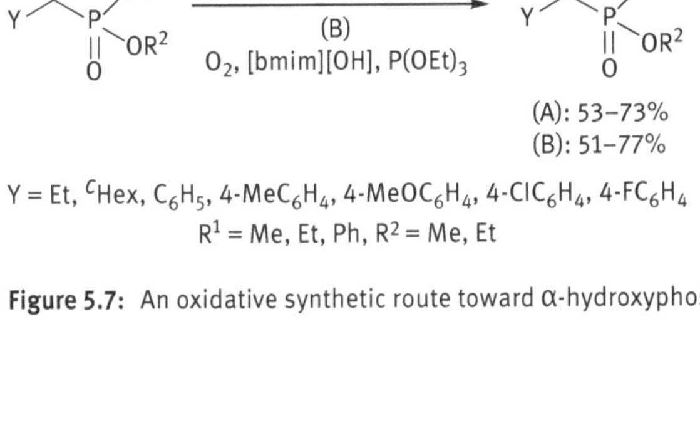

Y = Et, ᶜHex, C_6H_5, 4-MeC_6H_4, 4-$MeOC_6H_4$, 4-ClC_6H_4, 4-FC_6H_4
R^1 = Me, Et, Ph, R^2 = Me, Et

Figure 5.7: An oxidative synthetic route toward α-hydroxyphosphonates.

5.5 Synthesis of Optically Active α-hydroxyphosphonates *via* the Pudovik Reaction

As the α-hydroxyphosphonates may have a chirality center on the α carbon atom, the selective preparation of one of the enantiomers is a real challenge. The easiest way to obtain optically active α-hydroxyphosphonates is the synthesis starting from chiral components [73]. The Pudovik reaction may be performed in the presence of an optically active catalyst, e.g. chiral organocatalysts [74–77], or metal complex catalysts with chiral ligands [78–84].

According to an elegant and simple method, squaramide was applied as the organocatalyst to enhance the asymmetric reaction of aldehydes and diphenyl phosphite [74]. The catalytic effect of squaramide is explained by the formation of *H*-bonds between the amino functions of the catalyst and the hydroxy and oxo functions of the starting components. The method resulted in the formation of (*R*)-α-hydroxyphosphonates in enantiomeric excesses (*ee*) up to 88% (Figure 5.8).

A variation of the above method is the cinchona diaminomethylene-malonitrile-catalyzed Pudovik reaction. This procedure led to the selective formation of the (*S*) enantiomer of the products in yields of 70–99% and *ee* values of 90–96% [75].

Japanese researchers performed the addition of dimethyl phosphite to aromatic aldehydes in the presence of an organophosphorus catalyst formed in situ from a chiral tetraaminophosphonium salt and KO^{t}Bu [76]. The reaction was performed at −98 °C, and the completion required 2–13 h. Excellent yields (91–99%) and *ee* values (94–99%) proved the high activity and selectivity of the catalyst used (Figure 5.9).

A series of chiral aziridinyl phosphonates were tried out as catalysts in the asymmetric reaction of aromatic aldehydes and triethyl phosphite [77]. Almost all of the tested aziridine-based compounds provided enantioselectivity to some extent, however, even with the more active catalyst shown below, only low to moderate *ee* values (17–44%) were attained (Figure 5.10).

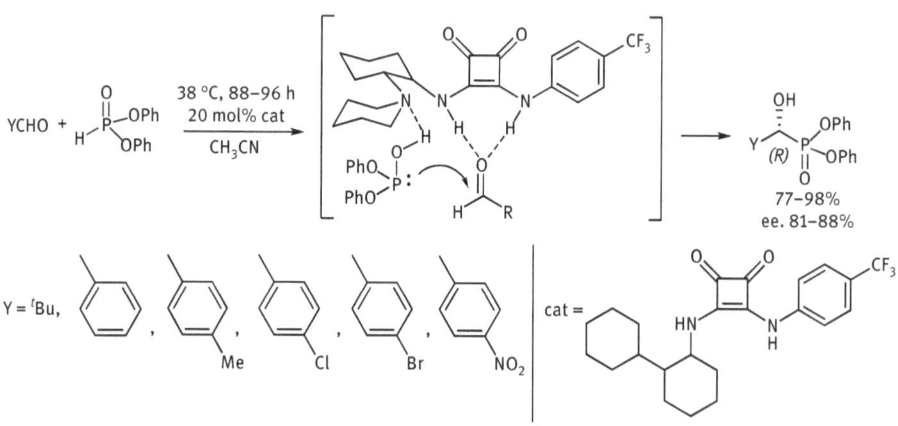

Figure 5.8: Asymmetric synthesis of α-hydroxyphosphonates catalyzed by squaramide.

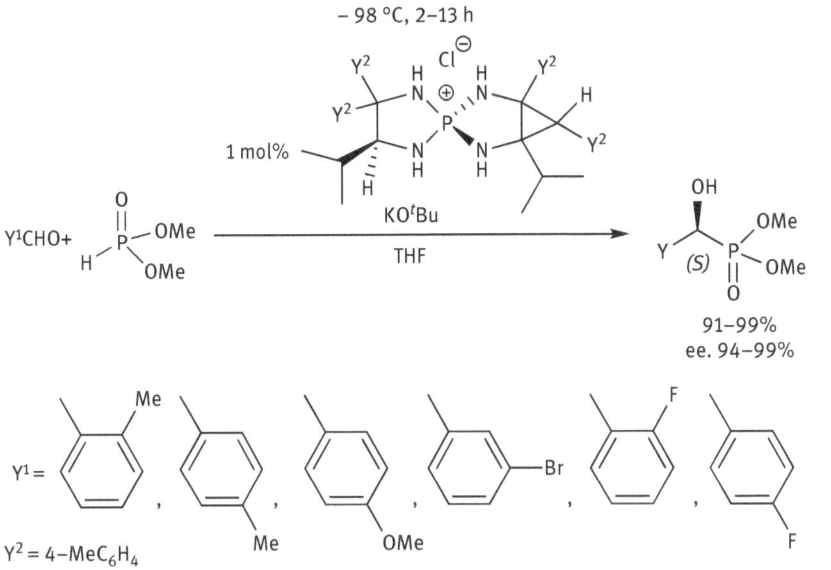

Figure 5.9: Asymmetric Pudovik reaction in the presence of a chiral aminophosphonium salt.

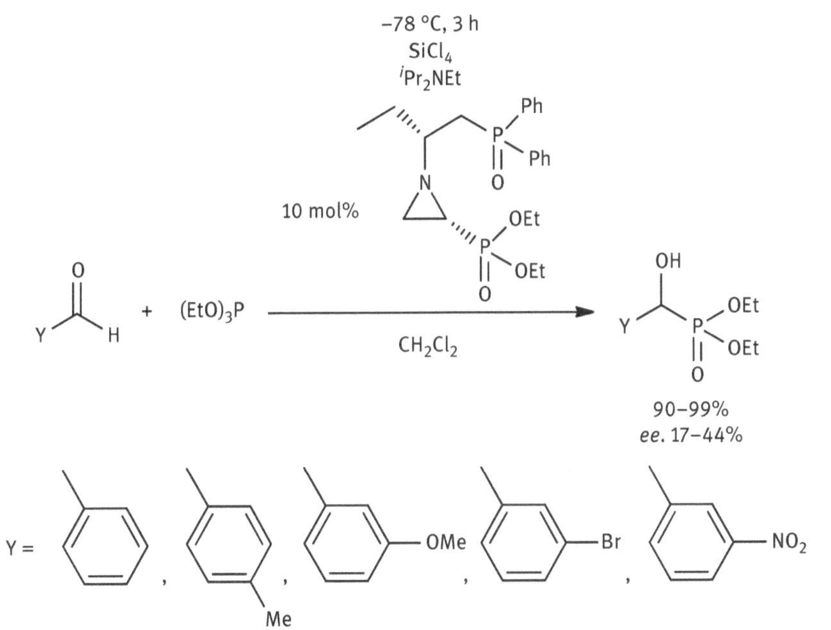

Figure 5.10: Chiral aziridinylphosphonates as catalysts in the Pudovik reaction.

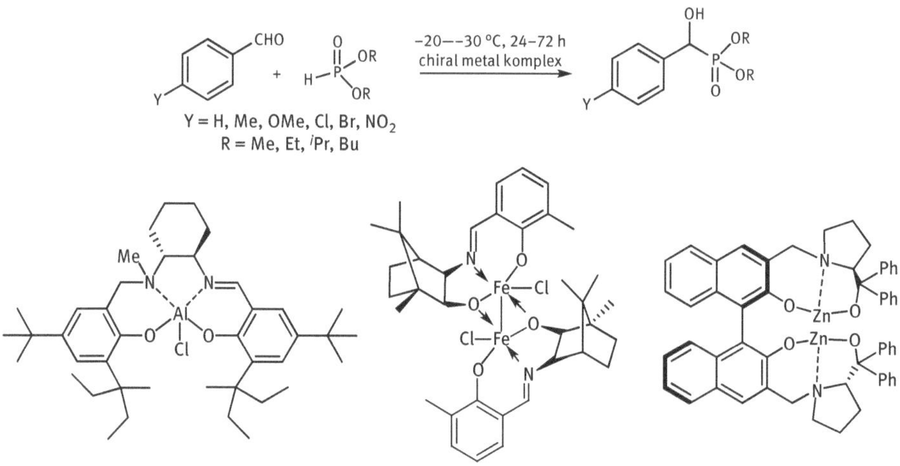

9 *ee*. 93–97%

10 *ee*. 85–99%

11 *ee*. 81–94%

Figure 5.11: Selected examples of chiral metal complexes used in asymmetric Pudovik reactions.

The next strategy for the synthesis of optically active α-hydroxyphosphonates *via* the Pudovik reaction involved chiral metal complexes. This method is more widespread than the application of chiral organocatalysts. A few examples are shown in Figure 5.11. Most of the procedures were carried out in the presence of Schiff base metal complexes (e.g. **9** and **10**), where the Al [78, 79] or Fe [80–82] was coordinated to the nonbonding electrons of the nitrogen atom. Catalyst **9** was applied in the amount of 10 mol% in order to enhance the enantioselective Pudovik reaction [83]. By adding one equivalent of potassium carbonate to the reaction mixture, the amount of the chiral catalyst could be reduced to 2 mol% affording the corresponding α-hydroxyphosphonates with high (93–97%) *ee* values. Camphor-based Schiff base Fe complexes (e.g. **10**) were found to be promising catalysts in the Pudovik reaction [81, 82]. Catalyst **10** promoted the formation of the (*R*) isomer of the corresponding α-hydroxyphosphonates with excellent *ee* values (up to 99%) [82]. The asymmetric Pudovik reaction of substituted aldehydes and dimethyl or diethyl phosphite was also performed in the presence of 5 mol% of BINOL-based Zn complex **11** [84]. Surprisingly, stirring the equimolar mixture of the starting compounds at –20 °C, an *ee* of only 16% was attained after 48 h, while by increasing the relative quantity of the *P* reagent to 10 equivalents, the *ee* values dramatically increased.

5.6 Miscellaneous Syntheses of Optically Active α-hydroxyphosphonates

α-Hydroxyphosphonates were acylated with 0.5 equivalents of achiral carboxylic acids in the presence of an optically active agent, (*R*)-benzotetramisole ((*R*)-BTM) and

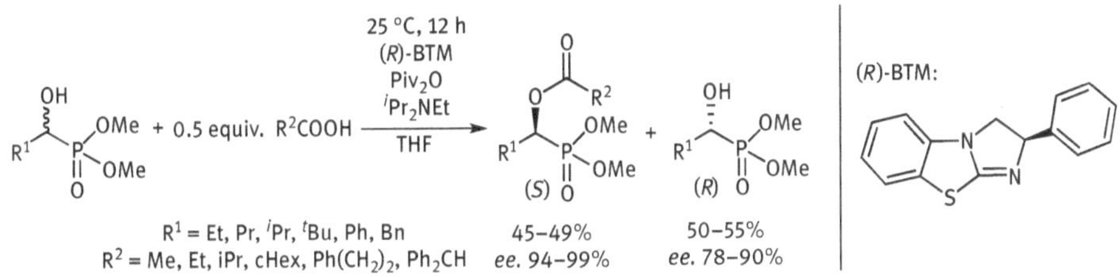

Figure 5.12: Kinetic resolution of α-hydroxyphosphonates through acylation.

pivalic anhydride (Figure 5.12) [85]. The method provided the (*S*) enantiomer of the acylated hydroxyphosphonate selectively with *ee* values up to 99%, while the unreacted α-hydroxyphosphonate was enriched in the (*R*) isomer. The *ee* values for the latter species fell in the range of 78–90%.

The organocatalytic synthesis of optically active α-hydroxy-β-nitrophosphonates was performed starting from the corresponding α-ketophosphonates that were reacted with nitromethane, where 5 mol% of cupreine or 9-*O*-benzylcupreine was the catalyst (Figure 5.13) [86]. High yields (71–93%) and *ee* values (91–99%) were obtained with both catalysts. The absolute configuration of the major enantiomer was not determined.

According to an elegant method, the oxidation of the corresponding benzyl phosphonates to α-hydroxyphosphonates was carried out in the presence of a chiral camphorsulfonic acid-based oxidant [87]. The first step involved deprotonation of the methylene group of the starting compound with sodium tetramethyl disilazane followed by the treatment with (+)-8,8-(dichlorocamphor)sulfonyloxaziridine to afford the corresponding oxidized products in variable (25–47%) yields with high (86–99%) enantioselectivities (Figure 5.14).

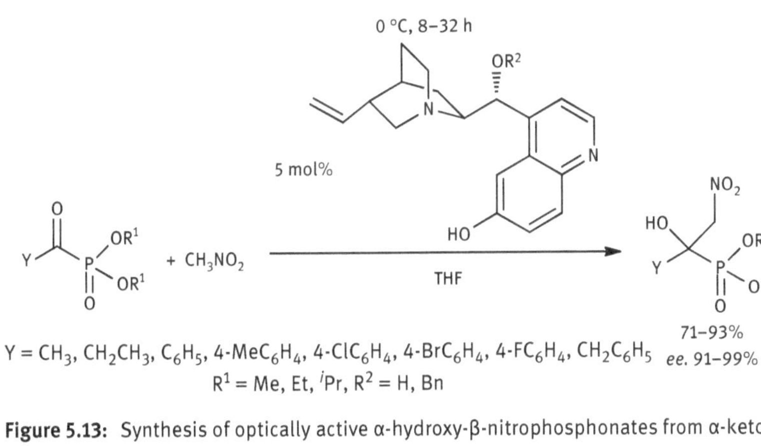

Figure 5.13: Synthesis of optically active α-hydroxy-β-nitrophosphonates from α-ketophosphonates.

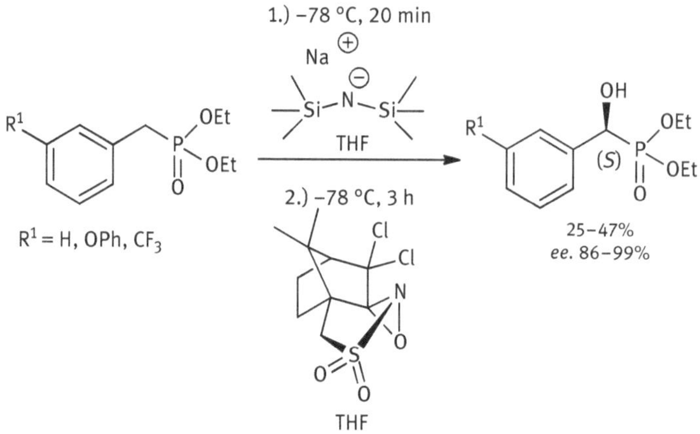

Figure 5.14: Oxidation of benzyl phosphonates to α-hydroxyphosphonates with a chiral oxidizing agent.

5.7 Summary

The synthetic methods to provide α-hydroxyphosphonates relevant as biologically active substrates were surveyed. The most common procedures involve the reaction of oxo compounds with dialkyl phosphites or trialkyl phosphites. The former variation is an attractive model for green chemists. The best procedure involves not only the addition of a dialkyl phosphite to an oxo compound, but also an environmentally friendly work-up. Finally, a few reductive and oxidative approaches were discussed followed by enantioselective methods.

Acknowledgments: This work was supported by the National Research Development and Innovation Fund of Hungary in the frame of projects K119202 and TÉT_16-1-2016-0111. Z. Rádai is grateful for the fellowship provided by Chinoin–Sanofi Pharmaceuticals.

References

[1] W. Lorenz, A. Henglein, G. Schrader, *J. Am. Chem. Soc.*, 1955, **77**, 2554–2556.
[2] H. Song, H. Mao, D. Shi, *Chin. J. Chem.*, 2010, **28**, 2020–2024.
[3] R. U. Pokalwar, R. V. Hangarge, P. V. Maske, M. S. Shingare, *Arkivoc*, 2006, **11**, 196–204.
[4] A. H. Kategaonkar, R. U. Pokalwar, S. S. Sonar, V. U. Gawali, B. B. Shingate, M. S. Shingare, *Eur. J. Med. Chem.*, 2010, **45**, 1128–1132.

[5] G. S. Reddy, C. S. Sundar, S. S. Prasad, E. D. Dadapeer, C. N. Raju, C. S. Reddy, *Der Pharma Chem.*, 2012, **4,** 2208–2213.
[6] S. Sampath, C. N. Raju, C. V. Rao, *Phosphorus, Sulfur, Silicon*, 2016, **191,** 95–99.
[7] R. M. N. Kalla, H. R. Lee, J. Cao, J. W. Yoo, I. Kim, *New J. Chem.*, 2015, **39,** 3916–3922.
[8] K. U. M. Rao, C. S. Sundar, S. S. Prasad, C. R. Rani, C. S. Reddy, *Bull. Korean Chem. Soc.*, 2011, **32,** 3343–3347.
[9] K. R. M. Naidu, K. S. Kumar, P. Arulselvan, C. B. Reddy, O. Lasekan, *Arch. Pharm. Chem. Life Sci.*, 2012, **345,** 957–963.
[10] D. V. Patel, K. Rielly-Gauvin, D. E. Ryono, C. A. Free, W. L. Rogers, S. A. Smith, J. M. DeForrest, R. S. Oehl, Jr. E. W. Petrillo, *J. Med. Chem.*, 1995, **38,** 4557–4569.
[11] A. N. Pudovik, G. A. Zametaeva, *Izvestiya Akademii Nauk. S.S.S.R., Seriya Khimicheskaya* 1952, 932–939.
[12] A. N. Pudovik, I. V. Konovalova, *Synthesis*, 1979, 81–96.
[13] M. S. Kharasch, R. A. Mosher, I. S. Bengelsdorf, *J. Org. Chem.*, 1960, **25,** 1000–1006.
[14] H. R. Hudson, R. O. Yusuf, R. W. Matthews, *Phosphorus, Sulfur, Silicon*, 2008, **183,** 1527–1540.
[15] C. Jin, H. He, *Phosphorus, Sulfur, Silicon*, 2011, **186,** 1397–1403.
[16] S. Sebti, A. Rhihil, A. Saber, M. Laghrissi, S. Boulaajaj, *Tetrahedron Lett.*, 1996, **37,** 3999–4000.
[17] A. R. Sardarian, B. Kaboudin, *Synth. Commun.*, 1997, **27,** 543–551.
[18] A. Smahi, A. Solhy, R. Tahir, S. Sebti, J. A. Mayoral, J. I. García, J. M. Fraile, M. Zahouily, *Catal. Commun.*, 2008, **9,** 2503–2508.
[19] A. Solhy, S. Sebti, R. Tahir, J. Sebti, M. O. Abba, M. Bousmina, S. Vaudreuil, M. Zahouily, *Curr. Org. Chem.*, 2010, **14,** 1517–1522.
[20] R. G. de Noronha, P. J. Costa, C. C. Romão, M. J. Calhorda, A. C. Fernandes, *Organometallics*, 2009, **28,** 6206–6212.
[21] S. Santhisudha, P. Sreelakshimi, S. H. Jayaprakash, B. V. Kumar, C. S. Reddy, *Phosphorus, Sulfur, Silicon*, 2015, **190,** 1479–1488.
[22] K. U. M. Rao, S. H. Jayaprakash, S. K. Nayak, C. S. Reddy, *Catal. Sci. Technol.*, 2011, **1,** 1665–1670.
[23] M. Sheykhan, H. Mohammadnejad, J. Akbari, A. Heydari, *Tetrahedron Lett.*, 2012, **53,** 2959–2964.
[24] A. Rostami, B. Atashkar, D. Moradi, *Appl. Catal., A.*, 2013, **467,** 7–16.
[25] T. K. Olszewski, *Synthesis*, 2014, **46,** 403–429.
[26] M. Pandi, P. K. Chanani, S. Govindasamy, *Appl. Catal., A.*, 2012, **441–442,** 119–123.
[27] K. P. Nandre, J. P. Nandre, V. S. Patil, S. V. Bhosale, *Chem. Biol. Interface*, 2012, **2,** 314–321.
[28] M. A. Kulkarni, U. P. Lad, U. V. Desai, S. D. Mitragotri, P. P. Wadgaonkar, *C. R. Chimie*, 2013, **16,** 148–152.
[29] R. M. N. Kalla, Y. Zhang, I. Kim, *New J. Chem.*, 2017, **41,** 5373–5379.
[30] K. S. Kumar, C. B. Reddy, M. V. N. Reddy, C. R. Rani, C. S. Reddy, *Org. Commun.*, 2012, **5,** 50–57.
[31] D. L. Kong, R. D: Liu, G. Z. Li, P. W. Zhang, M. S. Wu, *Asian J. Chem.*, 2014, **26,** 1246–1248.
[32] G. Keglevich, V. R. Tóth, L. Drahos, *Heteroat. Chem.*, 2011, **22,** 15–17.
[33] S. Kumari, A. Shekhar, D. D. Pathak, *Chem. Sci. Trans.*, 2014, **3,** 45–54.
[34] B. Kaboudin, R. Nazari, *J. Chem. Res. (S)*, 2002, 291–292.
[35] S. S. Prasad, S. H. Jayaprakash, K. U. Rao, N. B. Reddy, P. C. R. Kumar, C. S. Reddy, *Org. Commun.*, 2014, **7,** 98–105.
[36] S. Zhou, Z. Wu, J. Rong, S. Wang, G. Yang, X. Zhu, L. Zhang, *Chem. Eur. J.*, 2012, **18,** 2653–2659.
[37] L. Zhao, H. Ding, B. Zhao, C. Lu, Y. Yao, *Polyhedron*, 2014, **83,** 50–59.
[38] S. Yang, X. Zhu, S. Zhou, S. Wang, Z. Feng, Y. Wei, H. Miao, L. Guo, F. Wang, G. Zhang, X. Gu, X. Mu, *Dalton Trans.*, 2014, **43,** 2521–2533.
[39] H. Miao, S. L. Zhou, S. W. Wang, L. J. Zhang, Y. Wei, S. Yang, F. H. Wang, Z. Chen, Y. Chen, Q. B. Yuan, *Sci. China Chem.*, 2013, **56,** 329–336.

[40] B. Liu, J. F. Carpentier, Y. Sarazin, *Chem. Eur. J.*, 2012, **18**, 13259–13264.
[41] V. Thottempudi, K. H. Chung, *Bull. Korean Chem. Soc.*, 2008, **29**, 1781–1783.
[42] I. Aouani, K. Lahbib, S. Touil, *Medicinal Chem.*, 2015, **11**, 206–213.
[43] C. Wang, J. Zhou, X. Lv, J. Wen, H. He, *Phosphorus, Sulfur, Silicon*, 2013, **188**, 1334–1339.
[44] H. R. Ramananarivo, A. Solhy, J. Sebti, A. Smahi, M. Zahouily, J. Clark, S. Sebti, *ACS Sustain. Chem. Eng.*, 2013, 403–409.
[45] C. Liu, Y. Zhang, Q. Qian, D. Yuan, Y. Yao, *Org. Lett.*, 2014, **16**, 6172–6175.
[46] X. Zhou, Y. Liu, L. Chang, J. Zhao, D. Shang, X. Liu, L. Lin, X. Feng, *Adv. Synth. Catal.*, 2009, **351**, 2567–2572.
[47] G. Keglevich, Z. Rádai, N. Z. Kiss, *Green Process Synth.*, 2017, **6**, 197–201.
[48] G. Consiglio, S. Failla, P. Finocchiaro, *Phosphorus, Sulfur, Silicon*, 1996, **117**, 37–54.
[49] P. V. Shinde, A. H. Kategaonkar, B. B. Shingate, M. S. Shingare, *Tetrahedron Lett.*, 2011, **52**, 2889–2892.
[50] A. S. Sadaphal, S. S. Sonar, R. U. Pokalwar, N. V. Shitole, M. S. Shingare, *J. Korean Chem. Soc.*, 2009, **53**, 536–541.
[51] S. M. Vahdat, R. Baharfar, M. Tajbakhsh, A. Heydari, S. M. Baghbanian, S. Khaksar, *Tetrahedron Lett.*, 2008, **49**, 6501–6504.
[52] V. B. Vangala, H. N. Pati, *Synth. Commun.*, 2016, **46**, 374–378.
[53] F. Jahani, B. Zamenian, S. Khaksar, M. Tajbakhsh, *Synthesis*, 2010, **19**, 3315–3318.
[54] P. Raju, G. G. Rajeshwaran, M. Nandakumar, A. K. Mohanakrishnan, *Eur. J. Org. Chem.*, 2015, **2015**, 3513–3523.
[55] A. Kumar, S. Jamwal, S. Khan, N. Singh, V. K. Rai, *Phosphorus, Sulfur, Silicon*, 2016, **192**, 381–385.
[56] R. U. Pokalwar, S. A. Sadaphal, A. H. Kategaonkar, B. B. Shingate, M. S. Shingare, *Green Chem. Lett. Rev.*, 2010, **3**, 33–38.
[57] N. Azizi, M. R. Saidi, *Phosphorus, Sulfur, Silicon*, 2003, **178**, 1255–1259.
[58] A. S. Mane, V. P. Chavan, B. K. Karale, R. V. Hangarge, M. S. Gaikwad, M. S. Shingare, *Chem. Commun.*, 2002, **32**, 2633–2636.
[59] T. Bhattacharya, B. Majumdar, T. K. Sarma, *Chem. Sel. Commun.*, 2016, **1**, 5265–5269.
[60] Q. Wu, J. Zhou, Z. Yao, F. Xu, Q. Shen, *J. Org. Chem.*, 2010, **75**, 7498–7501.
[61] L. Nagarapu, R. Mallepalli, U. N. Kumar, P. Venkateswarlu, R. Bantu, L. Yeramanchi, *Tetrahedron Lett.*, 2012, **53**, 1699–1700.
[62] H. S. Wang, J. E. Zeng, *Phosphorus, Sulfur, Silicon*, 2010, **185**, 1425–1428.
[63] K. Ramesh, B. Madhav, S. N. Murthy, Y. V. D. Nageswar, *Synth. Commun.*, 2012, **42**, 258–265.
[64] A. Bouzina, N. E. Aouf, M. Berredjem, *Res. Chem. Intermed.*, 2016, **42**, 5993–6002.
[65] P. G. Mandhane, R. S. Joshi, D. R. Nagargoje, C. H. Gill, *Tetrahedron Lett.*, 2010, **51**, 1490–1492.
[66] A. Heydari, A. Arefi, S. Khaksar, M. Tajbakhsh, *Catal. Commun.*, 2006, **7**, 982–984.
[67] M. Tajbakhsh, K. Samad, T. Zahra, B. Ahmadreza, *Chin. J. Chem.*, 2012, **30**, 827–829.
[68] S. S. Sonar, A. H. Kategaonkar, M. N. Ware, C. H. Gill, B. B. Shingate, M. S. Shingare, *Arkivoc*, 2009, **2**, 138–148.
[69] O. Seven, S. Polat-Cakir, M. S. Hossain, M. Emrullahoglu, A. S. Demir, *Tetrahedron*, 2011, **67**, 3464–3469.
[70] H. Maeda, K. Takahashi, H. Ohmori, *Tetrahedron*, 1998, **54**, 12233–12242.
[71] L. Gu, C. Jin, H. Zhang, *New J. Chem.*, 2015, **39**, 1579–1582.
[72] X. Li, C. Jin, L. Gu, *J. Org. Chem.*, 2015, **80**, 2443–2447.
[73] M. Drąg, E. Wieczerzak, M. Pawełczak, L. Berlicki, Z. Grzonka, P. Kafarski, *Biochimie*, 2013, **95**, 1640–1649.
[74] J. V. Alegre-Requena, E. Marqués-López, P. J. S. Miguel, R. P. Herrera, *Org. Biomol. Chem.*, 2014, **12**, 1258–1264.

[75] S. Hirashima, R. Arai, K. Nakashima, N. Kawai, J. Kondo, Y. Koseki, T. Miura, *Adv. Synth. Catal.*, 2015, **357,** 3863–3867.
[76] D. Uraguchi, T. Ito, T. Ooi, *J. Am. Chem. Soc.*, 2009, **131,** 3836–3837.
[77] Ö. Dogan, M. Isci, M. Aygun, *Tetrahedron: Asymmetry*, 2013, **24,** 562–567.
[78] X. Zhou, X. Liu, X. Yang, D. Shang, J. Xin, X. Feng, *Angew. Chem. Int. Ed.*, 2008, **47,** 392–394.
[79] C. Wang, C. Xu, X. Tan, H. Peng, H. He, *Org. Biomol. Chem.*, 2012, **10,** 1680–1685.
[80] P. Muthupandi, G. Sekar, *Org. Biomol. Chem.*, 2012, **10,** 5347–5352.
[81] F. Xu, Y. Liu, J. Tu, C. Lei, G. Li, *Tetrahedron: Asymmetry*, 2015, **26,** 891–896.
[82] R. Boobalan, C. Chen, *Adv. Synth. Catal.*, 2013, **355,** 3443–3450.
[83] K. Suyama, Y. Sakai, K. Matsumoto, B. Saito, T. Katsuki, *Angew. Chem. Int. Ed.*, 2010, **49,** 797–799.
[84] L. Sun, Q. P. Guo, X. Li, L. Zhang, Y. Y. Li, C. S. Da, *Asian J. Org. Chem.*, 2013, **2,** 1031–1035.
[85] I. Shiina, K. Ono, T. Nakahara, *Chem. Commun.*, 2013, **49,** 10700–10702.
[86] T. Mandal, S. Samanta, G. G. Zhao, *Org. Lett.*, 2007, **9,** 943–945.
[87] D. Skropeta, R. R. Schmidt, *Tetrahedron:Asymmetry*, 2003, **14,** 265–273.

Erika Bálint, Anna Tripolszky and Ádám Tajti

6 Synthesis of α-aminophosphonates by the Kabachnik–Fields reaction and by the Pudovik reaction

Abstract: α-Aminophosphonates are of significant importance due to their biological activity. The most widely applied synthetic routes towards α-aminophosphonates are the Kabachnik–Fields reaction involving the condensation of amines, oxo compounds and >P(O)H species, such as dialkyl phosphites, and the Pudovik reaction of imines and >P(O)H reagents. By the double Kabachnik–Fields reaction, bis(aminophosphonates) have also became available. This chapter summarizes the synthesis of α-aminophosphonates and related derivatives through the two main routes as described in the literature over the last five years.

Keywords: Kabachnik–Fields reaction, Pudovik reaction, α-aminophosphonates, bis(aminophosphonates).

6.1 Introduction

α-Aminophosphonic acids are considered as bioisosteres of the corresponding α-aminocarboxylic acids, in which the planar carboxylic group is replaced by a tetrahedral phosphonic acid functionality. Due to their versatile biological activity, they are important targets in biochemistry [1, 2], medicinal chemistry [3–6] and pesticide chemistry [7–9]. The synthesis and application of α-aminophosphonic acid derivatives attracted considerable interest, comprising more than one thousand five hundred papers since 1952 (Figure 6.1).

The most widely used synthetic routes to α-aminophosphonates are the Kabachnik–Fields condensation and the (aza-)Pudovik reaction. This chapter is aimed at giving insights into the synthesis of α-aminophosphonates and related derivatives surveying the literature data over the last five years.

6.2 Kabachnik–Fields Reaction

The Kabachnik–Fields (phospha-Mannich) reaction is a three-component condensation of a primary or secondary amine, a carbonyl compound (aldehyde or ketone) and a >P(O)H reagent, such as a dialkyl phosphite or a secondary phosphine oxide (Figure 6.2) [10–13].

https://doi.org/10.1515/9783110535839-006

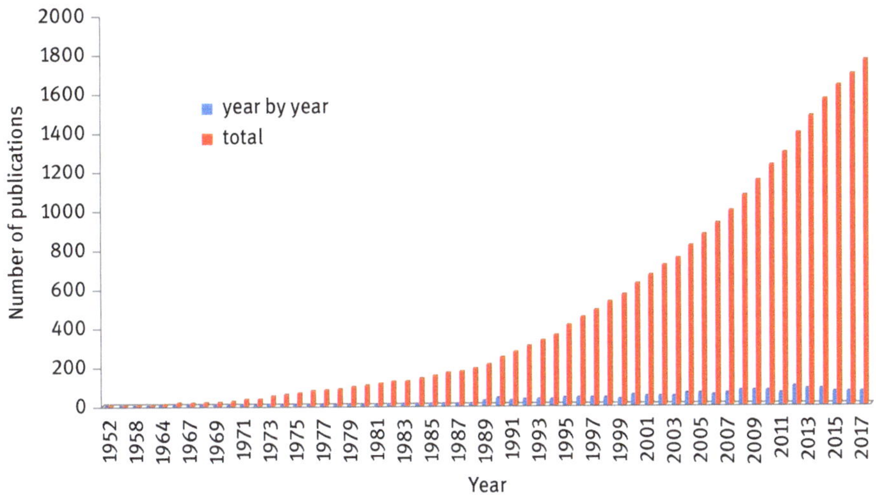

Figure 6.1: The number of publication on α-aminophosphonic acid derivatives (1952–2017). ScienceDirect keyword search on "α-aminophosphonic acid" and "α-aminophosphonate".

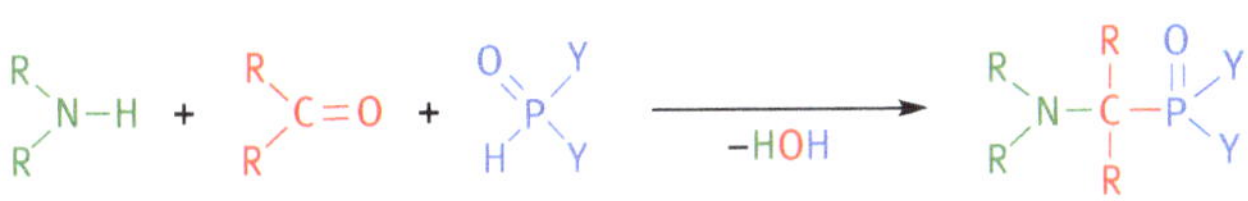

Figure 6.2: General scheme for the Kabachnik–Fields reaction.

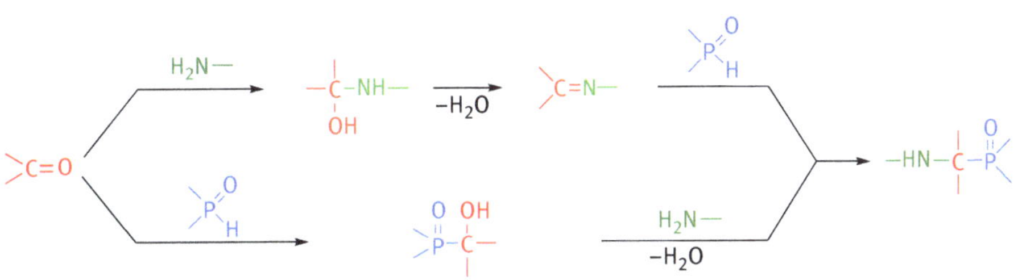

Figure 6.3: Two possible mechanistic pathways for the Kabachnik–Fields reaction.

In general, there are two possible reaction pathways for the Kabachnik–Fields reaction (Figure 6.3). One is when the carbonyl compound and the primary amine react with each other resulting in the formation of an imine (Schiff base) intermediate, and then the P-reagent is added on the C=N unit. According to the other route, the first step is the addition of dialkyl phosphites to the carbonyl group of the oxo component, to provide an α-hydroxyphosphonate, which undergoes substitution by the amine to furnish the α-aminophosphonate. On the basis of kinetic studies, it was concluded that the mechanism depends on the nature of the reactants [12–14].

Kabachnik–Fields reactions may be accomplished in many variations. These condensations are usually carried out in the presence of various catalysts and/or solvents [13]. They can be carried out under green chemical conditions, including microwave (MW)-assisted synthesis, which are also of special interest [15].

6.2.1 Kabachnik–Fields Reactions in the Presence of Catalysts and Solvents

Most of the papers published in the field of Kabachnik–Fields reaction since 2002 suggest the use of special catalysts [16–25], such as metal triflates [17], lanthanide triflates [17, 18], gallium(III) iodide [19], bismuth(I) nitrate [20], magnesium perchlorate [21, 22], samarium(II) iodide [23], indium(III) chloride [24] and phthalocyanine–$AlCl_3$ complex [25] etc. in solvents or under neat conditions. According to the literature reports, catalytic approaches still dominated over the last five years (Table 6.1). The main goal is to find efficient, cost-effective, reusable and environmentally benign catalysts for the three-component synthesis of α-aminophosphonates. In the case of the condensation of aromatic amines, benzaldehyde derivatives and dialkyl phosphites, the use of triflates was still a common practice (Table 6.1, entries 1–3). In one case, the reaction was carried out with ytterbium triflate [$Yb(OTf)_3$] in water using polyoxyethanyl α-tocopheryl sebacate (PTS) as an amphiphilic species (Table 6.1, entry 1), while in another instance, copper triflate was applied with a fluorous bis(oxazoline) ligands using acetonitrile as the solvent (Table 6.1, entry 2). Pentafluorophenylammonium triflate (PFPAT) was also efficient as a new organocatalyst for the synthesis of α-aminophosphonates at room temperature (Table 6.1, entry 3). Various diethyl(3,5-dibromo-4-hydroxyphenylamino) methylphosphonates were synthesized in a household MW oven using cerium(III) chloride heptahydrate as the catalyst in THF (Table 6.1, entry 4). A series of α-aminophosphonates were prepared *via* a Kabachnik–Fields reaction catalyzed by a gold–bipyridine complex (Table 6.1, entry 5). The three-component condensation of anilines, benzaldehydes and diphenyl phosphite was elaborated in dichloromethane applying zinc(II) di(L-prolinate) as the catalyst (Table 6.1, entry 6). An efficient method has been developed for the synthesis of α-aminophosphonates using a heterogeneous, reusable silica-supported dodecatungstophosphoric acid (DTP/SiO_2) catalyst at room temperature (Table 6.1, entry 7). Polyaniline-methanesulfonic acid salt (PANI-MSA)-coated glass slide in hexane was also an effective catalyst in the reaction of various amines, aldehydes and dimethyl phosphite (Table 6.1, entry 8). A combination of [bmim][$AlCl_4$] ionic liquid and ultrasonic irradiation was used as an alternative to conventional acid catalysts in the Kabachnik–Fields reaction (Table 6.1, entry 9). The ultrasound-assisted synthesis of α-aminophosphonates is reviewed in detail by Bubun and Keglevich in Chapter 13. In another case, 1,4-diazabicyclo[2.2.2]octane hydrochloride

([H-DABCO]Cl) quaternary ammonium salt was applied as the catalyst in methanol (Table 6.1, entry 10). Finally, a nonionic surfactant (Tween-20) catalyzed process was also reported in aqueous media (Table 6.1, entry 11).

Table 6.1: Catalytic Kabachnik–Fields reactions using solvents.

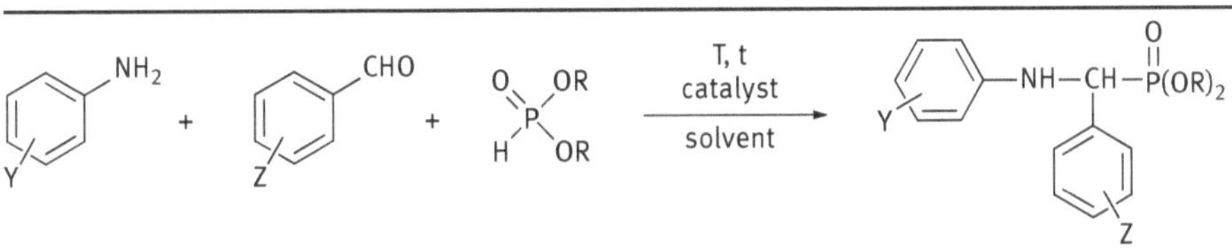

Entry	Y	Z	R	Catalyst	Solvent	*T*, *t*	Yield (%)	Ref.
1	H, 4-Me, 4-MeO, 2-Cl, 3-Cl, 4-Cl, 4-F, $4\text{-}NO_2$	H, 4-Me, 4-MeO, 4-Cl, $2\text{-}NO_2$, $4\text{-}NO_2$	Et	$Yb(OTf)_3$ (1%)	PTS/H_2O	25 °C, 1 h	83–96	[26]
2	H, 4-Me, 4-MeO, 4-Cl, $4\text{-}NO_2$	H, 4-Me, 4-MeO, 3-Cl, 4-Cl, 2-Br, 4-F, $3\text{-}NO_2$, $4\text{-}NO_2$, $3\text{-}CF_3$, $4\text{-}CF_3$, $4\text{-}NMe_2$	Et	$Cu(OTF)_2$ (5%) with fluorous bis(oxazoline) ligands (5%)	DCM	25 °C, 6 h	73–94	[27]
3	H, 4-MeO, 4-HO, 4-Cl, 4-Br	H, 4-Cl, 4-Br	Me	$C_6F_5NH_3(OTf)$ (PFPAT) (10%)	ACN	25 °C, 1–2 h	85–95	[28]
4[a]	3,5-diBr-4-OH	4-MeO, 4-Cl, 4-HO, $3\text{-}NO_2$, $4\text{-}NMe_2$, 2,4-diCl, 2,4-diMeO	Et	$CeCl_3{\cdot}7H_2O$ (5%)	THF	490 W, 8–11 min	89–94	[29]
5	H, 4-Me, 4-MeO, 4-Cl	H, 4-Me, 4-MeO, 4-Cl, 4-Br, $4\text{-}NO_2$	Et	$[AubpyCl_2]Cl$ (5%)	ACN	40 °C, 3–10 h	85–95	[30]
6	H, 4-Me, 2,6-diMe, 4-MeO, 4-Cl, 4-Br, 4-I, 4-F, $4\text{-}NO_2$	H, 4-Me, 4-MeO, 4-Cl, 4-Br, 4-F, $3\text{-}NO_2$, 3-HO, $4\text{-}CF_3$, $4\text{-}NMe_2$	Ph	$Zn(\text{L-Pro})_2$ (10%)	DCM	25 °C, 20–60 min	87–98	[31]
7	H, 3-Cl, 4-Cl, 2,4,6-triMe, 4-MeO, $4\text{-}NO_2$	H, 4-Me, 4-MeO, 2,5-diMeO, 4-Cl,	Me, Et, Bn	DTP/SiO_2 (20%)	ACN	25 °C, 1 h	93–98	[32]
8	H, 4-MeO, 4-Br, $4\text{-}NO_2$	4-Me, 4-MeO, 4-Cl, 4-OH, $4\text{-}NO_2$	Me	PANI-MSA-coated glass slide	Hexane	25 °C, 3 h	73–98	[33]

(continued)

Table 6.1: (Continued)

Entry	Y	Z	R	Catalyst	Solvent	*T*, *t*	Yield (%)	Ref.
9[b]	H, 4-Me	H, 2-Me, 4-MeO, 2-Cl, 3-NO_2, 4-NO_2	Me	[bmim][$AlCl_4$] (10%)	MeOH	25 °C, 5–10 min	87–93	[34]
10	H, 4-Me, 4-MeO, 4-F	H	Et	[H-DABCO]Cl (1 equiv.)	MeOH	25 °C, 10–30 min	90–93	[35]
11	H, 2-Me, 4-MeO, 4-Cl, 4-Br, 2-NO_2, 4-NO_2	6-Niroben-zo[*d*]-[1,3] dioxole-5-car-baldehyde	Et	Tween-20 (5%)	Water	60 °C, 25–60 min	82–91	[36]

[a]Under MW irradiation in a household MW oven. [b]Under ultrasonic irradiation.

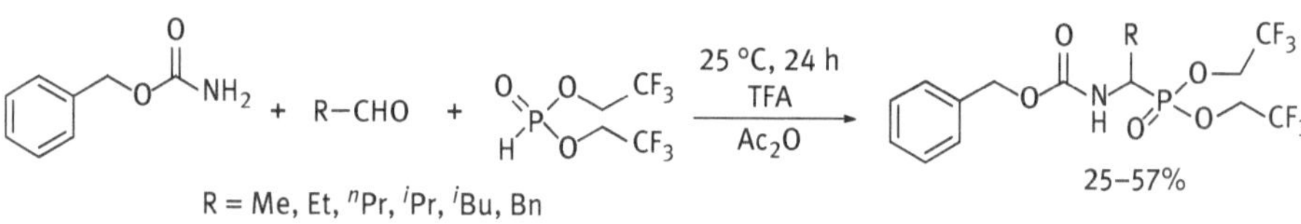

Figure 6.4: Synthesis of bis(trifluoroethyl) esters of α-aminophosphonic acids.

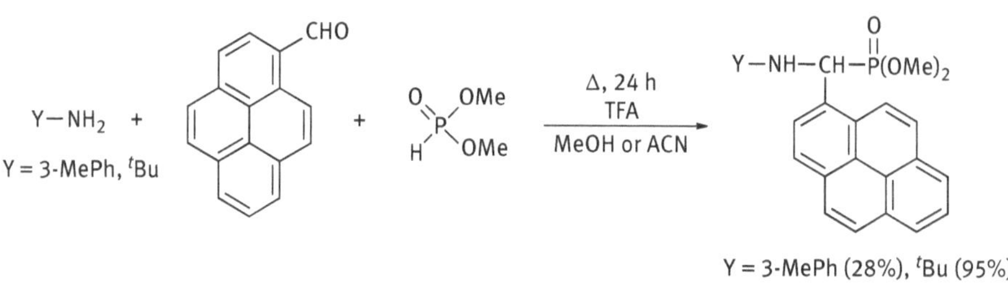

Figure 6.5: Synthesis of pyrene-derived α-aminophosphonates.

Bis(trifluoroethyl) esters of α-aminophosphonic acids were prepared as inhibitors of serine proteases by the reaction of benzyl carbamate, aldehydes and bis(2,2,2-trifluoroethyl) phosphite in the presence of trifluoroacetic acid (TFA) and acetic anhydride (Figure 6.4) [37].

The reaction of 3-methylaniline or *tert*-butylamine with pyrene-1-carboxaldehyde and dimethyl phosphite was also described in the presence of a catalytic amount of TFA (Figure 6.5) [38]. The pyrene-derived α-aminophosphonates obtained showed fluorescent properties.

2-Cyclopropylpyrimidin-4-yl-aryl- and 2-cyclopropylpyrimidin-4-yl-benzothiazole-derived α-aminophosphonates were obtained by the condensation of anilines or *N*-benzothiazole amines, 2-cyclopropylpyrimidin-4-carbaldehyde and dialkyl phosphites using phosphomolybdic acid ($H_3PMo_{12}O_{40}$) as the catalyst in dichloromethane (Figure 6.6) [39].

The three-component condensation of amines, 2-alkynylindole-3-carbaldehydes and dimethyl phosphite was carried out applying a catalytic amount of $BF_3{\cdot}OEt_2$ as a Lewis acid catalyst (Figure 6.7) [40].

Another example for the $BF_3{\cdot}OEt_2$-mediated Kabachnik–Fields reaction involves the preparation of a series of α-aminophosphonates containing a pyrazole moiety (Figure 6.8) [41].

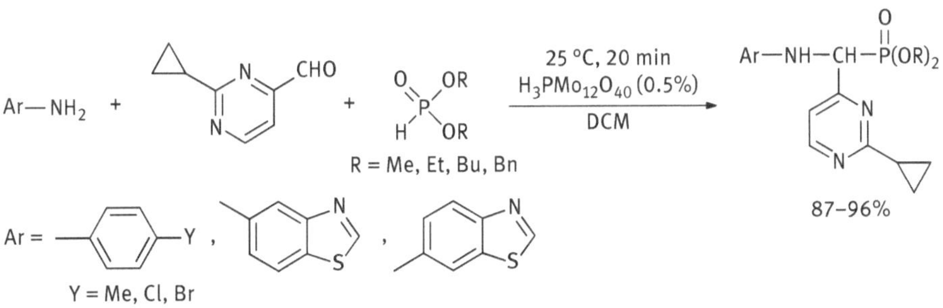

Figure 6.6: Kabachnik–Fields reaction of 2-cyclopropyl pyrimidine-4-carbaldehyde.

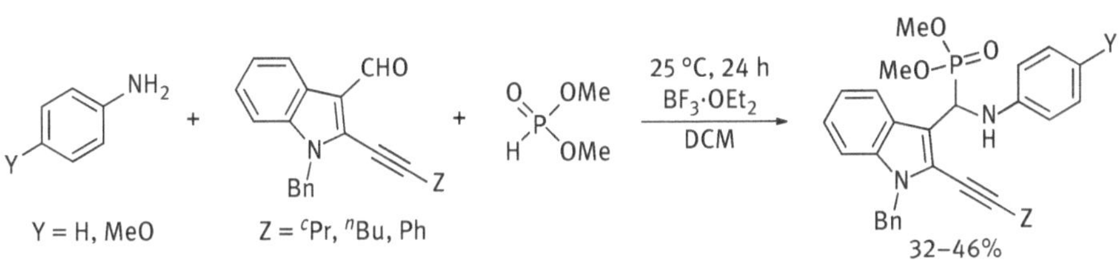

Figure 6.7: Kabachnik–Fields reaction of 2-alkynylindole-3-carbaldehydes.

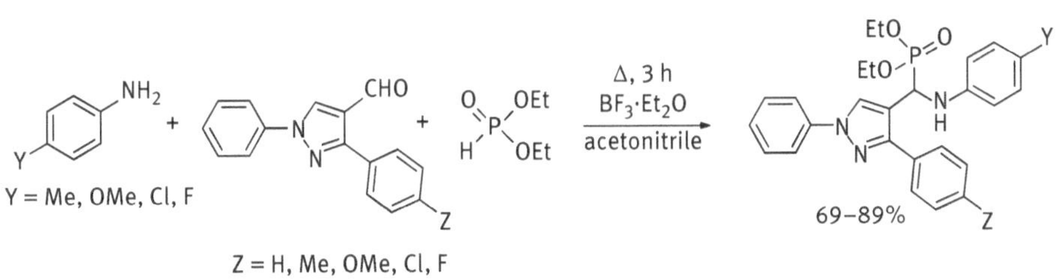

Figure 6.8: Synthesis of α-aminophosphonates with a pyrazole moiety.

In other instances, trialkyl phosphite was used instead of dialkyl phosphite as the P-component in the Kabachnik–Fields condensation. Only a few examples are introduced in the following paragraphs. A series of α-aminophosphonates were synthesized by the condensation of amines, aldehydes and trialkyl phosphites using hafnium(IV) chloride in ethanol (Figure 6.9) [42].

The synthesis of α-aminophosphonates with an isoxazole ring was accomplished using trialkyl phosphites in the presence of iron(III) chloride in THF (Figure 6.10) [43].

A comparative study was reported by our group, where the possibilities for the Kabachnik–Fields reaction of benzylamine, benzaldehyde and triethyl phosphite or diethyl phosphite were investigated in water (Figure 6.11) [44]. It was found that in the case of trialkyl phosphites, *p*-toluenesulfonic acid (PTSA) had to be used as the catalyst. On the other hand, the reaction could be performed without any catalyst, in a solvent-free manner when diethyl phosphite was the P component.

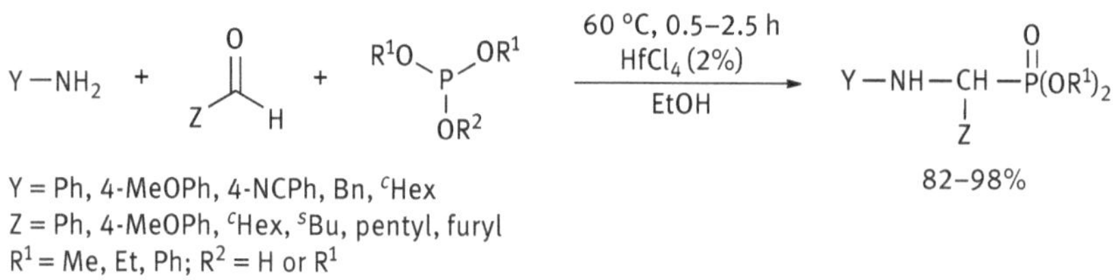

Figure 6.9: $HfCl_4$-catalyzed Kabachnik–Fields reaction with trialkyl phosphites.

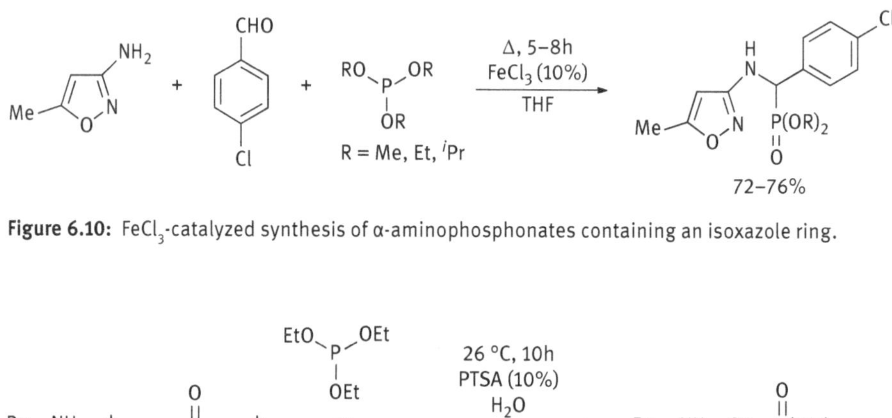

Figure 6.10: $FeCl_3$-catalyzed synthesis of α-aminophosphonates containing an isoxazole ring.

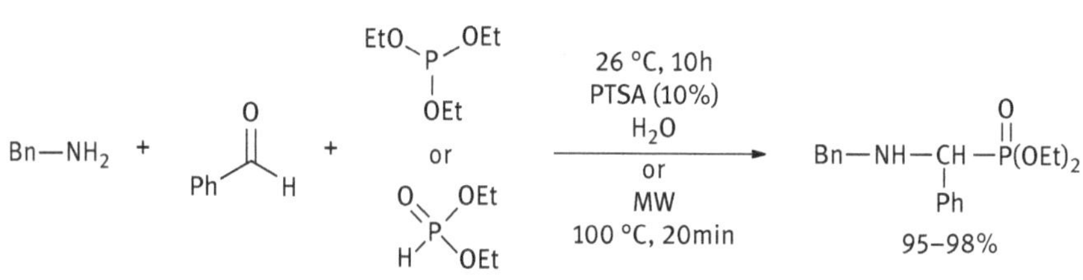

Figure 6.11: Kabachnik–Fields reaction with triethyl phosphite or diethyl phosphite.

6.2.2 Kabachnik–Fields Reactions in the Presence of Catalysts under Solvent-free Conditions

More examples from the literature cover catalytic Kabachnik–Fields reactions under solvent-free conditions (Table 6.2). A series of new fluorinated α-aminophosphonates was synthesized starting from fluorinated aniline derivatives, aldehydes and diethyl phosphite in the presence of $BF_3 \cdot OEt_2$ (Table 6.2, entry 1). In most cases, metal-containing catalysts were applied in the solvent-free Kabachnik–Fields reactions. For example, the three-component reactions of various amines, aldehydes and dialkyl phophites were performed with zinc acetate dihydrate (Table 6.2, entry 2). In another instance, a series of α-aminophosphonates was prepared in the presence of cadmium perchlorate hydrate (Table 6.2, entry 3). Nickel(II) sulfate hexahydrate was an efficient and reusable catalyst for the synthesis of α-aminophosphonates under mild conditions (Table 6.2, entry 4). Micron-particulate aluminium nitride (AlN/Al) was also applied as a new heterogeneous catalyst in the Kabachnik–Fields condensation (Table 6.2, entry 5). Acid supported magnetite nanoparticles (Fe_3O_4) also acted as novel and effective catalysts for the solvent-free synthesis of *α*-aminophosphonates (Table 6.2, entries 6 and 7). In one case, dehydroascorbic acid (DHAA; Table 6.2, entry 6), while in another case, phosphotungstic acid (PTA) was the acid component of the catalyst (Table 6.2, entry 7). Several α-aminophosphonates were synthesized in MW-assisted reactions of 2-naphthyl- or 2-fluorenylamine, different aldehydes and dimethyl phosphite using a titanium dioxide-silica catalyst (Table 6.2, entry 8). Air-stable zirconocene bis(perfluorobutanesulfonate) was prepared and applied as a catalyst in the solvent-free Kabachnik–Fields reaction of amines, aldehydes/ketones and diethyl phosphite (Table 6.2, entry 9). A series of new pyrenyl-α-aminophosphonates was synthesized by the one-pot reaction of aryl/heteroaryl amines, pyrene aldehyde and diethyl phosphite in the presence of silica-supported polyacrylic acid (PAA; Table 6.2, entry 10). The preparation of quinoline-containing α-aminophosphonate derivatives was elaborated using a polyethyleneimine-grafted mesoporous nanomaterial as the catalyst (Table 6.2, entry 11). Acidic Amberlite-IR 120 resin was also an effective catalyst in the Kabachnik–Fields condensations, which were carried out in a household MW oven (Table 6.2, entry 12). Another acidic catalyst, such as triflic acid supported carbon was found to be suitable for the synthesis of various α-aminophosphonates (Table 6.2, entry 13). Phenylboronic acid-catalyzed Kabachnik–Fields reactions of benzyl amine, aliphatic or aromatic aldehydes and dimethyl phosphite were accomplished in the absence of a solvent (Table 6.2, entry 14). The condensation was extended to aliphatic and aromatic ketones (Table 6.2, entry 15). A facile method was developed for the synthesis of tertiary and quaternary α-aminophosphonates using phenylphosphonic acid as the catalyst (Table 6.2, entry 16). 2,3-Dihydro-1*H*-inden-5-amine and 2-aminofluorene were reacted with various aldehydes and dimethyl phosphite under MW conditions (Table 6.2, entries 17 and 18). In the first case, the condensations were catalyzed with molybdate sulfuric acid (MSA) (Table 6.2, entry 17), while in the latter

Table 6.2: Solvent-free Kabachnik–Fields reactions in the presence of catalysts.

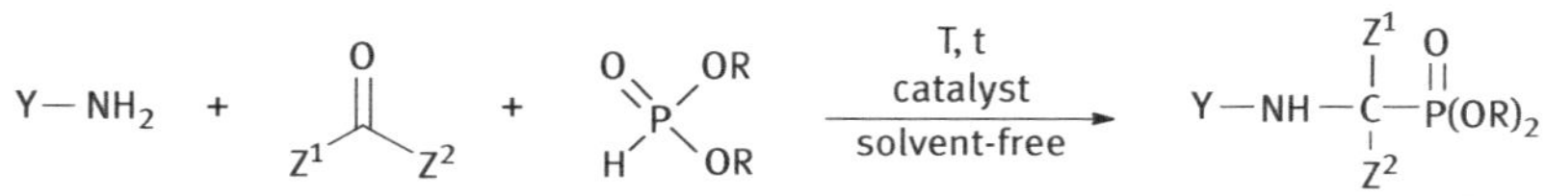

Entry	Y	Z^1	Z^2	R	Catalyst	*T*, *t*	Yield (%)	Ref.
1	4-FPh, 3,4-diFPh	4-FPh, 2-F-,4-ClPh	H	Et	$BF_3 \cdot Et_2O$	70 °C, 2 h	83–96	[45]
2	2,4-diClPh, 4-BrPh, 4-NO_2Ph	4-HOPh, 4-MeOPh	H	Me, Et, Bn	$Zn(OAc)_2 \cdot H_2O$ (12%)	50 °C, 15–22 min	85–94	[46]
3	Ph, 4-BrPh, 4-FPh, 4-NO_2Ph	Ph, 4-^{i}PrPh, 4-MeOPh, 4-EtOPh, 4-ClPh, 4-BrPh, 4-HOPh, 4-NO_2Ph	H	Me, Et	$Cd(ClO_4)_2 \cdot H_2O$ (5%)	25 or 40 °C, 10–360 min	52–98	[47]
4	Ph, 4-MePh, 4-CF_3Ph	Ph, 4-MeOPh, 4-ClPh, 4-NO_2Ph, 4-biphenyl, 3-furyl	H	Et	$NiSO_4 \cdot 6H_2O$ (5%)	25 °C, 10–20 min	92–98	[48]
5	4-FPh, 4-^{i}PrPh	Ph, 3,4-diMeOPh, 4-MeSPh, 4-HOPh, indole-3-carbaldehyde, fluorene-3-carbaldehyde	H	Et, ^{n}Bu	AlN/Al (5%)	50 °C, 1 h	84–94	[49]
6	Ph, 4-MePh, 4-BrPh	Ph, 4-MePh, 4-MeOPh, 4-ClPh, PhCH=CH, 2-thienyl	H	Me	DHAA–Fe_3O_4 (0.9%)	40 °C, 1–2 h	75–95	[50]
7	Ph, Bn	Ph, 2-MePh, 2-MeOPh, 4-MeOPh, 2-ClPh, 4-ClPh, 2-NO_2Ph, 4-NO_2Ph, 2-CF_3Ph	H	Me, Et	Fe_3O_4@SiO_2-PTA (5%)	25 °C, 3–6 h	72–97	[51]
8[a]	2-Naphthyl, 2-Fluorenyl	4-MePh, 4-HOPh, 4-MeOPh, 4-EtOPh, 4-ClPh 4-BrPh, 4-FPh, 4-NO_2Ph, Et, ^{n}Pr, ^{n}Bu	H	Me	TiO_2–SiO_2 (5%)	70 °C, 3–5 min	85–97	[52]
9	Ph, Bn, 4-MePh, 4-ClPh, 4-NO_2Ph, 4-CF_3Ph, 1-naphthyl, benzidinyl, 2-pyridinyl, 4-pyridinyl, cyclohexyl	Ph, 4-MePh, 4-MeOPh, 4-HOPh, 3-NO_2Ph, 2-CF_3Ph, 4-CF_3Ph, 4-ClPh, 2-BrPh, 4-BrPh, 2-FPh, ^{n}Pr, cyclohexanone	H, Me	Et	$[Cp_2Zr(OSO_2C_4F_9)_2 \cdot 2H_2O]$ (5%)	25 °C, 2.5–12 h	49–98	[53]

10	Ph, 4-MePh, 4-MeOPh, 4-HOPh, 3-ClPh, 4-ClPh, 4-FPh, 4-NO_2Ph, 4-pyridinyl, 6-methyl-2-pyridinyl, 3-isoxazolyl	1-Pyrenyl	H	Et	PAA-SiO_2 (30%)	70 °C, 90–130 min	81–96	[54]
11	Ph, 4-FPh, CH_2CH_2Ph	4-Quinoline-carboxaldehyde		Et, Ph	MCM-41@ PEI (20%)	50–80 °C, 5–6 h	86–91	[55]
12[a]	^{n}Pr, Bn, Ph, 2-HOPh, 4-HOPh, 2-MePh, 2,6-diMePh, 3-ClPh, 3-FPh	Piperonyl, H, 4-MePh, 2-MeOPh, 4-FPh, 3,4-diMeOPh, 3-NO_2Ph, 4-NO_2Ph	H	Et, ^{n}Bu, Ph, allyl	Amberli-te-IR 120 (H^+) (10%)	900 W, 1–5 h	79–98	[56]
13	Ph, 4-BrPh, 3-ClPh, 4-ClPh, 3-NO_2Ph, 4-NO_2Ph	Ph, 3-MePh, 4-MePh, 4-^{i}PrPh, 4-MeOPh, 4-HOPh, 2,5-$(MeO)_2$Ph, 3,4-diHOPh, 4-ClPh, 4-FPh	H	Et	TfOH/C (10%)	25 °C, 1–9 h	85–98	[57]
14	Bn	^{i}Pr, ^{s}Bu, ^{t}Bu, 2-MeOPh, 4-MeOPh, 3,4-diMeOPh, 4-ClPh, 3-HOPh	H	Me	$PhB(OH)_2$ (10%)	50 °C, 15–45 min	62–93	[58]
15	Bn	Acetophenone, prophiophenone, acetone, diethyl ketone, methyl ethyl ketone, 1-indanone, 2-indanone, cyclohexanone		Me	$PhB(OH)_2$ (10%)	50 °C, 0.5–8 h	28–93	[58]
16	Bn	3-HOPh, 3,4-diHOPh, 2-MeOPh, 4-MeOPh, 4-ClPh, 4-CF_3Ph, 4-biphenyl, indole-3-yl, pyrrole-2-yl, 2-furyl, Me, Et, ^{n}Pr, ^{i}Pr, ^{i}Bu, ^{t}Bu	H, Me, Et	Me	PhP(O) $(OH)_2$ (10%)	50 °C, 25–100 min	47–98	[59]
17[a]	2,3-Dihydro-1*H*-indenyl	3-MePh, 4-MePh, 4-^{i}PrPh, 4-MeOPh, 4-EtOPh, 4-ClPh, 3-BrPh, 4-BrPh, 2-FPh, 4-FPh, 4-NO_2Ph	H	Me	MSA (5%)	60 °C, 4–10 min	90–97	[60]
18[a]	2-Fluorenyl	4-MePh, 4-MeOPh, 2-ClPh, 4-ClPh, 4-BrPh, 4-FPh, 2-NO_2Ph, 4-NO_2Ph, Et, ^{n}Pr, ^{i}Pr, ^{n}Bu, ^{c}Hex, piperonyl	H	Me	PS/PTSA (5%)	70 °C, 3–5 min	84–98	[61]

(continued)

Table 6.2: (Continued)

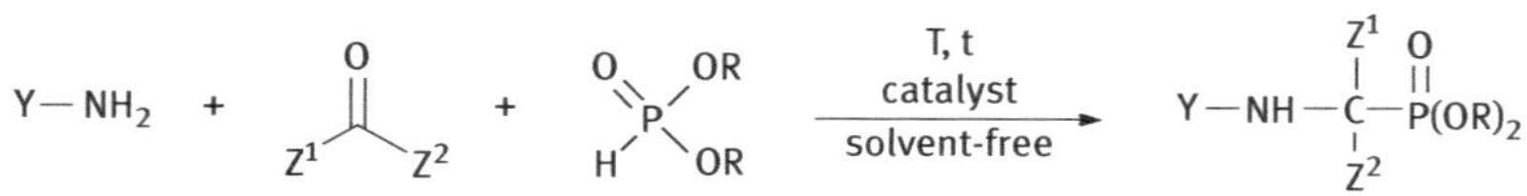

Entry	Y	Z^1	Z^2	R	Catalyst	*T, t*	Yield (%)	Ref.
19	Ph, Bn, 4-ClPh, 4-BrPh, 4-FPh	Ph, 4-MeOPh, 2-ClPh, 3-ClPh, 3-BrPh, 4-BrPh, 4-HOPh	H	Et	$C_5H_{14}ClNO \cdot 2ZnCl_2$ IL (15%)	25 °C, 30–120 min	70–98	[62]
20	Ph, 4-MePh, 4-NO_2Ph, ^{n}Pr, Bn	Ph, 4-Ph, 4-MeOPh, 4-HOPh, 4-NO_2Ph, 4-ClPh, 4-BrPh Terephthalaldehyde, 1- or 2-naphthaldehyde, furan-2-carbaldehyde, acetophenone, cyclohexanone	H	Et	MWCNT-[mpIm] HSO_4 (7%)	25 °C, 33–80 min	89–96	[63]
21	Ph, 4-ClPh	Ph, 4-MePh, 4-MeOPh, 3-ClPh, 4-ClPh, 3-NO_2Ph, 4-NO_2Ph	H	Et	Acidic ionic liquid (10%)	50 °C, 1–2.5 h	85–96	[64]
22	Bn, Ph, 2-MeOPh, 2-HOPh, 3-FPh	Ph, 4-MePh, 2-HOPh, 4-HOPh, 3-MeOPh, 4-MeOPh, 4-NO_2Ph 4-MeOPh, 5-HOPh	H	Et	Bakers' yeast, phosphate buffer (pH 7.0), D-glucose	25 °C, 36–48 h	55–83	[65]

[a]Under microwave irradiation in a household MW oven.

instance, polystyrene-supported PTSA was applied (Table 6.2, entry 18). Ionic liquids may also be important as catalysts in Kabachnik–Fields reactions. In one example, the condensations were performed using choline chloride·2$ZnCl_2$ ionic liquid (Table 6.2, entry 19). In the next two instances, the utilization of 1-methyl-3-(ethoxysilylpropyl)-imidazolium hydrogensulfate anchored on multiwalled carbon nanotube (MWCNT) and benzimidazolium-based dicationic acidic ionic liquid as catalysts was reported (Table 6.2, entries 20 and 21). A biocatalyst-, such as bakers' yeast-mediated synthesis of α-aminophosphonates was also described (Table 6.2, entry 22).

A SiO_2–$ZnBr_2$-catalyzed variation of the Kabachnik–Fields reaction of 4-(4-chlorophenoxy)aniline, various aldehydes and diethyl phosphite was developed (Figure 6.12) [66]. The condensations were carried out under conventional heating, as well as ultrasonic and MW irradiation. The latter one proved to be more efficient, than the others, and the corresponding α-aminophosphonates were obtained in yields of 93–98%.

A wide range of α-aminophosphonates containing a 1,3,4-thiadiazole moiety was prepared by the three-component condensation of 2-amino-5-ethyl-1,3,4-thiadiazole, aldehydes and diethyl phosphite (Figure 6.13) [67]. The reactions were performed in a household MW oven using phosphosulfonic acid, as a reusable heterogeneous solid acid catalyst.

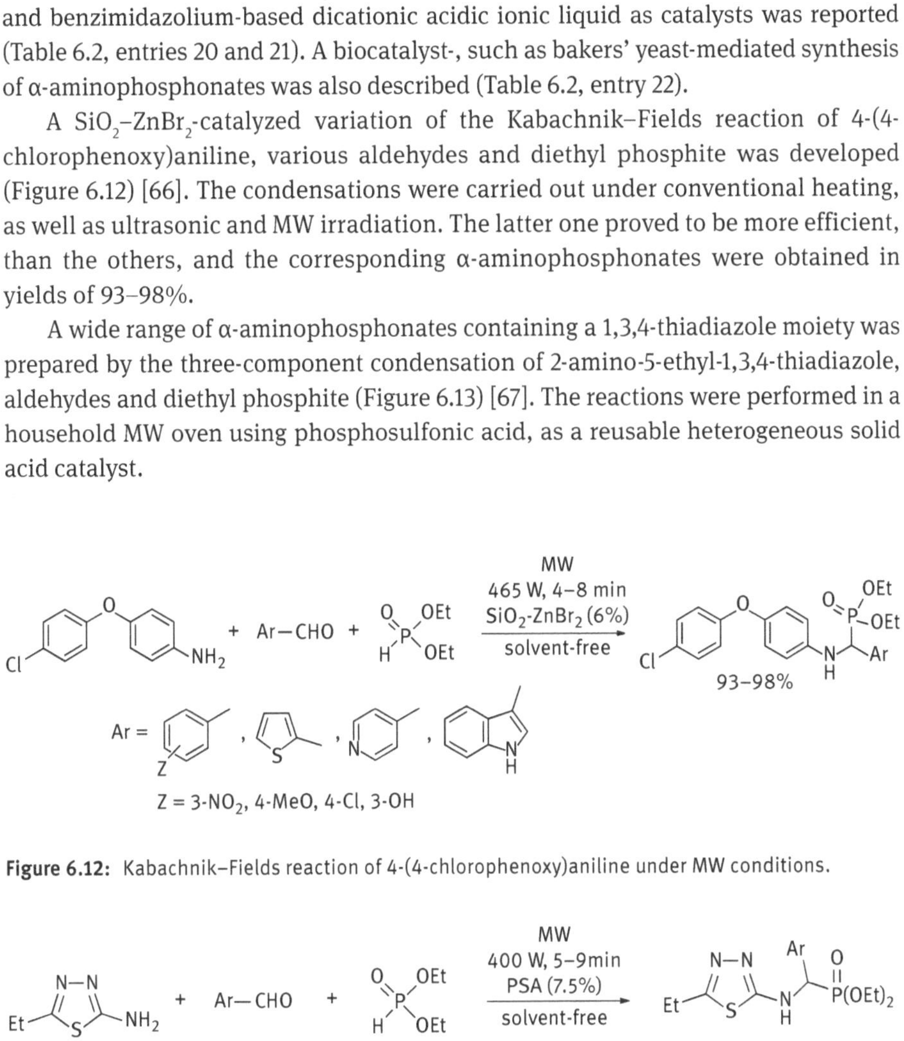

Figure 6.12: Kabachnik–Fields reaction of 4-(4-chlorophenoxy)aniline under MW conditions.

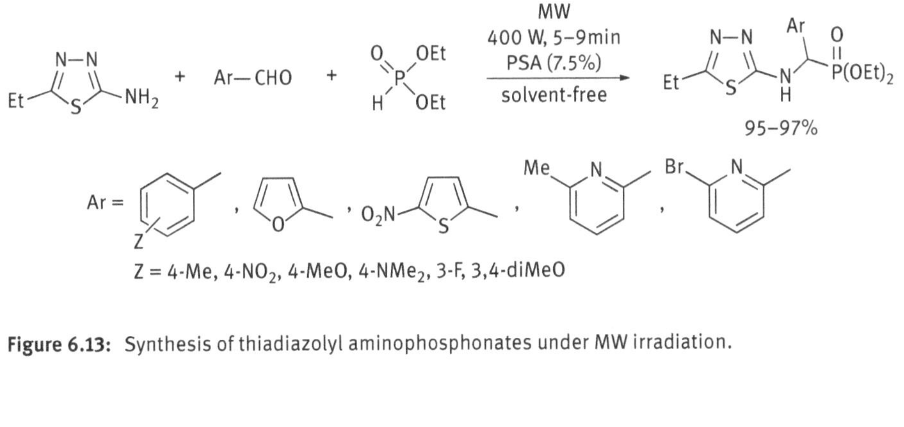

Figure 6.13: Synthesis of thiadiazolyl aminophosphonates under MW irradiation.

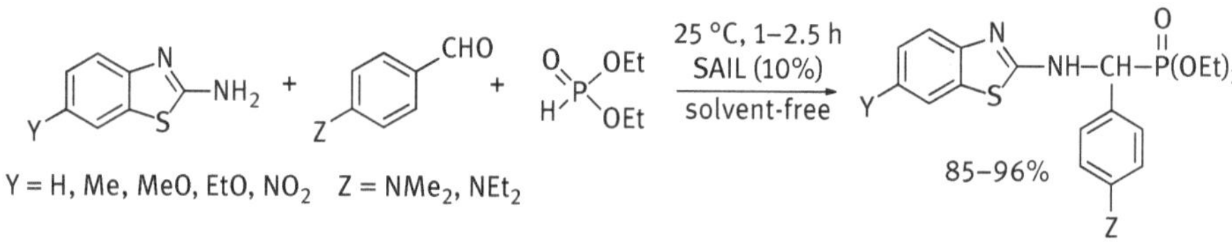

Figure 6.14: Kabachnik–Fields reaction of 2-aminobenzothiazoles in the presence of an acidic ionic liquid.

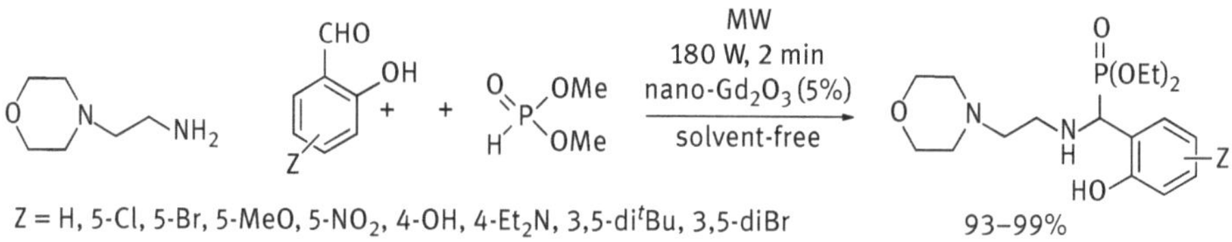

Figure 6.15: Nano-Gd_2O_3-catalyzed synthesis of α-aminophosphonates incorporating an *N*-morpholinoethyl moiety.

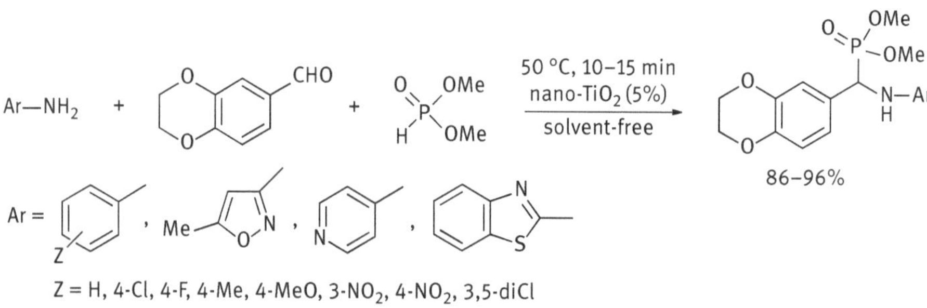

Figure 6.16: Synthesis of dimethyl (2,3-dihydrobenzodioxinyl)-(arylamino)-methylphosphonates.

A sulfonic acid-functionalized ionic liquid was synthesized and applied as the catalyst in the solvent-free synthesis of *N*-benzothiazolyl-α-aminophosphonates (Figure 6.14) [68].

2-Morpholinoethanamine was reacted with salicylaldehydes and dimethyl phosphite using gadolinium oxide nanopowder (nano-Gd_2O_3) under neat conditions (Figure 6.15) [69]. The condensations were performed in a household MW oven, and the corresponding α-aminophosphonates were obtained in yields of 93–99%.

The reaction of various amines, 2,3-dihydrobenzo[*b*][1,4]dioxine-6-carbaldehyde and dimethyl phosphite was carried out in the presence of nano-TiO_2 at 50 °C for 10–15 min (Figure 6.16) [70].

L-Cysteine-functionalized magnetic nanoparticles were used as magnetic reusable catalysts in the synthesis of α-aminophosphonates incorporating benzimidazole, theophylline or adenine nucleobases (Figure 6.17) [71].

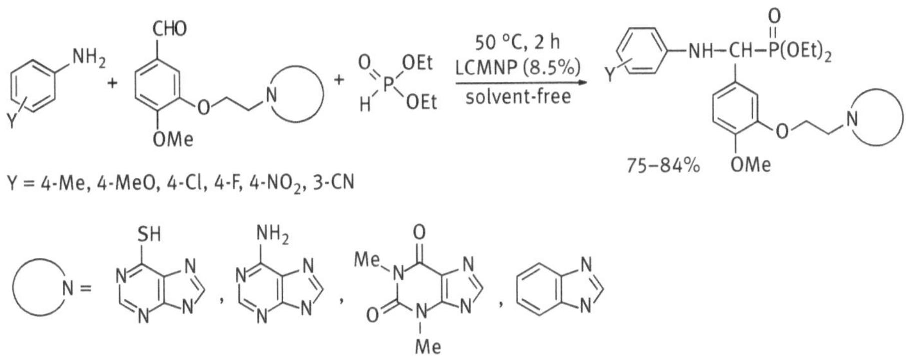

Figure 6.17: Synthesis of α-aminophosphonates containing a benzimidazole, theophylline or adenine moiety

6.2.3 Catalyst-free Kabachnik–Fields Reactions

Only a few examples have been reported on the catalyst-free Kabachnik–Fields condensations. A number of carbazole- and phenothiazine-based α-aminophosphonates were prepared by the reaction of aniline derivatives, bromo-9-ethyl-9*H*-carbazole-3-carbaldehyde or 10-ethyl-10*H*-phenothiazine-3-carbaldehyde and diethyl phosphite using polyethylene glycol (PEG) as a green reaction media (Figure 6.18) [72, 73].

An MW-assisted catalyst- and solvent-free Kabachnik–Fields reaction of amines, aldehydes and dimethyl phosphite was also described (Figure 6.19) [74]. The condensations were carried out in a multimode MW reactor at 80 °C, and the corresponding α-aminophosphonates were obtained in yields of 40–98%.

An eco-friendly accomplishment for the synthesis of various α-aminophosphonates and α-aminophosphine oxides was developed by the Keglevich group (Figure 6.20) [75].

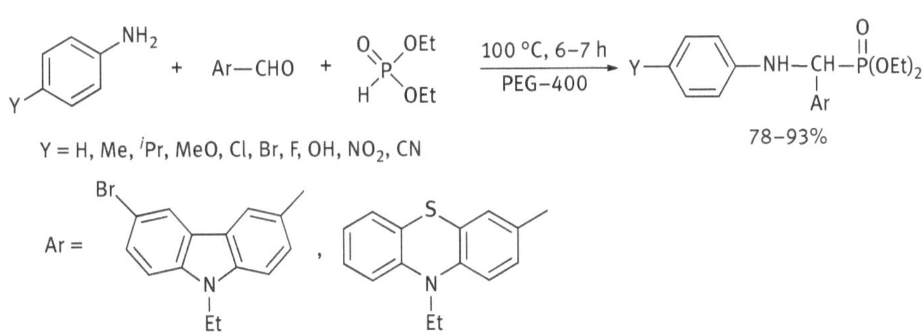

Figure 6.18: Synthesis of carbazole- and phenothiazine-based α-aminophosphonates.

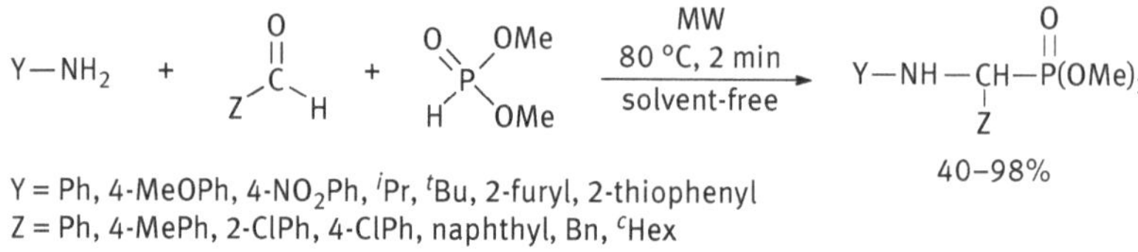

Figure 6.19: MW-assisted catalyst- and solvent-free Kabachnik–Fields reactions.

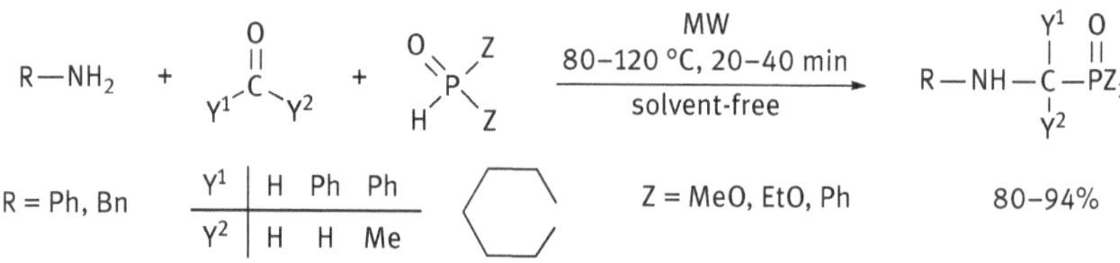

Figure 6.20: MW-assisted synthesis of α-aminophosphonates in the absence of catalyst and solvent.

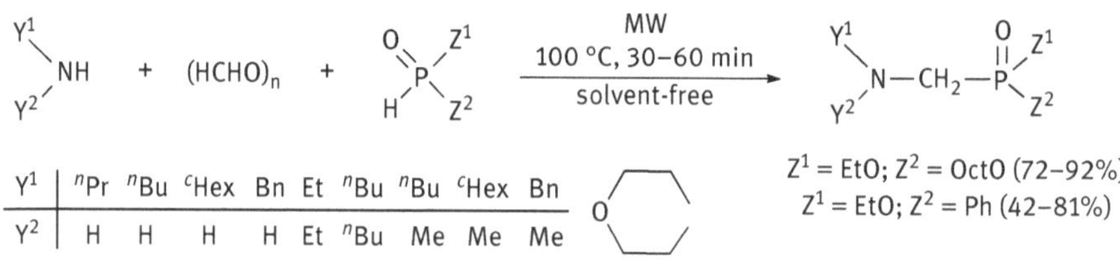

Figure 6.21: Green accomplishment of Kabachnik–Fields reactions with ethyl octyl phosphite and alkyl phenyl-*H*-phosphinates.

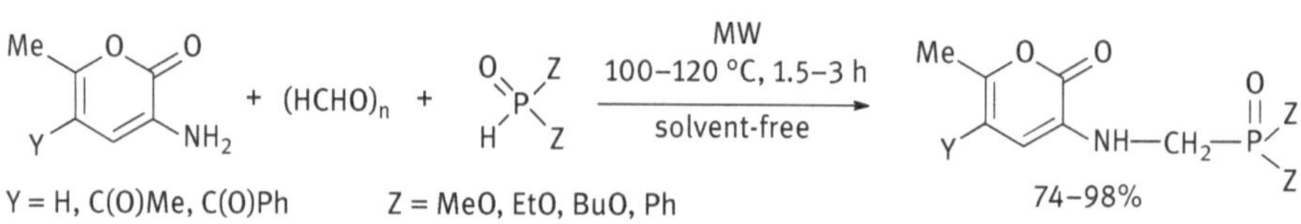

Figure 6.22: Kabachnik–Fields reaction of 3-amino-6-methyl-2*H*-pyran-2-ones under green conditions.

The MW-assisted Kabachnik–Fields reactions of primary or secondary amines, paraformaldehyde and ethyl octyl phosphite or alkyl phenyl-*H*-phosphinates were investigated by our research group (Figure 6.21) [76, 77].

A series of new *N*-(2*H*-pyranonyl)-α-aminophosphonates and α-aminophosphine oxides was obtained in high yields under catalyst-free MW conditions (Figure 6.22) [78]. When dialkyl phosphites were used, the condensations were carried out in the

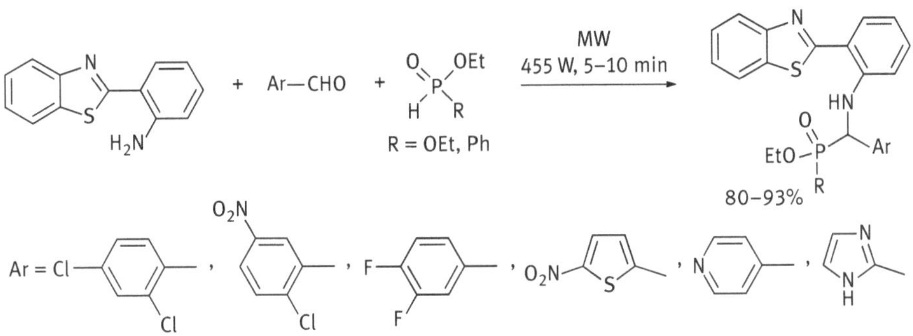

Figure 6.23: Kabachnik–Fields reaction of 2-(2-aminophenyl)benzothiazole.

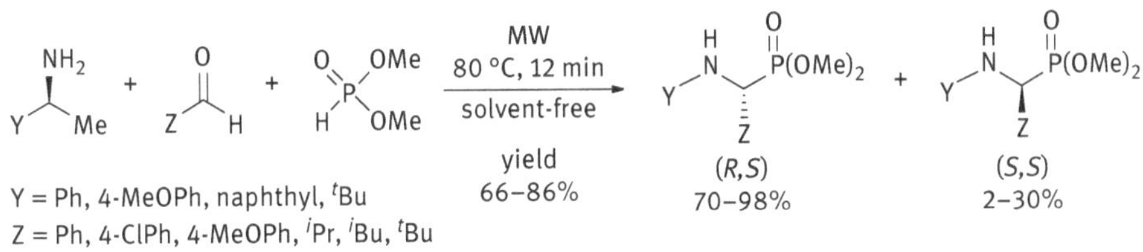

Figure 6.24: Diastereoselective synthesis of α-aminophosphonates.

absence of a solvent. On the other hand, acetonitrile had to be used when diphenyl phosphine oxide was the P component.

The Kabachnik–Fields reaction of 2-(2-aminophenyl)benzothiazole, aromatic or heteroaromatic aldehydes and diethyl phosphite or ethyl phenyl-*H*-phosphinate was performed in a household MW oven without any catalyst and solvent (Figure 6.23) [79].

Ordónez and co-workers reported a catalyst- and solvent-free MW-assisted, highly diastereoselective synthesis of α-aminophosphonates by the condensation of chiral amines, alkyl or aryl aldehydes and dimethyl phosphite (Figure 6.24) [80].

(*S*)-α-Phenylethylamine was applied as a chiral building block in MW-assisted Kabachnik–Fields condensations with paraformaldehyde and various >P(O)H reagents affording the formation of optically active α-aminophosphonate derivatives (Figure 6.25) [81].

A convenient approach was elaborated for the synthesis of novel heterocyclic α-aminophosphonates starting from primary amines, 2-hydroxybenzaldehyes or 2-hydroxyacetophenones and dialkyl phosphites (Figure 6.26) [82].

A wide range of α-ureidophosphonates was synthesized by the catalyst-free condensation of urea, benzaldehyde derivatives and diethyl phosphite in toluene (Figure 6.27) [83].

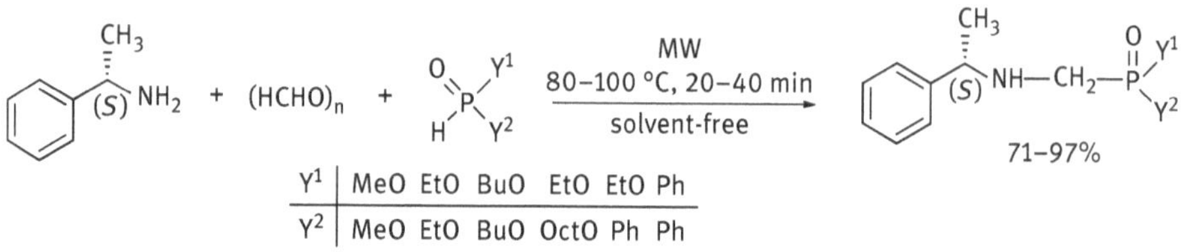

Figure 6.25: Synthesis of optically active α-aminophosphonate derivatives.

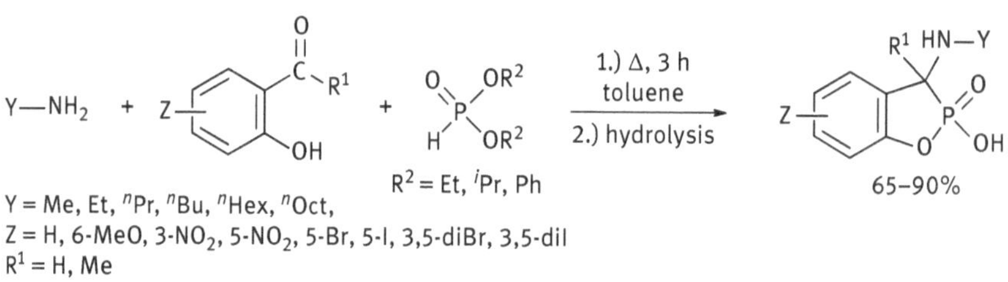

Figure 6.26: Synthesis of a dihydro-oxaphosphole oxide by the Kabachnik–Fields condensation followed by subsequent transformation.

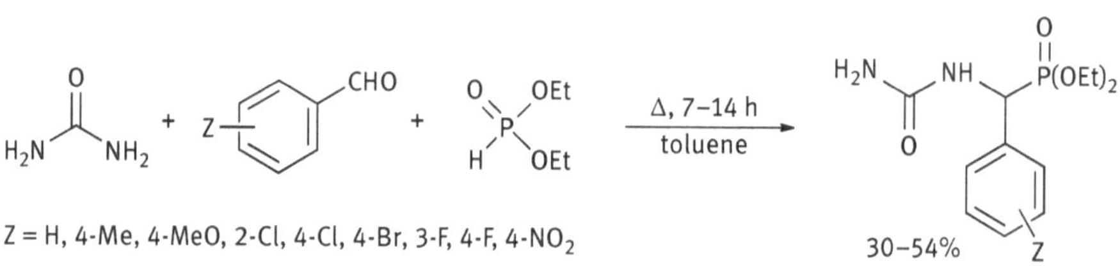

Figure 6.27: Synthesis of α-ureidophosphonates.

6.3 Double Kabachnik–Fields Reactions

In the double Kabachnik–Fields (bis(phospha-Mannich)) condensation, a primary amine reacts with 2 equivalents of a carbonyl compound (aldehyde or ketone) and with 2 equivalents of a P-reagent, such as dialkyl phosphite or secondary phosphine oxide (Figure 6.28). This reaction is an elegant synthetic route for the preparation of bis(α-aminophosphonates) and related derivatives.

According to the literature, double Kabachnik–Fields reactions are usually carried out in various solvents, without any catalyst [84–91]. Nowadays, green chemical approaches, such as the catalyst- and solvent-free MW-assisted syntheses, have also became more and more popular [77, 92–96].

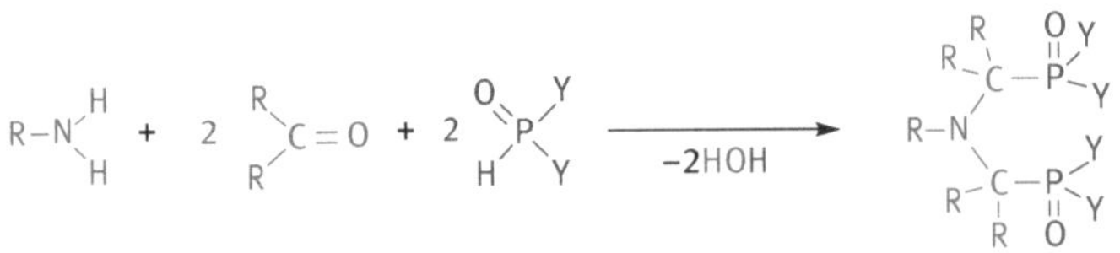

Figure 6.28: General scheme for the double Kabachnik–Fields reaction.

6.3.1 Double Kabachnik–Fields Reactions of Primary Amines

Most of the papers published in the field of double Kabachnik–Fields reaction deal with the condensation of various primary amines, 2 equivalents of formaldehyde or paraformaldehyde and the same amount of dialkyl phosphites or secondary phosphine oxides (Table 6.3). In two examples, acidic or basic catalyst was used (Table 6.3, entries 1 and 2). In one case, the three-component reaction was carried out in the presence of PTSA as the catalyst in toluene (Table 6.3, entry 1) [97]. In another instance, potassium carbonate was applied as the catalyst without any solvent (Table 6.3, entry 2) [98]. The catalyst-free reaction of benzylamine with 2 equivalents of paraformaldehyde and diethyl phosphite was carried out in D_2O (Table 6.3, entry 3) [84]. The catalyst-free double Kabachnik–Fields reactions were also performed using propargyl amine, formaldehyde or paraformaldehyde and dimethyl phosphite (Table 6.3, entry 4) [85, 86].

Table 6.3: Double Kabachnik–Fields reactions in the presence of catalysts and/or solvents.

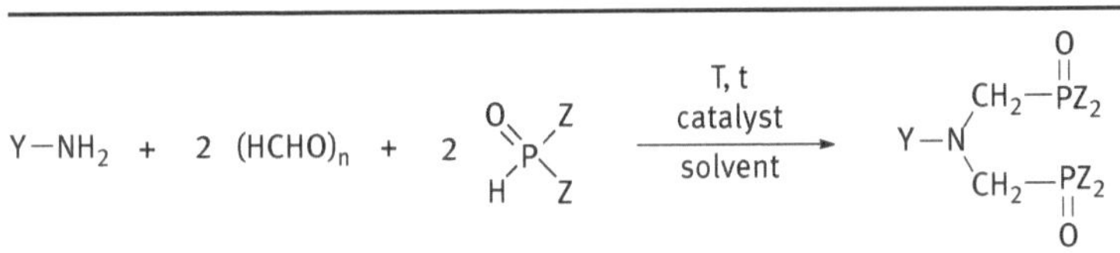

Entry	Y	Z	Catalyst	Solvent	*T, t*	Yield (%)	Ref.
1	Bn	BuO	PTSA (15%)	Toluene	110 °C, 24 h	72	[97]
2	CH_2=$CHCH_2$, ⁿBu, Bn, ᶜHex	EtO	K_2CO_3 (1.4%)	–	100 °C, 4 h	78–84	[98]
3[a]	Bn	EtO	–	D_2O	0 °C → 110 °C, 72 h	57	[84]
4	HC≡CCH_2	MeO	–	THF	25 °C → 70 °C, 12 h	60/76	[85, 86]

[a]Deuterated paraformaldehyde $(DCDO)_n$ was applied.

The three-component reactions of propylamine or 4-aminopyridine with 2 equivalents of both the heterocyclic aldehydes and the diethyl phosphite were elaborated without any catalyst in toluene (Figure 6.29) [87]. The corresponding bis(aminophosphonates) were synthesized in 73–95% yields.

The MW-assisted catalyst- and solvent-free double Kabachnik–Fields condensation of primary amines, paraformaldehyde and various >P(O)H reagents, such as dialkyl phosphites, alkyl phenyl-*H*-phosphinates or oxaphosphorine oxide derivatives, was described by our research group (Figure 6.30) [77, 92, 93].

Amino acid derivatives may also be the starting materials in the Kabachnik–Fields reaction. The condensation of 4-(aminomethyl)benzoic acid, 2 equivalents of paraformaldehyde and the same amount of dimethyl phosphite was carried out in THF, and 4-[[bis[(dimethoxyphosphinyl)-methyl]-amino]-methyl]-benzoic acid was obtained in excellent yield (Figure 6.31) [88].

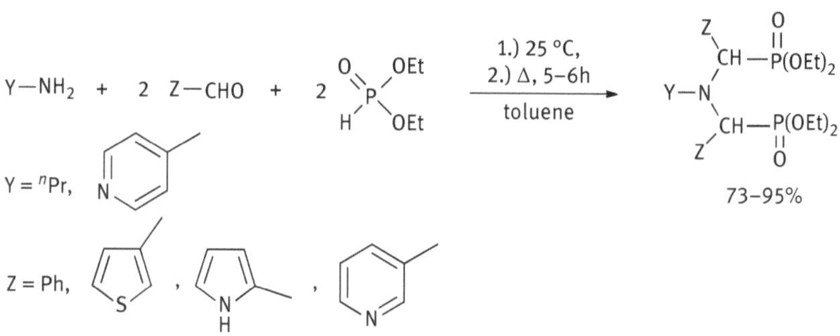

Figure 6.29: Double Kabachnik–Fields reaction of heterocyclic aldehydes.

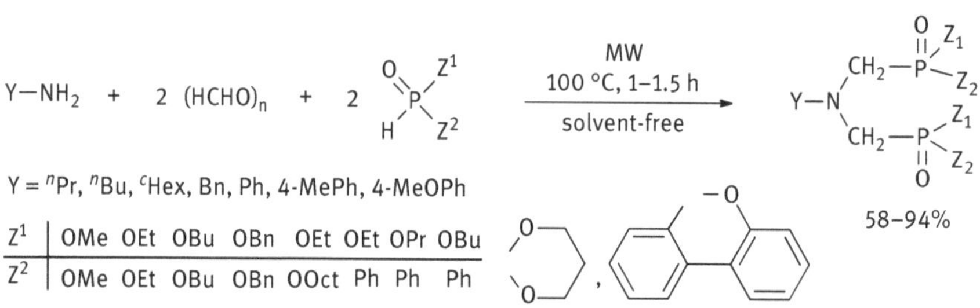

Figure 6.30: Catalyst- and solvent-free double Kabachnik–Fields reaction.

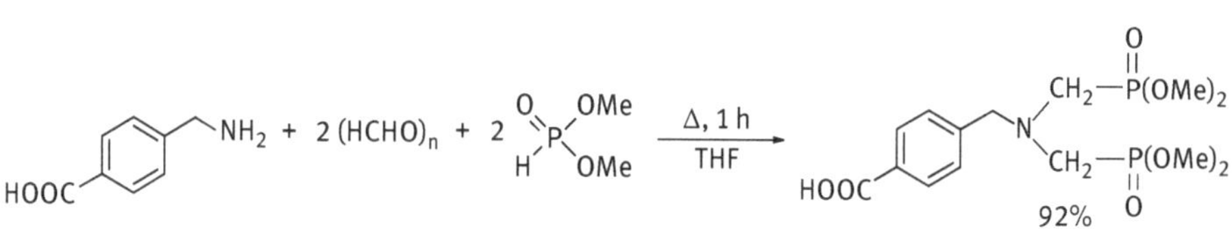

Figure 6.31: Kabachnik–Fields reaction using 4-(aminomethyl)benzoic acid as the starting material.

Table 6.4: Double Kabachnik–Fields reactions of long-chain amino acids.

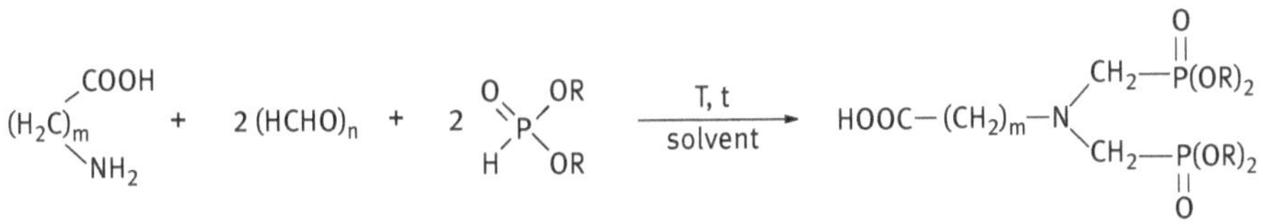

Entry	m	R	Solvent	T, t	Yield (%)	Ref.
1	5, 10	Me	THF	66 °C, 2 h	92, 95	[88]
2	5, 10	Oct	Acetonitrile	82 °C, 3 h	81, 84	[89]

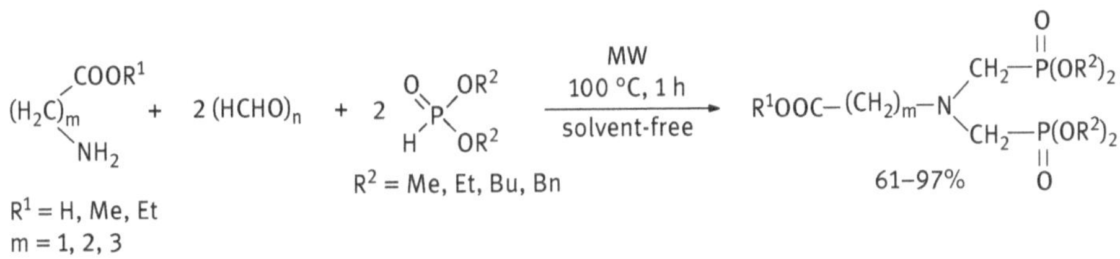

Figure 6.32: Kabchnik–Fields reactions of α-, β- and γ-amino acid derivatives under green conditions.

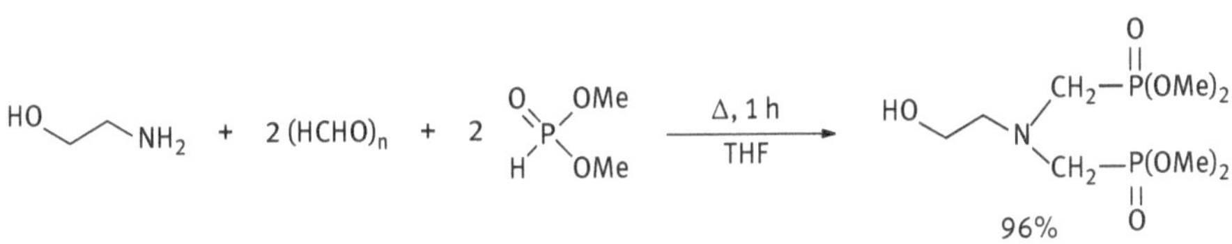

Figure 6.33: Double Kabachnik–Fields reaction of ethanolamine.

The reactions of long-chain amino acids with 2 equivalents of paraformaldehyde and dimethyl or dioctyl phosphite were also described (Table 6.4) [88, 89]. The condensations were performed in THF or in acetonitrile, and the corresponding bis(α-aminophosphonates) were obtained in 81–95% yields.

The catalyst- and solvent-free MW-assisted condensation of α-, β- and γ-amino acid derivatives with 2 equivalents of paraformaldehyde and 2 equivalents of dialkyl phosphites was studied in our group (Figure 6.32) [94, 95].

Ethanolamine was reacted with 2 equivalents of paraformaldehyde and 2 equivalents of dimethyl phosphite in THF, and the corresponding bis(α-aminophosphonate) was obtained in a yield of 96% (Figure 6.33) [88].

The double condensation of amino-terminated polyethylene glycol was accomplished with 2 equivalents of formalin and dimethyl phosphite in THF (Figure 6.34) [90].

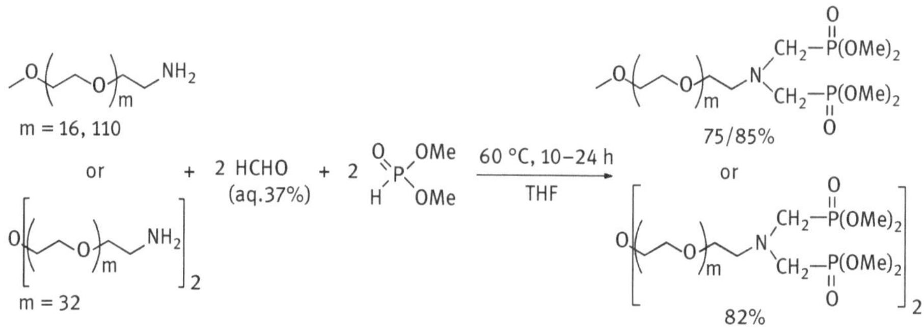

Figure 6.34: Preparation of bis(aminophosphonates) incorporating polyethylene glycol moiety.

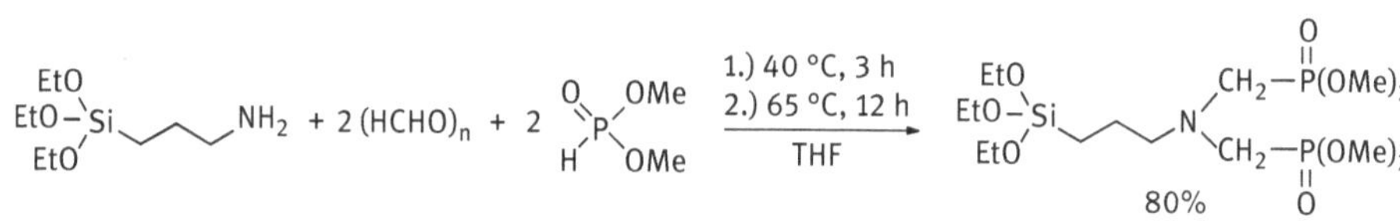

Figure 6.35: Synthesis of 3-(triethoxysilyl)propylazanediyl containing bis(aminophosphonate).

(3-Aminopropyl)triethoxysilane was also applied as the starting material in the double Kabachnik–Fields reaction, and the bis(aminophosphonate) obtained was applied as a novel flame retardant (Figure 6.35) [91].

6.3.2 Double Kabachnik–Fields Reactions of Diamines and Dialdehydes

According to the literature, there are several particular examples for the Kabachnik–Fields condensation of diamines and dialdehydes also affording bis(aminophosphonate) derivatives. Milen and co-workers studied the one-pot reaction of *p*-phenylenediamine, 2 equivalents of aromatic aldehydes and the same amount of diethyl phosphite in the presence of propylphosphonic anhydride (T3P®) yielding diamine bisphosphonate derivatives (Figure 6.36) [99].

The three-component condensation of *p*-phenylenediamine or benzidine, various aldehydes and diethyl phosphite was also performed using a benzimidazolium dicationic ionic liquid as the catalyst (Figure 6.37) [100]. The advantages of this method were the high yields and the reusability of the catalyst.

Cerium(III) chloride heptahydrate supported silica ($CeCl_3 \cdot 7H_2O$-SiO_2) was used as a catalyst for the double Kabachnik–Fields reaction of a 4,4'-sulfonyldianiline, aldehydes and diethyl phosphite (Figure 6.38) [101]. The condensation was carried out under conventional heating and MW irradiation. The latter method proved to be more efficient.

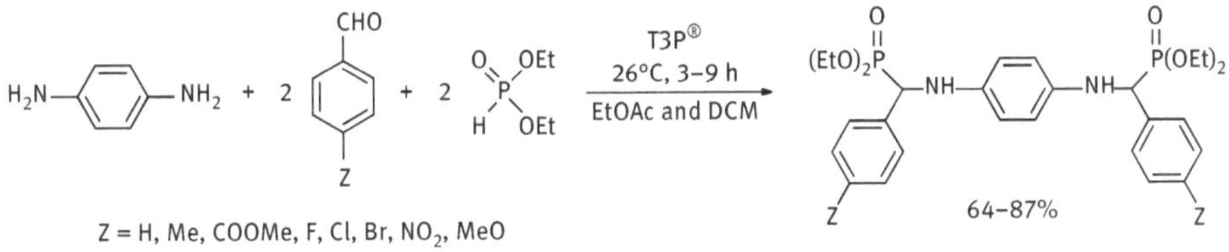

Figure 6.36: T3P®-catalyzed Kabachnik–Fields reaction of *p*-phenylenediamine.

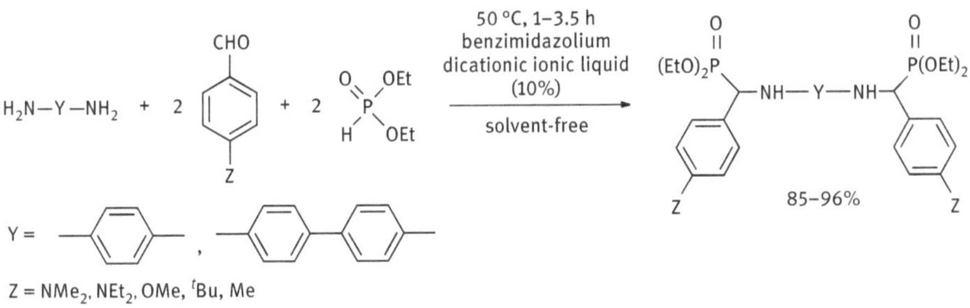

Figure 6.37: Condensation of diamines, aromatic aldehydes and diethyl phosphite.

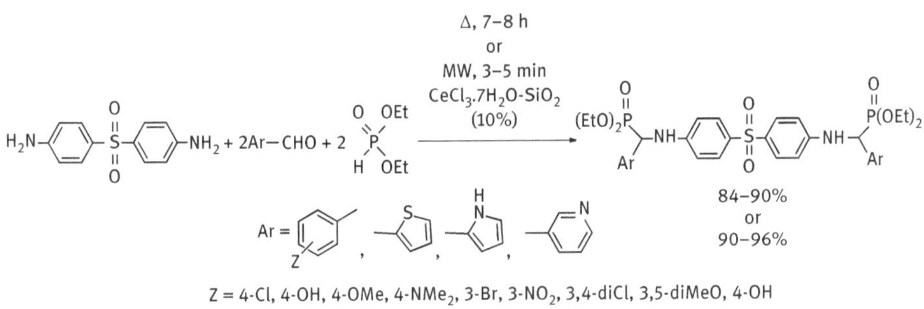

Figure 6.38: Synthesis of α-diaminophosphonates in the presence of $CeCl_3 \cdot 7H_2O$-SiO_2.

Diamines were also reacted with 2 equivalents of 9-anthracenecarboxaldehyde and the same amount of diethyl phosphite in benzene without the use of catalyst (Figure 6.39) [102].

PEG-600 in water, as a green reaction media, was used in the three-component condensation of 4,4′-dioxyaniline, 2 equivalents of aromatic aldehyde and 2 equivalents of diphenyl phosphite (Figure 6.40) [103].

The catalyst- and solvent-free condensation of 2,6-di(aminomethyl)pyridine with formaldehyde and diethyl phosphite was studied at 100 °C. The tetramethylphosphonic octaethyl ester was obtained in a yield of 60% (Figure 6.41) [104].

A solvent- and catalyst-free MW-assisted method was developed for the one-pot reaction of a dialdehyde, 2 equivalents of aromatic amines and 2 equivalents of diethyl- or dibutyl phosphite (Figure 6.42) [105].

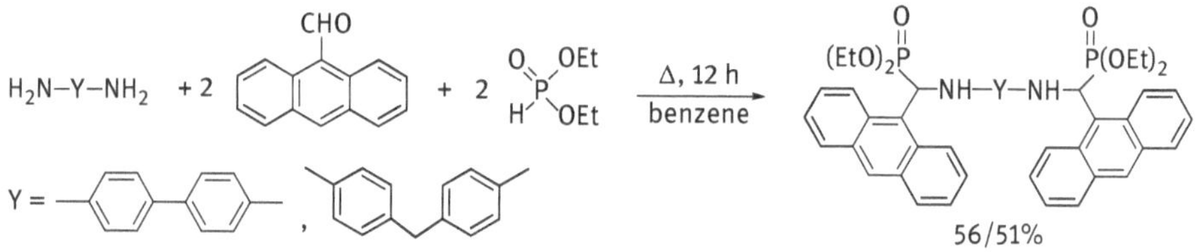

Figure 6.39: Kabachnik–Fields reaction using 9-anthracenecarboxaldehyde.

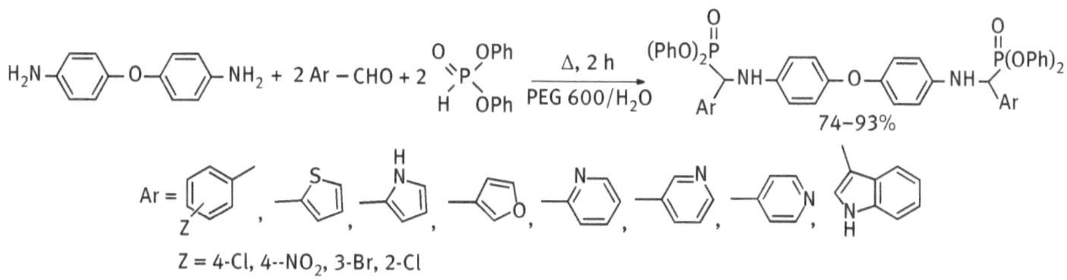

Figure 6.40: Double Kabachnik–Fields reaction applying PEG-600 in water.

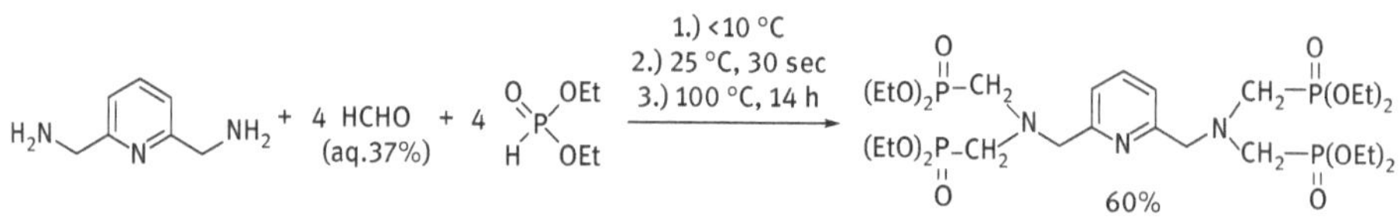

Figure 6.41: Catalyst- and solvent-free synthesis of a tetramethylphosphonic ethyl ester

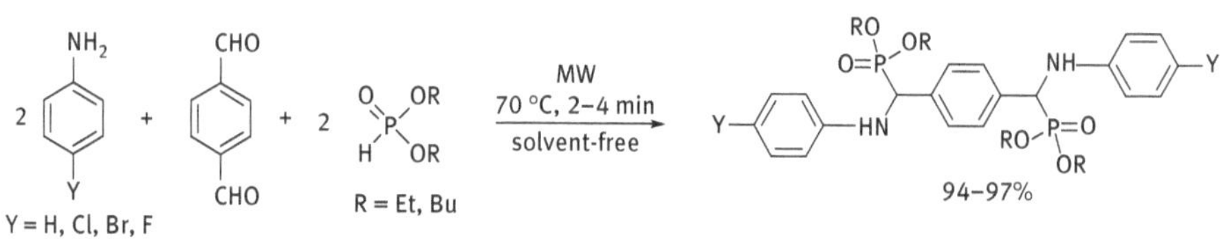

Figure 6.42: Synthesis of bis(α-aminophosphonates) under MW conditions.

6.3.3 An Insight into the Synthesis and Utilization of Bis(Aminophosphine Oxides)

In our research group, the catalyst-free MW-assisted double Kabachnik–Fields reaction of primary amines, 2 equivalents of paraformaldehyde and the same amount of secondary phosphine oxides in acetonitrile was elaborated for the synthesis of bis(aminophosphine oxides) (Figure 6.43) [93, 96].

The bis(phosphinoylmethyl)amines were utilized as precursors of bidentate phosphine ligands [93, 96]. The bisphosphines obtained after double deoxygenation

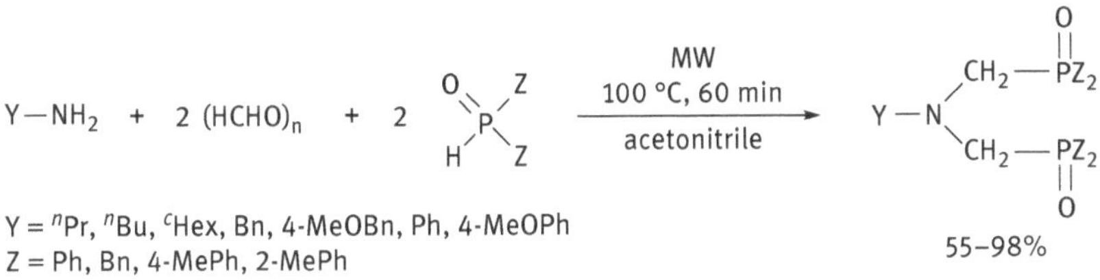

Figure 6.43: The MW-assisted synthesis of bis(aminophosphine oxides).

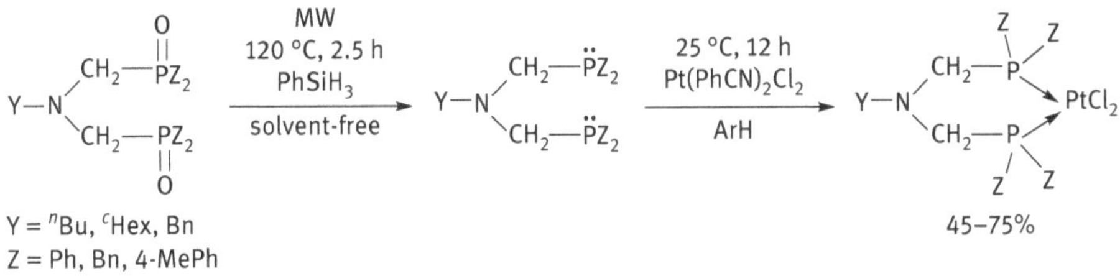

Figure 6.44: Utilization of bis(phosphinoylmethyl)amines as precursors of bidentate P-ligands.

were converted to cyclic platinum complexes, which were tested as catalysts in the hydroformylation of styrene (Figure 6.44).

6.4 The Pudovik Reaction

The Pudovik (also known as aza-Pudovik) reaction of imines and >P(O)H reagents (such as dialkyl phosphites, alkyl phenyl-*H*-phosphinates or secondary phosphine oxides) represents a common synthetic route towards α-aminophosphonates, α-aminophosphinates and α-aminophosphine oxides [106]. In most cases, α-(aryl or hetaryl)-α-aminophosphonate derivatives were prepared by this reaction, starting from α-aryl or α-hetaryl imines, while aliphatic Schiff bases are used rather rarely [107, 108]. Asymmetric syntheses of α-aminophosphonate derivatives were also described applying chiral starting materials or chiral catalysts. The most recent developments of the Pudovik reaction are summarized in the following section.

6.4.1 Pudovik Reactions of α-aryl Imines

6.4.1.1 Pudovik Reactions in the Presence of Catalysts and Solvents

α-Aminophosphonates bearing an *N*-indazole moiety were prepared by the addition of diethyl phosphite on the C=N unit of *N*-indazole imines. The reactions were performed in ethanol with tetramethyl guanidine (TMG) as the catalyst (Figure 6.45) [109].

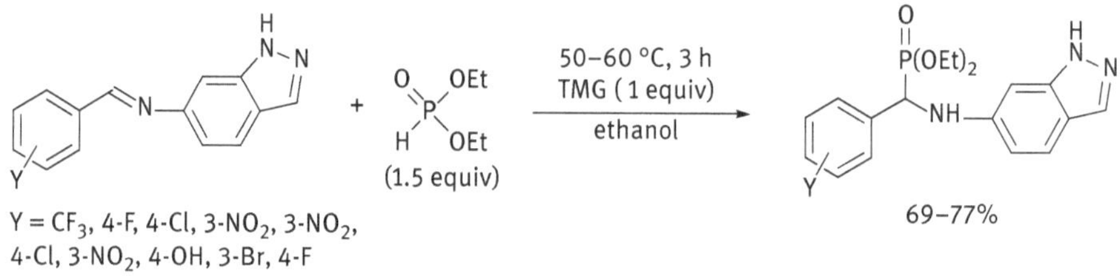

Figure 6.45: Addition of diethyl phosphite to *N*-indazole imines catalyzed by TMG.

1,8-Diazabicyclo[5.4.0]undec-7-ene (DBU) was also checked as a catalyst in Pudovik reactions. The products were obtained in yields of 75–88%. The use of a large excess (4 equivalents) of both diethyl phosphite and DBU were necessary for the reaction completion at 60 °C after 24 h (Figure 6.46) [110].

New α-aminophosphonates were prepared by the reaction of an α-anthryl Schiff base with dimethyl phosphite in toluene applying CdI_2 as the catalyst. The product showed only weak genotoxic and cytotoxic activity in *in vivo* tests (Figure 6.47) [111].

A nucleophilic carbene, such as 1,3-bis(2,6-diisopropylphenyl)imidazol-2-ylidene was also applied as an organocatalyst in the Pudovik reaction of (*N*-benzilidene)butylamine and dimethyl phosphite in THF (Figure 6.48) [112].

A complex with Yb and Li central atoms was synthesized and used as a catalyst in the Pudovik reaction of aromatic imines. After a reaction at 40 °C for 6 h in acetonitrile, the target compounds were obtained in yields of 42–99% (Figure 6.49) [113].

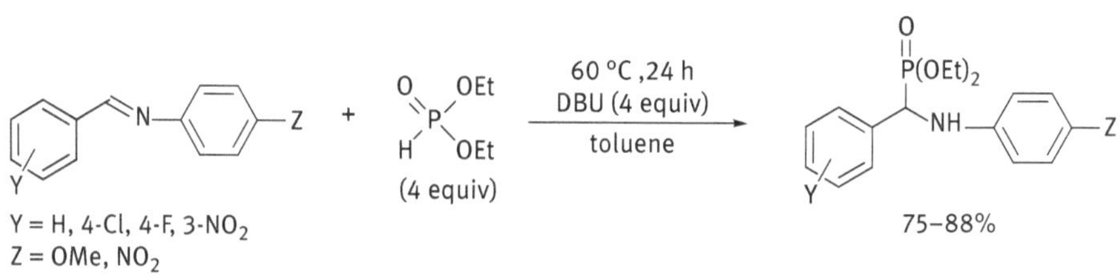

Figure 6.46: DBU-promoted synthesis of α-aminophosphonates.

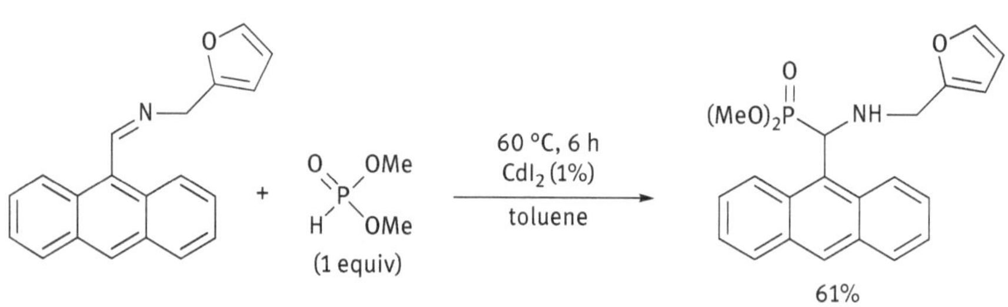

Figure 6.47: Pudovik reaction catalyzed by CdI_2.

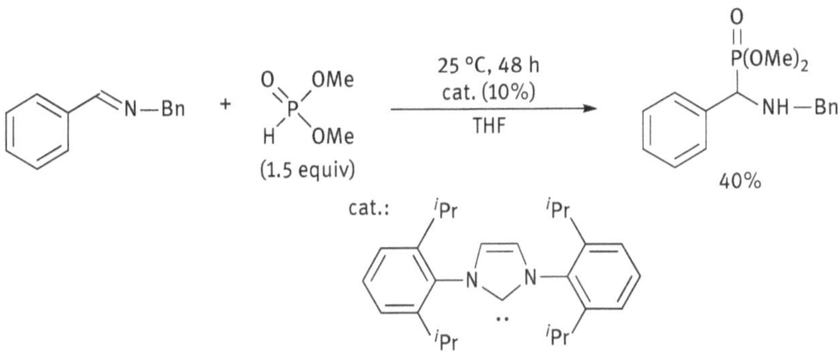

Figure 6.48: A nucleophilic carbene used as an organocatalyst in the Pudovik reaction.

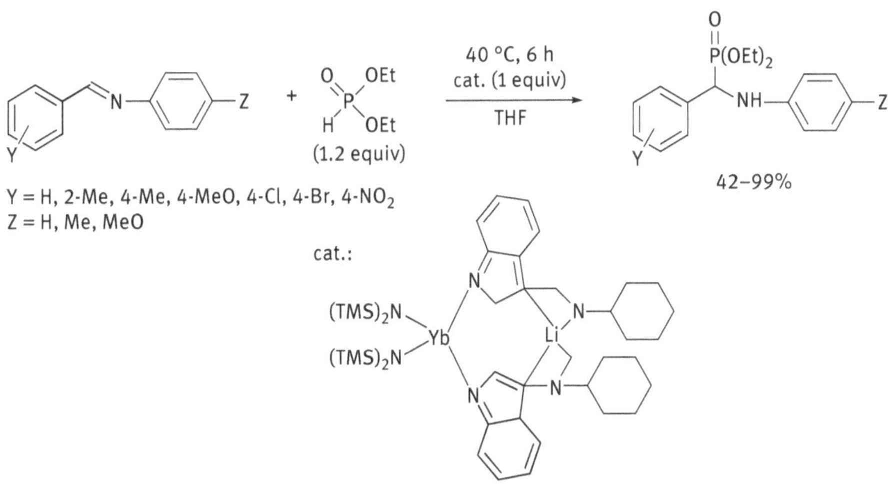

Figure 6.49: Pudovik reaction catalyzed by a Yb- and Li-containing complex.

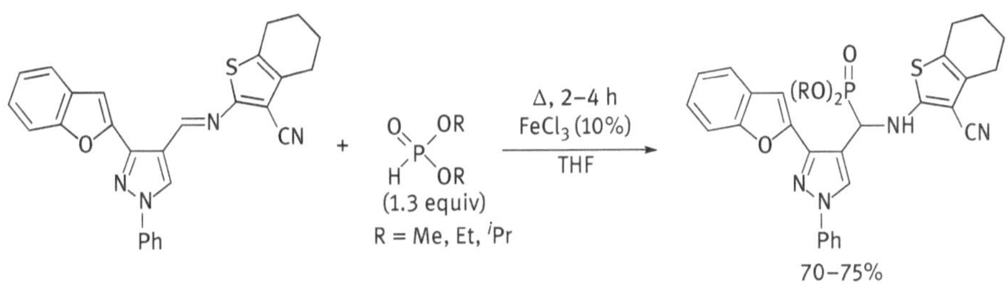

Figure 6.50: Hydrophosphonylation of a pyrazole-containing imine catalyzed by $FeCl_3$.

A pyrazole-containing Schiff base was converted to the corresponding α-aminophosphonates by hydrophosphonylations using dialkyl phosphites and $FeCl_3$ catalyst in THF. The target compounds were obtained in yields of 70–75% (Figure 6.50) [114].

TMG was a useful catalyst in the preparation of α-aminophosphonates with α-thiophene and *N*-carbazole moieties. The addition of dialkyl phosphites led to the products in yields of 78–86% (Figure 6.51) [115].

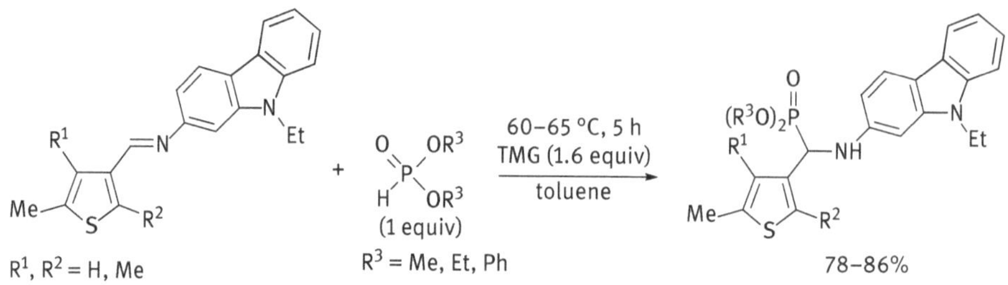

Figure 6.51: TMG-mediated Pudovik reaction of imines with α-thiophene and *N*-carbazole moieties.

6.4.1.2 Pudovik Reactions in the Presence of Catalysts under Solvent-free Conditions

The addition of dialkyl phosphites to the C=N double bond of imines derived from amlodipine was performed in the presence of $SnCl_2{\cdot}2H_2O$ as the catalyst. The ultrasound-mediated procedure was compared with a control experiment carried out using conventional heating. In the case of sonication, the reaction times of 4–6 h could be shortened to 20–24 min. The products showed antibacterial, antiviral and antifungal activities (Figure 6.52) [116].

MoO_2Cl_2 was also used as an efficient catalyst in the Pudovik reactions of aromatic Schiff bases. After 5–60 min of reaction times at 80 °C, the α-aminophosphonates were obtained in yields of 80–95% (Figure 6.53) [117].

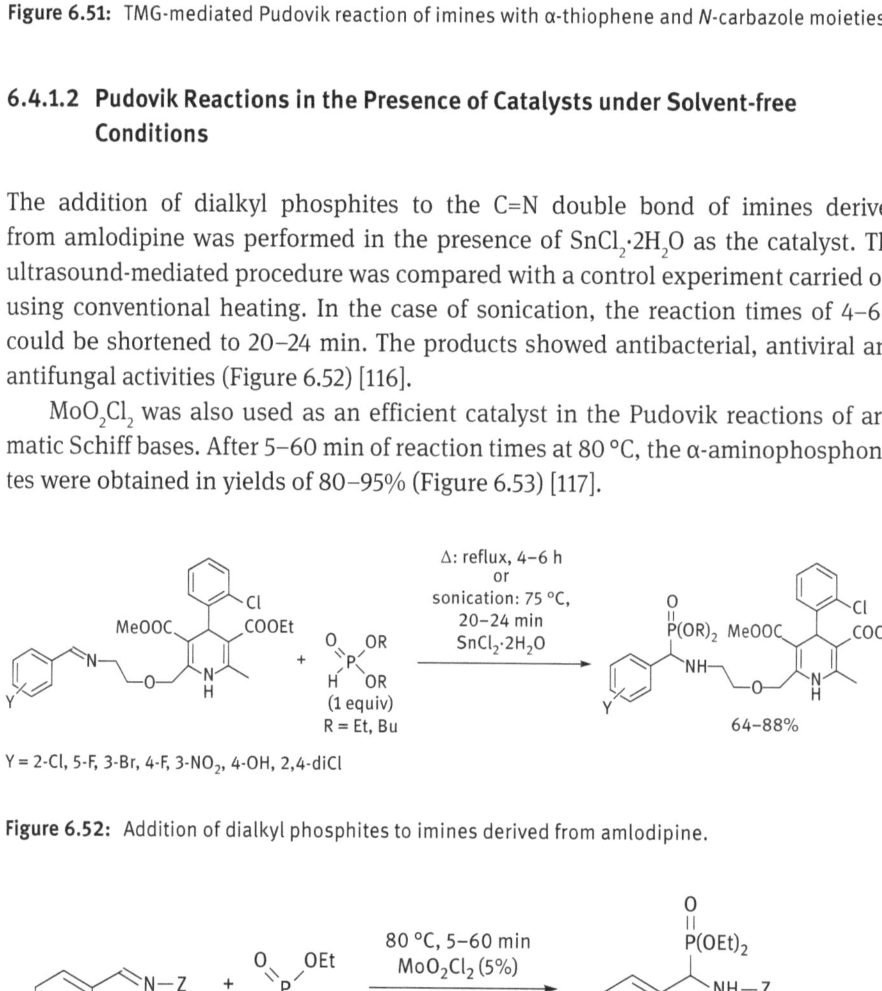

Figure 6.52: Addition of dialkyl phosphites to imines derived from amlodipine.

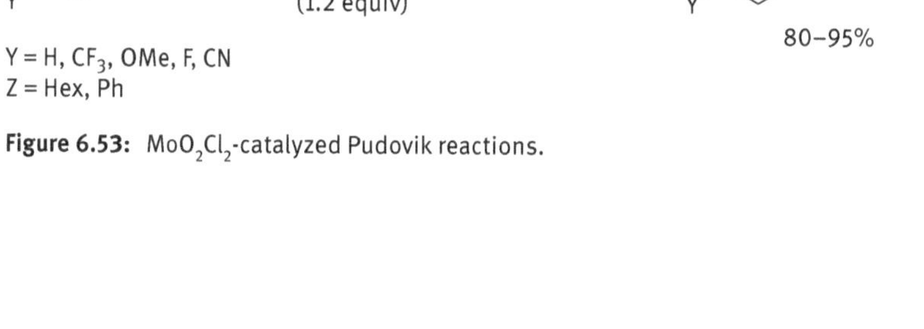

Figure 6.53: MoO_2Cl_2-catalyzed Pudovik reactions.

6.4.1.3 Catalyst-free Pudovik Reactions in Solvents

A number of α-aryl and *N*-aryl aminophosphonates were prepared by the reaction of the corresponding imines with diethyl phosphite in ethanol. Despite the rather long (12–24 h) reaction times, only moderate yields (30–64%) could be reached (Figure 6.54) [118, 119].

A fluorescent aminophosphonate library was synthesized by the Pudovik reaction. The Schiff bases incorporating potentially fluorescent groups were reacted with dibenzyl and diphenyl phosphites in THF (Figure 6.55) [120].

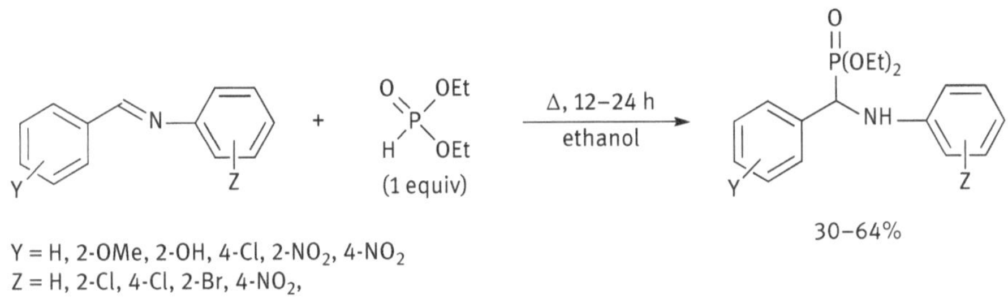

Figure 6.54: Pudovik reaction of aromatic imines with diethyl phosphite.

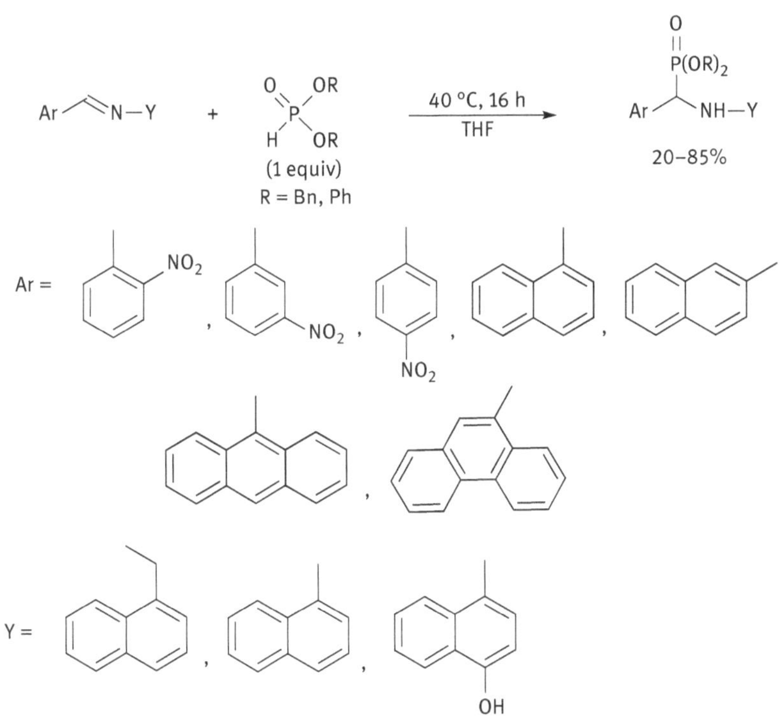

Figure 6.55: Synthesis of aminophosphonates with fluorescent properties.

Imines derived from anthracene carbaldehyde were reacted with dialkyl phosphites in refluxing diethyl ether or benzene. The α-anthryl-aminophosphonates were obtained in yields of 66–74% (Figure 6.56) [121].

Various P-reagents (diethyl phosphite, ethyl phenyl-*H*-phosphinate and diphenylphosphine oxide) were added to imines incorporating an α-imidazole ring in boiling toluene (Figure 6.57) [122].

Schiff bases, prepared from quinoline derivatives, also served as starting materials in Pudovik reactions. Completion of the addition of P-species on the C=N bond required 2 h in boiling toluene (Figure 6.58) [123].

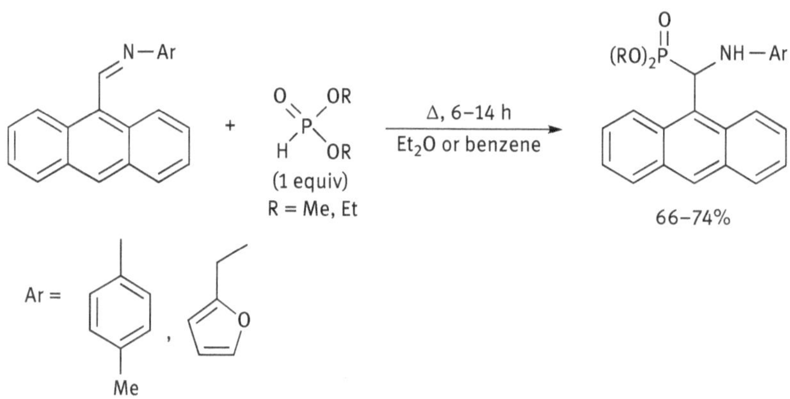

Figure 6.56: Pudovik reaction of anthryl-containing imines.

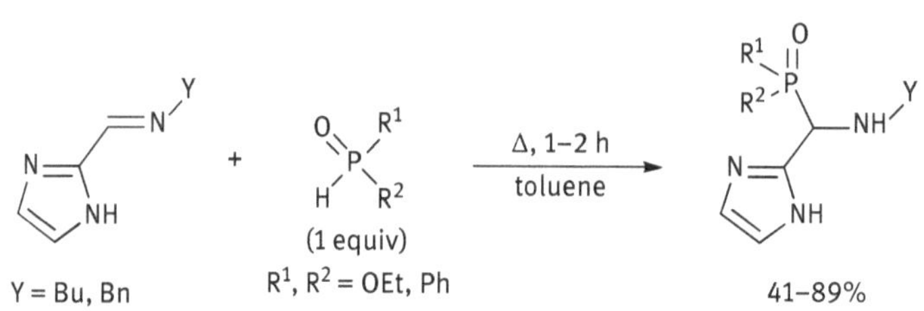

Figure 6.57: Addition of >P(O)H reagents to imines with an α-imidazole ring.

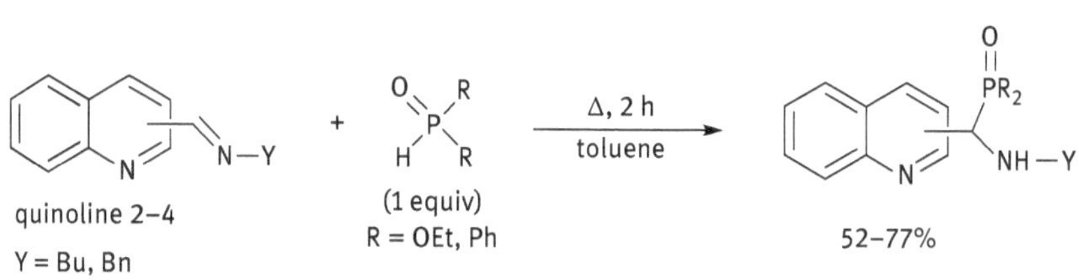

Figure 6.58: Preparation of quinoline-containing α-aminophosphonate derivatives.

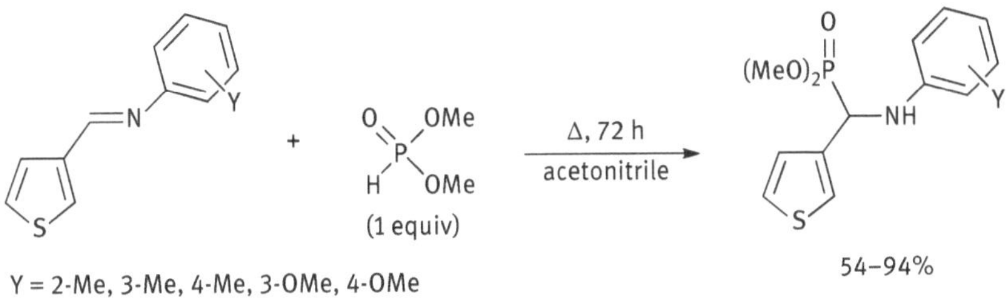

Figure 6.59: Pudovik reaction of imines with an α-thiophene ring.

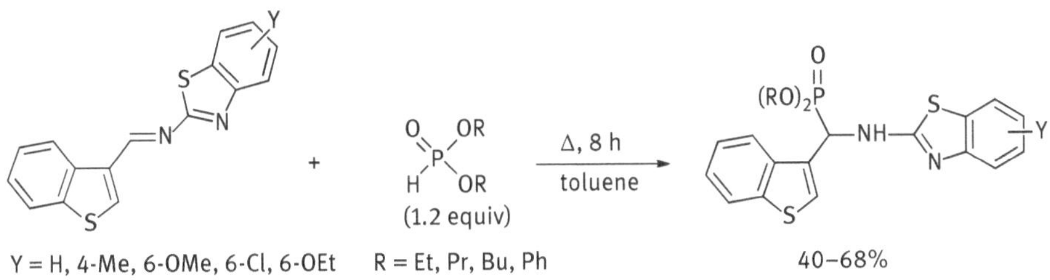

Figure 6.60: Imines with a benzothiazole ring as starting materials in the Pudovik reaction.

α-Thienyl-aminophosphonates were obtained by the Pudovik reaction of imines with α-thiophene ring and dimethyl phosphite. The components were refluxed in acetonitrile for a rather long (72 h) reaction time (Figure 6.59) [124].

Imines bearing a benzothiazole moiety were reacted with dialkyl phosphites or diphenyl phosphite in toluene at reflux temperature for 8 h (Figure 6.60) [125].

6.4.1.4 Catalyst- and Solvent-free Pudovik Reactions

The addition of >P(O)H reagents to α-aryl imines took place under MW-assisted solvent- and catalyst-free conditions at 80–100 °C to afford α-aminophosphonates and α-aminophosphine oxides in yields of 68–97% (Figure 6.61) [126].

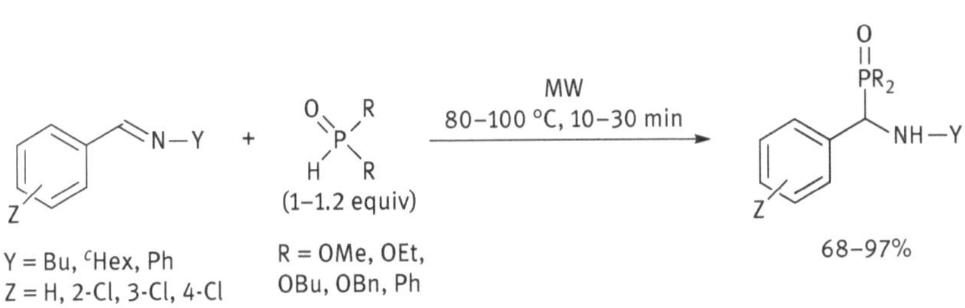

Figure 6.61: MW-assisted solvent- and catalyst-free Pudovik reaction of α-aryl imines.

Diethyl phosphite was added to a 2-hydroxyphenyl Schiff base; the reaction was carried out at 110 °C for 1 h. The corresponding 2-hydroxyphenyl α-aminophosphonate was reacted further without purification (Figure 6.62) [127].

In the synthesis of aminophosphonates containing an *N*-pyrazole moiety, unusually large excess (130 equivalents) of diethyl phosphite was applied. In fact, this method cannot be considered as a green procedure. The products showed satisfactory insecticidal activities in bioassay studies (Figure 6.63) [128].

An imine derived from furan-3-carbaldehyde and propargyl amine was converted to an aminophosphonate by a hydrophosphonylation reaction with dimethyl phosphite. The catalyst- and solvent-free addition was complete at 60 °C after 8 h (Figure 6.64) [129].

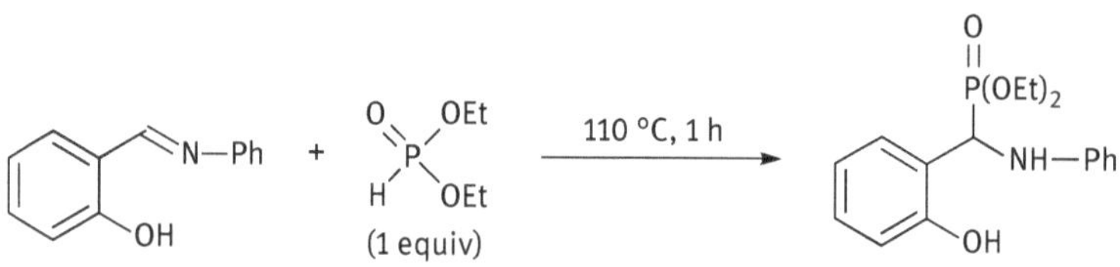

Figure 6.62: Synthesis of a 2-hydroxyphenyl α-aminophosphonate without catalyst and solvent.

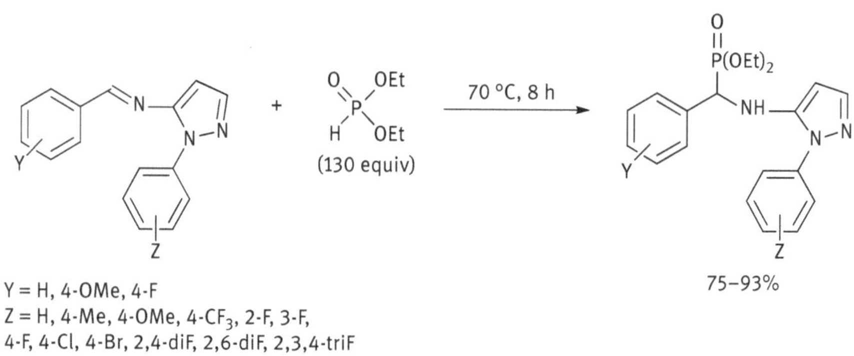

Figure 6.63: Preparation of *N*-pyrazol-containing α-aminophosphonates.

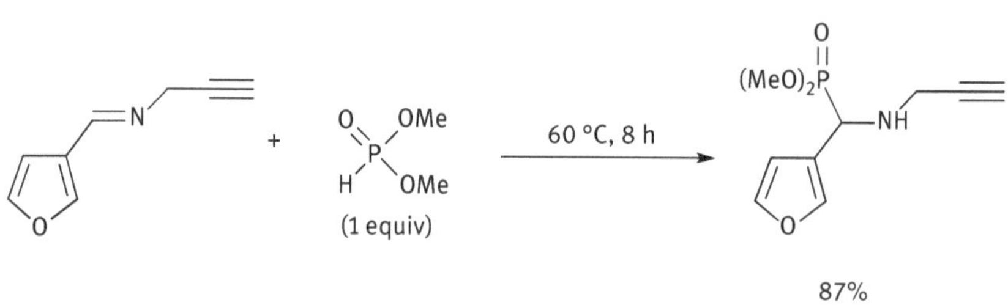

Figure 6.64: Pudovik reaction of an α-furyl *N*-propargyl imine.

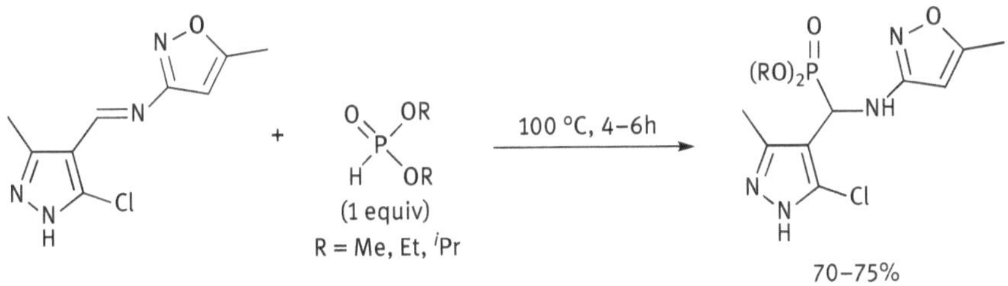

Figure 6.65: Addition of dialkyl phosphites to a Schiff base containing α-pyrazol and *N*-isoxazole moieties.

An α-pyrazol and *N*-isoxazole containing Schiff base was reacted with dialkyl phosphites at 100 °C for 4–6 h. The aminophosphonates were obtained in yields of 70–75% (Figure 6.65) [130].

6.4.2 Pudovik Reactions of α-aliphatic Imines

α-Fluoroalkyl-aminophosphonates were also synthesized *via* a Pudovik reaction. Schiff bases derived from fluoroalkyl aldehydes were reacted with diethyl phosphite in the presence of $BF_3 \cdot Et_2O$ in dichloromethane at room temperature (Figure 6.66) [131].

Pudovik reactions of α-aliphatic imines can also be catalyzed by MoO_2Cl_2. The reaction with diethyl phosphite was performed at 80 °C for 5–30 min. The desired α-aminophosphonates were obtained in yields of 95% (Figure 6.67) [117].

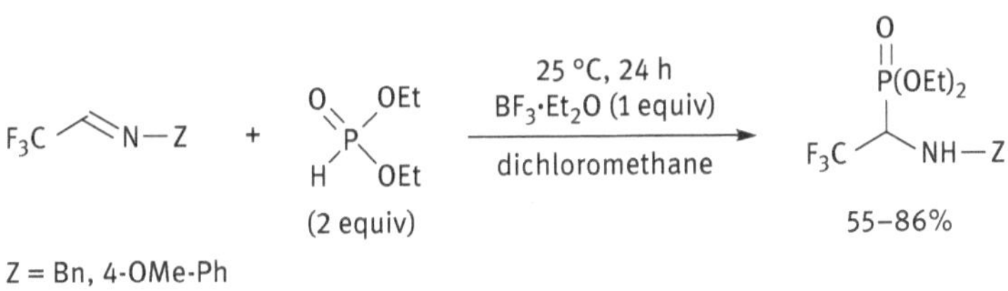

Figure 6.66: Synthesis of fluoroalkyl-aminophosphonates.

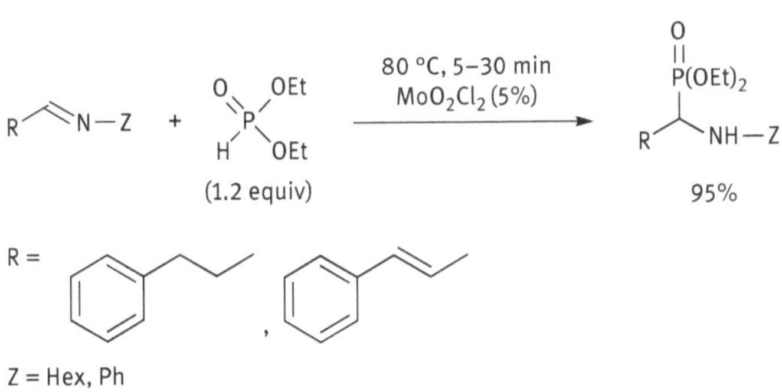

Figure 6.67: Pudovik reaction of α-aliphatic imines catalyzed by MoO_2Cl_2.

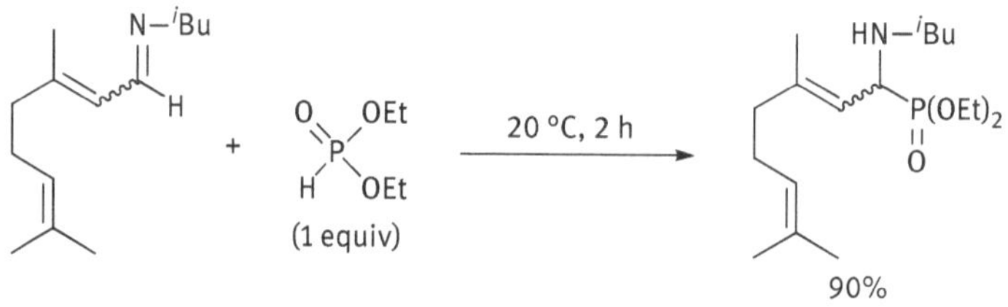

Figure 6.68: Pudovik reaction of a terpene-type imine.

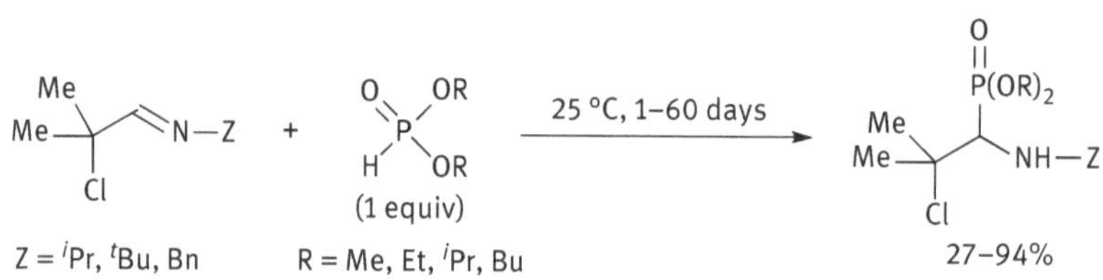

Figure 6.69: The hydrophosphonylation of (*N*-2-methyl-2-chloropropylidene)alkylamines with dialkyl phosphites.

A terpene-type Schiff base was converted to an α-aminophosphonate by reaction with diethyl phosphite at 20 °C for 2 h (Figure 6.68) [132].

The hydrophosphonylation of (*N*-2-methyl-2-chloropropylidene)alkylamines with dialkyl phosphites was performed at room temperature. Depending on the substituents, prolonged (60 days) reaction times were necessary in a few cases (Figure 6.69) [133].

N-Sulfonyl and *N*-carboxylic imine derivatives were also applied as starting materials in the Pudovik reaction to prepare the corresponding α-aminophosphonates [134–137].

6.4.3 Stereoselective Pudovik Reactions

A phosphoric acid with 1,1'-bi-2-naphthol (BINOL) moiety was applied as the catalyst in the asymmetric α,β-hydrophosphonylation of conjugated imines. The reactions were performed at room temperature for 24 h in xylene. The corresponding products were obtained in moderate yields (30–65%) and enantiomer excess values (8–62%; Figure 6.70) [138].

Another example for the asymmetric Pudovik reaction was carried out in the presence of chiral Ti-complexes. The additions needed rather long (20–30 h) reaction times at room temperature in toluene to afford the chiral aminophosphonates in good yields (82–92%) and high enantiomeric purities (92–98%; Figure 6.71) [139].

Chiral quinine derivatives were also utilized as catalysts in the asymmetric hydrophosphonylation of aromatic imines. The desired products were obtained in excellent yields (93–99%) and *ee* values (91–99%) after the reaction at room temperature for 12 h (Figure 6.72) [140].

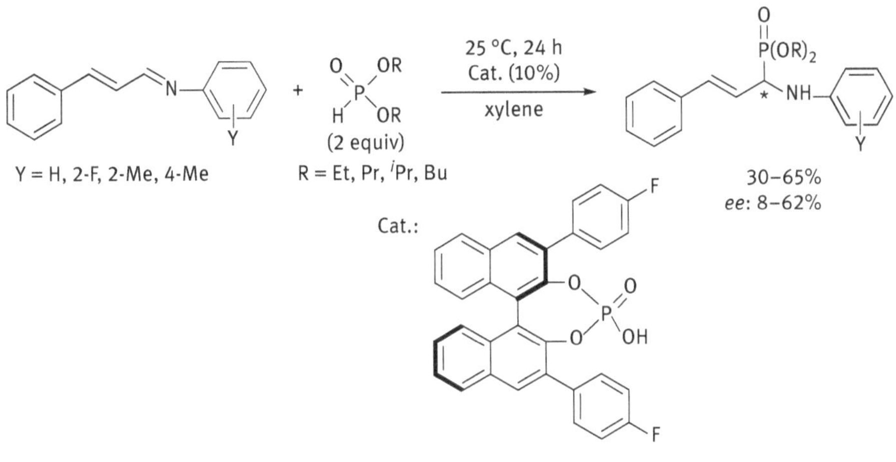

Figure 6.70: Asymmetric addition of dialkyl phosphites catalyzed by a chiral phosphoric acid derivative.

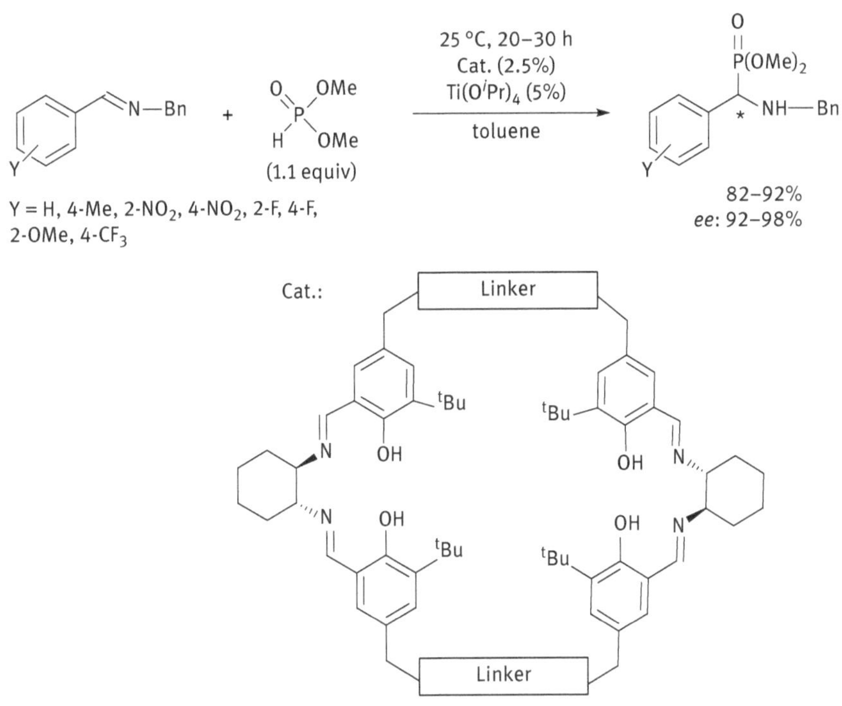

Figure 6.71: A Pudovik reaction catalyzed by chiral Ti-complexes.

In another case, (*R*)-menthyl-phenyl-*H*-phosphinate was added to the double bond of *O*-pivaloylated-D-galactosyl imines. In the presence of $BF_3{\cdot}Et_2O$ in THF at room temperature, the addition afforded the desired products in yields of 45–88% and with *de* values of 75–90% (Figure 6.73) [141].

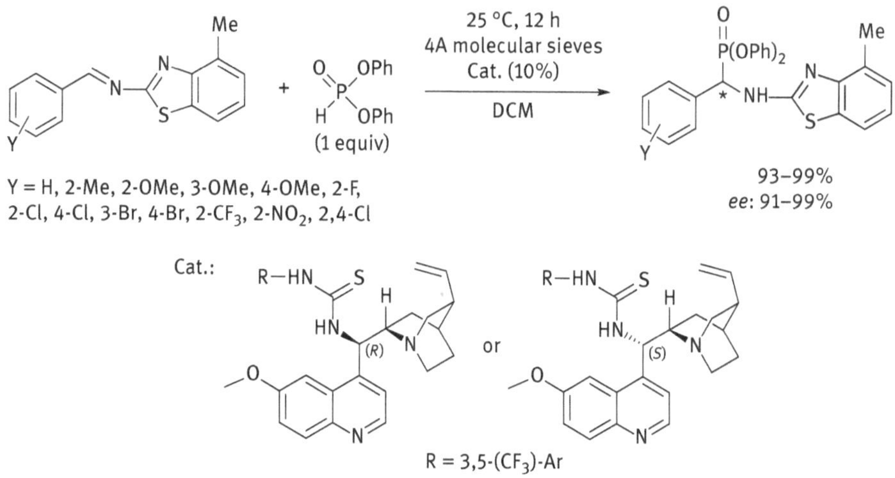

Figure 6.72: Asymmetric hydrophosphonylation of aromatic imines mediated by chiral quinine derivatives.

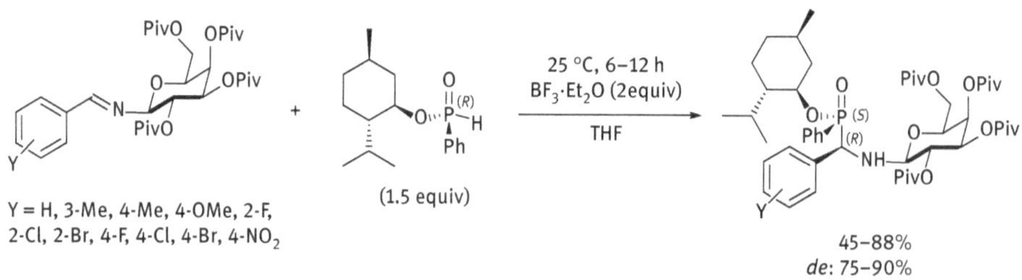

Figure 6.73: Diastereoselective hydrophosphonylation of *O*-pivaloylated-D-galactosyl imines by a chiral phosphinate.

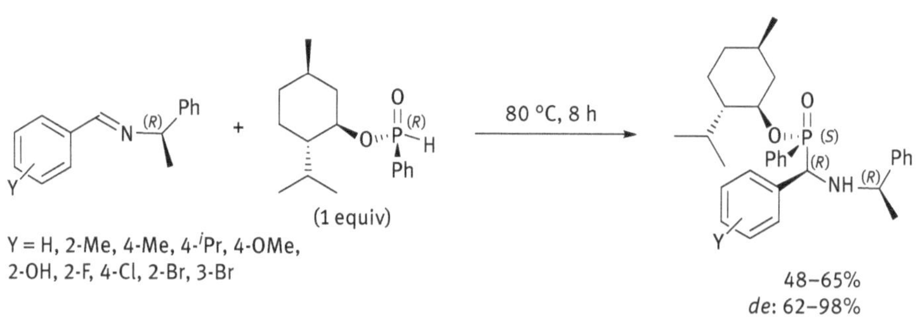

Figure 6.74: Reaction of optically pure chiral aromatic imines with (*R*)-menthyl-phenyl-*H*-phosphinate.

The reaction of the same chiral phosphinate with chiral aromatic imines led to one predominant diastereomer. After a reaction at 80 °C for 8 h, the products were obtained in yields of 48–65%. The diastereomer excess was 62–98% (Figure 6.74) [142].

6.5 Conclusion

In this chapter, the synthesis of α-aminophosphonates and related derivatives by the single and the double Kabachnik–Fields condensations, as well as by the Pudovik reaction was summarized. This field of organophosphorus chemistry has attracted much attention over the last five years. Although there is a wide variety of catalytic methods still being applied, the green accomplishments, such as the catalyst- and solvent-free reactions and the MW-assisted synthesis, come to the forefront of contemporary research.

Acknowledgment: This work was supported by the Hungarian Scientific Research Fund (PD111895) and in part (E. B.) by the János Bolyai Research Scholarship of the Hungarian Academy of Sciences. Ádám Tajti thanks Gedeon Richter Talentum Foundation and Pro Progressio Foundation for financial support.

References

[1] V. P. Kukhar, H. R. Hudson, *Aminophosphonic and Aminophosphinic Acids: Chemistry and Biological Activity*; Wiley: Chichester, 2000.
[2] A. Mucha, P. Kafarski, L. Berlicki, *J. Med. Chem.*, 2011, **54**, 5955.
[3] J. G. Allen, F. G. Atherton, M. J. Hall, C. H. Hassall, S. W. Holmes, R. W. Lambert, L. J. Nisbet, P. S. Ringrose, *Nature*, 1978, **272**, 56.
[4] G. Lavielle, P. Hautefaye, C. Schaeffer, J. A. Boutin, C. A. Cudennec, A. Pierre, *Med. Chem.*, 1991, **34**, 1998.
[5] J. Grembecka, A. Mucha, T. Cierpicki, P. Kafarski, *J. Med. Chem.*, 2003, **46**, 2641.
[6] M. Sienczyk, J. Oleksyszyn, *Curr. Med. Chem.*, 2009, **16**, 1673.
[7] G. Forlani, L. Berlicki, M. Duo, G. Dziedziola, S. Giberti, M. Bertazzini, P. Kafarski, *J. Agric. Food Chem.*, 2013, **61**, 6792.
[8] G. Forlani, A. Occhipinti, L. Berlicki, G. Dziedziola, A. Wieczorek, P. Kafarski, *J. Agric. Food Chem.*, 2008, **56**, 3193.
[9] N. Long, X. J. Cai, B. A. Song, S. Yang, Z. Chen, P. S. Bhadury, D. Y. Hu, L. H. Jin, W. Xue, *J. Agric. Food Chem.*, 2008, **56**, 5242.
[10] M. I. Kabachnik, T. Y. Medved, *Dokl. Akad. Nauk SSSR*, 1952, **83**, 689.
[11] E. K. Fields, *J. Am. Chem. Soc.*, 1952, **74**, 1528.
[12] R. A. Cherkasov, V. I. Galkin, *Russ. Chem. Rev.*, 1998, **67**, 857.
[13] G. Keglevich, E. Bálint, *Molecules*, 2012, **17**, 12821.
[14] E. D. Matveeva, N. S. Zefirov, *Dokl. Chem.*, 2008, **420**, 137.
[15] P. Kafarski, M. G. Górniak, I. Andrasiak, *Curr. Green Chem.*, 2015, **5**, 218.
[16] H. Firouzabadi, N. Iranpoor, S. Sobhani, *Synthesis*, 2004, **2004** 2692.
[17] S. Lee, J. H. Park, J. Kang, J. K. Lee, *Chem. Commun.*, 2001, 1698.
[18] S. Lee, J. K. Lee, C. E. Song, D.-C. Kim, *Bull. Korean Chem. Soc.*, 2002, **23**, 667.
[19] P. Sun, Z. Hu, Z. Huang, *Synth. Commun.*, 2004, **34**, 4293.

[20] A. K. Bhattacharya, T. Kaur, *Synlett*, 2007, 745.
[21] S. Bhagat, A. K. Chakraborti, *J. Org. Chem.*, 2007, **72**, 1263.
[22] J. Wu, W. Sun, H.-G. Xia, X. Sun, *Org. Biomol. Chem.*, 2006, **4**, 1663.
[23] F. Xu, Y. Luo, M. Deng, Q. Shen, *Eur. J. Org. Chem.*, 2003, **24**, 4728.
[24] B. C. Ranu, A. Hajra, U. Jana, *Org. Lett.*, 1999, **1**, 1141.
[25] E. D. Matveeva, T. A. Podrugina, E. V. Tishkovskaya, L. G. Tomilova, N. S. Zefirov, *Synlett*, 2003, 2321.
[26] M. Shen, S. Shang, Z. Song, D. Wang, X. Rao, H. Gao, J. Wang, *Synth. Commun.*, 2013, **44**, 361.
[27] H. Wang, T. Deng, C. Cai, *J. Fluor. Chem.*, 2014, **168**, 144.
[28] F. Malamiri, S. Khaksar, *J. Chem. Sci.*, 2014, **126**, 807.
[29] M. Varalakshmi, D. Srinivasulu, D. Rajasekhar, C. Naga Raju, S. Sreevani, *Phosphorus, Sulfur, Silicon*, 2014, **189**, 106.
[30] Y. Zhang, C. Zhu, *Catal. Commun.*, 2012, **28**, 134.
[31] A. R. de Oliveira, R. Katla, M. P. D. Rocha, T. B. Albuquerque, C. D. G. da Silva, V. L. Kupfer, A. W. Rinaldi, N. L. C. Domingues, *Synthesis*, 2016, **48**, 4489.
[32] S. A. R. Mulla, M. Y. Pathan, S. S. Chavan, S. P. Gampleb, D. Sarkarb, *RSC Adv.*, 2014, **4**, 7666.
[33] B. Rajender, G. Ramesh, S. Palaniappan, *Catal. Commun.*, 2014, **43**, 93.
[34] S. Rostamnia, M. Amini, *Chem. Pap.*, 2014, **68**, 834.
[35] Y.-Q. Yu, D.-Z. Xu, *Synthesis*, 2015, **47**, 1869.
[36] C. S. Sundar, D. Srinivasulu, S. K. Nayak, C. S. Reddy, *Phosphorus, Sulfur, Silicon*, 2012, **187**, 523.
[37] M. Skorenski, J. Oleksyszyn, M. Sienczyk, *Tetrahedron Lett.*, 2013, **54**, 1566.
[38] J. Lewkowski, M. R. Moya, A. Wrona-Piotrowicz, J. Zakrzewski, R. Kontek, G. Gajek, *Beilstein J. Org. Chem.*, 2016, **12**, 1229.
[39] P. S. Reddy, P. V. G. Reddy, S. M. Reddy, *Tetrahedron Lett.*, 2014, **55**, 3336.
[40] R. Bukšnaitienė, A. Urbanaitė, I. Čikotienė, *J. Org. Chem.*, 2014, **79**, 6532.
[41] Z.-L. Ren, J. Zhang, H.-D. Li, M.-J. Chu, L.-S. Zhang, X.-K. Yao, Y. Xia, X.-H. Lv, H.-Q. Cao, *Chem. Pharm. Bull.*, 2016, **64**, 1755.
[42] X.-C. Li, S.-S. Gong, D.-Y. Zeng, Y.-H. You, Q. Sun, *Tetrahedron Lett.*, 2016, **57**, 1782.
[43] A. A. Shady, S. M. A. Bakr, M. D. Khidre, J. *Heterocycl. Chem.*, 2017, **54**, 71.
[44] G. Keglevich, E. Bálint, R. Kangyal, M. Bálint, M. Milen, *Heteroat. Chem.*, 2014, **25**, 282.
[45] R. M. Abdel-Rahman, T. E. Ali, *Monatsh. Chem.*, 2013, **144**, 1243.
[46] U. R. S. Arigala, C. Matcha, K. R. Yoon, *Heteroat. Chem.*, 2012, **23**, 160.
[47] S. D. Dindulkar, M. V. Reddy, Y. T. Jeong, *Catal. Commun.*, 2012, **17**, 114.
[48] S. G. Lakoud, M. Merabet-Khelassi, L. Aribi-Zouioueche, *Res. Chem. Intermed.*, 2016, **42**, 4403.
[49] S. U. Tekale, S. S. Kauthale, R. U. Shaikh, R. P. Marathe, R. B. Nawale, R. P. Pawar, *J. Iran. Chem. Soc.*, 2014, **11**, 717.
[50] D. Saberi, S. Cheraghi, S. Mahdudi, J. Akbari, A.Heydari, *Tetrahedron Lett.*, 2013, **48**, 6403.
[51] M. Nazish, S. Saravanan, N. H. Khan, P. Kumari, R. I. Kureshy, S. H. R. Abdi, H. C. Bajaj, *ChemPlusChem.*, 2014, **79**, 1753.
[52] R. R. Chinthaparthi, I. Bhatnagar, C. S. R. Gangireddy, S. C. Syama, S. R. Cirandur, *Arch. Pharm. Chem. Life Sci.*, 2013, **346**, 667.
[53] N. Li, X. Wang, R. Qiu, X. Xu, J. Chen, X. Zhang, S. Chen, S. Yin, *Catal. Commun.*, 2014, **43**, 184.
[54] S. H. Jayaprakash, K. U. M. Rao, B. S. Krishna, S. S. Prasad, C. S. Sundar, C. S. Reddy, *Phosphorus, Sulfur, Silicon*, 2015, **190**, 449.
[55] J. Taran, A. Ramazani, H. Aghahosseini, F. Gouranlou, R. Tarasi, M. Khoobi, S. W. Joo, *Phosphorus, Sulfur, Silicon*, 2017, **192**, 776.
[56] A. K. Bhattacharya, D. S. Raut, K. C. Rana, I. K. Polanki, M. S. Khan, S. Iram, *Eur. J. Med. Chem.*, 2013, **66**, 146.
[57] A. A. Jafari, S. Amini, F. Tamaddon, *J. Iran. Chem. Soc.*, 2013, **10**, 677.

[58] G. D. Tibhe, M. Bedolla-Medrano, C. Cativiela, M. Ordónez, *Synlett*, 2012, **23**, 1931.
[59] M. Bedolla-Medrano, E. Hernández-Fernández, M. Ordónez, *Synlett*, 2014, **25**, 1145.
[60] M. V. Reddy, G. C. S. Reddy, Y. T. Jeong, *RSC Adv.*, 2014, **4**, 24089.
[61] C. S. R. Gangireddy, R. R. Chinthaparthi, V. R. Mudumala, M. Mamilla, U. R. S. Arigala, *Heteroat. Chem.*, 2014, **25**, 147.
[62] S. T. Disale, S. R. Kale, S. S. Kahandal, T. G. Srinivasan, R. V. Jayaram, *Tetrahedron Lett.*, 2012, **53**, 2277.
[63] K. P. Boroujeni, E. R. Shirazi, M. M. Doroodmand, *Phosphorus, Sulfur, Silicon*, 2016, **191**, 683.
[64] H. Eshghi, M. Mirzaei, M. Hasanpour, M. Mokaber-Esfahani, *Phosphorus, Sulfur, Silicon*, 2015, **190**, 1606.
[65] A. K. Bhattacharya, M. Mujahid, *Synth. Commun.*, 2013, **43**, 2583.
[66] M. R. Sivala, S. R. Devineni, M. Golla, V. Medarametla, G. K. Pothuru, N. R. Chamarthi, *J. Chem. Sci.*, 2016, **128**, 1303.
[67] M. Gundluru, S. Sarva, M. K. R. Kandula, V. R. Netala, V. Tartte, S. R. Cirandur, *Heteroat. Chem.*, 2016, **27**, 269.
[68] M. Mirzaei, H. Eshghi, M. Hasanpour, R. Sabbaghzadeh, *Phosphorus, Sulfur, Silicon*, 2016, **191**, 1351.
[69] K. M. K. Reddy, S. Santhisudha, G. Mohan, K. Peddanna, C. A. Rao, C. S. Reddy, *Phosphorus, Sulfur, Silicon*, 2016, **191**, 933.
[70] C. S. Sundar, N. B. Reddy, S. S. Prasad, K. U. M. Rao, S. H. J. Prakash, C. S. Reddy, *Phosphorus, Sulfur, Silicon*, 2014, **189**, 551.
[71] F. Bahrami, F. Panahi, F. Daneshgar, R. Yousefi, M. B. Shahsavanib, A. Khalafi-Nezhad, *RSC Adv.*, 2016, **6**, 5915.
[72] M. A. Kumar, Y.-K. Park, K. D. Lee, *Chem. Pharm. Bull.*, 2012, **60** 1531.
[73] M. A. Kumar, K. D. Lee, *Phosphorus, Sulfur, Silicon*, 2012, **187**, 899.
[74] X.-J. Mu, M.-Y. Lei, J.-P. Zou, W. Zhang, *Tetrahedron Lett.*, 2006, **47**, 1125.
[75] G. Keglevich, A. Szekrényi, *Lett. Org. Chem.*, 2008, **5**, 616.
[76] Á. Tajti, E. Bálint, G. Keglevich, *Curr. Org. Synth.*, 2016, **13**, 638.
[77] E. Bálint, R. E. Tóth, G. Keglevich, *Heteroat. Chem.*, 2016, **27**, 323.
[78] E. Bálint, J. Takács, L. Drahos, A. Juranovič, M. Kočevar, G. Keglevich, *Heteroat. Chem.*, 2013, **24**, 221.
[79] S. T. Basha, H. Sudhamani, S. Rasheed, N. Venkateswarlu, T. Vijaya, C. N. Raju, *Phosphorus, Sulfur, Silicon*, 2016, **191**, 1339.
[80] G. D. Tibhe, M. A. Reyes-González, C. Cativiela, M. Ordóñez, *J. Mex. Chem. Soc.*, 2012, **56**, 183.
[81] E. Bálint, Á. Tajti, D. Kalocsai, B. Mátravölgyi, K. Karaghiosoff, M. Czugler, G. Keglevich, *Tetrahedron*, 2017, **73**, 5659.
[82] Z. Qu, X. Chen, J. Yuan, Y. Bai, T. Chen, L. Qu, F. Wang, X. Li, Y. Zhao, *Tetrahedron*, 2012, **68**, 3156.
[83] B. Kaboudin, M. B. Afsharinezhad, T. Yokomatsu, *Arkivoc*, 2012, **(iv)**, 70.
[84] N. Souri, P. Tian, C. Platas-Iglesias, K. L. Wong, A. Nonat, L. J. Charbonniere, *J Am. Chem. Soc.*, 2017, **139**, 1456.
[85] P. Cao, L. Tong, Y. Hou, G. Zhao, G. Guerin, M. A. Winnik, M. Nitz, *Langmuir*, 2012, **28**, 12861.
[86] E. Cavero, M. Zablocka, A.-M. Caminade, J. P. Majoral, *Eur. J. Org. Chem.*, 2010, **2010**, 2759.
[87] R. Damiche, S. Chafaa, *J. Mol. Struct.*, 2017, **1130**, 1009.
[88] K. Chougrani, G. Niel, B. Boutevin, G. David, *Beilstein J. Org. Chem.*, 2011, **7**, 364.
[89] N. V. Davletshina, S. A. Koshkin, A. R. Garifzyanov, R. R. Davletshin, M. A. Filimonova, R. A. Cherkasov, *Russ. J. Gen. Chem.*, 2015, **85**, 1791.
[90] C.-O. Turrin, A. Hameau, A.-M. Caminade, *Synthesis*, 2012, **44**, 1628.
[91] S. Jiang, B. Yu, K. Zhou, H. Yang, Y. Shi, S. Lo, Y. Hu, Z. Gui, *J. Sol-Gel. Sci. Technol.*, 2013, **69**, 418.

[92] G. Keglevich, A. Szekrényi, Á. Szöllősy, L. Drahos, *Synt. Commun.*, 2011, **41**, 2265.
[93] E. Bálint, E. Fazekas, P. Pongrácz, L. Kollár, L. Drahos, T. Holczbauer, M. Czugler, G. Keglevich, *J. Organomet. Chem.*, 2012, **717**, 75.
[94] E. Bálint, E. Fazekas, L. Drahos, G. Keglevich, *Heteroat. Chem.*, 2013, **24**, 510.
[95] E. Bálint, E. Fazekas, J. Kóti, G. Keglevich, *Heteroat. Chem.*, 2015, **26**, 106.
[96] E. Bálint, A. Tripolszky, E. Jablonkai, K. Karaghiosoff, M. Czugler, Z. Mucsi, L. Kollár, P. Pongrácz, G. Keglevich, *J. Organomet. Chem.*, 2016, **801**, 111.
[97] M. Iqbal, R. G. Struijk, J. Huskens, M. Sypula, A. Wilden, G. Modolo, W. Verboom, *New J. Chem.*, 2012, **36**, 2048.
[98] A. A. Prishchenko, M. V. Livantsov, O. P. Novikova, L. I. Livantsova, V. S. Petrosyan, *Heteroat. Chem.*, 2010, **21**, 430.
[99] M. Milen, P. Ábrányi-Balogh, R. Kangyal, A. Dancsó, D. Frigyes, G. Keglevich, *Heteroat. Chem.*, 2014, **25**, 245.
[100] H. Eshghi, M. Mirzaei, M. Hasanpour, M. Mokaber-Esfahani, *Phosphorus, Sulfur, Silicon*, 2015, **190**, 1606.
[101] S. R. Devineni, S. Doddaga, R. Donka, N. R. Chamarthi, *Chin. Chem. Lett.*, 2013, **24**, 759.
[102] I. Kraicheva, E. Vodenicharova, E. Tashev, T. Tosheva, I. Tsacheva, K. Troev, *Phosphorus, Sulfur, Silicon*, 2012, **187**, 660.
[103] C. Sampath, P. Harika, *Phosphorus, Sulfur, Silicon*, 2015, **190**, 1893.
[104] S. Abada, A. Lecointre, M. Elhabiri, L. J. Charbonniere, *Dalton Trans.*, 2010, **39**, 9055.
[105] G. C. S. Reddy, C. R. Rani, M. V. N. Reddy, C. S. Reddy, *Arch. Pharm.*, 2014, **347**, 819.
[106] Z. Wang, Comprehensive Organic Name Reactions and Reagents; Wiley: Hoboken, 2009.
[107] T. E. Ali, S. M. Abdel-Kariem, *Arkivoc*, 2015, **(vi)**, 246.
[108] R. M. Abdel-Rahman, T. E. Ali, S. M. Abdel-Kariem, *Arkivoc*, 2016, **(i)**, 183.
[109] C. Subramanyam, S. T. Basha, G. Madhava, S. N. Rasool, S. Adam, S. D. S. Murthy, C. N. Raju, *Phosphorus, Sulfur, Silicon*, 2017, **192**, 267.
[110] S. Motevalli, N. Iranpoor, E. Etemadi-Davan, K. R. Moghadam, *RSC Adv.*, 2015, **5**, 100070.
[111] I. Kraicheva, I. Tsacheva, R. Nikolova, M. Topashka-Ancheva, I. Stoineva, B. Shivachev, *Phosphorus, Sulfur, Silicon*, 2017, **192**, 403.
[112] Z.-H. Cai, G.-F. Du, B. Dai, L. He, *Synthesis*, 2012, **44**, 694.
[113] X. Zhu, S. Wang, S. Zhou, Y. Wei, L. Zhang, F. Wang, Z. Feng, L. Guo, X. Mu, *Inorg. Chem.*, 2012, **51**, 7134.
[114] W. M. Abdou, R. F. Barghash, M. S. Bekheit, *Monatsh. Chem.*, 2011, **142**, 649.
[115] S. S. Reddy, V. K. Rao, B. S. Krishna, C. S. Reddy, P. V. Rao, C. N. Raju, *Phosphorus, Sulfur, Silicon*, 2011, **186**, 1411.
[116] S. Rasheed, K. Venkataramana, K. Chandrasekhar, G. Fareeda, C. N. Raju, *Arch. Pharm. Chem., Life Sci.*, 2012, **345**, 294.
[117] R. G. de Noronha, C. C. Romão, A. C. Fernandes, *Chem. Commun.*, 20011, **12**, 337.
[118] Q. Wang, L. Yang, H. Ding, X. Chen, H. Wang, X. Tang, *Bioorg. Chem.*, 2016, **69**, 132.
[119] Q. Wang, M. Zhu, R. Zhu, L. Lu, C. Yuan, S. Xing, X. Fu, Y. Mei, Q. Hang, *Eur. J. Med. Chem.*, 2012, **49**, 354.
[120] M. G. Górniak, P. Kafarski, *Phosphorus, Sulfur, Silicon*, 2016, **191**, 511.
[121] I. Kraicheva, I. Tsacheva, E. Vodenicharova, E. Tashev, T. Tosheva, A. Kril, M. Topashka-Ancheva, I. Iliev, T. Gerasimova, K. Troev, *Bioorg. Med. Chem.*, 2012, **20**, 117.
[122] B. Boduszek, T. K. Olszewski, W. Goldeman, K. Grzegolec, P. Blazejewska, *Tetrahedron*, 2012, **68**, 1223.
[123] J. Michalska, B. Boduszek, T. K. Olszewski, *Heteroat. Chem.*, 2011, **22**, 617.
[124] J. Lewkowski, Z. Malinowski, A. Matusiak, M. Morawska, D. Rogacz, P. Rychter, *Molecules*, 2016, **21**, 694.

[125] P. Zhang, C. Tang, Z. Chen, B. Wang, X. Wang, L. Jin, S. Yang, D. Hu, *Phosphorus, Sulfur, Silicon*, 2013, **189**, 530.
[126] E. Bálint, Á. Tajti, A. Ádám, I. Csontos, K. Karaghiosoff, M. Czugler, P. Ábrányi-Balogh, G. Keglevich, *Beilstein J. Org. Chem.*, 2017, **13**, 76.
[127] L.-K. Lin, C.-S. Wu, W.-C. Su, Y.-L. Liu, *J. Polym. Sci., Part A: Polym. Chem.*, 2013, **51**, 3523.
[128] G. Chun-wei, S.-H. Wu, C. Fu-li, H. Zhen-yu, F. Xiao-huan, R. Wan, *Phosphorus, Sulfur, Silicon*, 2016, **191**, 1250.
[129] T. T. T. N'Guyen, H. T. T. Duong, J. Basuki, V. Montembault, S. Pascual, C. Guibert, J. Fresnais, C. Boyer, M. R. Whittaker, T. P. Davis, L. Fontaine, *Angew. Chem. Int. Ed.*, 2013, **52**, 14152.
[130] A. A. Shady, S. M. A. Bakr, M. D. Khidre, *J. Heterocycl. Chem.*, 2017, **54**, 71.
[131] G.-V. Röschenthaler, V. P. Kukhar, I. B. Kulik, A. E. Sorochinsky, V. A. Soloshonok, *J. Fluorine Chem.*, 2011, **132**, 834.
[132] I. S. Nizamov, F. D. Yambushev, I. D. Nizamov, A. D. Voloshina, V. A. Alfonsov, *Heteroat. Chem.*, 2013, **24**, 36.
[133] M. B. Gaziziov, R. A. Khairullin, A. A. Minnikhanova, A. I. Alekhina, A. A. Bashkirtsev, O. G. Sinyashin, *Dokl. Chem.*, 2010, **433**, 186.
[134] G.-V. Röschenthaler, V. P. Kukhar, I. B. Kulik, M. Y. Belik,A. E. Sorochinsky, E. B. Rusanov, V. A. Soloshonok, *Tetrahedron Lett.*, 2012, **53**, 539.
[135] P. P. Onys'ko, K. A. Zamulko, O. I. Kyselyova, *Phosphorus, Sulfur, Silicon*, 2016, **191**, 274.
[136] P. P. Onys'ko, K. A. Zamulko, O. I. Kyselyova, *Phosphorus, Sulfur, Silicon*, 2016, **191**, 1574.
[137] P. P. Onys'ko, K. A. Zamulko, O. I. Kyselyova, I. P. Yelenich, Y. V. Rassukana, *J. Fluorine Chem.*, 2016, **185**, 191.
[138] W. Xu, S. Zhang, S. Yang, L.-H. Jin, P. S. Bhadury, D.-Y. Hu, Y. Zhang, *Molecules*, 2010, **15**, 5782.
[139] M. Nazish, A. Jakhar, N. H. Khan, S. Verma, R. I. Kureshy, S. H. R. Abdi, H. C. Bajaj, *Appl. Catal., A.*, 2015, **515**, 116.
[140] G. Zhang, G.-F. Hao, J. Pan, J. Zhang, D. Hu, B.-A. Song, *J. Agric. Food Chem.*, 2016, **64**, 4207.
[141] S. Liu, Y. Li, Z. Yin, Q. Yu, Z. Miao, *J. Org. Chem.*, 2017, **82**, 2481.
[142] M. Yang, H. Xu, Z.-Y. Zhou, H. Zhang, L.-J. Liu, Y.-M. Sun, S.-Z. Nie, C.-Q. Zhao, *Tetrahedron: Asymmetry*, 2016, **27**, 815.

Réka Henyecz, Mátyás Milen, Károly Kánai and György Keglevich

7 The use of the T3P® reagent in the synthesis of phosphinic and phosphonic derivatives

Abstract: Cyclic tri(propylphosphonate), the T3P® reagent, made a significant progress in the last decade. After utilizing it in the synthesis of peptides, it was applied in a series of organic reactions. This chapter summarizes the possible use of the T3P® reagent in organophosphorus syntheses. The T3P® reactant was useful in the esterification and amidation of phosphinic acids, as well as in different variations of the Kabachnik–Fields condensation to afford α-aminophosphonates. The scope and limitations of the use of the T3P® reagent is also summarized.

Keywords: T3P® reagent, phosphinates, phosphinic amides, α-aminophosphonates.

7.1 Introduction

Propylphosphonic anhydride (T3P®) is one of the most important organophosphorus reagents. T3P® was discovered and introduced in synthetic organic chemistry by Wissmann and Kleiner in 1980 [1]. Originally, it was applied as a coupling agent in the synthesis of simple peptide derivatives. Some advantageous properties of the T3P® reagent have been soon recognized. The model peptides were obtained in good yields and high optical purity. In addition, the anhydride has a good solubility in various organic solvents; solutions with more than 50% concentration can be prepared from the T3P® reagent in dichloromethane, 1,4-dioxane, tetrahydrofuran, ethyl acetate and *N*,*N*-dimethylformamide [1, 2]. The excess of the reagent and the "T3P®+H_2O" adduct formed as the byproduct or even smaller fragments may be easily removed from the reaction mixture by alkaline extraction. The T3P® reagent shows low allergenic potential and low toxicity. There were no symptoms of intoxication in mice at a more than 2000 mg/kg oral dose according to biological examinations [3].

The T3P® reagent may be synthesized from propylphosphonic dichloride (**Method A**) or propyl phosphonic acid (**Method B**) (Figure 7.1). In the first case (**Method A**), propylphosphonic dichloride was treated with 1 equivalent of water, and the mixture was heated to 80 °C. Then, the heating was followed at 80 °C under 17 torr for 3 h, and under 5 torr for another 3 h. The crude polypropylphosphonic acid was obtained in an almost quantitative yield, whose distillation at 200 °C under 0.3 torr afforded the final product as colorless oil [1]. In the second instance (**Method B**), propylphosphonic acid was heated with acetic anhydride at the boiling point for 24 h, followed by the removal of acetic acid formed and the excess acetic anhydride. The

https://doi.org/10.1515/9783110535839-007

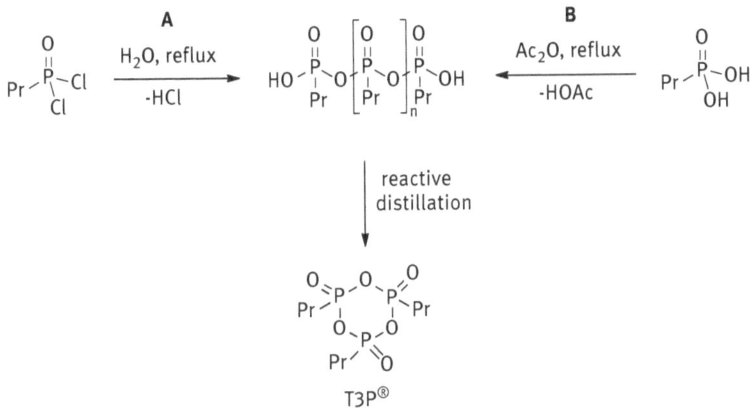

Figure 7.1: Synthesis of the T3P® reagent.

oligomeric residue was Kugelrohr distilled (300 °C/0.3 mbar) to give the desired trimer product in a yield of 72% [4].

The freshly prepared T3P® should be taken up immediately in an inert organic solvent to avoid hydrolysis.

Beyond peptide chemistry, the T3P® reagent was successfully applied in different water elimination or condensation reactions. For example, preparation of alkenes from primary, secondary or tertiary alcohols under mild conditions [5], the direct conversion of carboxylic acid amides into nitriles, or formamides into isocyanides [6], the N-benzylation of phenothiazine with benzyl alcohols [7], esterification of carboxylic acids with alcohols [8] and the synthesis of heterocyclic compounds via cyclization [9–11]. Furthermore, T3P® has also been utilized as a promoter in multicomponent reactions [12, 13], rearrangements [14, 15], synthesis of pharmaceuticals [16], spirocycles [17] and natural products [17, 18].

Hitherto, the synthetic application of propylphosphonic anhydride (T3P®) has been summarized in two comprehensive articles [19, 20] and three short reviews [21–23].

This chapter focuses on the recently discovered T3P®-assisted preparation of organophosphorus compounds, such as phosphinates, phosphinic amides and a series of α-aminophosphonates.

7.2 T3P®-mediated Synthesis of Organophosphorus Compounds

7.2.1 T3P®-assisted Synthesis of Phosphinic Derivatives

Phosphinates are important organophosphorus scaffolds. It is known that the direct esterification of phosphinic acids does not take place on conventional heating. For this, alternative methods had to be developed for the esterification of phosphinic acids [24].

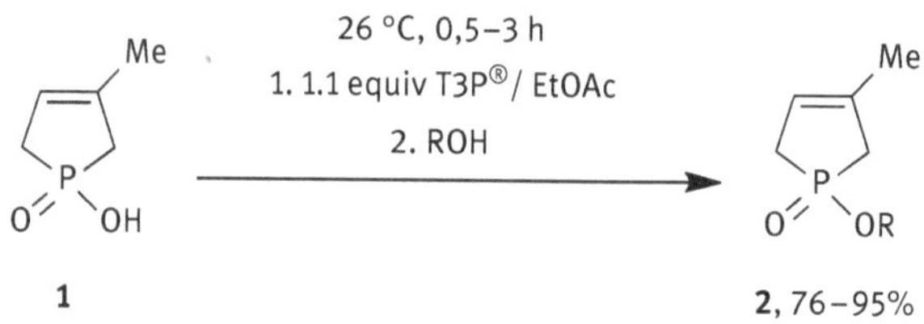

R = Me, Et, ^{n}Pr, ^{i}Pr, ^{n}Bu, ^{i}Bu, ^{s}Bu, Pent, ^{i}Pent, 3-pentyl, ^{c}Hex, Bn, 3-phenylethyl, 2-(1-naphtyl)ethyl, mentyl

Figure 7.2: T3P®-promoted esterification of 1-hydroxy-3-methyl-3-phospholene 1-oxide.

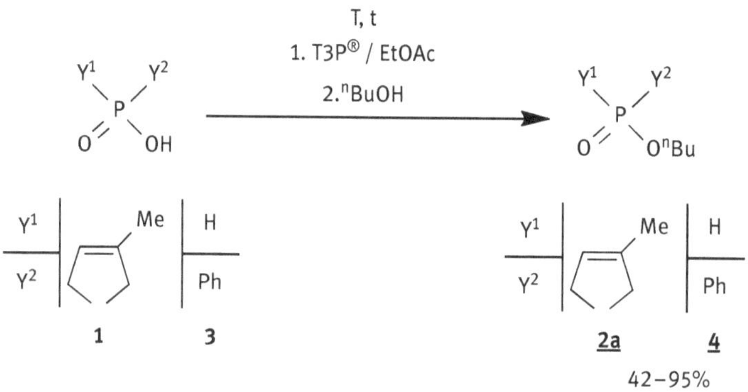

Figure 7.3: A detailed study of the T3P®-assisted esterification of selected phosphinic acids.

Keglevich and Jablonkai elaborated the efficient esterification of 1-hydroxy-3-methyl-3-phospholene 1-oxide (**1**) with a series of acyclic alcohols under mild conditions applying 1.1 equivalents of T3P® (Figure 7.2) [25].

Then, two model compounds (**1** and **3**) were chosen for the detailed study of the T3P®-mediated esterification (Figure 7.3) [26].

It can be seen from Table 7.1 that the use of 1.1 equivalents of the T3P® reagent led practically to quantitative conversions. The utilization of a second unit of the T3P® reagent is also possible, but in the case of the less reactive hydroxy-methyl-3-phospholene 1-oxide (**1**) using 0.66 equivalents of the T3P® reagent, MW irradiation was necessary. Applying only 0.44 equivalents of the T3P® reagent, the esterification is possible only with the more reactive phenyl-*H*-phosphinic acid (**3**), again under MW irradiation.

The esterification of other five-membered cyclic phosphinic acids was also studied (Figure 7.4). The reaction of 1-hydroxy-3,4-dimethyl-3-phospholene 1-oxide (**5A**) with simple C1–C4 alcohols could be performed using only 0.66 equivalents of the T3P® reagent under MW conditions. However, in the case of the less reactive hydroxy-methylphospholane oxides (**5B** and **5C**), it is better to use 1.1 equivalents of the T3P® reagent to obtain the phosphinates (**6**) in high yields.

Table 7.1: The optimization of the esterification in the presence of the T3P® reagent.

Product	T3P® (equiv.)	Mode of heating	Temp. (°C)	Time (min)	Yield (%)
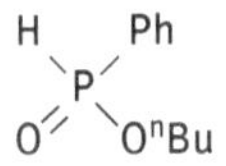 **4**	1.1	–	25	30	95
	0.66	**–**	**25**	**30**	**91**
	0.44	–	25	30	79
	0.44	Δ	85	15	78
	0.44	**MW**	**85**	**15**	**89**
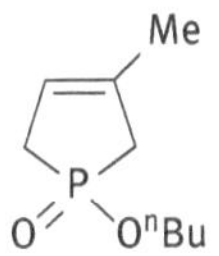 **2a**	1.1	–	25	30	82
	0.66	–	25	30	42
	0.66	Δ	85	30	56
	0.66	**MW**	**85**	**30**	**79**

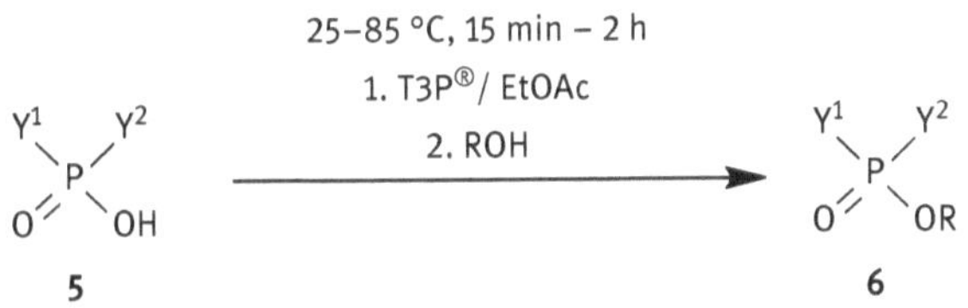

R = Me, Et, ⁿPr, ⁱPr, ⁿBu, ⁱBu, ˢBu, ᵗBu

	Y1 Y2	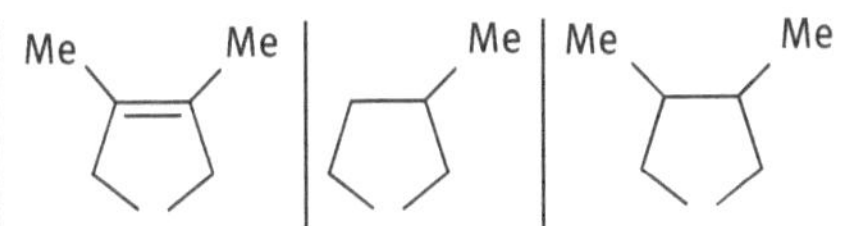		
		A	**B**	**C**
Yield (%)	1.1 equiv. T3P®	59–85	67–81	67–75
	0.66 equiv. T3P®	57–83	57–67	55–65

Figure 7.4: Esterification of five-membered cyclic phosphinic acids in the presence of the T3P® reagent.

Application of the T3P® reagent was also extended to the amidation of phosphinic acids. 1-Hydroxy-3-methylphospholene oxide (**1**) was converted to the corresponding amides (**7**) by reaction with primary and secondary amines using 1.1 equivalents of the T3P® reagent at room temperature (Figure 7.5).

It is worth noting that this was the first synthesis of phosphinic amides directly from phosphinic acids.

The efficiency of these derivatization techniques lies in the fact that in the first step, the phosphinic acid is converted to a more reactive mixed anhydride (**8**) by reaction with the T3P® reagent (Figure 7.6) [27]. Then, the reactive intermediate (**8**) is attacked by the nucleophile that may be butyl alcohol or butylamine. It was

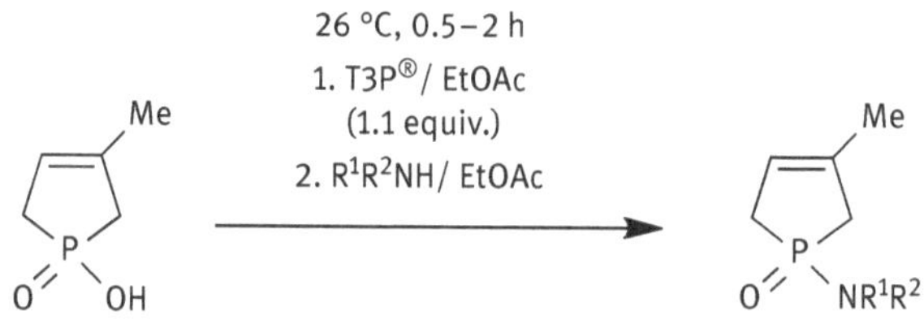

R^1	Pr	^{i}Pr	^{n}Bu	^{i}Bu	^{s}Bu	^{n}Hex	^{c}Hex	Bn	Et	Pr	Bu
R^2	H	H	H	H	H	H	H	H	Et	Pr	Bu
Yield (%)	61	60	65	68	62	86	82	87	65	61	70

Figure 7.5: Amidation of 1-hydroxy-3-methyl-3-phospholene 1-oxide using the T3P® reagent.

found that the amidation of intermediate **8** is a two-step process, while the esterification follows a one-step protocol. The driving force for the esterification is higher (ΔH = –59.2 kJ mol^{-1}) than that for the amidation (ΔH = –31.6 kJ mol^{-1}), which is in accord with the experimental results.

7.2.2 The T3P®-assisted Synthesis of α-aminophosphonates

The α-aminophosphonates are potentially bioactive molecules as they are structural analogues of the α-amino acids. They can be used as antibiotics, antibacterial, anticancer and antiviral drugs [28–30]. They can also be applied as herbicides, fungicides and insecticides in the agriculture industry [31–34]. Therefore, the synthesis of these valuable compounds is nowadays in the focus.

Milen et al. described a T3P®-promoted efficient synthesis of α-aminophosphonates under mild conditions (Figure 7.7) [35]. The one-pot three-component reaction of benzaldehyde, aromatic amines and triethyl phosphite at room temperature in the presence of 1 equivalent of the T3P® reagent led to the corresponding products (**9**) in high yields (80–96%).

The T3P®-assisted Kabachnik–Fields reaction was extended to several benzaldehyde derivatives as well (Figure 7.8). The different substituents did not have an impact on the yields (84–95%) of the products (**10**).

According to the mechanism suggested, T3P® is involved in the first step of the reaction as a dehydrating agent (Figure 7.9). The formation of the imine intermediate (**11**) from the aldehyde and the aniline leads to tripropyl-triphosphonic acid ("T3P®+H_2O") as the byproduct. In the next step, triethyl phosphite attacks the carbon atom of the imine function. Then, "T3P®+H_2O" protonates the adduct so formed, and

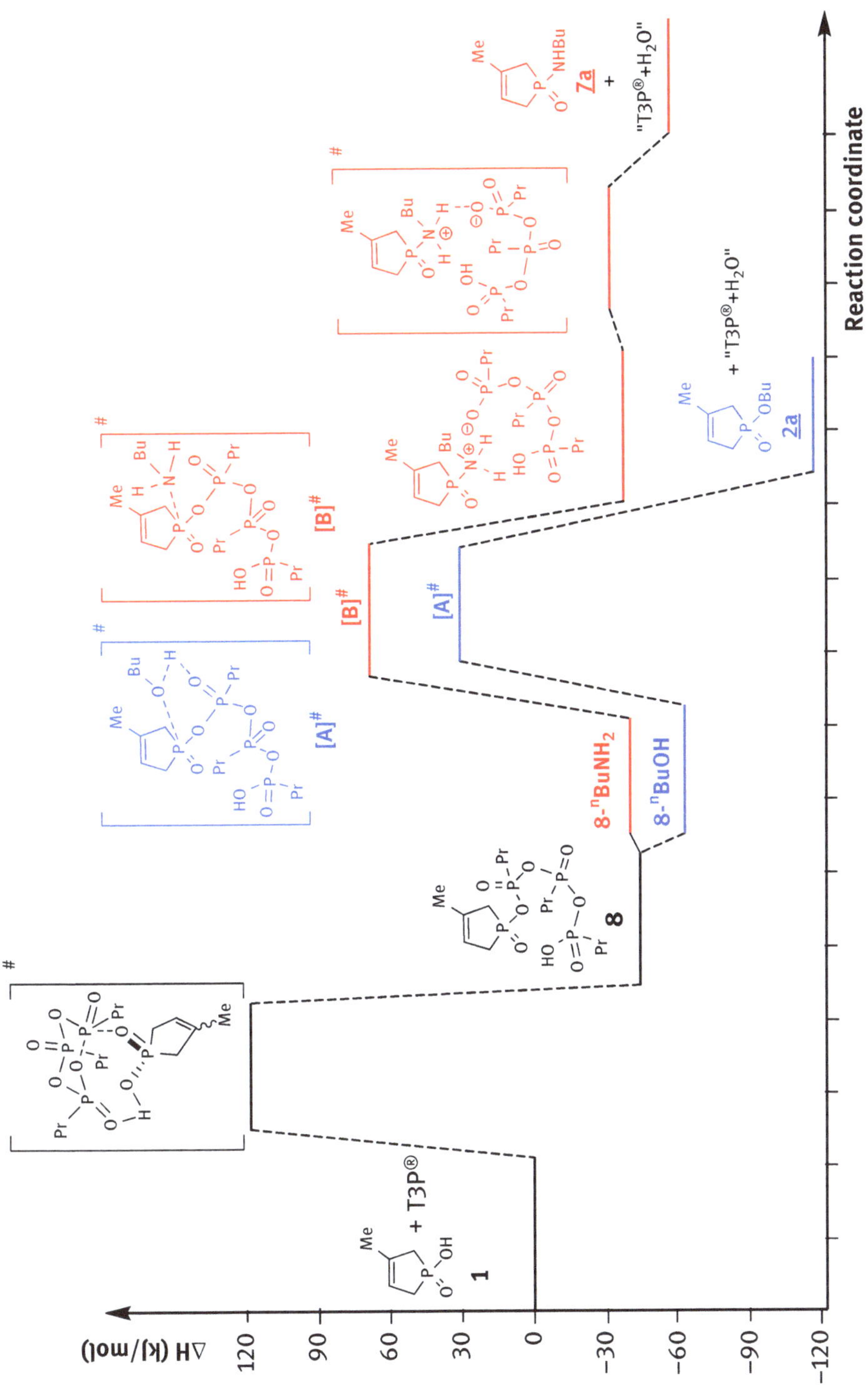

Figure 7.6: The mechanism of the T3P®-promoted derivatization calculated by the B3LYP/6-311G(2d,2p) method.

the phosphonium salt (**12**) undergoes an Arbuzov fission to afford the α-aminophosphonate (**13**) as the product and "T3P®+EtOH" as the byproduct.

Bis(α-aminophosphonates) (**14**) were synthesized in a similar way (Figure 7.10) [36]. 4-Phenylenediamine was reacted with 2 equivalents of an aromatic aldehyde and 2 equivalents of triethyl phosphite in the presence of 2 equivalents of the T3P®

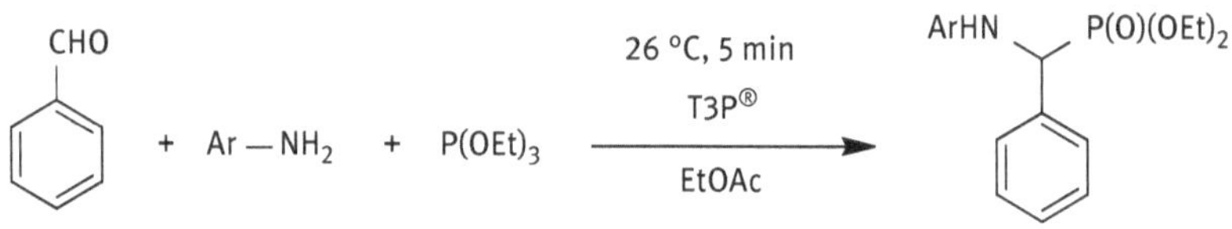

9, 80–96%

Ar = Ph, 4-MeC$_6$H$_4$, 4-PhC$_6$H$_4$, 4-FC$_6$H$_4$, 4-ClC$_6$H$_4$, 4-BrC$_6$H$_4$, 4-O$_2$NC$_6$H$_4$, 4-EtO$_2$CC$_6$H$_4$, 4-MeOC$_6$H$_4$, 3,4-(MeO)$_2$C$_6$H$_3$, 4-MeSC$_6$H$_4$, 2-Ph(OC)C$_6$H$_4$, 3-PhOC$_6$H$_4$, 3-F$_3$CC$_6$H$_4$, 3-F$_3$COC$_6$H$_4$, 1-naphthyl

Figure 7.7: T3P®-promoted condensation of benzaldehyde, aromatic amines and triethyl phosphite.

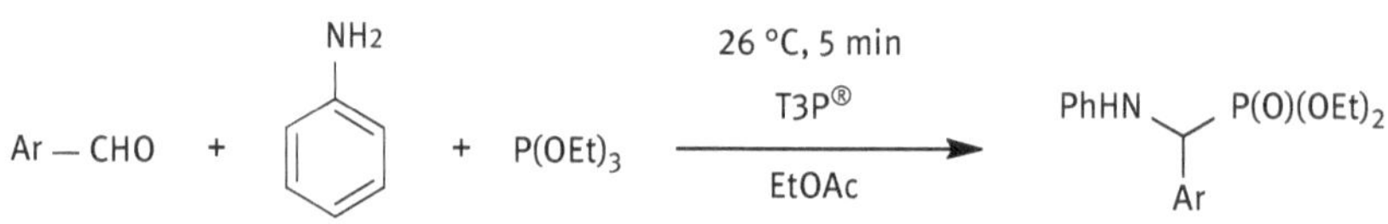

10, 84–95%

Ar = 4-MeC$_6$H$_4$, 4-FC$_6$H$_4$, 4-ClC$_6$H$_4$, 4-O$_2$NC$_6$H$_4$, 4-F$_3$CC$_6$H$_4$, 4-MeOC$_6$H$_4$, 4-BnOC$_6$H$_4$, 3-NCC$_6$H$_4$, 1-naphthyl

Figure 7.8: Kabachnik–Fields reaction of substituted benzaldehydes, aniline and triethyl phosphite in the presence of the T3P® reagent.

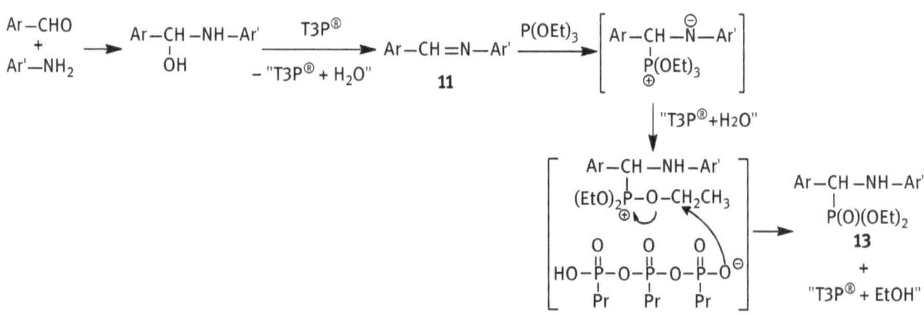

Figure 7.9: Proposed mechanism for the T3P®-assisted Kabachnik–Fields reaction of an aromatic aldehyde, aromatic amine and triethyl phosphite.

reagent at 25 °C. It was found that using diethyl phosphite instead of triethyl phosphite, the condensation was slightly more efficient. Compare the yields of 66–87% versus 64–80%.

Terephtalaldehyde was also converted to bis(α-aminophosphonates) in reaction with 2 equivalents of the aniline derivatives and the P-reagents. Both triethyl phosphite and diethyl phosphite were suitable reactants (Figure 7.11). The corresponding products (**15**) were obtained in variable yields (46–95%) under mild conditions.

The T3P® reagent was successfully applied in the synthesis of dialkyl (3-oxo-2,3-dihydro-1*H*-isoindol-1-yl)phosphonates (**16**) (Figure 7.12) [37]. The reaction of 2-formylbenzoic acid, amines and triethyl phosphite or diethyl phosphite required 1.5 equivalents of the T3P® reagent. The one-pot condensation was carried out in boiling ethyl acetate to give the products (**16**) in yields of 52–92%. The substituents of the amines greatly influenced the yields.

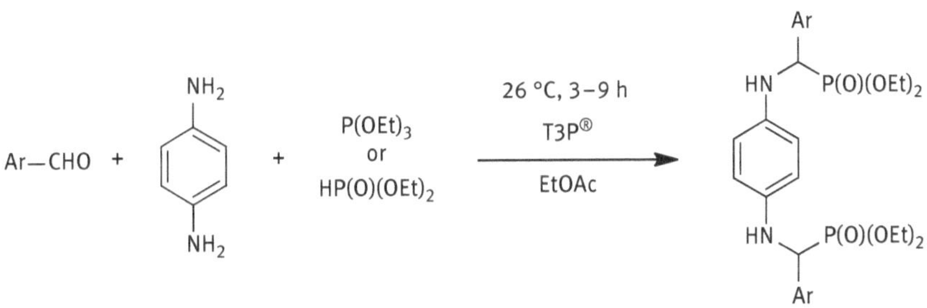

14, 64–87%

Ar = Ph, 4-MeC_6H_4, 4-$MeO_2CC_6H_4$, 4-FC_6H_4, 4-ClC_6H_4, 4-BrC_6H_4, 4-$O_2NC_6H_4$, 4-$MeOC_6H_4$

Figure 7.10: The synthesis of bis(α-aminophosphonates) starting from 4-phenylenediamine using the T3P® reagent.

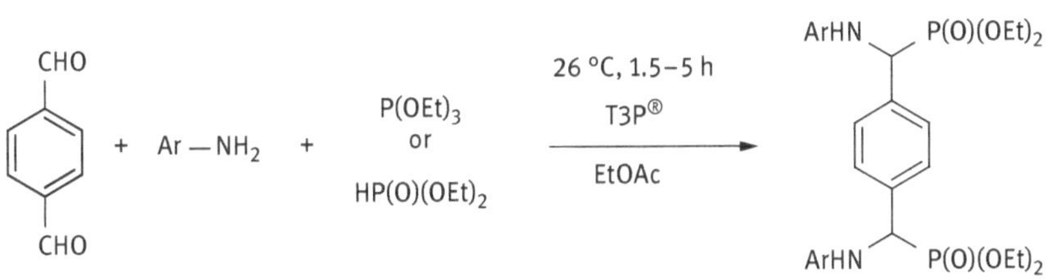

15, 46–95%

Ar = Ph, 4-MeC_6H_4, 4-$EtO_2CC_6H_4$, 4-FC_6H_4, 4-BrC_6H_4, 4-$O_2NC_6H_4$, 4-$MeOC_6H_4$, 2-$MeSC_6H_4$, 3-$F_3CC_6H_4$

Figure 7.11: T3P®-promoted bis-condensation of terephtaldehyde with substituted anilines and P-reagents.

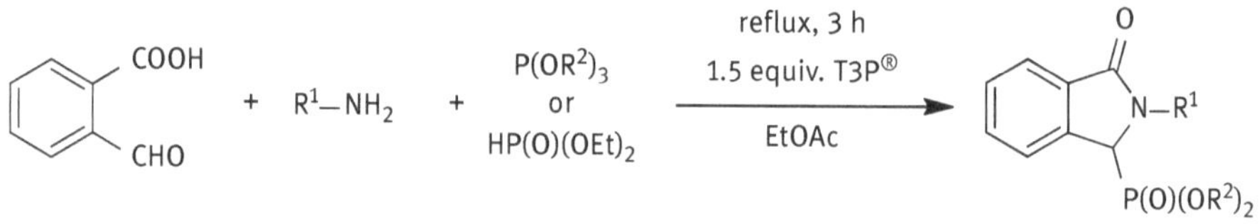

16, 52–92%

R^1 = Ph, 3-ClC_6H_4, 2-$^iPrC_6H_4$, 2-ClC_6H_4, 2-$O_2NC_6H_4$, 2-$MeSC_6H_4$, 2-$PhSC_6H_4$, 4-FC_6H_4, 2-NCC_6H_4, 4-$O_2NC_6H_4$, 2,6-$Cl_2C_6H_4$, 2,5-$F_2C_6H_4$, 3,4-$(MeO)_2C_6H_4$, 2,3-dihydro-1*H*-inden-5-yl, $PhCH_2$, $Ph(CH_2)_2$, $Ph(CH_2)_3$, $Me(CH_2)_7$, $EtMe_2C$, naphthalen-1-yl, 5-chloropyridin-2-yl, quinolin-5-yl
R^2 = Me, Et, ^iPr

Figure 7.12: T3P®-assisted one-pot synthesis of (oxoisoindolin-1-yl)phosphonates.

7.3 Scope and Limitations of the Use of the T3P® Reagent

The T3P® reagent is indeed a powerful condensing agent. There are only two shortcomings. One of them is the relatively high price of this reactant (100 ml of a 1.63 mmol/ml ethyl acetate solution costs 207 Euros [38]). The other disadvantage is the relatively low atomic efficiency that is the consequence of the fact that only one –PrP(O)O– unit is utilized from among the three P moieties of the reactant. It was shown in Section 7.2 that the more reactive phosphinic acids, such as 1-hydroxy-3-methyl- and 3,4-dimethyl-3-phospholene oxides **1** and **5A,** could be esterified efficiently in the presence of only 0.66 equivalents of the condensing agent. However, there was a need for a temperature of 85 °C under MW irradiation. In special cases, such in the reaction of the rather reactive phenyl-*H*-phosphinic acid (**3**), already 0.44 equivalents of the T3P® reagent is enough at 85 °C on MW irradiation [26].

7.4 Summary

The T3P® reagent received a number of applications in organic chemistry. This chapter summarized the recent results on the utilization of this elegant activating agent in the field of organophosphorus reactions, as exemplified by the derivatization of phosphinic acids, and the different variations of the Kabachnik–Fields reactions.

Acknowledgment: This work was supported by the National Research, Development and Innovation Office (K119202).

References

[1] H. Wissmann, H.-J. Kleiner, *Angew. Chem. Int. Ed. Engl.*, 1980, **19**, 133.
[2] J. Glauder, *Speciality Chem. Mag.*, 2014, **25**, 277.
[3] H. Wissmann, *Phosphorus, Sulfur Silicon Relat. Elem.*, 1987, **30**, 645.
[4] H. Pizova, P. Bobal, *Tetrahedron Lett.*, 2015, **56**, 2014.
[5] A. Meudt, S. Scherer, C. Böhm, PCT Int. Appl. WO 2005123632, 2005; *Chem. Abstr.*, 2005, **144**, 68544.
[6] A. Meudt, S. Scherer, S. Nerdinger, PCT Int. Appl. WO 2005070879, 2005; *Chem. Abstr.*, 2005, **143**, 172649.
[7] L. Levi, S. Scheuren, T. J. J. Müller, *Synthesis*, 2014, **46**, 3059.
[8] M. Wedel, A. Walter, F.-P. Montforts, *Eur. J. Org. Chem.*, 2001, 1681.
[9] M. Milen, P. Ábrányi-Balogh, A. Dancsó, G. Simig, B. Volk, *Tetrahedron*, 2014, **70**, 465.
[10] A. Ilangovan, S. Saravanakumar, S. Umesh, *J. Chem. Sci.*, 2015, **127**, 797.
[11] P. Ábrányi-Balogh, T. Földesi, A. Grün, B. Volk, G. Keglevich, M. Milen, *Tetrahedron Lett.*, 2016, **57**, 1953.
[12] S. S. Reddy, B. R. P. Reddy, P. V. G. Reddy, *Chin. Chem. Lett.*, 2015, **26**, 739.
[13] M. Milen, A. Dancsó, T. Földesi, B. Volk, *Tetrahedron*, 2017, **73**, 70.
[14] J. K. Augustine, A. Bombrun, A. B. Mandal, P. Alagarsamy, R. N. Atta, P. Selvam, *Synthesis*, 2011, 1477.
[15] B. Vasantha, H. P. Hemantha, V. V. Sureshbabu, *Synthesis*, 2010, 2990.
[16] J. R. Dunetz, J. Magano, G. A. Weisenburger, *Org. Process Res. Dev.*, 2016, **20**, 140.
[17] W. P. Unsworth, R. J. K. Taylor, *Synlett*, 2016, **27**, 2051.
[18] C. Kitsiou, W. P. Unsworth, G. Coulthard, R. J. K. Taylor, *Tetrahedron*, 2014, **70**, 7172.
[19] Basavaprabhu, T. M. Vishwanatha, N. R. Panguluri, V.V. Sureshbabu, *Synthesis*, 2013, **45**, 1569.
[20] A. A. Waghmare, R. M. Hindupur, H. N. Pati, *Rev. J. Chem.* 2014, **4**, 53.
[21] R. Amitrano, P. Koch, M. Neuber, D. Neumann, P, Petrovic, M. Roos, C. Vedder, *Pept. Youth: Adv. Exp. Med. Biol.*, 2009, **611**, 193.
[22] M. Schwarz, *Synlett*, 2000, 1369.
[23] A. L. L. García, *Synlett*, 2007, 1328.
[24] G. Keglevich, N. Z. Kiss, Z. Mucsi, E. Jablonkai, E. Bálint, *Green Process. Synth.*, 2014, **3**, 103.
[25] E. Jablonkai, M. Milen, L. Drahos, G. Keglevich, *Tetrahedron Lett.*, 2013, **54**, 5873.
[26] E. Jablonkai, R. Henyecz, M. Milen, J. Kóti, L. Drahos, G. Keglevich, *Tetrahedron*, 2014, **70**, 8280.
[27] P. Ábrányi-Balogh, E. Jablonkai, R. Henyecz, M. Milen, G. Keglevich, *Curr. Org. Chem.*, 2016, **20**, 1135.
[28] M. C. Allen, W. Fuhrer, B. Tuck, R. Wade, *J. Med. Chem.*, 1989, **32**, 1652.
[29] A. Mucha, P. Kafarski, L. Berlicki, *J. Med. Chem.*, 2011, **54**, 5955.
[30] J. Bird, R. C. De Mello, G. P. Harper, D. J. Hunter, E. H. Karran, R. E. Markwell, A. J. Miles-Williams, S. S. Rahman, R. W. Ward, *J. Med. Chem.*, 1994, **37**, 158.
[31] M. Sienczyk, J. Oleksyszyn, *Curr. Med. Chem.*, 2009, **16**, 1673.
[32] S. C. Fields, *Tetrahedron*, 1999, **55**, 12237.
[33] B. A. Song, Y. L. Wu, S. Yang, D. Y. Hu, X. Q. He, L. H. Jin, P. Lu, *Moculecules*, 2003, **8**, 186.
[34] J. Emsley, D. Hall, *The Chemistry of Phosphorus*; Harper and Row: London, 1976.
[35] M. Milen, P. Ábrányi-Balogh, A. Dancsó, D. Frigyes, L. Pongó, G. Keglevich, *Tetrahedron Lett.*, 2013, **54**, 5430.
[36] M. Milen, P. Ábrányi-Balogh, R. Kangyal, A. Dancsó, D. Frigyes, G. Keglevich, *Heteroat. Chem.*, 2014, **25**, 245.
[37] M. Milen, A. Dancsó, T. Földesi, Slégel P., B. Volk, *Tetrahedron*, 2016, **72**, 5091.
[38] http://www.sigmaaldrich.com/catalog/product/aldrich/431303?lang=hu&region=HU

Réka Henyecz and György Keglevich

8 P–C couplings by the Hirao reaction

Abstract: The Hirao reaction that comprises a P–C coupling between an aryl halide and a dialkyl phosphite in the presence of a Pd(0) catalyst to afford the corresponding arylphosphonate ran a nice career in its 37 years. This chapter summarizes the bases, and the most important variations/extensions showing the synthesis of phosphinates and tertiary phosphine oxides besides phosphonates. The green chemical accomplishments get a special stress.

Keywords: Hirao reaction, P–C coupling, phosphonate, phosphinate, phosphine oxide, green chemistry.

8.1 Introduction

The P–C coupling reaction of aryl, hetaryl and vinyl halides (or other derivatives) with $Y^1Y^2P(O)H$ reagents, such as dialkyl phosphites (Y^1,Y^2 = RO), alkyl *H*-phosphinates (Y^1 = RO, Y^2 = R, Ar) and secondary phosphine oxides (Y^1,Y^2 = R, Ar), means an important protocol for the synthesis of phosphonates, phosphinates and phosphine oxides, respectively [1, 2]. The original Hirao reaction involves the preparation of arylphosphonates. This was followed by different variations applying a series of >P(O)H reactants, preformed or in situ formed Pd(0) catalysts from Pd(II) salts, or other metal precursors (Ni and Cu) and mono- and bidentate P-ligands, bases (mostly amines), and solvents. In this chapter, the Pd-catalyzed accomplishments are in the focus. The green chemical approaches comprising phase transfer catalytic and microwave-assisted variations, along with P-ligand-free and solvent-free protocols are also surveyed.

8.2 Hirao Reaction using the Pd(0) Catalyst

Hirao reported the first P–C cross-couplings comprising the reactions of aryl and vinyl halides with dialkyl phosphites in the presence of tetrakis(triphenylphosphine)palladium (Figure 8.1). The phosphonates (**1**) were obtained in most cases in high yields, using triethylamine, tributylamine and pyridine as the base along with toluene as the solvent [3–5].

https://doi.org/10.1515/9783110535839-008

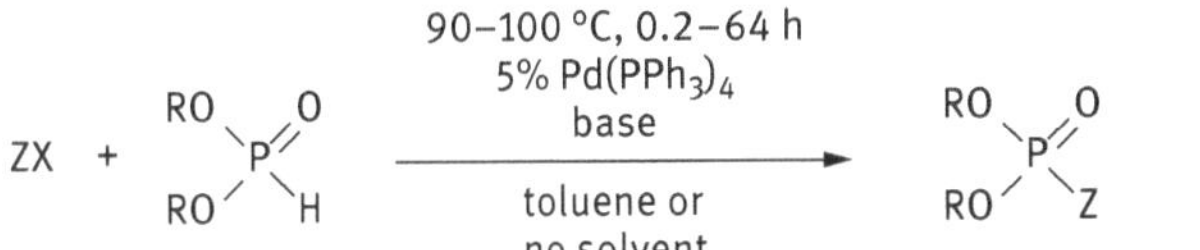

1, 3–98%

base: Et_3N, Bu_3N, pyridine
Z = Ph, 4-$NH_2C_6H_4$, 4-HOC_6H_4, 2-$(CH_2OH)C_6H_4$, 4-$MeOC_6H_4$, 4-MeC_6H_4, 2-MeC_6H_4, 4-NHC(O)MeC_6H_4, 4-ClC_6H_4, 4-BrC_6H_4, 2-BrC_6H_4, 2-$HO_2CC_6H_4$, 2-$MeO_2CC_6H_4$, 4-C(O)MeC_6H_4, 4-CNC_6H_4, 4-$NO_2C_6H_4$, 1-naphthyl, 3-pyridyl, allyl, 1-cyclohexenyl, styryl, 1-phenylvinyl
X = I, Br
R = Et, ^{i}Pr, Bu

Figure 8.1: The first examples of the Hirao reaction.

It is generally believed that the mechanism of the Hirao reaction is similar to the C–C coupling reactions (Figure 8.2). The first step is the oxidative addition of the aryl derivative (ArX) to the Pd(0) complex. The next step involves the ligand exchange, when the deprotonated $(RO)_2P(O)H$ replaces the X^- anion of adduct **2**. Finally, the corresponding phosphonate (**4**) is formed by reductive elimination from intermediate **3**.

The Hirao reaction was then extended to other models. The most important results are summarized in Figure 8.3 [6–37]. Substituted aryl and vinyl halide or triflates were reacted with dialkyl phosphites, alkyl *H*-phosphinates and secondary phosphine oxides utilizing the $Pd(PPh_3)_4$ catalyst in the presence of triethylamine,

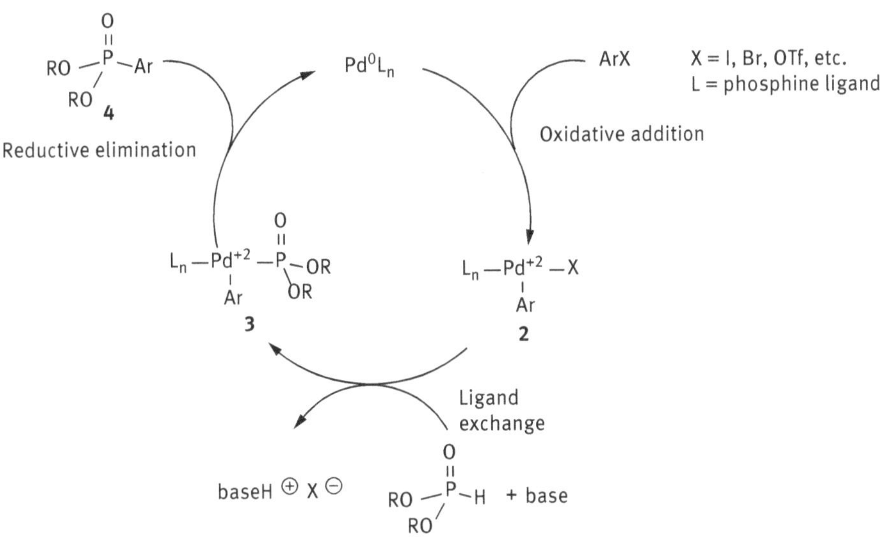

Figure 8.2: Mechanism of the classical Hirao reaction.

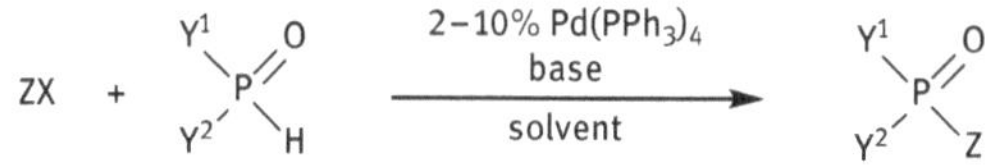

5, 18–98%

base: Et_3N, $^iPr_2NEt$, Me-morpholine
solvent: toluene, acetonitrile, DMF, THF, benzene
Z = Ph, 4-$NH_2C_6H_4$, 4-$MeOC_6H_4$, 4-NHC(O)MeC_6H_4, 4-ClC_6H_4, 4-BrC_6H_4, 2-BrC_6H_4, 2-$MeO_2CC_6H_4$, 4-C(O)MeC_6H_4, 4-CNC_6H_4, 4-$NO_2C_6H_4$, 2-naphthyl, allyl, 1-(4-^{t}Bu-cyclohexenyl), styryl, 1-heptenyl, etc.
X = I, Br, OTf
Y^1, Y^2 = Ph [28,35], 4-BrC_6H_4 [25], OMe [15], OEt [6-21,36,37], O^{i}Pr [6], OBu [6], O^{t}Bu [11,29]

Y^1	Ph	Ph	Ph	Ph	Ph	Ph	Ph	Me	Me	Et	Pr	Bu	$CH(OEt)_2$	CH_2OBn
Y^2	Bu	Hex	Bn	OMe	OEt	OPr	OBu	O^{i}Pr	OBu	OEt	OBu	OEt	OEt	OEt
ref.	34	34	34	33	28,36	36	30	22-24	31	36	36	31	26,27	32

Figure 8.3: Extension of the Hirao reaction.

diizopropylethylamine or methylmorpholine in toluene, acetonitrile, DMF, THF or benzene as the solvent. The yields fell in the range of 18–98%, but were mostly high.

It is also possible that a Pd(0) precursor is used together with bidentate P-ligands. Typical combinations are $Pd_2(dba)_3$/dppf [38, 39], $Pd_2(dba)_3$/dppp [40, 41], $Pd(dba)_2$/dppp [42–45], $Pd_2(dba)_3$/BINAP, Xantphos, Josiphos and dppf [46].

8.3 P–C Couplings using a Pd Precursor/Ligand System

Instead of $Pd(PPh_3)_4$, Pd salts in combination with P-ligands could also be used. In this case, the active Pd(0) catalyst is formed in situ in the reaction mixture. Stawinski and coworkers compared different Pd precursor/PPh_3 systems in the reaction of iodo- and bromobenzene with diethyl phosphite (Figure 8.4) [47]. From among the two Pd salts tested, $Pd(OAc)_2$ was found to be the best. In the catalytic systems,

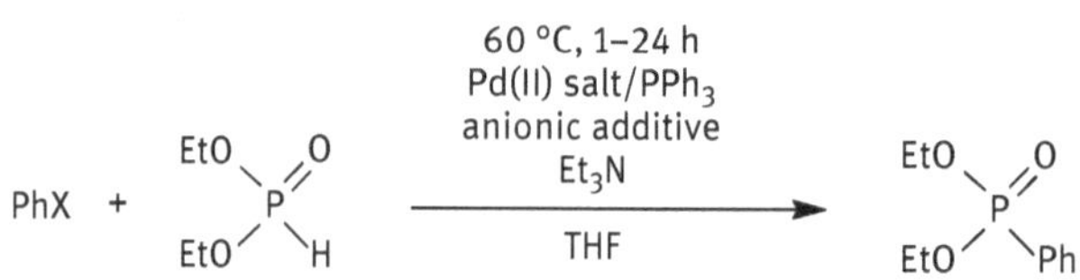

6, ~100%

X = I, Br
Pd(II) salt: $Pd(OAc)_2$, $PdCl_2$
anionic additives: $Bu_4N(OAc)$, Bu_4NBr, Bu_4NCl

Figure 8.4: Testing Pd(III) salts as Pd(0) precursors.

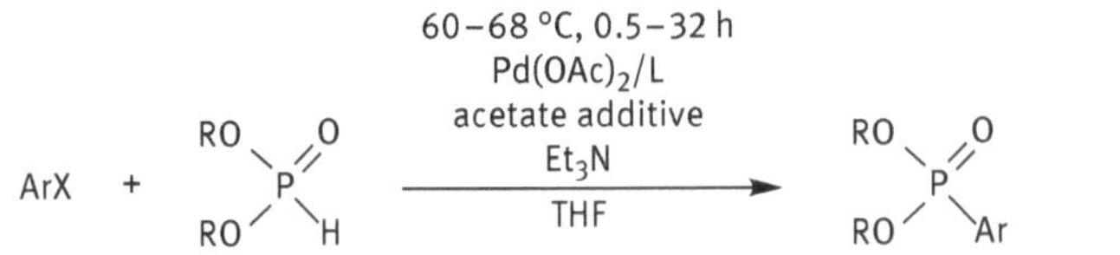

4, 63–98%

Ar = Ph, 4-MeOC$_6$H$_4$, 2-MeC$_6$H$_4$, 4-FC$_6$H$_4$, 4-NO$_2$C$_6$H$_4$, 1-naphthyl, 2-naphthyl, 9-phenantryl, 3-pyridyl, 4-pyridyl
X = I, Br, OTf
R = Me, Et, ^{i}Pr
L = PPh$_3$, dppp, dppb, dppf, BINAP
acetate additive: NaOAc, KOAc, LiOAc, CsOAc, Bu$_4$N(OAc)

Figure 8.5: Testing acetate additives in the Hirao reaction.

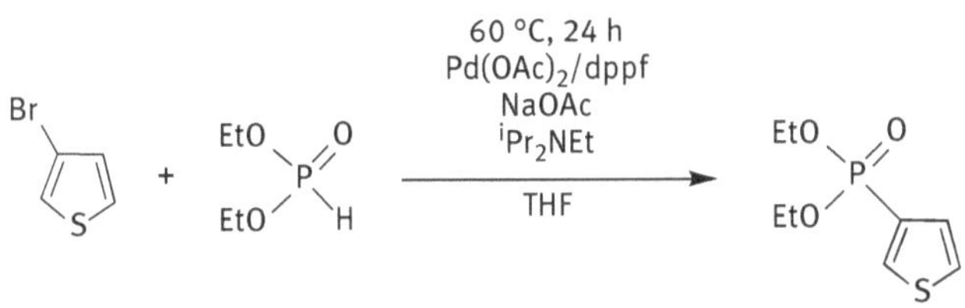

7, 72%

Figure 8.6: Phosphonation of bromothiophene.

ammonium salts were tested as additives, and it was found that the acetate ions promoted the P–C coupling.

The same research group investigated the reaction of aryl halides and triflates with dialkyl phosphites in the presence of Pd(OAc)$_2$ using, this occasion, different P-ligands together with acetate additives (Figure 8.5) [48]. It was found that a smaller amount of the catalyst was enough in the case of KOAc additive.

As a heterocyclic phosphonate, 3-phosphonothiophene (**7**) was synthesized by the P–C coupling of 3-bromothiophene with diethyl phosphite using Pd(OAc)$_2$ and dppf as the catalyst system, and NaOAc as an acetate additive in THF (Figure 8.6) [49].

The Pd(OAc)$_2$/ligand system was often applied in the synthesis of arylphosphonates, and this catalyst system was, in most cases, found more efficient than the classical Pd(PPh$_3$)$_4$ catalyst [50–60].

Aryl-imidazolium sulfuric derivatives were also suitable starting materials in coupling with dialkyl phosphites to afford the arylphosphonates (**4**) in high yields (Figure 8.7) [61]. However, the use of aryl-imidazolium sulfuric derivatives is not too practical from the point of view of atomic efficiency.

Montchamp and coworkers synthesized a series of phosphinates (**8**) by the reaction of aryl derivatives and *H*-phosphinates in solvent/co-solvent systems (Figure 8.8) [62, 63]. It was assumed that the role of the co-solvent is to facilitate the tautomerization

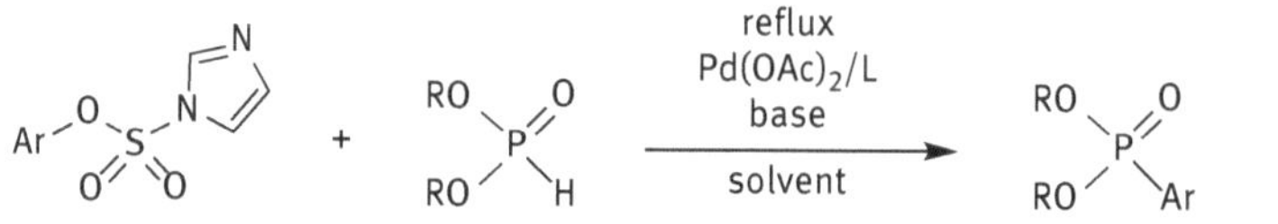

4, 85–99%

Ar = Ph, 4-$MeOC_6H_4$, 3-$MeOC_6H_4$, 4-MeC_6H_4, 4-PhC_6H_4, 4-ClC_6H_4, 4-$EtO_2CC_6H_4$, 4-MeC(O)C_6H_4, 1-naphthyl
R = Et, ^{i}Pr
L = dppp, dppf, XPhos
base: Et_3N, $^iPr_2NEt$, Na_2CO_3, Cs_2CO_3, K_3PO_4
solvent: toluene, DMF, DMA, acetonitrile, THF, dioxane

Figure 8.7: Aryl-imidazolium sulfuric derivatives as "aryl" sources in the Hirao reaction.

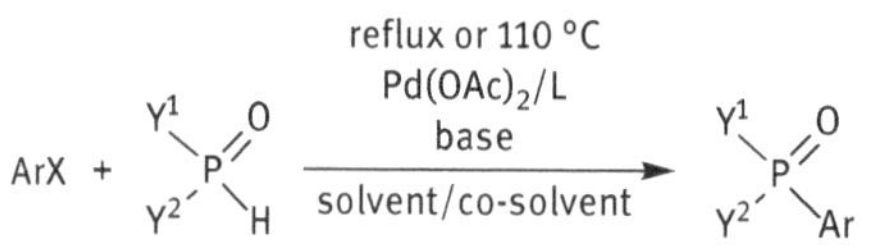

Y^1	octyl	^{c}hexyl	Me	$(CH_2)_4Ph$	$CMe(OEt)_2$
Y^2	OEt	OEt	OBu	OEt	OEt

8, 47–96%

Ar = Ph, 3,4,5-tri$MeOC_6H_2$, 2-ClC_6H_4, 4-$MeO_2CC_6H_4$, 2-$EtO_2CC_6H_4$, 4-$CF_3C_6H_4$, 4-CNC_6H_4, 4-$NO_2C_6H_4$, 2-$NO_2C_6H_4$, C_6F_5, 2-pyridyl, 2-pyrimidyl, 2-Cl-pyrazine
X = I, Cl, Br, OTf
L = PPh_3, dppf, dppe, dppp, BINAP, DPEphos, xantphos, polystyrene-supported Pd-nixantphos
base: $^iPr_2NEt$, pyridine, propylene oxide
solvent: toluene, DMF, ethanol, tertamyl alcohol, acetonitrile, ethylene glycol, dimethoxyethane
co-solvent: butanol, ethylene glycol, polyethylene glycol, polypropylene glycol, dimethoxyethane, tetramil alcohol, 1,3-diphenylurea, methylcellulose, diglyme

Figure 8.8: Utilization of the Hirao reaction in the synthesis of phosphinates.

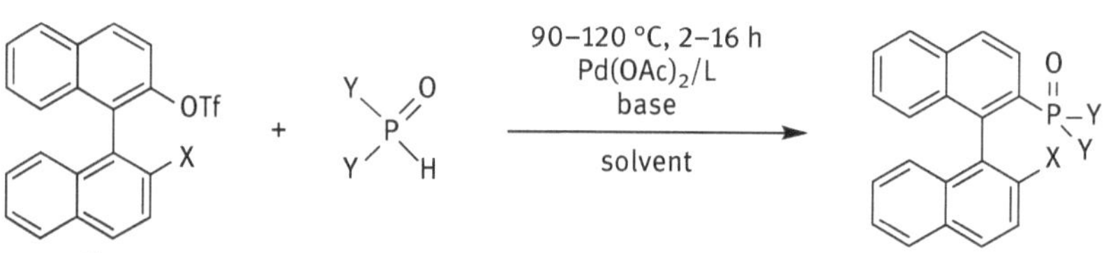

9

10, 10–90%

X = H, OH, OTf, NMe_2, CO_2Me, NHC(O)Me
Y = Ph, 4-$MeOC_6H_4$, 4-FC_6H_4, 3,5-diFC_6H_3, Me, ^{i}Pr, ^{c}Hex
L = dppp, dppb
base: $^iPr_2NEt$, $NaHCO_3$
solvent: DMSO, toluene

Figure 8.9: Utilization of the Hirao reaction in the synthesis of tertiary phosphine oxides.

of the >P(O)H species to the >P(OH) form. It was assumed that it is the trivalent form that reacts with the aryl derivatives. It is noteworthy that aryl chlorides could also be used in the coupling reactions.

In combination with bidentate P-ligands, $Pd(OAc)_2$ was used in the synthesis of 1,1-binaphthyl derivatives with a phosphinoyl function (**9**) (Figure 8.9) [64–74].

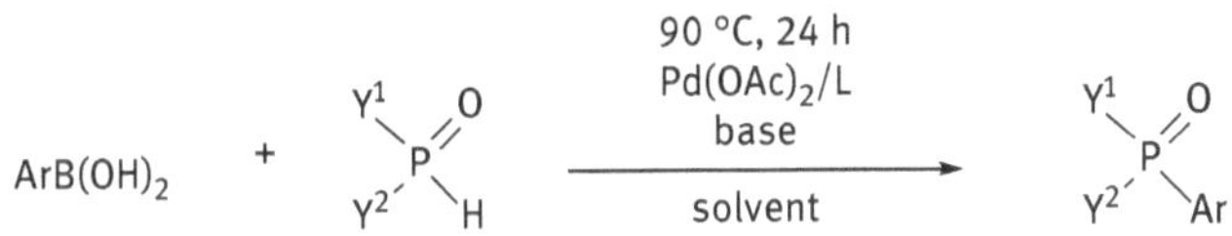

8, 20–95%

Ar = Ph, 4-$SMeC_6H_4$, 4-$MeOC_6H_4$, 4-$CF_3OC_6H_4$, 4-PhC_6H_4, 4-MeC_6H_4, 4-$MeO_2CC_6H_4$, 3-$CF_3C_6H_4$, 4-CNC_6H_4, 3-$NO_2C_6H_4$, 4-$HOCH_2C_6H_4$, 3-MeC(O)NH, 4-$NPh_2C_6H_4$, 2-naphthyl, 3-pyridyl

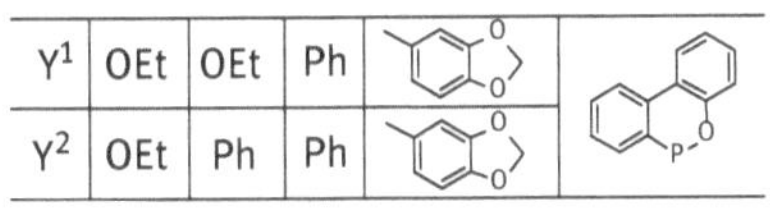

L = Ph_3, dppp, dppb, bpy
base: Et_3N, K_2CO_3, Na_2CO_3, Cs_2CO_3, K_3PO_4
solvent: dioxane, DMF, toluene, acetonitrile, dichloroethane

Figure 8.10: The use of arylboronic acids in the Hirao reaction.

Among others, diethyl phosphite, ethyl phenyl-*H*-phosphinate and diphenylphosphine oxide were reacted with arylboronic acids in the presence of $Pd(OAc)_2$ and different P-ligands (Figure 8.10) [75]. The best results were obtained with dppp as the P-ligand and K_2CO_3 as the base.

8.4 P–C Couplings in the Presence of Cu and Ni Catalysts

Other transition metal catalysts, such as Cu and Ni catalysts, were also applied in P–C coupling reactions. Regarding cost, this variation is cheaper than the Pd-catalyzed accomplishment. Cu and Ni catalysts can be used in the Hirao reaction, when the aryl derivative and/or the >P(O)H species are reactive enough.

8.4.1 Cu-catalyzed P–C Coupling Reactions

Beletskaya et al. synthesized diethyl arylphosphonates (**11**) by the coupling of aryl iodides and diethyl phosphite in the presence of CuI as the catalyst precursor and N-ligands (Figure 8.11) [76].

The Cu-catalyzed P–C coupling of 2-haloacetanilides and dialkyl phosphites or diphenylphosphine oxide was described in the presence of CuI and N-methylpyrrolidine-2-carboxamide as the N-ligand (Figure 8.12) [77].

Another example may be the Cu-catalyzed coupling of oxazoline-substituted aryl bromides with different diarylphosphine oxides (Figure 8.13) [78].

Zhao and coworkers studied the P–C coupling reaction of aryl halides with an *H*-phosphinate, a secondary phosphine oxide and dialkyl phosphites in the presence of the CuI/N-ligand catalyst systems to provide the phosphorylated products

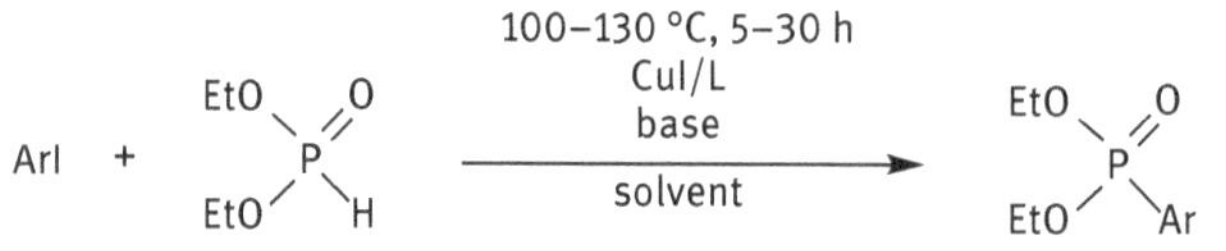

11, 42–100%

Ar = Ph, 4-MeC_6H_4, 4-$MeOC_6H_4$, 4-BrC_6H_4, 4-$MeC(O)C_6H_4$
L = bpy, phenanthroline, Me-imidazole, TMEDA, benzoin-oxime, salicylaldehyde-oxime, salicylaldehyde-phenylhydrazone
base:K_2CO_3, Cs_2CO_3
solvent toluene, dioxane, DMF

Figure 8.11: Example for a Cu-catalyzed Hirao reaction.

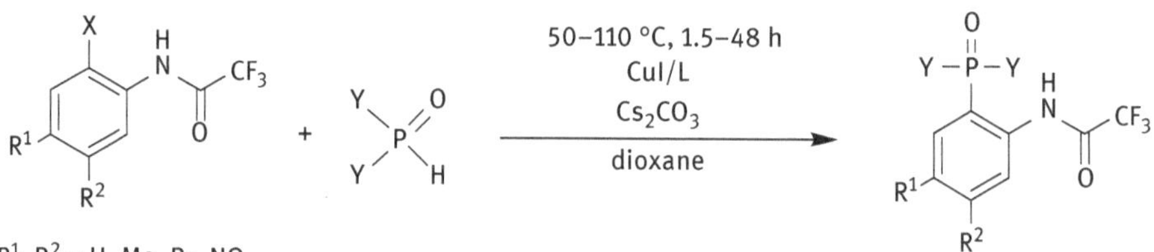

12, 50–93%

R^1, R^2 = H, Me, Br, NO_2
X = I, Br, Cl
Y = O^iPr, OBu, Ph
L = *N*-methylpyrrolidine-2-carboxamide

Figure 8.12: Hirao reactions starting from 2-haloacetanilides.

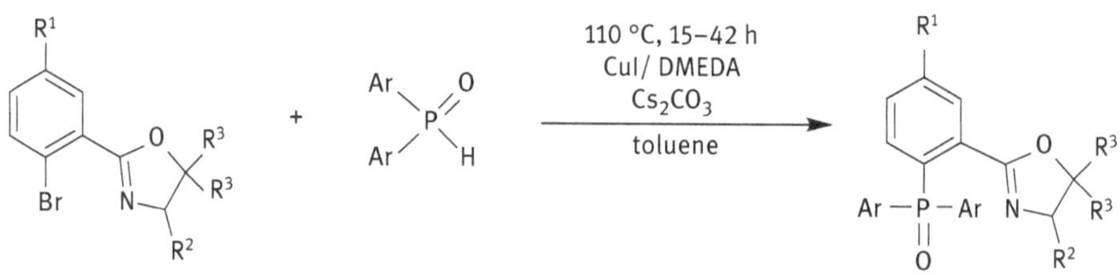

13 , 36–85%

R^1, R^2, R^3 = H, ^tBu, ^iPr, CF_3, Ph
Ar = 4-$CF_3C_6H_4$, 3,5-di$CF_3C_6H_3$

Figure 8.13: Hirao reactions providing triarylphosphine oxides.

(**8**) in variable yields (Figure 8.14) [79, 80]. The reactivity of aryl bromides could be enhanced via halogen exchange by the addition of KI.

Arylboronic acids may be suitable starting materials in the Cu-catalyzed coupling with dialkyl phosphites to furnish arylphosphonates (**6**) in yields of 54–96% (Figure 8.15) [81].

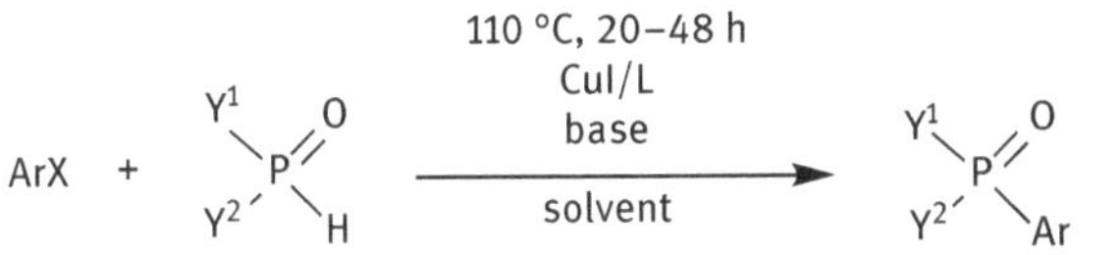

Y^1	OEt	Ph	OEt	O^iPr
Y^2	Ph	Ph	OEt	O^iPr

8, 24–91%

Ar = Ph, 4-MeC_6H_4, 4-ClC_6H_4, 4-BrC_6H_4, 4-$NO_2C_6H_4$
X = I, Br
L = pyrrolidine 2-phosphonic acid phenyl monoester, proline, pipecolinic-acid
base: Cs_2CO_3, K_2CO_3, DMAP
solvent: toluene, DMF

Figure 8.14: Additional examples for Cu-catalyzed P–C couplings.

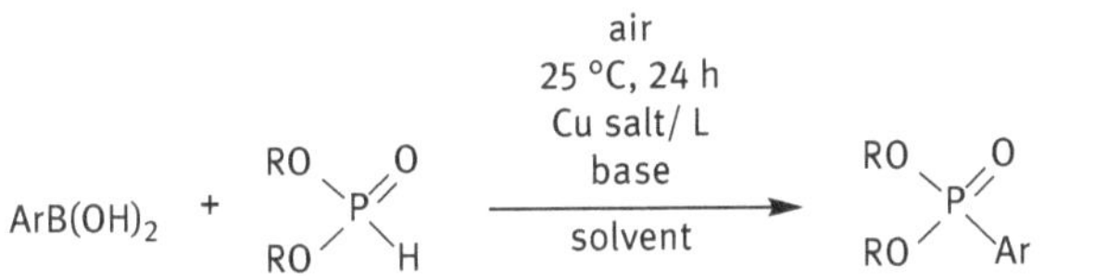

6, 54–96%

Ar = Ph, 4-MeC_6H_4, 3-MeC_6H_4, 2-MeC_6H_4, 4-$MeOC_6H_4$, 4-PhC_6H_4, 4-FC_6H_4, 4-$MeO_2CC_6H_4$, 4-$MeC(O)C_6H_4$, 4-$CF_3C_6H_4$, 4-CNC_6H_4, 3-$NO_2C_6H_4$, 2-MeO-4-BrC_6H_3, 4-vinyl-C_6H_4, 1-naphthyl, 2-naphthyl
R = Me, Et, ^{i}Pr, Bn
Cu salt: CuO, Cu2O, $Cu(OAc)_2$, $Cu(OTf)_2$, $CuCl_2$, $CuBr_2$, CuCl, CuBr, CuI
L = 1,10-phenantroline, TMEDA, DMEDA, hydroxyquinoline, 2-aminoaniline
base: Et_3N, $^iPr_2NEt$, K_2CO_3, Cs_2CO_3, pyridine
solvent: toluene, acetonitrile, CH_2Cl_2, THF, DME, DMSO

Figure 8.15: Arylboronic acids as starting materials in Cu-catalyzed P–C coupling reactions.

8.4.2 Ni-catalyzed P–C Coupling Reactions

Zhang et al. described the coupling of aryl mesylates and tosylates with >P(O)H reagents, such as diarylphosphine oxides, *H*-phosphinates and diethyl phosphite in the presence of the $NiCl_2$(dppf) catalyst system (Figure 8.16) [82]. Zn was added to the reaction mixture to promote the reduction of Ni(II) to Ni(0).

Aryl bromides were reacted with dimethyl phosphite and diphenylphosphine oxide in a similar way applying Ni catalysts (Figure 8.17) [83]. These reactions gave the best results in the presence of $NiCl_2$(dppp) and K_3PO_4 in dioxane.

Phenol derivatives could also be involved in P–C coupling reactions after converting the hydroxy function to a better leaving group by reaction with bromotripyrrolidinophosphonium hexafluorophosphate (PyBroP) (Figure 8.18) [84].

The reactions of aryl halides with diphenylphosphine oxides were also carried out using the Ni(Zn) catalyst together with N-ligands in water (Figure 8.19) [85].

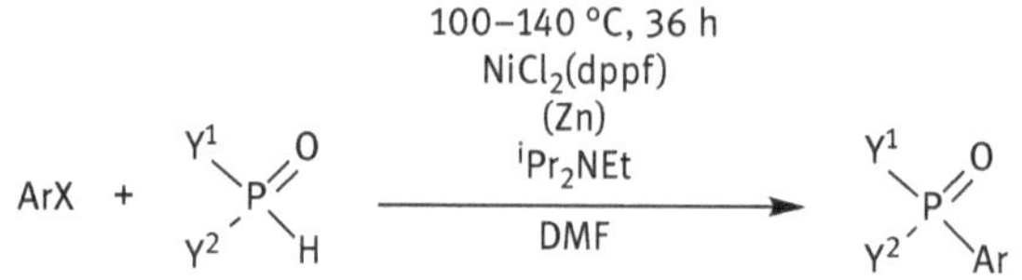

8, 29–92%

Ar = Ph, 4-MeC_6H_4, 3-MeC_6H_4, 2-MeC_6H_4, 4-MeOC_6H_4, 3-MeOC_6H_4, 2-MeOC_6H_4, 4-MeO_2CC_6H_4, 1-naphthyl, 2-naphthyl
X = OMs, OTs

Y^1	Ph	3,5-diMeC_6H_3	4-MeOC_6H_4	4-MeC_6H_4	4-$CF_3C_6H_4$	OEt	OEt
Y^2	Ph	3,5-diMeC_6H_3	4-MeOC_6H_4	4-MeC_6H_4	4-$CF_3C_6H_4$	Ph	OEt

Figure 8.16: Ni-catalyzed Hirao reactions including a bidentate P-ligand.

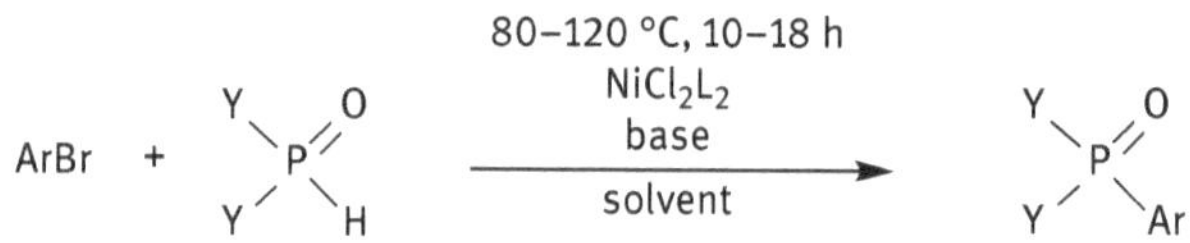

8, 28–97%

Ar = 4-MeOC_6H_4, 4-MeC(O)C_6H_4, 3-MeC(O)C_6H_4,4-CNC_6H_4, 2-CNC_6H_4, 1-naphthyl, 2-naphthyl, 3-quinolinyl, 5-indolyl, 2-pyrazinyl, 2-pyridyl
Y = OMe, Ph
L = PPh_3, P(^{c}hexyl)$_3$, dppe, dppp, dppf, $AsPh_3$
base: K_2CO_3, Na_2CO_3, Cs_2CO_3, K_3PO_4, NaOAc, ^{t}BuOK, ^{i}Pr$_2$NEt
solvent: dioxane, toluene, acetonitrile, DMF, THF

Figure 8.17: Additional P–C couplings in the presence of the Ni(II)/P-ligand catalyst system.

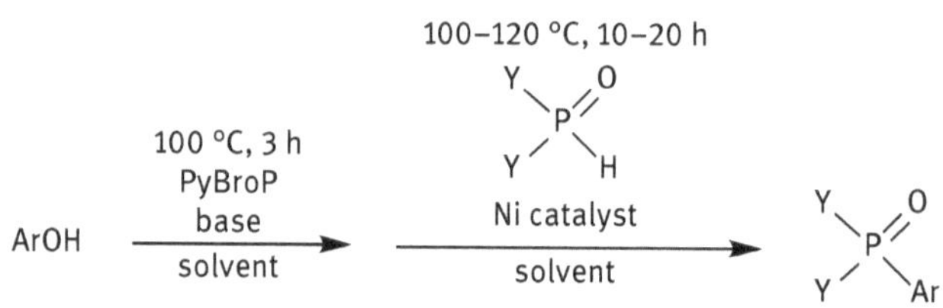

8, 30–98%

Ar = 4-MeOC_6H_4, 4-PhC_6H_4, 4-MeC(O)C_6H_4, 4-CNC_6H_4, 1-naphthyl, 2-naphthyl, 2-quinolinyl, 8-quinolinyl, 5-phenanthridinyl, 4-carbazolyl, 2-pyridyl
Y = Ph, OMe, OEt
Ni catalyst: $NiCl_2$(dppp), $NiCl_2$(dppe), $NiCl_2$(dppf), $NiCl_2$(P(^{c}hexyl)$_3$)$_2$
base: K_3PO_4, K_2CO_3
solvent: dioxane, toluene, acetonitrile

Figure 8.18: Hirao reaction after activating the hydroxy group of phenols.

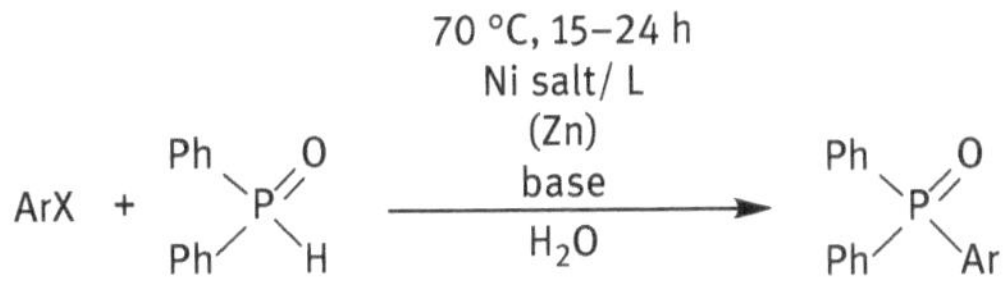

14, 75–97%

Ar = Ph, 4-$NH_2C_6H_4$, 3-$NH_2C_6H_4$, 2-$NH_2C_6H_4$, 4-OHC_6H_4, 4-PhC_6H_4, 4-MeC_6H_4, 4-$MeOC_6H_4$, 3-$MeOC_6H_4$, 4-ClC_6H_4, 4-BrC_6H_4, 4-IC_6H_4, 4-MeC(O)C_6H_4, 1-naphthyl, 2-naphthyl
X = I, Br

Y^1	Ph	3,5-diMeC_6H_3	4-$MeOC_6H_4$	4-MeC_6H_4	4-$CF_3C_6H_4$	OEt	OEt
Y^2	Ph	3,5-diMeC_6H_3	4-$MeOC_6H_4$	4-MeC_6H_4	4-$CF_3C_6H_4$	Ph	OEt

Ni salt: $NiCl_2$*$6H_2O$, $Ni(OAc)_2$*$6H_2O$, $Ni(NO_3)_2$*$6H_2O$
L = bpy, pyridine, dibenzylethylenediamine, dianilinoethane, TMEDA

Figure 8.19: Ni-catalyzed Hirao reaction including N-ligands.

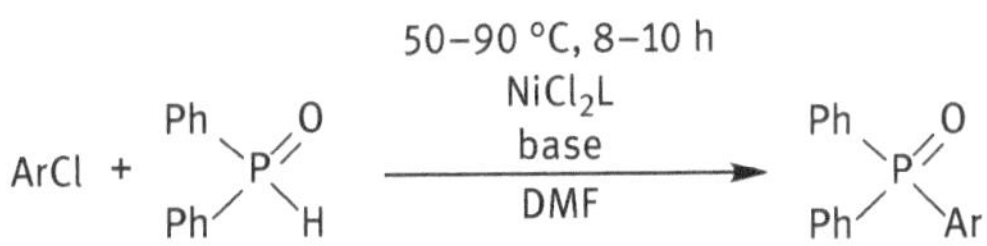

14, 46–87%

Ar = 4-$MeOC_6H_4$, 4-MeC_6H_4, 4-$CF_3C_6H_4$, 4-CNC_6H_4, 3-CNC_6H_4, 2-CNC_6H_4, 4-NHC(O)MeC $_6H_4$, 1-naphthyl, 3-(*N*-Me-indolyl), 8-quinolyl, 2-benzothiazolyl
L = dimethoxyethane
base: K_2CO_3, ^{t}BuONa

Figure 8.20: Hirao reaction in the presence of $NiCl_2$-dimethoxyethane.

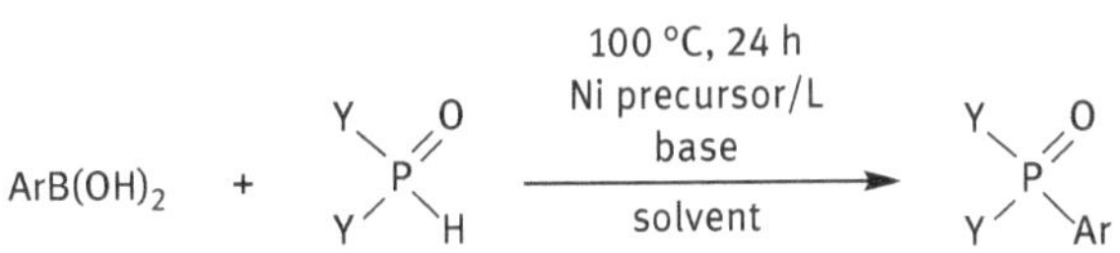

8, 50–97%

Ar = Ph, 4-PhC_6H_4, 4-$MeOC_6H_4$, 4-MeC_6H_4, 2-MeC_6H_4, 4-$CF_3OC_6H_4$, 4-$CF_3C_6H_4$, 4-FC_6H_4, 4-ClC_6H_4, 4-MeC(O)C_6H_4, 4-$MeO_2CC_6H_4$, 2-naphthyl
Y = Ph, 4-MeC_6H_4, 4-$MeOC_6H_4$, 4-$NMeC_6H_4$, 4-FC_6H_4, Pent, OEt

Ni precursor: $NiBr_2$, $NiCl_2$, $Ni(OAc)_2$, $Ni(acac)_2$, $Ni(PPh_3)_2Cl_2$
L = pyridine, bpy, dppp, TMEDA, 1,10-phenantroline
base: K_2CO_3, Cs_2CO_3, pyridine, Et_3N
solvent: dichloroethane, dioxane, toluene, DMF

Figure 8.21: Ni-catalyzed P–C coupling reactions starting from arylboronic acids.

A special Ni catalyst comprising dimethyoxyethane as the ligand was used in the P–C coupling of aryl/hetaryl chlorides and diphenylphosphine oxide. The triarylphosphine oxides (**14**) were obtained in acceptable yields (Figure 8.20) [86].

Zhao et al. performed the reaction of arylboronic acids with different >P(O)H reagents in the presence of Ni catalysts comprising different N- and P-ligands (Figure 8.21) [87]. $NiBr_2$ with pyridine as the ligand is a good choice for the catalyst, and the corresponding products (**8**) were obtained in variable yields.

8.5 Green Chemical Approaches of the Hirao Reaction

8.5.1 Microwave-assisted and Phase Transfer Catalyzed P–C Coupling Reactions

"Greener" and more effective accomplishments of the Hirao reaction were elaborated in the past two decades. The widely used microwave (MW) and phase transfer catalytic techniques offered advantages also in the coupling reactions under discussion.

The first MW-assisted Hirao reaction was reported to take place between aryl halides/triflates and diethyl phosphite using bis(triphenylphosphine)palladium-dichloride as the catalyst, triethylamine as the base, and triethylsilane as the reducing agent (Figure 8.22) [88]. The coupling reaction was carried out in a Teflon autoclave using a domestic MW oven.

The cross-coupling of dialkyl phosphites with aryl and vinyl halides/triflates was also investigated in the presence of tetrakis(triphenylphosphine)palladium under MW irradiation (Figure 8.23) [89]. The best results (72–96%) were obtained using Cs_2CO_3 as the base in THF.

The MW technique was successfully applied in the synthesis of progesterone receptor antagonist P-containing 11β-aryl-substituted steroids (Figure 8.24) [90].

Several arylboronic acids and arylfluoroborates were reacted with dialkyl phosphites employing the $Pd(OAc)_2$ or $Pd(O_2CCF_3)_2$/dmphen catalyst system and *p*-benzoquinone without the addition of a base (Figure 8.25) [91]. The authors assumed a

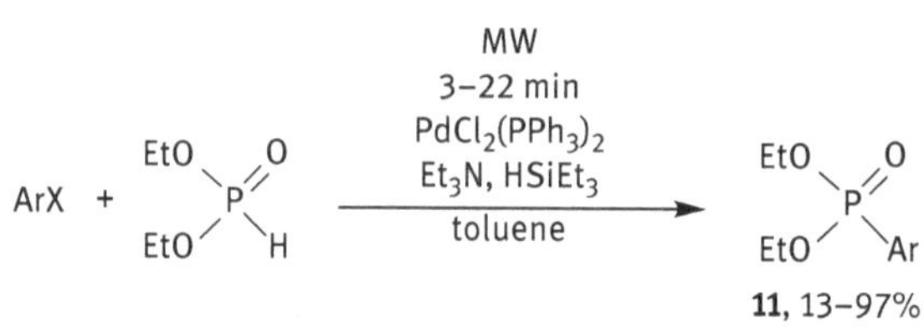

Ar = 4-MeC_6H_4, 4-$MeOC_6H_4$, 4-BrC_6H_4, 4-ClC_6H_4, 4-$HO_2CC_6H_4$, 4-$MeO_2CC_6H_4$, 3-$MeO_2CC_6H_4$, 2-$MeO_2CC_6H_4$, 2-$MeO_2CCH_2C_6H_4$, 2-$MeOCH_2C_6H_4$, 4-MeC(O)NHC_6H_4, 4-$MeSO_3C_6H_4$
X = I, Br, Cl, OTf

Figure 8.22: The first Hirao reaction under MW conditions.

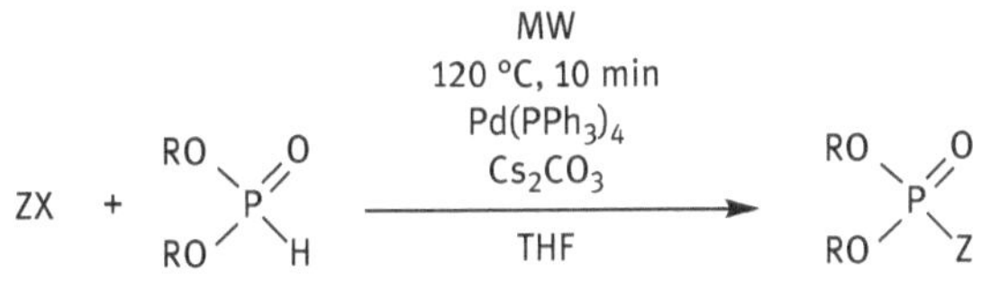

1, 72–96%

Z = Ph, 4-$MeOC_6H_4$, 4-$NO_2C_6H_4$, 1-naphthyl, 2-naphthyl, 9-phenanthryl, 3-pyridyl, vinyl, 1-methylvinyl, 2-methylvinyl
X = I, Br, OTf
R = Me, Et, ^{i}Pr, Bn

Figure 8.23: MW-assisted P–C coupling in the presence of $Pd(PPh_3)_4$.

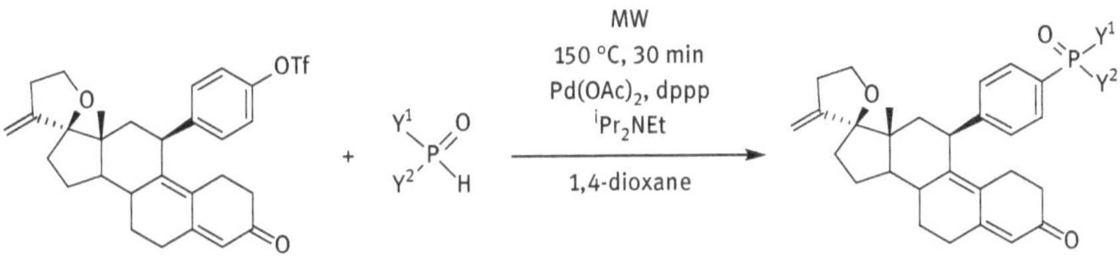

Y = Me, OMe, OEt, OCH_2CF_3, Ph, 4-ClC_6H_4, OPh

15, 19–52%

Figure 8.24: The synthesis of 11β-aryl-substituted steroids utilizing the Hirao reaction.

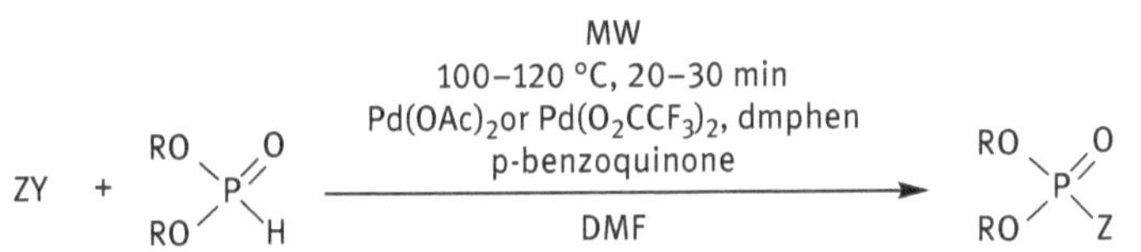

1, 37–90%

Z = Ph, 4-MeC_6H_4, 3,4-$(MeO)_2C_6H_3$, 3-PhC_6H_4, 4-BrC_6H_4, 3-BrC_6H_4, 4-$MeO_2CC_6H_4$, 4-MeC(O)C_6H_4, 1-naphthyl
Y = $B(OH)_2$, BF_3K
R = Me, Et

Figure 8.25: P–C coupling of arylboronic acids and arylfluoroborates under MW conditions.

mechanism involving a "reoxidant for the regeneration of Pd." The authors of this chapter believe that the use of an oxidant in the Hirao reaction is mistaken. Rather, a reductive agent should be used.

A new cyclodiphosphazane-containing Pd catalyst was developed and tested in the synthesis of triarylphosphine oxides from aryl bromides and diphenylphosphine oxide (Figure 8.26) [92]. The use of this exotic catalyst and Cs_2CO_3 as a base in acetonitrile led to yields of 46–95% under MW conditions.

Beletskaya et al. investigated solid–liquid phase transfer catalyzed P–C coupling reactions (Figure 8.27) [93–96]. The beneficial effect of the triethylbenzylammonium chloride (TEBAC) catalyst was proved. The experimental data are listed in Table 8.1.

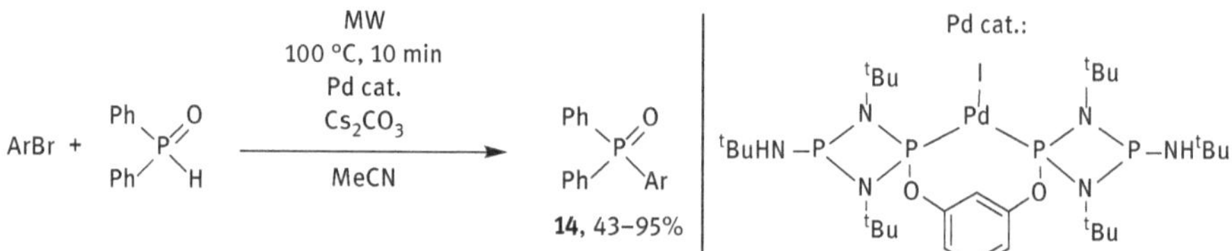

Ar = Ph, 2-MeC_6H_4, 4-MeC_6H_4, 3-$MeOC_6H_4$, 4-$MeOC_6H_4$, 4-CNC_6H_4, 2-$HOCH_2C_6H_4$, 2-naphthyl

Figure 8.26: A Hirao reaction catalyzed by a cyclodiphosphazane-containing Pd complex.

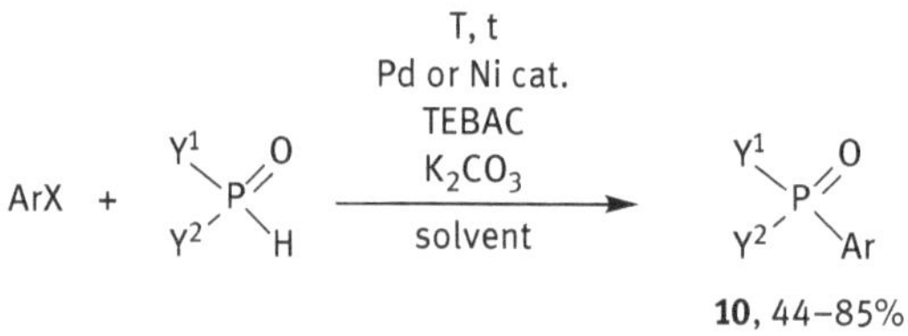

Ar = Ph, 4-$MeOC_6H_4$, 4-MeC_6H_4, 4-$ClOC_6H_4$, 4-$HO_2C_6H_4$, 4-$EtO_2CC_6H_4$, 4-CNC_6H_4, 9-Br-10-antracenyl
X = I, Br
For Y^1Y^2 P(O)H, catalyst and solvent, see Table 8.1.

Figure 8.27: A phase transfer catalyzed P–C coupling protocol.

Table 8.1: Hirao reaction in the presence of TEBAC as the phase transfer catalyst.

$Y^1Y^2P(O)H$	Catalyst	Solvent	Temp. (°C)	Time (h)	Yield (%)	Ref.
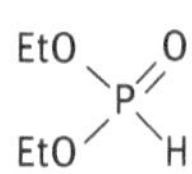	$PdCl_2$, $Pd(OAc)_2$	– or MeCN	60–80	6–64	52–86	[93, 94]
	$Pd(OAc)_2/PPh_3$ $Pd(OAc)_2/P(2\text{-}furyl)_3$	MeCN	70–80	6–40	69–85	[94]
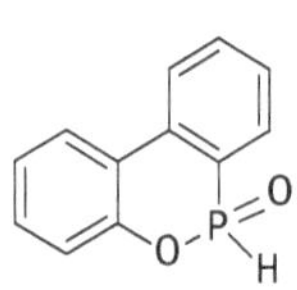	$PdCl_2$				73	
	$PdCl_2(PPh_3)_2$, $NiCl_2(PPh_3)_2$	Dioxane	100	6	50–75	[95]
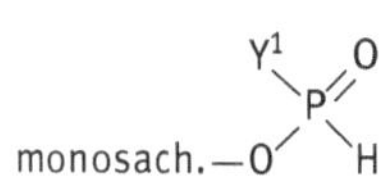 Y^1 = OEt, OBu, Ph, cyclo-$C_6H_{11}O$,SugO	$PdCl_2(PPh_3)_2$	Dioxane	90–100	1.5–5[a]	55–67	[96]

[a] In the case of the coupling with butyl glucofuranosyl phosphite, combination of the phase transfer catalytic and the MW techniques led to reduced (from 3 h to 14 min) reaction times.

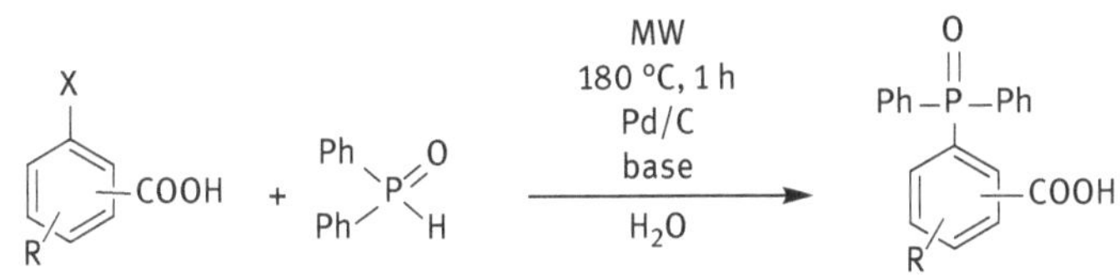

17, 18–87%

X = 3-I, 4-I, 3-Br, 4-Br, 4-Cl
R = H, 3-NH_2, 4-MeO, 4-Me
base: Na_2CO_3, K_2CO_3, NaOH, Et_3N

Figure 8.28: P–C coupling in aqueous medium applying Pd/C under MW conditions.

The $Pd(OAc)_2$-catalyzed phosphonylation of aryl bromides with diethyl phosphite required a reaction time of 64 h in the absence of added P-ligands, while the coupling with the more reactive iodides was complete after 6 h under the same conditions. In the presence of P-ligands, the P–C couplings were more efficient [94]. The arylation of 6*H*-dibenzo[*c*,*e*][1,2]oxaphosphinine 2-oxide was performed in the presence of different Pd and Ni catalysts in dioxane. Both the P-ligand-free and the P-ligand included versions were efficient [95]. "Phosphorylated" monosaccharides were also applied as >P(O)H reagents [96].

Interestingly, the reaction of iodo- and bromobenzoic acids with diphenylphosphine oxide took place in water applying the Pd/C catalyst under MW irradiation (Figure 8.28) [97]. In this instance, the use of tetrabutylammonium bromide had no effect on the reaction.

8.5.2 "P-ligand-free" Hirao Reactions

Keglevich and Jablonkai realized that the transition metal-catalyzed P–C coupling reactions may take place in the presence of $Pd(OAc)_2$ without any added P-ligand under solvent-free and MW-assisted conditions (Figure 8.29) [98, 99].

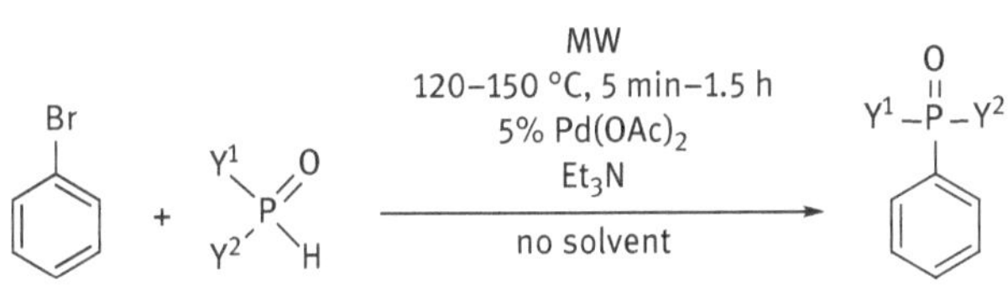

(1,5 equiv.)

18, 41–92%

Y^1	OEt	OBu	OEt	O^nPr	O^iPr	O^nBu	O^iBu	O^iPent	Ph
Y^2	OEt	OBu	Ph	Ph	Ph	Ph	Ph	Ph	Ph

Figure 8.29: MW-assisted Hirao reaction using $Pd(OAc)_2$ without an added P-ligand.

The reaction of bromobenzene and 1.5 equivalents of the >P(O)H species (dialkyl phosphites, phenyl-*H*-phosphinates and diphenylphosphine oxide) was performed in the presence of 5% $Pd(OAc)_2$ as the catalyst, and 1.1 equivalents of triethylamine as the base. The relevance of the MW technique was shown by comparative thermal experiments. In the case of diethyl phosphite, the conversion was almost complete at 120°C, but the best results (yields of 84–92%) were obtained at 150°C. The MW-assisted P–C couplings were also efficient in respect of reaction time (5 min to 1.5 h).

Then, the $Pd(OAc)_2$-catalyzed "P-ligand-free" P–C coupling reactions were extended to substituted bromobenzenes (Figure 8.30). It was found that both electron-donating and electron-withdrawing substituents decrease the reactivity, as in these cases higher temperatures (175–200°C) were necessary to obtain arylphosphonates in good yields (69–92%) (Table 8.2). The methoxy- and alkyl-substituted bromoarenes turned out to be the least reactive; in these instances, occasionally, a temperature of 200°C, and 10% of the $Pd(OAc)_2$ catalyst had to be applied.

Hirao et al. described another "P-ligand-free" variation (Figure 8.31) [100]. Diethyl phosphite was reacted with 2-nitro-5-bromoanisole applying $Pd(OAc)_2$ as the catalyst and Na_2CO_3 as the base in xylene. After a reaction time of 24 h, the corresponding arylphosphonate was isolated in a yield of 69%.

Xiao and his coworkers described the Pd-catalyzed coupling of an arylsulfinate salt and dialkyl phosphites using $PdCl_2$ without the addition of a P-ligand (Figure 8.32) [101]. 4-Methylphenylphosphonates were obtained in good yields (60–96%) applying silver carbonate as the oxidant under MW irradiation. Here, we note again the probable unnecessity of an oxidant during the Hirao reaction.

The Keglevich group found that $NiCl_2$ may also be a suitable catalyst in the P–C coupling of brombenzene and different >P(O)H species (Figure 8.33) [102].

The experiments were carried out at 150°C under MW irradiation. Applying 1.5 equivalents of triethylamine in the absence of any solvent, the reaction of diethyl phosphite and brombenzene required 2 h, but the formation of byproducts reduced the yield (67%). The use of K_2CO_3 as the base in acetonitrile was more advantageous: in the presence of 5% $NiCl_2$ and after a reaction time of 45 min, the required product could be obtained in a yield of 92%. Using $NiCl_2$ at 150°C, the alkyl diphenylphosphi-

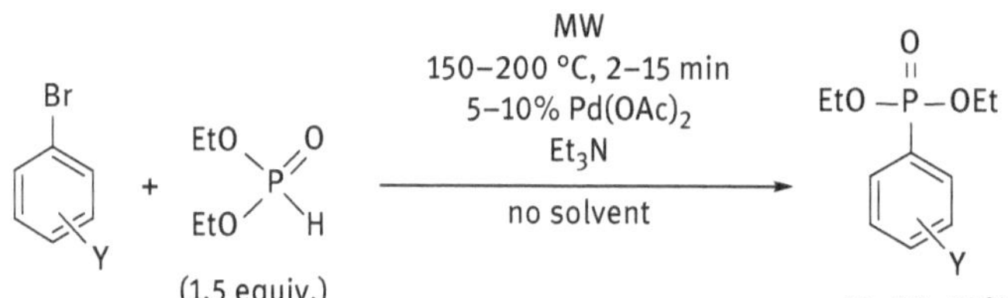

19, 60–95%

Y = H, 4-MeO, 3-MeO, 4-Pr, 4-Et, 4-Me, 4-Cl, 3-Cl, 4-F, 3-F, 4-CO_2Et, 3-CO_2Et, 4-C(O)Me, 3-C(O)Me

Figure 8.30: P–C coupling of aryl bromides using $Pd(OAc)_2$ as the catalyst.

Table 8.2: P–C coupling of bromoarenes and diethyl phosphite.

Y	$Pd(OAc)_2$ (%)	Temp. (°C)	Time (perc)	Conversion (%)	Yield (19) (%)
H	5	150	5	99	93
4-MeO	10	200	2	80	69
3-MeO	10	200	2	93	79
4-Pr	10	200	2	86	71
4-Et	10	175	15	93	85
4-Me	10	175	10	86	73
4-Cl	10	175	10	95	83
3-Cl	10	175	10	95	87
4-F	5	175	5	99	91
3-F	5	175	10	100	88
4-CO_2Et	5	175	15	100	89
3-CO_2Et	10	200	2	93	81
4-C(O)Me	5	175	5	96	71
3-C(O)Me	10	175	5	100	92

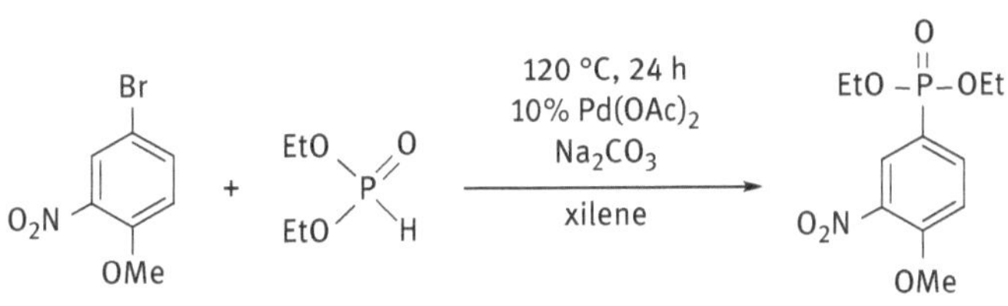

20, 69%

Figure 8.31: Another $Pd(OAc)_2$-catalyzed "P-ligand-free" Hirao reaction.

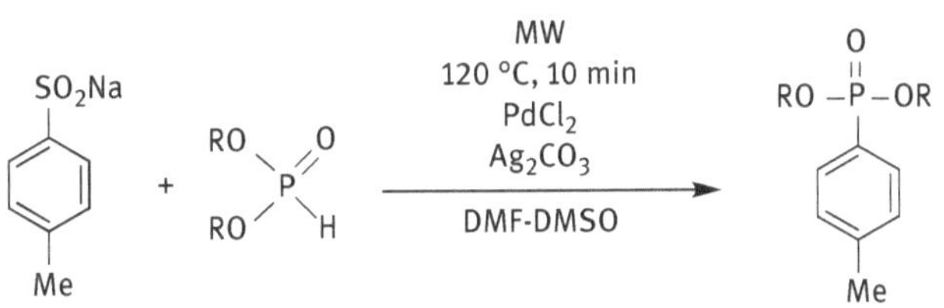

R = Me, Et, ^{n}Pr, ^{i}Pr, ^{n}Bu, ^{i}Bu, CH_2CH_2OMe

16, 60–96%

Figure 8.32: An MW-assisted P-ligand-free desulfitative phosphonation.

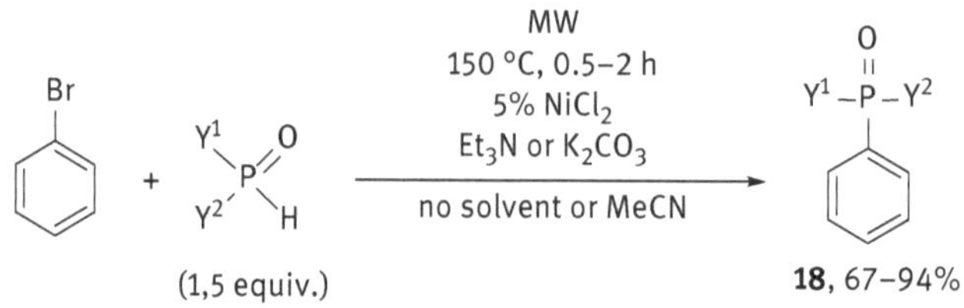

Y^1	OEt	OBu	OEt	O^nPr	O^iPr	O^nBu	O^iBu	O^iPent
Y^2	OEt	OBu	Ph	Ph	Ph	Ph	Ph	Ph

Y^1	Ph	4-$MeOC_6H_4$	4-$^tBuC_6H_4$	4-MeC_6H_4	3-MeC_6H_4	4-ClC_6H_4	4-FC_6H_4
Y^2	Ph	4-$MeOC_6H_4$	4-$^tBuC_6H_4$	4-MeC_6H_4	3-MeC_6H_4	4-ClC_6H_4	4-FC_6H_4

Figure 8.33: P–C couplings using $NiCl_2$ as the catalyst without added P-ligands under MW conditions.

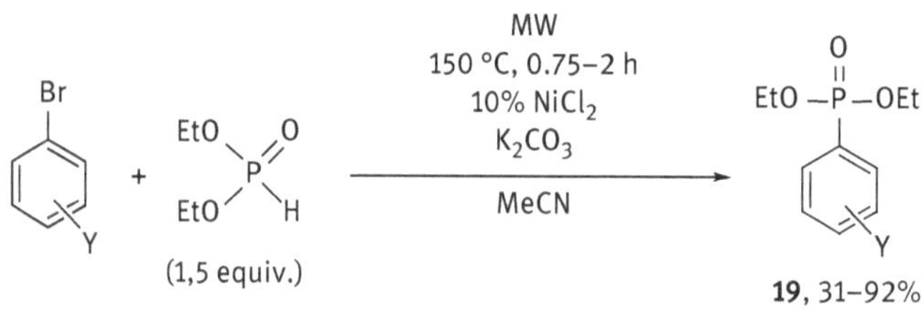

Y = H, 4-MeO, 3-MeO, 4-^{t}Bu, 4-Pr, 4-Et, 4-Me, 3-Me, 3-Cl, 4-F, 3-F

Figure 8.34: $NiCl_2$-catalyzed Hirao reactions of bromoarenes with diethyl phosphite.

nates were obtained in high yields (84–89%) from phenyl-*H*-phosphinates after an irradiation of 30 min. Diphenylphosphine oxide and other aryl-substituted secondary phosphine oxides were also suitable reagents in the "P-ligand-free" Hirao reaction under discussion. The $NiCl_2$-catalyzed phosphonylation of substituted bromoarenes led to similar results as in the presence of $Pd(OAc)_2$, but the scope of the aryl bromides was somewhat limited (Figure 8.34) [102].

Both Pd- and Ni-catalyzed "P-ligand-free" reactions are suitable for the coupling of bromoarenes and various >P(O)H species. Considering the reaction conditions, costs and safety issues, one can conclude that the use of $Pd(OAc)_2$ is favorable, but the application of $NiCl_2$ may be a good alternative too. These recent developments and extensions of the Hirao reaction mean a big step further, since there is no need for expensive and sensitive P-ligands. The authors of this chapter proved that in the "P-ligand-free" accomplishment, the excess of the >P(O)H reagent existing under a tautomeric equilibrium may serve as the P-ligand. The excess of the >P(O)H species may also promote the reduction of Pd(II) before entering the catalytic cycles.

Finally, a new catalyst-free method was developed for the reaction of halobenzoic acids and secondary phosphine oxides in aqueous medium (Figure 8.35) [103]. 4-Iodo-, 3-bromo- and 4-bromobenzoic acids were reacted with diaryl phosphine

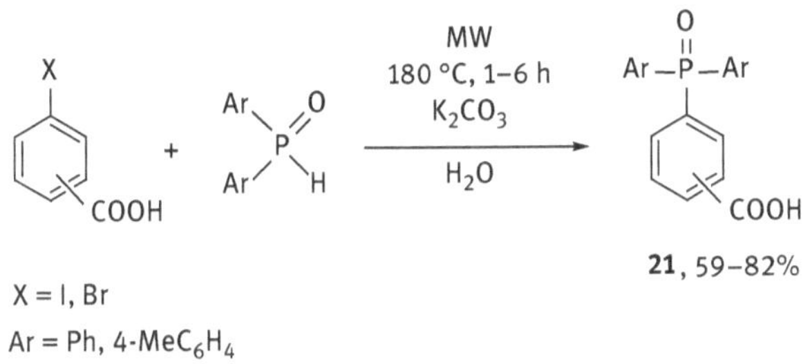

Figure 8.35: Catalyst-free P–C coupling reaction in water.

oxides in the presence of K_2CO_3 at 180°C under MW conditions. The only limitation of this green P–C coupling reaction is the low water solubility of the secondary phosphine oxides.

8.6 Summary

This chapter may give a good overview on the present state of the art of the Hirao reaction belonging to the important group of coupling reactions. The Hirao reaction utilizing a series of aryl derivatives and >P(O)H reagents along with a Pd, Ni or Cu catalyst gives access to arylphosphonates, tertiary phosphine oxides and related compounds that may be useful intermediates. The segment of chemistry outlined hides interesting green chemical challenges, such as MW activation, solvent- and catalyst-free accomplishments, and even phase transfer catalysis.

Acknowledgment: This work was supported by the National Research, Development and Innovation Office (K119202).

References

[1] E. Jablonkai, G. Keglevich, *Curr. Org. Synth.*, 2014, **11**, 429.
[2] E. Jablonkai, G. Keglevich, *Org. Prep. Proceed. Int.*, 2014, **46**, 281.
[3] T. Hirao, T. Masunaga, Y. Ohshiro, T. Agawa, *Tetrahedron Lett.*, 1980, **21**, 3595.
[4] T. Hirao, T. Masunaga, N. Yamada, Y. Ohshiro, T. Agawa, *Bull. Chem. Soc. Jpn.*, 1982, **55**, 909.
[5] T. Hirao, T. Masunaga, Y. Ohshiro, T. Agawa, *Synthesis*, 1981, 56.
[6] X. Lu, J. Zhu, *Synthesis*, 1987, 726.
[7] P. Machnitzki, T. Nickel, O. Stelzer, C. Landgrafe, *Eur. J. Inorg. Chem.*, **1998**, 1029.
[8] R. F. Ziessel, L. J. Charbonnière, S. Mameri, F. Camerel, *J. Org. Chem.*, 2005, **70**, 9835.
[9] M. Kant, S. Bischoff, R. Siefken, E. Gründemann, A. Köckritz, *Eur. J. Org. Chem.*, 2001, 477.
[10] J. J. Bulot, E. E. Aboujaoude, N. Collignon *Phosphorus, Sulfur*, 1984, **21**, 197.

[11] S. S. Chauhan, A. Varshney, B. Verma, M. W. Pennington, *Tetrahedron*, 2007, **48**, 4051.
[12] J. A. Bennett, E. G. Hope, K. Singh, A. M. Stuart, *J. Fluorine Chem.*, 2009, **130**, 615.
[13] I. **Řehoř**, V. Kubiček, J. Kotek, P. Hermann, I Lukeš, J. Száková, L. V. Elst, R. N. Muller, J. A. Peters, *J. Mater. Chem.*, 2009, **19**, 1494.
[14] Y.-C. Kim, S. G. Brown, T. K. Harden, J. L. Boyer, G. Dubyak, B. F. King, G. Burnstock, K. A. Jacobson, *J. Med. Chem.*, 2001, **44**, 340.
[15] D. A. Holt, J. M. Erb, *Tetrahedron Lett.*, 1989, **30**, 5393.
[16] K. S. Petrakis, T. L. Nagabhushan, *J. Am. Chem. Soc.*, 1987, **109**, 2831.
[17] M. A. Kazankova, I. G. Trostyanskaya, S. V. Lutsenko, I. P. Beletskaya, *Tetrahedron Lett.*, 1999, **40**, 569.
[18] P. Zhong, Z. X. Xiong, X. Huang, *Synth. Commun.*, 2000, **30**, 273.
[19] Y. Kobayashi, A. D. William, *Adv. Synth. Catal.*, 2004, **346**, 1749.
[20] M. Ghalib, B. Niaz, P. G. Jones, J. W. Heinicke, *Tetrahedron Lett.*, 2012, **53**, 5012.
[21] Y. Kobayashi, A. William, Y. Tokoro, *J. Org. Chem.*, 2001, **66**, 7903.
[22] J. Zhang, Y. Xu, G. Huang, H. Guo, *Tetrahedron Lett.*, 1988, **29**, 1955.
[23] Y. Xu, J. Zhang, *J. Chem. Soc.*, 1986, 1606.
[24] Y. Xu, H. Wei, J. Zhang, G. Huang, *Tetrahedron Lett.*, 1989, **30**, 949.
[25] B. M Trost, R. Radinov, *J. Am. Chem. Soc.*, 1997, **119**, 5962.
[26] S. N. L. Bennett, R. G. Hall, *J. Chem. Soc.*, 1995, 1145.
[27] R. G. Hall, P. Riebli, *Synlett*, 1999, 1633.
[28] J.-H. Xie, L.-X. Wang, Y. Fu, S.-F Zhu, B.-M Fan, H.-F. Duan, Q.-L. Zhou, *J. Am. Chem. Soc.*, 2003, **125**, 4404.
[29] K. Muthukumaran, R. S. Loewe, A. Ambroise, S.-i. Tamaru, Q. Li, G. Mathur, D. F. Bocian, V. Misra, J. S. Lindsey, *J. Org. Chem.*, 2004, **69**, 1444.
[30] Y. Xu, Z. Li, Xia, J. H. Guo, Y. Huang, *Synthesis*, 1983, 377.
[31] Y. Xu, J. Zhang, *Synthesis*, 1984, 778.
[32] H.-J. Cristau, A. Hervé, F. Loiseau, D. Virieux, *Synthesis*, 2003, **14**, 2216.
[33] M. Schuman, X. Lopez, M. Karplus, V. Gouvemeur, *Tetrahedron*, 2001, **57**, 10299.
[34] Y. Xu, Z. Li, J. Xia, H. Guo, Y. Huang, *Synthesis*, 1984, 781.
[35] D. A. Rankic, M. Parvez, B. A. Keay, *Tetrahedron: Asymmetry*, 2012, **23**, 754.
[36] G. Németh, Z. Greff, A. Sipos, Z. Varga, R. Székely, M. Sebestyén, Z. Jászay, Sz. Béni, Z. Nemes, J.-L. Pirat, J.-N. Volle, D. Virieux, Á. Gyuris, K. Kelemenics, É. Áy, J. Minarovits, S. Szathmary, G. Kéri, L. Őrfi, *J. Med. Chem.*, 2014, **57**, 3939.
[37] L. Li, A. Li, L. Song, Z.-H. Wang, X.-H. Zhou, T. Yang, W. Huang, *J. Mol. Struct.*, 2014, **1067**, 37.
[38] G. P. Luke, W. C. Shakespeare, *Synth. Commun.*, 2002, **32**, 2951.
[39] S. Montel, C. Midrier, J.-N. Volle, R. Braun, K. Haaf, L. Willms, J.-L. Pirat, D. Virieux, *Eur. J. Org. Chem.*, 2012, 3237.
[40] K. Matsumura, H. Shimizu, T. Saito,Kumobayashi, H. *Adv. Synth. Catal.*, 2003, **345**, 180.
[41] C. Wang, G. Yang, J. Zhuang, Zhang, W. *Tetrahedron Lett.*, 2010, **51**, 2044.
[42] T. Oshiki, T. Imamoto, *J. Am. Chem. Soc.*, 1992, *114*, 3975.
[43] T. Imamoto, T. Oshiki, T. Onozawa, M. Matsuo, T. Hikosaka, M. Yanagawa, *Heteroat. Chem.*, 1992, **3**, 563.
[44] J. R. Moncarz, T. J. Brunker, J. C. Jewett, M. Orchowski, D. S. Glueck, *Organometallics*, 2003, **22**, 3205.
[45] B. H. Lipshutz, D. J. Buzard, C. Y. Yun, *Tetrahedron Lett.*, 1999, **40**, 201.
[46] A. J. Bloomfield, S. B. Herzon, *Org. Lett.*, 2012, **14**, 4370.
[47] M. Kalek, J. Stawinski, *Organometallics*, 2007, **26**, 5840.
[48] M. Kalek, M. Jezowska, J. Stawinski, *Adv. Synth. Catal.*, 2009, **351**, 3207.
[49] M. Lilley, B. Mambwe, R. F. W. Jackson, R. Muimo, *Chem. Commun.*, 2014, **50**, 9343.
[50] L. J. Gooßen, M. K. Dezfuli, *Synlett*, 2005, 445.

[51] A. Bessmertnykh, C. M. Douaihy, R. Guilard, *Chem. Lett.*, 2009, **38**, 738.
[52] Y. Belabassi, S. Alzghari, J.-L. Montchamp, *J. Organomet. Chem.*, 2008, **693**, 3171.
[53] I. Bonnaventure, A. B. Charette, *J. Org. Chem.*, 2008, **73**, 6331.
[54] M. Nandi, J. Jin, T. V. RajanBabu, *J. Am. Chem. Soc.*, 1999, **121**, 9899.
[55] Y.-Y. Yan, T. V. RajanBabu, *J. Org. Chem.*, 2000, **65**, 900.
[56] L. H. Davies, B. Stewart, R. W. Harrington, W. Clegg, L. J. Higham, *Agnew. Chem. Int. Ed.*, 2012, **51**, 4921.
[57] S. P. Cummings, J. Savchenko, P. E. Fanwick, A. Kharlamova, T. Ren, *Organometallics*, 2013, **32**, 1129.
[58] B. Reisinger, N. Kuzmanovic, P. Löffler, R. Merkl, B. König, R. Sterner, *Angew. Chem. Int. Ed.*, 2014, **53**, 595.
[59] M. Bazaga-García, R. M. P. Colodrero, M. Papadaki, P. Garczarek, J. Zoń, P. Olivera-Pastor, E. R. Losilla, L. León-Reina, M. A. G. Aranda, D. Choquesillo-Lazarte, K. D. Demadis, A. Cabeza, *J. Am. Chem. Soc.*, 2014, **136**, 5731.
[60] N. D. Contrella, J. R. Sampson, R. F. Jordan, *Organometallics*, 2014, **33**, 3546.
[61] Y. Luo, J. Wu, *Organometallics*, 2009, **28**, 6823.
[62] E. L. Deal, C. Petit, J.-L. Montchamp, *Org. Lett.*, 2011, **13**, 3270.
[63] O. Berger, C. Petit, E. L. J.-L. Montchamp, *Adv. Synth. Catal.*, 2013, **355**, 1361.
[64] L. Kurz, G. Lee, D. Morgans Jr., M. J. Waldyke, T. Ward, *Tetrahedron Lett.*, 1990, **31**, 6321.
[65] T. Hamada, A. Chieffi, J. Åhman, S. L. Buchwald, *J. Am. Chem. Soc.*, 2002, **124**, 1261.
[66] S. Y. Cho, M. Shibasaki, *Tetrahedron Lett.*, 1998, **39**, 1773.
[67] **Š.** Vyskočil, M. Smrčina, V. Hanuš, M. Polášek, P. Kočovský, *J. Org. Chem.*, 1998, **63**, 7738.
[68] Y. Uozumi, N. Suzuki, A. Ogiwara, T. Hayashi, *Tetrahedron*, 1994, **50**, 4293.
[69] Y. Uozumi, A. Tanahashi, S.-Y. Lee, T. Hayashi, *J. Org. Chem.*, 1993, **58**, 1945.
[70] M. Shi, C.-Q. Li, *Tetrahedron: Asymmetry*, 2005, **16**, 1385.
[71] T. Hayashi, *Acc. Chem. Res.*, 2000, **33**, 354.
[72] J.-X. Chen, J. F. Daeuble, J. M. Stryker, *Tetrahedron*, 2000, **56**, 2789.
[73] K. Ding, Y. Wang, H. Yun, J. Liu, Y. Wu, M. Terada, Y. Okubo, K. Mikami, *Chem. Eur. J.*, 1999, **5**, 1734.
[74] C. Anstiss, P. Karuso, M. Richardson, L. Liu, *Molecules*, 2013, **18**, 2788.
[75] T. Fu, H. Qiao, Z. Peng, G. Hu, X. Wu, Y. Gao, Y. Zhao, *Org. Biomol. Chem.*, 2014, *12*, 2895.
[76] N. B. Karlstedt, I. P. Beletskaya, *Russ. J. Org. Chem.*, 2011, *47*, 1011.
[77] D. Jiang, Q. Jiang, H. Fu, Y. Jiang, Y. Zhao, *Synthesis*, 2008, 3473.
[78] N. T. McDougal, J. Streuff, H. Mukherjee, S. C. Virgil, B. M. Stoltz, *Tetrahedron Lett.*, 2010, *51*, 5550.
[79] H. Rao, Y. Jin, H. Fu, Y. Jiang, Y. Zhao, *Chem. Eur. J.*, 2006, *12*, 3636.
[80] C. Huang, X. Tang, H. Fu, Y. Jiang, Y. Zhao, *J. Org. Chem.*, 2006, *71*, 5020.
[81] R. Zhuang, J. Xu, Z. Cai, G. Tang, M. Fang, Y. Zhao, *Org. Lett.*, 2011, *13*, 2110.
[82] C. Shen, G. Yang, W. Zhang, *Org. Biomol. Chem.*, 2012, *10*, 3500.
[83] Y.-L. Zhao, G.-J. Wu, Y. Li, L.-X. Gao, F.-S. Han, *Chem. Eur. J.*, 2012, *18*, 9622.
[84] Y.-L. Zhao, G.-J. Wu, F.-S. Han, *Chem. Commun.*, 2012, *48*, 5868.
[85] X. Zhang, H. Liu, X. Hu, G. Tang, J. Zhu, Y. Zhao, *Org. Lett.*, 2011, *13*, 3478.
[86] H.-Y Zhang, M. Sun, Y.-N. Ma, Q.-P. Tian, S.-D. Yang, *Org. Biomol. Chem.*, 2012, *10*, 9627.
[87] G. Hu, W. Chen, T. Fu, Z. Peng, H. Qiao, Y. Gao, Y. Zhao, *Org. Lett.*, 2013, *15*, 5362.
[88] D. Villemin, P.-A. Jaffrès, F. Simèon, *Phosphorus, Sulfur, Silicon*, 1997, *130*, 59.
[89] M. Kalek, A. Ziadi, J. Stawinski, *Org. Lett.*, 2008, *10*, 4637.
[90] W. Jiang, G. Allan, J. J. Fiordeliso, O. Linton, P. Tannenbaum, J. Xu, P. Zhu, J. Gunnet, K. Demarest, S. Lundeen, Z. Sui, *Bioorg. Med. Chem.*, 2006, *14*, 6726.
[91] M. Andaloussi, J. Lindh, J. Sävmarker, P. J. R. Sjöberg, M. Larhed, *Chem. Eur. J.*, 2009, *15*, 13069.

[92] G. S. Ananthnag, J. T. Mague, M S. Balakrishna, *Dalton Trans.*, 2015, *44*, 3785.
[93] Z. S. Novikova, N. N. Demik, A. Yu. Agarkov, I. P. Beletskaya, *Russ. J. Org. Chem.*, 1995, *31*, 129.
[94] M. M. Kabachnik, M. D. Solntseva, V. V. Izmer, Z. S. Novikova, I. P. Beletskaya, *Russ. J. Org. Chem.*, 1998, *34*, 93.
[95] I. P. Beletskaya, E. G. Neganova, Yu. A. Veits, *Russ. J. Org. Chem.*, 2004, **40**, 1782.
[96] I. P. Beletskaya, N. B. Karlstedt, E. E. Nifant'ev, D. V. Khodarev, T. S. Kukhareva, A. V. Nikolaev, A. J. Ross, *Russ. J. Org. Chem.*, 2006, **42**, 1780.
[97] S. M. Rummelt, M. Ranocchiari, J. A. van Bokhoven, *Org. Lett.*, 2012, *14*, 2188.
[98] E. Jablonkai, G. Keglevich, *Tetrahedron*, 2013, *54*, 4185.
[99] G. Keglevich, E. Jablonkai, L. B. Balázs, *RSC Adv.*, 2014, **4**, 22808.
[100] T. Amaya, Y. Abe, Y. Inada, T. Hirao, *Tetrahedron Lett.*, 2014, **55**, 3976.
[101] J. Li, X. Bi, H. Wang, J. Xiao, *RSC Adv.*, 2014, **4**, 19214.
[102] E. Jablonkai, L. B. Balázs, G. Keglevich, *Curr. Org. Chem.*, 2015, **19**, 197.
[103] E. Jablonkai, G. Keglevich, *Tetrahedron Lett.*, 2015, **56**, 1638.

Tamara Kovács and György Keglevich

9 Deoxygenation of phosphine oxides

Abstract: The deoxygenation of phosphine oxides to phosphines is important in the preparation of P-ligands, and in the regeneration of phosphine oxide wastes. The possibilities for the deoxygenation are surveyed in this chapter. Hydrides may be suitable reagents, but silanes, such as $PhSiH_3$, Cl_3SiH, $Me_2SiH-O-HSiMe_2$ (TMDS) and $[-MeSiH-]_n$ (PMHS) form the real choice for reducing agents. TMDS and PMHS are cheap and user-friendly reagents, whose lower reactivity may be overcome by using them under microwave-assisted and solvent-free conditions. Mechanistic aspects are also discussed besides the special methods for deoxygenation.

Keywords: phosphine oxides, deoxygenation, phosphines, silanes.

9.1 Application of tertiary phosphines

Tertiary phosphines (PR_3) are widely used as ligands in transition metal complexes and as extractants, or as starting materials and reagents in various organic chemical syntheses [1, 2]. Favourable features making possible their applications include good coordination ability, strong nucleophilicity and oxygen binding ability. Their use as ligands in transition metal (Pd, Pt and Rh) complexes is important, as these species may be applied in homogeneous catalytic hydrogenations and hydroformylations, as well as in hydrosilylation and palladium-catalyzed coupling reactions [3]. Tertiary phosphines may be the starting materials for quaternary phosphonium salts used as surfactants, ionic liquids or phase transfer catalysts. Typical examples for the application of tertiary phosphines as reagents in various organic reactions include the Wittig reaction (formation of C=C bonds), the Mitsunobu reaction (formation of C–O or C–N bonds) and Appel-type reactions (formation of C–halogen bonds) [4–6]. The driving force of the reactions is the high affinity of phosphines towards oxygen. When a P–O bond is formed, ~400 kJ/mol energy is released. The mentioned reactions are of poor atom economy that is the consequence of the formation of the phosphine oxide byproduct that is a waste. Phosphines should be regenerated and recycled. Reduction is the simplest method to regenerate phosphines from phosphine oxides.

https://doi.org/10.1515/9783110535839-009

9.2 Synthesis of tertiary phosphines

Basically, there are three major methods for the synthesis of phosphines. The first of these possibilities comprises the substitution reactions of carbon electrophiles with alkali metal phosphides (Figure 9.1-(1A)), or substitution reactions of carbon nucleophiles with P–Cl derivatives (Figure 9.1-(1B)) [7]. The second route is the hydrophosphination of alkenes or alkynes (Figure 9.1-(2)) [8], while the third possibility is the reduction of phosphine oxides (Figure 9.1-(3)) [9].

9.3 Deoxygenation of tertiary phosphine oxides

The reduction of phosphine oxides is the simplest method to obtain phosphines. The P=O bonds are highly stable thermodynamically. With a bond energy of around 502 kJ mol^{-1}, the P=O bonds are more stable than other functional groups (e.g. the C=O moiety that is easily reduced by catalytically activated hydrogen or sodium borohydride) [10]. Depending on the nature of the reducing agent and on the presence of additives, the reduction may occur with retention or inversion of the stereogenic phosphorus atom.

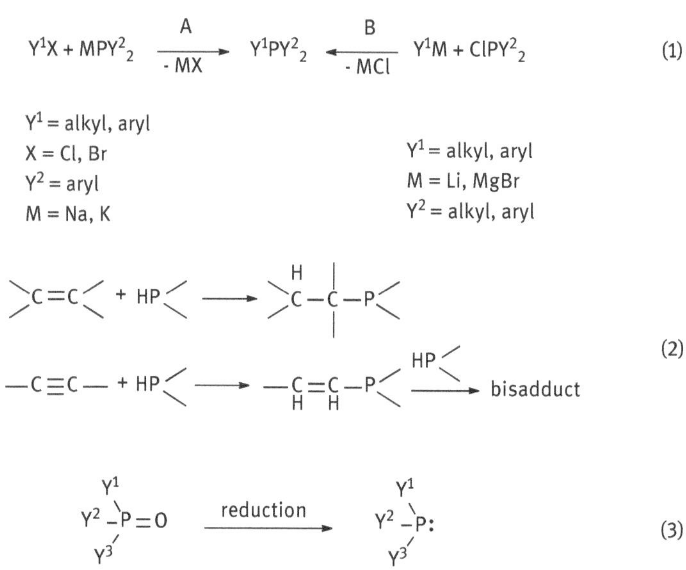

Figure 9.1: Synthesis routes to tertiary phosphines.

9.3.1 Deoxygenation of phosphine oxides by metal hydrides

Metal hydrides, such as aluminum hydrides (e.g. lithium aluminum hydride ($LiAlH_4$) alone, or combined with cerium chloride ($CeCl_3$), sodium aluminum hydride ($NaAlH_4$), sodium aluminum tetrachloride ($NaAlCl_4$) and di-isobutylaluminum hydride (DIBAL-H)) were employed for the deoxygenation of phosphine oxides [11]. The reduction of triphenylphosphine oxide using $LiAlH_4$ alone was not too efficient, as PPh_3 was obtained in only a yield of 54% (Figure 9.2-(1)) [12]. However, the application of additives, such as Lewis acids led to better conversions. The use of $CeCl_3$ together with $LiAlH_4$ resulted in a yield of 95% at a temperature of 40 °C after a reaction time of 30 min [13]. The $LiAlH_4$–$CeCl_3$ system associated with $NaBH_4$ allowed the one-pot preparation of phosphine–boranes from phosphine oxides (Figure 9.2-(2)). Hence, the air-sensitive phosphines were stabilized as boranes [14].

The reduction of tertiary phosphine oxides is also possible using amine-assisted aluminum hydrides. The method is characterized by mild conditions, short reaction times, high efficiency and a wide substrate scope [15].

The direct reduction of phosphine oxides by borane complexes is also possible. Borane (used as different complexes) is much more user friendly than the metal hydrides are, and its application leads to stable and easy to handle phosphine–boranes. Triphenylphosphine oxide may be deoxygenated by reaction with alkylboranes at a temperature of 120–180 °C [16], but the reduction of strained five-ring phosphine oxides may be achieved with borane–dimethyl sulfide complex under mild conditions (25–65 °C). If the molecule has an unsaturation, besides the expected deoxygenation and phosphine–borane formation, partial reduction of the double bond may also take place (Figure 9.3). Therefore, the incompatibility with double bonds may be a drawback of the borane reductions [17].

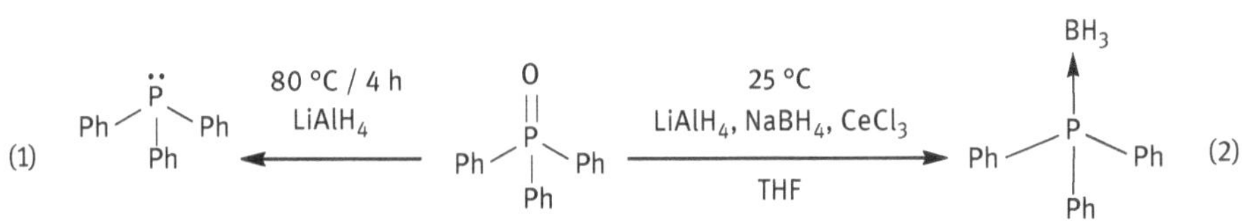

Figure 9.2: Deoxygenation of triphenylphosphine oxide by $LiAlH_4$.

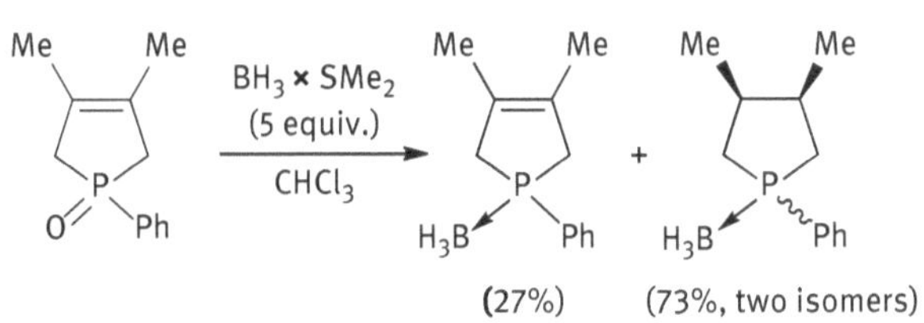

Figure 9.3: Reduction of five-ring phosphine oxides with the borane–dimethyl sulfide complex.

The reduction with borane is dependent on the ring strain. While the reduction of the more strained five-membered cyclic phosphine oxides may take place, there is no reaction with the six-ring phosphine oxides.

Secondary phosphine oxides can be converted expeditiously to secondary phosphine–boranes by treatment with an excess (3–10 equivalents) of $BH_3 \cdot SMe_2$ or $BH_3 \cdot THF$ at room temperature. The expected products were obtained in moderate to good yields (40–100%). In a few cases, the formation of phosphinous acid–boranes was also observed, even in the presence of the large excess of the reducing agent (Figure 9.4). Selectivity towards the formation of secondary phosphine–boranes could be improved by the addition of a small amount of water to the reaction mixture.

This method was extended to the reduction of phosphine oxides with proximal α- or β-hydroxy groups. The placement of a hydroxy group in the vicinity of the P=O function resulted in an intramolecular activating effect. The presence of a neighboring α- or β-hydroxy group in the molecule is a key factor enabling the reduction. The lack of a proximal hydroxy group or its longer distance from the phosphorus atom may cause unreactivity of the P=O functions towards BH_3 [18].

The mechanism of the deoxygenation of bridged phosphine oxides with boranes has been described. The oxygen atom of the bridging P=O function initiates a nucleophilic attack on the boron atom of borane. This is followed by a hydride migration leading to an intermediate containing a pentacoordinated phosphorus atom. Finally, the intermediate with trigonal–bipyramidal geometry is stabilized by the elimination of BH_2OH. After deprotonation, phosphine reacts with the excess of borane to furnish the phosphine–borane complex (Figure 9.5) [19].

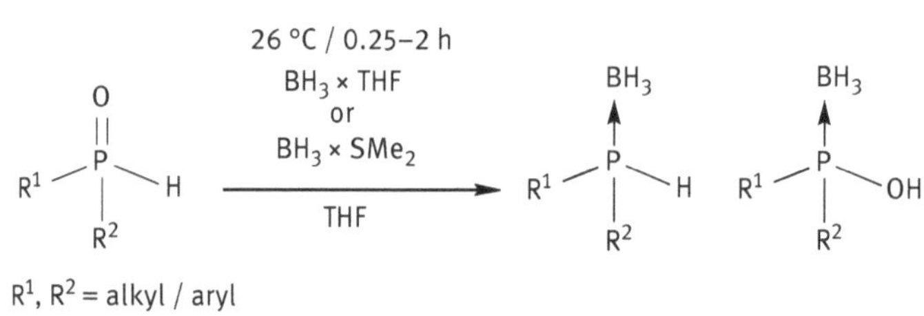

Figure 9.4: The reaction of secondary phosphine oxides with borane complexes.

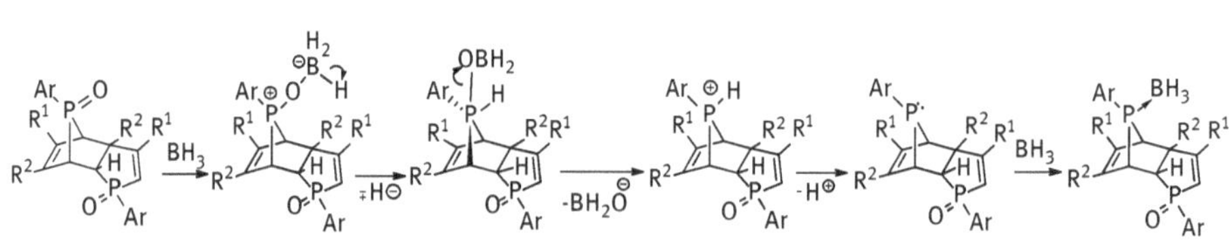

Figure 9.5: The mechanism of deoxygenation of phosphine oxides with borates.

9.3.2 Reduction of phosphine oxides by silanes

9.3.2.1 Deoxygenation of Phosphine Oxides by Phenylsilanes

The use of phenylsilanes, such as $PhSiH_3$, Ph_2SiH_2 and Ph_3SiH as reducing agents have a long history. They have been applied for the deoxygenation of phosphine oxides since 1964. All of them could be used without any solvent. $PhSiH_3$ was applied in the range of room temperature to 150 °C, while Ph_3SiH at 300 °C [20, 21]. Obviously, $PhSiH_3$ is the best choice that may utilize three hydrogen atoms during the deoxygenation. However, $PhSiH_3$ is rather expensive.

Phenylsilane reduced selectively the P=O bonds of a bis(secondary phosphine oxide) bearing two allyl groups. The resulting bisalkylphosphine underwent two consecutive intramolecular hydrophosphinations to form eventually the corresponding bicyclic tertiary phosphine (Figure 9.6) [22].

Unsaturated secondary phosphine oxides are important intermediates. The conventional method using phenylsilane can be applied well for the deoxygenation of secondary ethynylphosphine oxides to furnish the corresponding phosphines selectively in yields of 55–75% (Figure 9.7) [23].

Phosphonates may be reduced to primary phosphines with phenylsilane or diphenylsilane in the presence of tris(pentafluorophenyl)borane. The reaction was slow at room temperature. However, no change was observed at a higher temperature. The yields were generally better with phenylsilane than with diphenylsilane (ca. 95% vs. 25%) (Figure 9.8). The formation of unidentified byproducts could not be avoided [23].

Applying $PhSiH_3$ in the reduction of five- and six-membered ring phosphine oxides, phosphines were obtained in yields of ≥85% (Figure 9.9) [24].

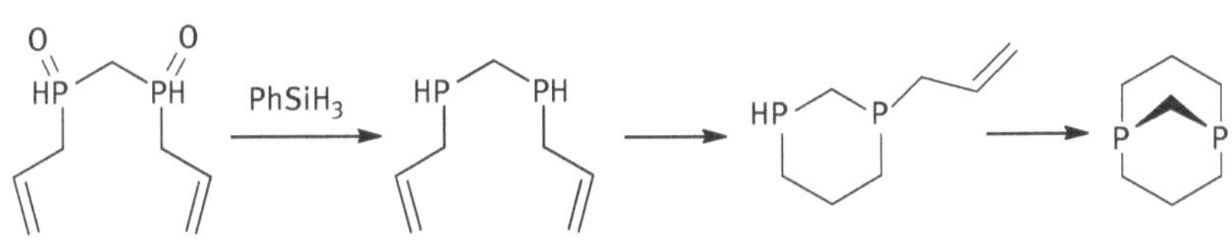

Figure 9.6: Reaction of a bis(secondary phosphine oxide) with phenylsilane.

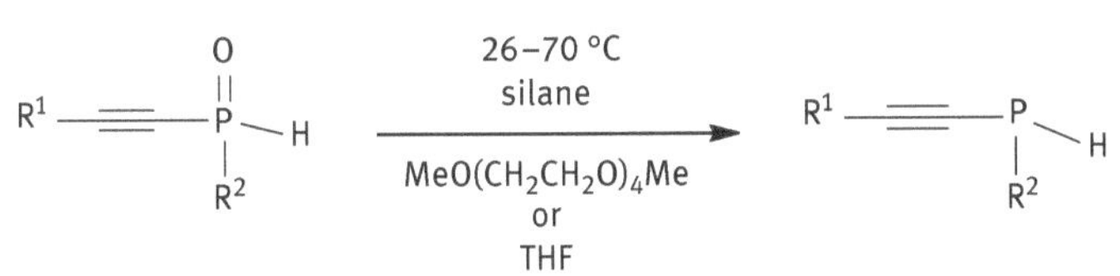

R^1	R^2	Yield (%)
H	*i*-Pr	65
H	*t*-Bu	55
Ph	*i*-Pr	70
Ph	*t*-Bu	70
Ph	Et	70

Figure 9.7: Deoxygenation of unsaturated secondary phosphine oxides with phenylsilane.

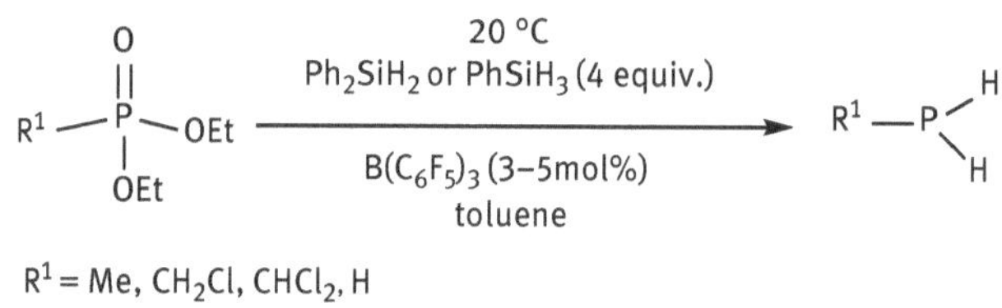

Figure 9.8: The reaction of phosphonates with phenylsilanes.

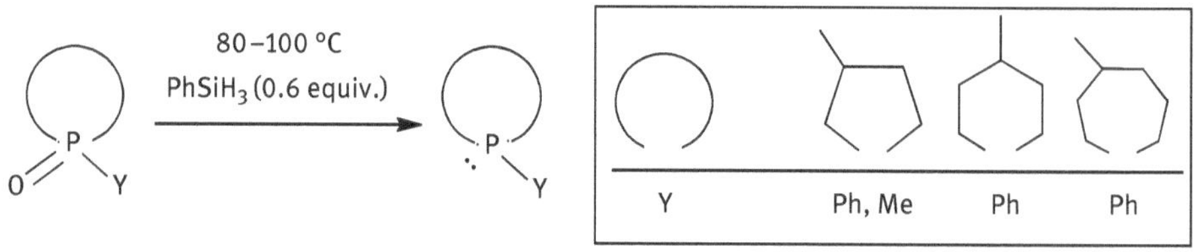

Figure 9.9: Deoxygenation of five- and six-membered ring phosphine oxides with phenylsilane.

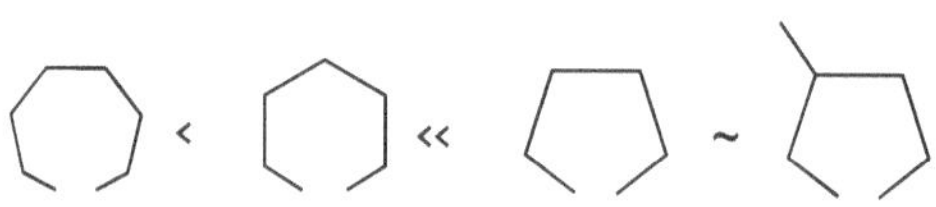

Figure 9.10: Reactivity of the phosphine oxides according to the ring size.

The effect of the ring size of cyclic phosphine oxides on the rate of the reduction by silanes was investigated, as well as the influence of a 3-methyl substituent in the ring. It was found that the remote 3-methyl substituent on 1-phenylphospholane oxide had practically no influence. Six- and seven-membered phosphine oxides were reduced rather slowly, with less than 20% conversion after a 3-h reaction time (Figure 9.10) [25].

From among phenylsilanes, $PhSiH_3$ is the reagent of choice that is a robust reducing agent. The only drawback is the higher price of this reagent.

9.3.2.2 Deoxygenation of Phosphine Oxides using Trichlorosilane

Within silanes, perhaps trichlorosilane (Cl_3SiH) is the most frequently used reagent in synthetic organic chemistry. Cl_3SiH is rather volatile (b.p.: 31.8 °C) and corrosive; hence, it is advisable to use it together with a tertiary amine. As one variation, the reduction of triphenylphosphine oxide with Cl_3SiH was performed in an autoclave at 200 °C without any solvent (Table 9.1, entry 1). The reaction was also efficient in boiling benzene using 2 equivalents of Cl_3SiH (Table 9.1, entry 2) [21]. It was observed that the need for 2 equivalents of Cl_3SiH may be avoided by the addition of triethylamine. In these cases, the use of 1.1 equivalents of Cl_3SiH was sufficient (Table 9.1, entry 3). Cl_3SiH was not found to be efficient in the deoxygenation of 1-phenyl-2-phospholene oxide in benzene solution together with triethylamine, as the yield was low (24%)

Table 9.1: Deoxygenation of different phosphine oxides by trichlorosilane.

Entry	Starting material	Product	Silane	Equiv.	Solvent	T (°C)	t (h)	Yield (%)	Ref.
1	Ph_3P=O	Ph_3P	Cl_3SiH	2	–	200	2	90	[21]
2	Ph_3P=O	Ph_3P	Cl_3SiH	2	PhH	78	2	98	[21]
3	Ph_3P=O	Ph_3P	Cl_3SiH^a	1.1	PhH	78	2	85	[21]
4	$(C_4H_9)_3P$=O	$(C_4H_9)_3P$	Cl_3SiH^a	1.1	PhH	80	2	92	[21]
5	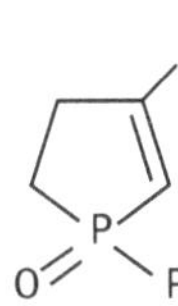 (A)	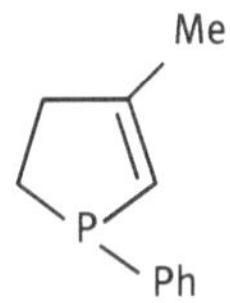	Cl_3SiH^a	1.1	PhH	78	2	24	[21]
6	(A)		Cl_3SiH^b	1.4	PhMe	110	2	91	[26]
7	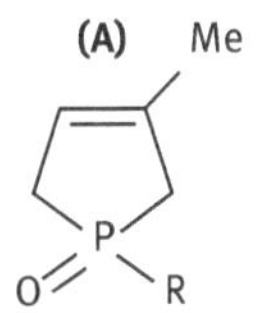 R = Ph, Et, Pr, Bu, ^{i}Bu, ^{i}Pent	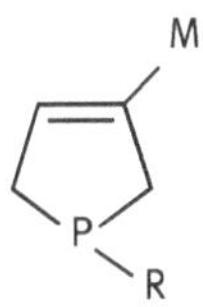	Cl_3SiH	5.9	PhH	0	3	~100	[26]
8	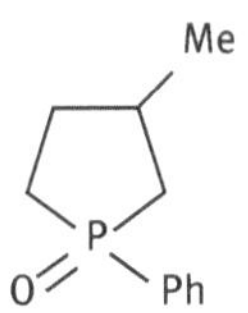	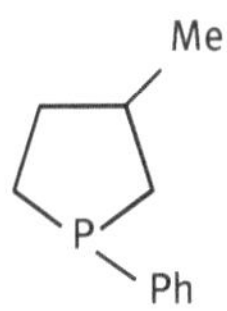	Cl_3SiH^b	1.4	PhMe	110	2	89	[26]
9	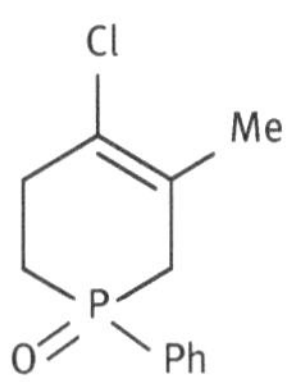	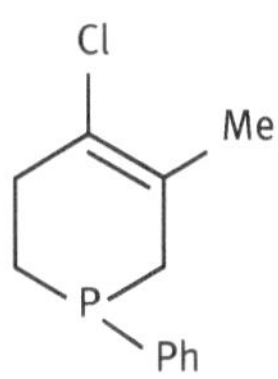	Cl_3SiH^b	1.4	PhMe	110	2	96	[26]
10	MePhBzP=O	MePhBzP	Cl_3SiH^a	–	PhH	180–190	2	69	[27]

[a] 1.1 equivalents of NEt_3 was measured in.
[b] 3 equivalents of C_6H_5N was measured in.

(Table 9.1, entry 5). It is noted that the low yield was probably due to preparative insufficiency, as Cl_3SiH may be used well in the reduction of tributylphosphine oxide under similar conditions (Table 9.1, entry 4). The reduction of 3-phospholene oxides, phospholane oxides and a 1,2,3,6-tetrahydrophosphinine oxide was also performed with Cl_3SiH together with 3 equivalents of C_6H_5N in boiling toluene for 2 h (Table 9.1, entries 6, 8 and 9) [21]. A phenyl-3-phospholene oxide was deoxygenated efficiently by 5.9 equivalents of Cl_3SiH at 0 °C, and phosphine so obtained reacted with dimethyl sulfide–borane to give the corresponding borane complex in a quantitative yield (Table 9.1, entry 7).

New platinum complexes of *cis*-Pt$(L)_2Cl_2$ type have been synthesized from these five- and six-membered cyclic phosphines [26]. Horner and Balzer reported the first study on the reduction of optically active phosphine oxides (Table 9.1, entry 10) [27].

Quin described the reduction of a 1-amino-3-phospholene oxide using trichlorosilane-pyridine to afford eventually a chlorophospholene formed after the fission of the P–N bond (Figure 9.11) [28].

Cl_3SiH is an especially useful reducing agent in the synthesis of P-chiral phosphines, as in these cases, phosphines may usually be obtained in high optical yields. The stereochemistry of the deoxygenation may depend on how Cl_3SiH is used. The application of Cl_3SiH alone, Cl_3SiH–pyridine or Cl_3SiH–*N,N*-diethylaniline resulted in retention of configuration at the phosphorus center, but the use of $Cl_3SiH–NEt_3$ led to phosphines formed with inversion of the P-configuration (Figure 9.12).

In the deoxygenation of phosphetane oxides using a mixture of $Cl_3SiH–NEt_3$ as the reagent, retention of the P-configuration was observed for both the *cis* and *trans* isomer (Figure 9.13). This result is in contrast with those described for acyclic phosphines. The steric hindrance around the P atom may prevent the attack of the hydride anion from the more crowded side [29].

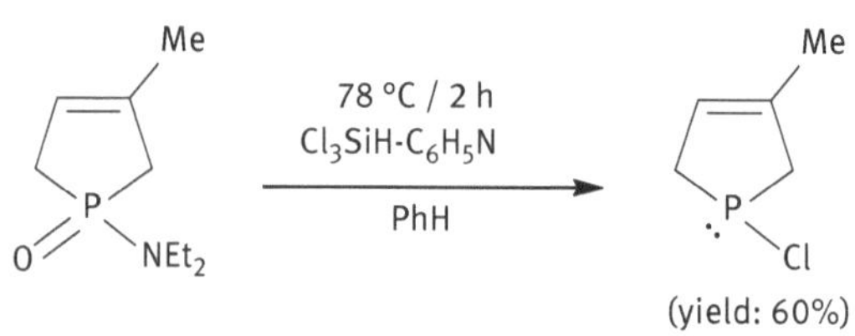

Figure 9.11: Reaction of a 1-amino-3-phospholene oxide using trichlorosilane-pyridine.

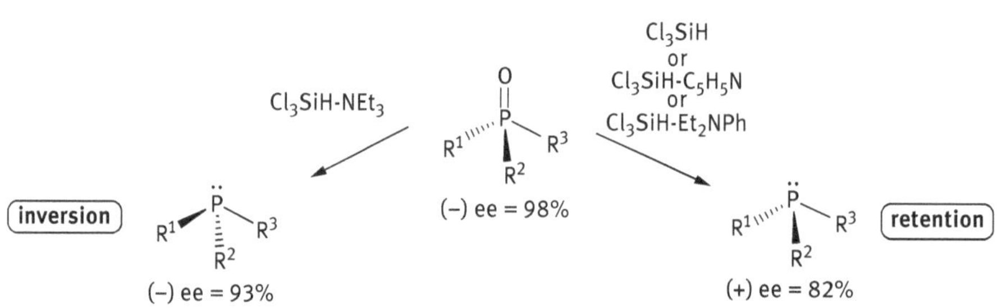

Figure 9.12: Stereochemistry of the reduction of P-chiral phosphine oxides.

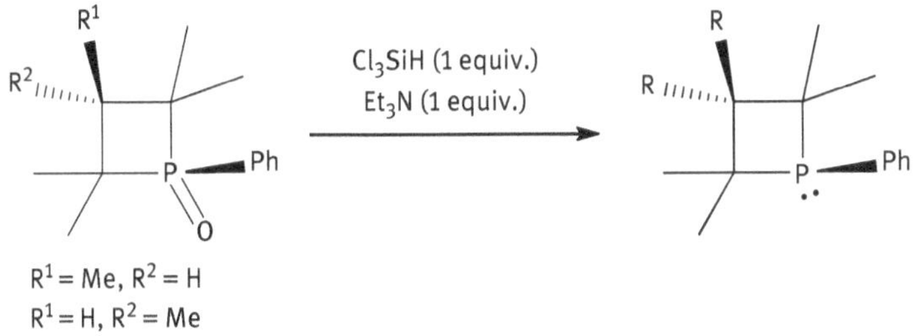

Figure 9.13: Reduction of chiral phosphetane oxides.

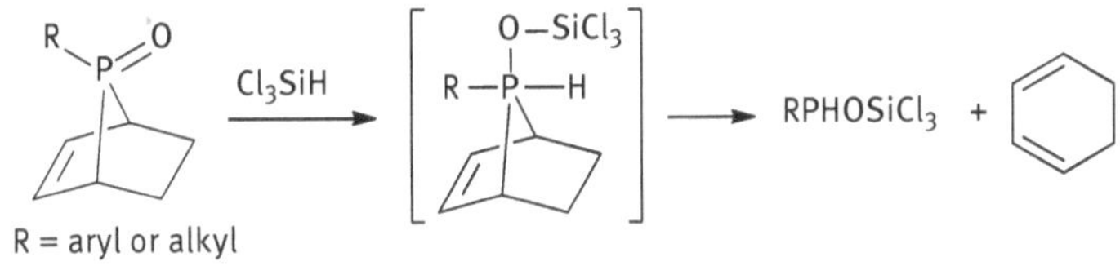

Figure 9.14: Reaction of bicyclic phosphine oxides with Cl_3SiH.

There is an example for the use of triphenylphosphine in addition to Cl_3SiH in the reduction of chiral phosphine oxides. The "electron-poor" phosphine used is suitable to capture the oxygen atom of the phosphine oxide deriving from an "electron-rich" phosphine. The reduction took place with retention of configuration at the P atom [30].

During the reduction of bridged phosphine oxides with Cl_3SiH, it was found that most of the reductions occurred with fragmentation releasing a phosphinite and cyclohexa-1,3-diene, that is a retro McCormack-like reaction (Figure 9.14). The ring strain in the pentacoordinated intermediate may be responsible for the phenomenon observed. The reactions performed in the presence of pyridine afforded the corresponding phosphines exclusively. The Cl_3SiH–pyridine complex does not act as the hydride donor, and the mechanism of this reduction does not involve a pentacoordinated P-intermediate [31].

Deoxygenation with Cl_3SiH under heterogeneous conditions was also possible [32]. The phosphine oxides were immobilized on silica, and then reduced to phosphines with good conversions. A phosphine oxide immobilized on polystyrene resin could be reduced quantitatively [33].

The reduction of triarylphosphine oxides by Cl_3SiH in the presence of NEt_3 can be promoted by microwave (MW) irradiation allowing a shorter reaction time of 10 min, and a yield of 83–95% [34].

As it was shown, Cl_3SiH is a widely applied reducing agent to prepare phosphines. It is better to use this corrosive and volatile reagent together with pyridine or NEt_3. The nature of the amine also has an impact on the stereochemistry of the reduction.

9.3.2.3 Deoxygenation of Phosphine Oxides using Perchlorosilanes

It is interesting to note that perchlorosilanes (e.g. hexachlorodisilane (Si_2Cl_6)) were also described as reducing agents [35]. The reduction of a chiral tertiary phosphine oxide with Si_2Cl_6 took place with inversion of configuration (Figure 9.15).

However, the Si_2Cl_6 reduction of phosphetane oxides proceeded with retention of configuration at phosphorus. [36].

The two diastereomers of the P-chiral binaphthyl monophosphine oxides showed different reactivity and stereoselectivity in silane reductions. Si_2Cl_6 reduced the diastereomers to give predominantly the corresponding phosphine with retention of configuration. In a sharp contrast, the use of $Cl_3SiH–NEt_3$ led to racemization at the phosphorus center (Figure 9.16) [37].

1-Chloro-3-phospholene may be prepared by the reduction of 1-chloro-3-phospholene oxide with Si_2Cl_6 at room temperature (Figure 9.17) [38].

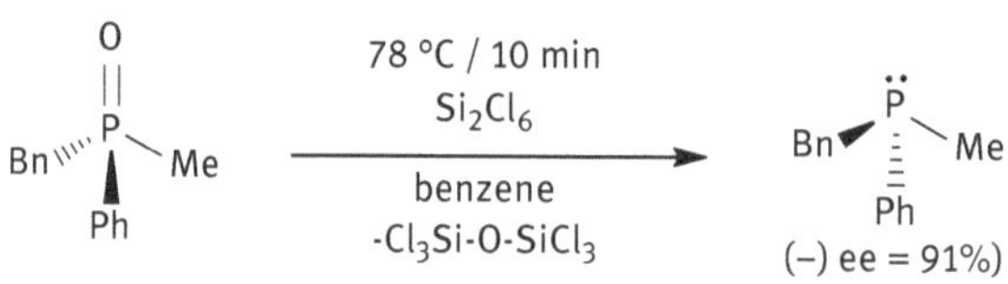

Figure 9.15: Reduction of a chiral phosphine oxide using hexachlorodisilane.

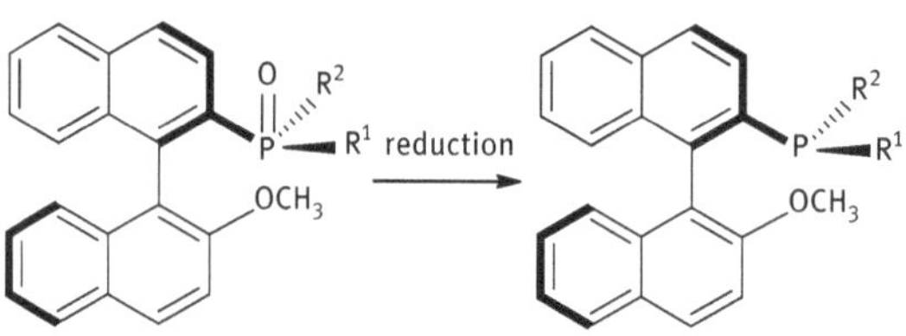

P=O	Silane	Phosphine (a)	Phosphine (b)
a	Si_2Cl_6	100	0
b	Si_2Cl_6	30	70
a	Cl_3SiH-NEt_3	50	50

R^1 = Me, R^2 = Ph (**a**)
R^1 = Ph, R^2 = Me (**b**)

Figure 9.16: Diastereoselectivity for the reduction of P-chiral binaphthyl monophosphine oxides.

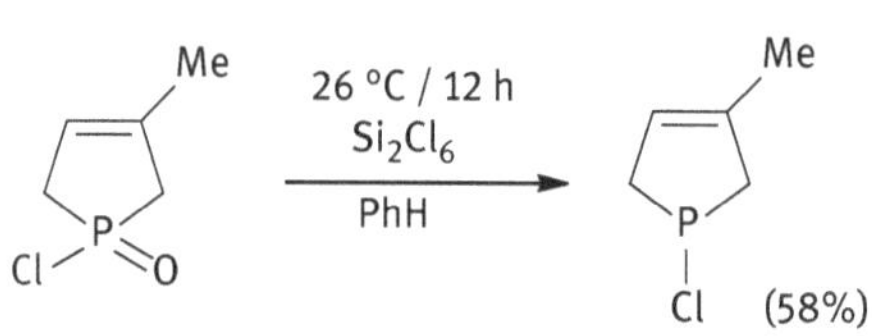

Figure 9.17: The deoxygenation of 1-chloro-3-phospholene oxide with Si_2Cl_6.

9.3.2.4 Deoxygenation of Phosphine Oxides by User-friendly Silanes

More user-friendly silanes are $(EtO)_3SiH$, $(EtO)_2MeSiH$, 1,1,3,3-tetramethyldisiloxane (TMDS) and polymethylhydrosiloxane called also as methylpolysiloxane ([PMHS or MPS] described by the formula $[-O-SiH(Me)-]_n$) [39, 40]. These silanes are, however, not too reactive.

Beller et al. studied the effect of different acids in the reduction of triphenylphosphine oxide with diethoxymethylsilane as the reducing agent (Figure 9.18) [39]. Without a catalyst, no reaction took place. Using benzoic acid as the catalyst, triphenylphosphine was obtained in a low yield. Surprisingly, the addition of the diphenyl ester of phosphoric acid as the catalyst resulted in PPh_3 in a yield of 75%. Furthermore, varying the electronic properties of the P-acid catalyst by introducing an electron-withdrawing substituent, such as NO_2 or CF_3 in the phenyl ring, the deoxygenation became quantitative [39].

Tertiary phosphine oxides could be converted to the corresponding phosphines using triethoxysilane together with a catalytic amount of titanium(IV) isopropoxide (Figure 9.19). The reductions were carried out in tetrahydrofuran, and proceeded slowly at room temperature, but became complete after a 1-h heating at 67 °C, if the silane was used in a 3 equivalent quantity [40].

Starting from aryl-diphenylphosphine oxides, the use of triethoxysilane in the presence of titanium tetraisopropoxide in boiling benzene gave the corresponding aryl-diphenylphosphine in a yield of around 90% after 30 min [41].

Catalyst	Yield (%)
–	<1
PhCOOH	6
$(PhO)_2P(O)OH$	75
$(4\text{-}NO_2C_6H_4O)_2P(O)OH$	>99
$(4\text{-}CF_3C_6H_4O)_2P(O)OH$	>99

Figure 9.18: Acid-catalyzed reduction of phosphine oxides with $(EtO)_2MeSiH$.

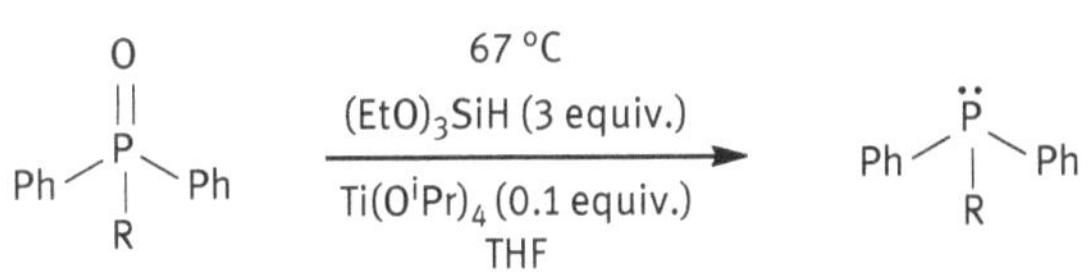

R	Yield (%)
Ph	85
$CH_2CH_2P(O)Ph_2$	90
Me	99
Et	95
iPr	81
iBu	41
Bn	98
CH_2Bn	83

Figure 9.19: Reductions with triethoxysilane.

The reduction of triphenylphosphine oxide was also carried out with TMDS in the presence of different catalysts. Copper(II) triflate was an efficient catalyst in the reduction of the P=O group at 100 °C in toluene [42]. Without a catalyst, even applying TMDS in a large excess (12 equivalents), no deoxygenation was observed. The use of 15 mol% of the diphenyl ester of phosphoric acid promoted the deoxygenation in boiling toluene [39]. Using 10 mol% of $Ti(OiPr)_4$ at 100 °C, triphenylphosphine was obtained in a high conversion. Later on, $InBr_3$ was also described as a catalyst for P=O-reductions. In the presence of 1% of $InBr_3$, the deoxygenation was quantitative at 100 °C [43] (Figure 9.20).

However, there is no need for any catalyst, as the deoxygenation of triphenylphosphine oxide with TMDS can be implemented under MW heating in the absence of any solvent. After an irradiation at 200 °C for 6.5 h, the deoxygenation was quantitative. On traditional heating at 175 °C, the completion required more than one day. As there is no need for any catalyst, the MW-assisted protocol means a green chemical approach (Figure 9.21) [44, 45].

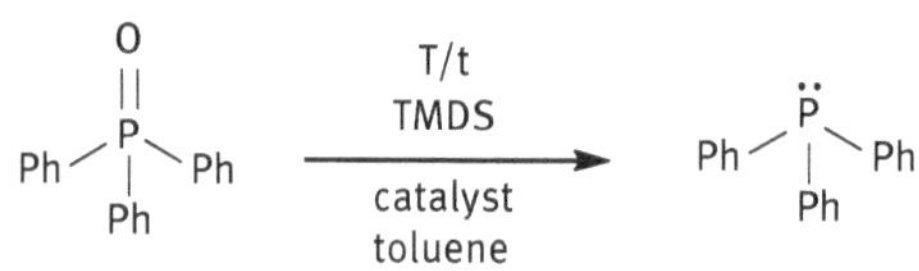

Equivalent of TMDS	Catalyst (mol %)	T (°C)	t (h)	Conv. (%)
12	–	100	2	<1
3	$Cu(OTf)_2$ (10)	100	15	96
3	$(PhO)_2P(O)OH$ (15)	110	24	62
1.2	$Ti(O^iPr)_4$ (10)	100	24	86
3	$InBr_3$ (1)	100	18	>99

Figure 9.20: The reduction of triphenylphosphine oxide with TMDS in the presence of different catalysts.

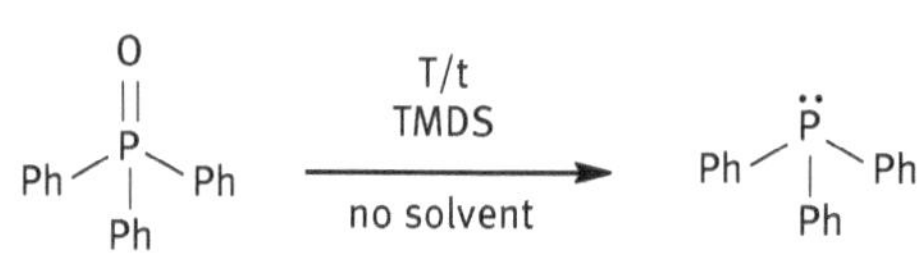

Equivalent of TMDS	Mode of Heating	T (°C)	t (h)	Conv. (%)
2	Δ	175	24	92
2	MW	200	6.5	100

Figure 9.21: Catalyst-free reduction of triphenylphosphine oxide with TMDS.

Deoxygenation of cyclic phosphine oxides, such as 3-methyl-1-phenyl-2-phospholene oxide, was studied with TMDS using 1% of $InBr_3$ catalyst in toluene at 100 °C [43]. This reaction could also be performed without the use of any catalyst in boiling toluene [46]. Under thermal- and MW-assisted conditions, the solvent- and catalyst-free reduction of 3-methyl-1-phenyl-3-phospholene-1-oxide was complete after a shorter reaction time (Figure 9.22) [44].

From among the "practical" reducing agents, PMHS was described for the deoxygenation of triphenylphosphine oxide. At 290 °C and without the use of any solvent, triphenylphosphine was obtained in a yield of 88% [20]. Regarding the application of PMHS, the relatively high temperature required is not too attractive. It was observed that even using a large excess (12 equivalents) of PMHS in toluene, practically no deoxygenation occurred at 100 °C after 2 h. However, using $Cu(OTf)_2$ as the catalyst at 100 °C for 15 h, the reduction of the P=O unit took place [42]. Applying 15 mol% of $(PhO)_2P(O)OH$ at 110 °C for 24 h, the reduction remained incomplete, and the phosphine was isolated in a yield of 35% [40]. Applying PMHS at a higher temperature of 175 °C without any catalyst and under solvent-free conditions, the deoxygenation was complete after 17 h on conventional heating. In the MW-assisted variation, 8 h was enough to attain a good conversion (Figure 9.23).

The first reduction of 3-methyl-1-phenyl-2-phospholene 1-oxide with PMHS was accomplished without any solvent at 250 °C [20]. Later on, this reaction was performed in boiling toluene for 6 h [46]. The reduction of 3-methyl-1-phenyl-3-phospholene 1-oxide was also studied under thermal- and MW-assisted conditions without the use of any solvent. These reductions were complete at 110 °C after 4 and 2 h, respectively (Figure 9.24) [45]. The outcome of the reduction of the dimethyl-3-phospholene oxide was better in boiling toluene.

TMDS and PMHS are user-friendly and cheap deoxygenating agents. Their relatively lower reactivity can be overcome by a solvent-free and MW-assisted accomplishment.

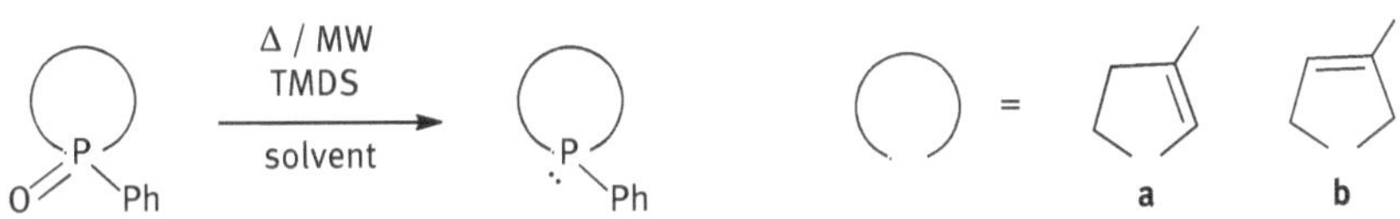

P=O	Equivalent of TMDS	Catalyst (mol %)	Solvent	Mode of heating	T (°C)	t (h)	Conv. (%)
a	3	$InBr_3$(1)	PhMe	Δ	100	40	95
a	2	–	PhMe	Δ	110	8	82
b	2	–	–	Δ	110	5	100
b	2	–	–	MW	110	3	100

Figure 9.22: Reduction of phospholene oxides with TMDS.

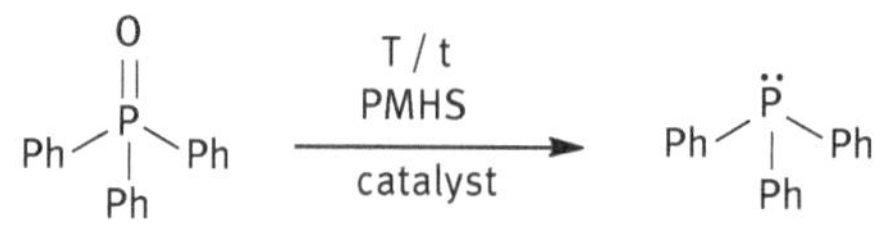

Equivalent of PMHS	Catalyst (mol %)	Mode of Heating	Solvent	T (°C)	t (h)	Yield (%)
5	–	Δ	–	290	2	86
12	–	Δ	PhMe	100	2	0
6	$Cu(OTf)_2$(10)	Δ	PhMe	100	15	88
4	$(PhO)_2P(O)OH$ (15)	Δ	PhMe	110	24	35
2	–	Δ	–	175	17	87
2	–	MW	–	175	8	90

Figure 9.23: The reduction of triphenylphosphine oxide with PMHS.

Phosphine oxide	Silane (equiv.)	Mode of Heating	Solvent	T (°C)	t (h)	Yield (%)
a	5	Δ	–	250	2	88
a	2	Δ	PhMe	110	6	85
b	2	Δ	–	110	4	91
b	2	MW	–	110	2	92
c	5	Δ	–	250	not provided	35
c	2	Δ	PhMe	110	6	92

Figure 9.24: The reduction of different phospholene oxides with PMHS.

$PhSiH_3$ ~ $NaphSiH_3$ > $BnSiH_3$ > $(4\text{-}MePh)_2SiH_2$ > PMHS > $(Ph_2SiH)_2$ ~ TMDS > $(1\text{-}napht)_2SiH_2$

Figure 9.25: The reactivity of the silanes investigated.

The deoxygenation of triphenylphosphine oxide was studied using a series of silanes comprising an aralkylsilane, arylsilanes, diarylsilanes, a disiloxane, a polysiloxane and a disilane under solvent-free thermal- or MW-assisted conditions [45]. The reactivity of the silanes was evaluated on the basis of the experimental data (Figure 9.25). The arylsilanes are somewhat more reactive than $BnSiH_3$. This group of reactive silanes bearing three hydrogen atoms is followed by those representatives bearing two hydrogen atoms, or only one proton per silicon moiety. The least reactive silane in this series is $(1\text{-}napht)_2SiH_2$. Its low reactivity is obviously the consequence of steric hindrance.

9.4 Mechanistic insights into the reduction of tertiary phosphine oxides by silanes

Three reaction pathways involving the reduction with Cl_3SiH, Cl_3SiH/Et_3N and Cl_3SiH/C_5H_5N were proposed [27].

Using Cl_3SiH alone, the reduction of the P=O moiety occurs through a four-membered transition state to afford an ion pair as the intermediate that is stabilized by a proton transfer. In this protocol, the configuration at the P atom is preserved (Figure 9.26).

The deoxygenation with phenylsilane follows a similar protocol [24]. The classical reduction scheme was refined by Pietrusiewicz et al. utilizing density functional theory calculations [47]. The first TS afforded a covalent P–O–Si species that is stabilized via a second TS by a proton transfer to furnish phosphine and the oxidated byproduct (Figure 9.27). It was proved that the model reaction between tributylphosphine oxide and phenylsilane involves a nonpolar mechanism.

When triethylamine was used as the base, the zwitterion reacts with the $HSiCl_3$–NEt_3 complex to form eventually phosphonium species with an inverted P atom through an intermolecular (S_N2) hydride transfer (Figure 9.28-(1)). [27]. Using pyridine, the internal hydride delivery is similar to that shown in Figure 9.26 (Figure 9.28-(2)) [48].

It was supposed that the difference in basicity between triethylamine and pyridine influences the reactivity of the corresponding complexes. Strong bases give phosphine with predominant inversion, while weak bases afford phosphine with predominant retention of configuration [49]. According to this theory, the $HSiCl_3$–NEt_3 complex is dissociated, and reacts with the phosphine oxide to give a phosphonium intermediate. Its pentavalent form is transformed to another phosphonium intermediate that is stabilized by the loss of Cl_3SiOH. As can be seen, the outcome is again inversion (Figure 9.29).

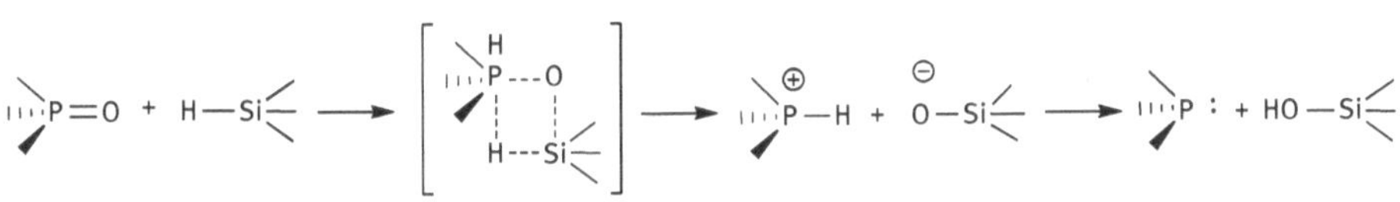

Figure 9.26: The reduction pathway using Cl_3SiH as the reducing agent.

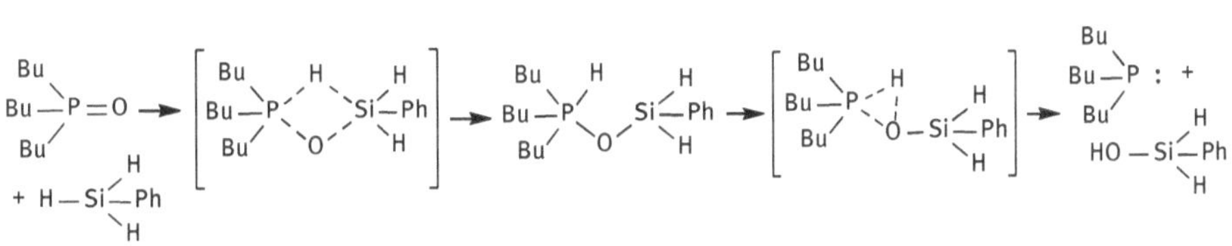

Figure 9.27: The reduction pathway using phenylsilane.

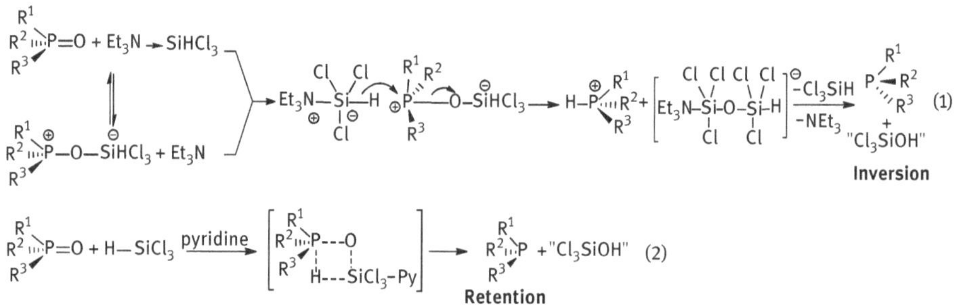

Figure 9.28: Reduction pathways using $HSiCl_3$–amine complexes.

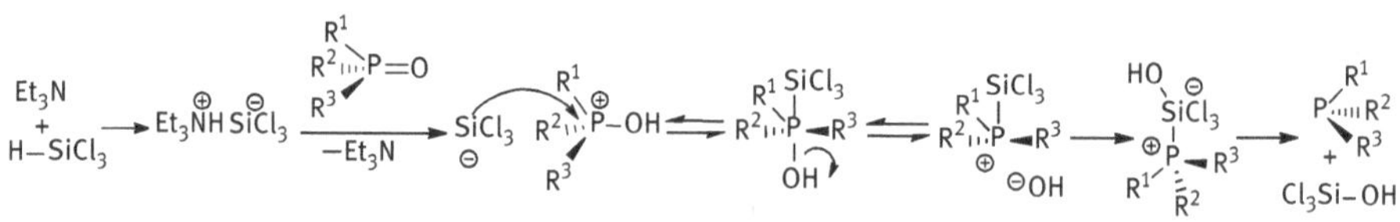

Figure 9.29: The reduction pathway using the $HSiCl_3$–NEt_3 complex.

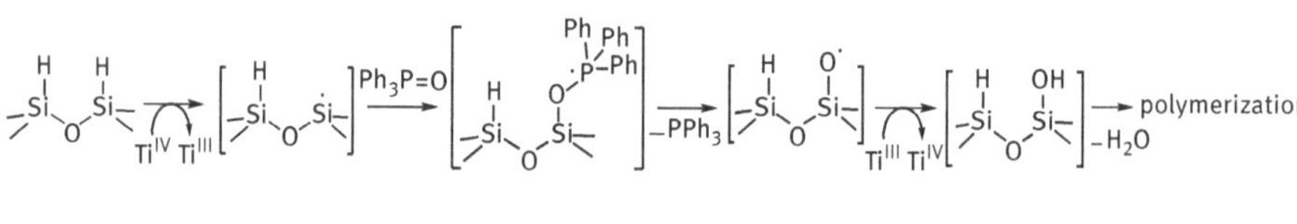

Figure 9.30: The reduction pathway using the TMDS-titanium(IV) catalyst.

Triethoxysilane and PMHS was investigated in the presence of titanium(IV) catalyst in the reduction of phosphine oxides. A mechanism resulting in retention of configuration was proposed [39].

The deoxygenation with TMDS was also investigated, and a mechanism based on electron paramagnetic resonance analysis and ^{29}Si NMR was suggested. The main result of these studies is the evidence for the presence of a Ti(III) intermediate, which may be formed by single electron transfer (Figure 9.30) [50, 51].

9.5 Deoxygenation of phosphine oxides with other reducing agents

Other reagents and combinations have also been developed for the direct reduction of phosphine oxides. Samarium iodide/hexamethylphosphoric triamide (SmI_2/HMPA) [52], $TiCp_2Cl_2$/Mg [53], Bi/TiO_2 [54], $SiCl_4$ along with a metal [55], sulfur derivatives

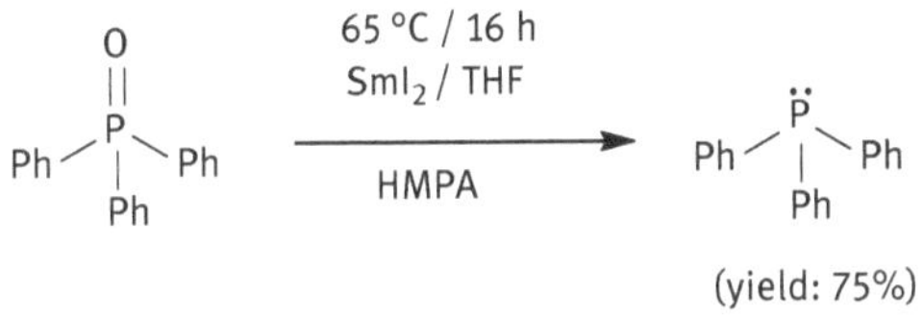

Figure 9.31: The deoxygenation of triphenylphosphine oxide by SmI_2.

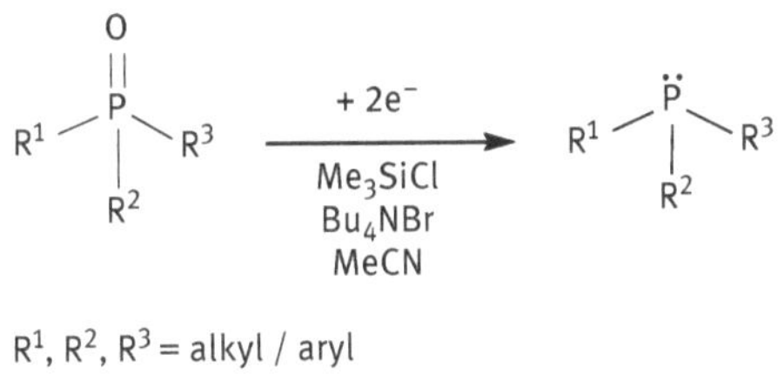

Figure 9.32: The deoxygenation of phosphine oxides by electroreduction.

[56] and hydrocarbons associated with activated carbon [57] have also been applied to reduction. However, the procedures mentioned are not too attractive, since the reagents are rather expensive and/or difficult to handle. Regarding the reduction of triphenylphosphine oxide, SmI_2 had to be applied in a large excess (22 equivalents) in HMPA as the solvent. Triphenylphosphine was obtained in a yield of 75% after a reaction time of 16 h (Figure 9.31) [52].

As a special method, the direct electroreduction of phosphine oxides in the presence of trimethylsilyl chloride was reported [58]. An electrochemical reduction that can be carried out under mild conditions is a highly promising procedure (Figure 9.32) [59].

9.6 Indirect reduction of phosphine oxides via phosphonium salt intermediates

The direct scheme of the title protocol may be described by two main steps: activation of the P=O bond by a suitable reagent to form a phosphonium or a radical intermediate, followed by a hydride transport. The reducing agents may be involved both as the activator and as the reductant itself. The transformation starting with chlorination by oxalyl chloride is a widely studied procedure [60, 61]. The reduction of triphenyldichlorophosphorane obtained by the reaction of triphenylphosphine oxide with $(COCl)_2$ was performed by hydrogenation at a high temperature and/or high pressure in the presence of a transition metal (Pt, Pd and Rh) catalyst (Figure 9.33) [62, 63].

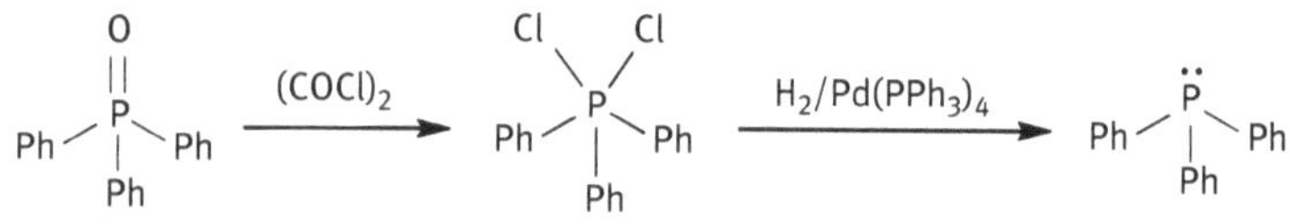

Figure 9.33: Indirect deoxygenation of phosphine oxides.

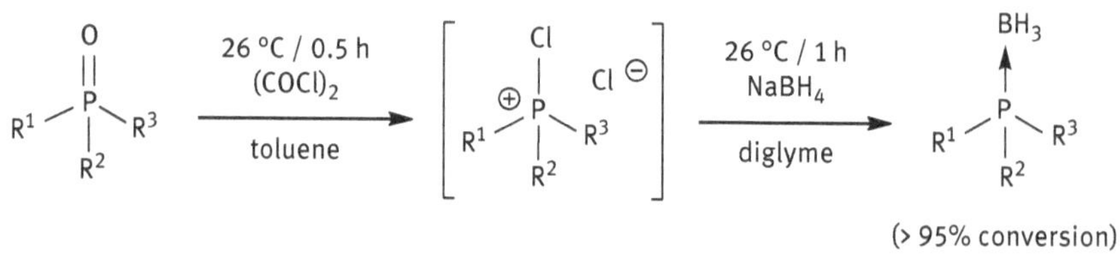

Figure 9.34: One-pot transformation of phosphine oxides to phosphine boranes

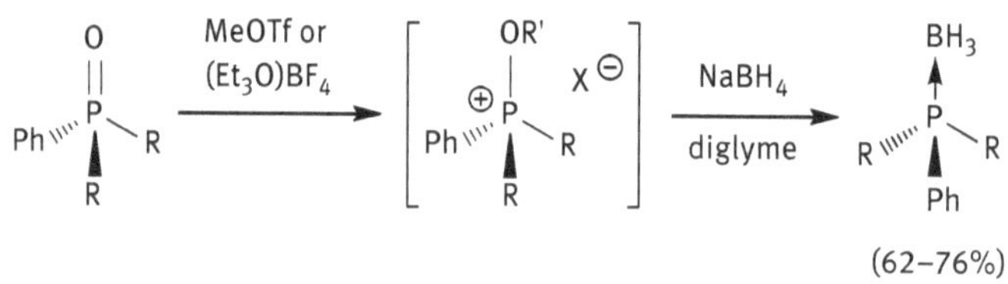

Figure 9.35: Indirect stereoselective conversion of phosphine oxides to phosphine boranes.

A one-pot transformation involving the treatment of various secondary and tertiary phosphine oxides with oxalyl chloride in toluene followed by reduction with $NaBH_4$ in diglyme was developed to form the corresponding phosphine boranes in good yields (Figure 9.34) [64]. The conversion of phosphine oxide to chlorophosphonium chloride may offer additional advantages beyond its easy reduction. The salt resulting from the reaction of the phosphine oxide and oxalyl chloride could be separated by filtration.

Interestingly, the use of alkylating reagents, such as methyl triflate or triethyloxonium tetrafluoroborate (Meerwein's salt) in the first step afforded alkoxy phosphonium salts, and after a subsequent reduction, phosphine boranes stereoselectively with inversion at the P-center (Figure 9.35) [65].

9.7 Conclusions

The main protocols for the deoxygenation of phosphine oxides to phosphines involve the application of different reducing agents, such as hydrides, silanes and special reagents. The group of silanes is the most important and widespread. The corrosive

Cl_3SiH and expensive $PhSiH_3$ may be replaced by user-friendly Me_2SiH–O–$HSiMe_2$ (TMDS) and $[-O-SiH(Me)-]_n$ (PMHS) under MW-assisted and solvent-free conditions. Possible mechanisms of the silane reductions and special methods for the deoxygenations were also discussed.

Acknowledgment: This work was supported by the National Research, Development and Innovation Office (K119202).

References

[1] D. H. Valentine, J. H. Hillhouse, *Synthesis*, 2003, 317.
[2] M. T. Honaker, J. M. Hovland, R. N. Salvatore, *Curr. Org. Synth.*, 2007, **4**, 30.
[3] S. L. Buchwald, C. Mauger, G. Mignani, U. Scholz, *Adv. Synth. Catal.*, 2006, **348**, 23.
[4] G. Wittig, G. Geissler, *Liebigs Ann. Chem.*, 1953, **580**, 44.
[5] O. Mitsunobu, M. Yamada, *Bull. Chem. Soc. Jpn.*, 1967, 2380.
[6] R. Appel, *Angew. Chem., Int. Ed.*, 1975, **14**, 801.
[7] A. Zapf, A. Ehrentraut, M. Beller, *Angew. Chem., Int. Ed.*, 2000, **39**, 4153.
[8] M. R. Douglass, C. L. Stern, T. J. Marks, *J. Am. Chem. Soc.*, 2001, **123**, 10221.
[9] D. Hérault, D. H. Nguyen, D. Nuel, G. Buono, *Chem. Soc. Rev.*, 2015, **44**, 2508.
[10] L. D. Quin, *A Guide to Organophosphorus Chemistry*; John Wiley & Sons: New York, 2000.
[11] F. Hein; K. Issleib; H. Rabold, *Z. Anorg. Allg. Chem.*, 1956, **287**, 208.
[12] K. Issleib, G. Grams, *Z. Anorg. Allg. Chem.*, 1959, **299**, 58.
[13] T. Imamoto, T. Takeyama, T. Kusumoto, *Chem. Lett.*, 1985, 1491.
[14] T. Imamoto, T. Oshiki, T. Onozawa, T. Kusumoto, K. Sato, *J. Am. Chem. Soc.*, 1990, **112**, 5244.
[15] S. Yang, X. Han, M. Luo, J. Gao, W. Chu, Y. Ding *Russ. J. Gen. Chem.*, 2015, **85**, 1156.
[16] R. Köster, Y. Morita, *Angew. Chem.*, 1965, 77, 589.
[17] G. Keglevich, M. Fekete, T. Chuluunbaatar, A. Dobó, V. Harmat, L. Tőke, *J. Chem. Soc., Perkin Trans. 1*, 2000, 4451.
[18] S. Sowa, M. Stankevič, A. Szmigielska, H. Małuszyńska, A. E. Kozioł, K. M. Pietrusiewicz *J. Org. Chem.*, 2015, **80**, 1672.
[19] G. Keglevich, T. Chuluunbaatar, K. Ludányi, L. Tőke, *Tetrahedron* 2000, **56**, 1.
[20] H. Fritzsche, U. Hasserodt, F. Korte, *Chem. Ber.*, 1964, **97**, 1988.
[21] H. Fritzsche, U. Hasserodt, F. Korte, *Chem. Ber.*, 1965, **98**, 171.
[22] F. Krech, K. Issleib, A. Zschunke, C. Muegge, S. Skvorcov, *Z. Anorg. Allg. Chem.*, 1991, **594**, 66.
[23] G. Keglevich, A. C. Gaumont, J. M. Denis, *Heteroatom Chem.*, 2001, **12**, 161.
[24] K. L. Marsi, *J. Org. Chem.*, 1974, **39**, 265.
[25] H. A Van Kalkeren, S. H. A. M. Leenders, C. R. A. Hommersom, F. P. J. T. Rutjes, F. L. van Delft, *Chem. Eur. J.*, 2011, **17**, 11290.
[26] P. Bagi, T. Kovács, T. Szilvási, P. Pongrácz, L. Kollár, L. Drahos, E. Fogassy, G. Keglevich, *J.* [20] *Organomet. Chem.*, 2014, **751**, 306.
[27] L. Horner, W. D. Balzer, *Tetrahedron Lett.*, 1965, 1157.
[28] L. D. Quin, J. Szewczyk, *Phosphorus, Sulfur*, 1984, **21**, 161.
[29] S. E. Cremer, R. J. Chorvat, *J. Org. Chem.*, 1967, **32**, 4066.
[30] H.-C. Wu, J-Q. Yu, J-B. Spencer, *Org. Lett.*, 2004, **6**, 4675.

[31] L. D. Quin, K. C. Caster, J. C. Kisalus, K. A. Mesch, *J. Am. Chem. Soc.*, 1984, **106**, 7021.
[32] L. Bemi, H. C. Clark, J. A. Davies, C. A. Fyfe, R. E. Wasylishen, *J. Am. Chem. Soc.*, 1982, **104**, 438.
[33] A. Fyfe, H. C. Clark, J. A. Davies, P. J. Hayes, *J. Am. Chem. Soc.*, 1983, **105**, 6577.
[34] K. Damian, M. L. Clarke, C. J. Cobley, *J. Mol. Catal. A: Chem.*, 2008, **284**, 46.
[35] E. H. Krenske, *J. Org. Chem.*, 2012, 77, 1.
[36] K. E. DeBruin, G. Zon, K. Naumann, Mislow, K. *J. Am. Chem. Soc.*, 1969, **91**, 7027.
[37] N. J. Kerrigan, E. C. Dunne, D. Cunningham, P. McArdle, K. Gilligan, D. G. Gilheany, *Tetrahedron Lett.*, 2003, **44**, 8461.
[38] D. K. Myers, L. D. Quin, *J. Org. Chem.*, 1971, **36**, 1285.
[39] Y. H. Li, L. Q. Lu, S. Das, S. Pisiewicz, K. Junge, M. Beller, *J. Am. Chem. Soc.*, 2012, **134**, 18325.
[40] T. Coumbe, N. J. Lawrence, F. Muhammad, *Tetrahedron Lett.*, 1994, **35**, 625.
[41] G. Onodera, H. Matsumoto, M. D. Milton, Y. Nishibayashi, S. Uemura, *Org. Lett.*, 2005, **7**, 4029.
[42] Y. H. Li, S. Das, S. L. Zhou, K. Junge, M. Beller, *J. Am. Chem. Soc.*, 2012, **134**, 9727.
[43] L. Pehlivan, E. Metay, D. Delbrayelle, G. Mignani, M. Lemaire, *Tetrahedron*, 2012, **68**, 3151.
[44] G. Keglevich, T. Kovács, F. Csatlós, *Heteroatom Chem.*, 2015, **26**, 199.
[45] T. Kovács, A. Urbanics, F. Csatlós, J. Binder, A. Falk, F. Uhlig, G. Keglevich, *Curr. Org. Synth.*, 2016, **13**, 148.
[46] G. Keglevich, T. Kovács, *Curr. Green Chem.*, 2014, **1**, 182.
[47] O. M. Demchuk, R. Jasiński, K. M. Pietrusiewicz, *Heteroatom Chem.* 2015, **26**, 441.
[48] K. Naumann, G. Zon, K. Mislow, *J. Am. Chem. Soc.*, 1969, **91**, 2788.
[49] K. Naumann, G. Zon, K. Mislow, *J. Am. Chem. Soc.*, 1969, **91**, 7012.
[50] M. Berthod, A. Favre-Réguillon, J. Mohamad, G. Mignani, G. Docherty, M. Lemaire, *Synlett*, 2007, 1545.
[51] C. Petit, E. Poli, A. Favre-Réguillon, L. Khrouz, S. Denis-Quanquin, L. Bonneviot, G. Mignani, M. Lemaire, *ACS Catal.*, 2013, 1431.
[52] Y. Handa, J. Inanaga, M. Yamaguchi, *J. Chem. Soc., Chem Commun.*, 1989, 298.
[53] F. Mathey, R. Maillet, *Tetrahedron Lett.*, 1980, **21**, 2525.
[54] A. Hagemeyer, C. W. Rieker, T. Lautensack, D. Hermeling, *U.S. Patent US 5689005A*, 1997; *Chem. Abstr.*, 1996, **125**, 141806.
[55] H. Fritzsche, U. Hasserodt, J. Van Olmen, *U.S. Patent US 3280195*, 1966; *Chem. Abstr.*, 1967, **66**, 28886.
[56] M. J. Petersson, W. A. Loughlin, I. D. Jenkins, *Chem. Commun.*, 2008, 4493.
[57] T. Dockner, *U.S. Patent US 4727193A*, 1988; *Chem. Abstr.*, 1987, **106**, 102561.
[58] H. Kawakubo, M. Kuroboshi, T. Yano, K. Kobayashi, S. Kamenoue, T. Akagi, H. Tanaka, *Synthesis*, 2011, **24**, 4091.
[59] T. Yano, M. Kuroboshi, H. Tanaka, *Tetrahedron Lett.*, 2010, **51**, 698.
[60] H. Li, Z. Chen, L. Wu, C. Wang, B. Hu, *CN 1660862*, 2005; *Chem. Abstr.*, 2006, **144**, 488805.
[61] M. Masaki, K. Fukui, *Chem. Lett.*, 1977, 151.
[62] M. Masaki, N. Kaketani, *Japanese Patent JP 53-034725*, 1978; *Chem. Abstr.*, 1978, **89**, 109953.
[63] K. Fukui, N. Kaketani, J. Kita, S. Fujimura, *Japanese Patent JP 55-149293*, 1980; *Chem. Abstr.*, 1981, **94**, 175259.
[64] L. J. Higham, E. F. Clarke, H. Müller-Bunz, D. G. Gilheany, *J. Organomet. Chem.*, 2005, **690**, 211.
[65] K. V. Rajendran, D. G. Gilheany, *Chem. Commun.*, 2012, **48**, 817.

Dávid Illés Nagy, Alajos Grün and György Keglevich

10 Dronic acid derivatives – An important group of phosphorus-containing drugs

Abstract: Hydroxy-methylenebisphosphonic acid (dronic acid) derivatives are important drugs against bone diseases. In this chapter, their way of action is summarized briefly, and their rationalized syntheses are surveyed together with the actual mechanisms. Preparations in the presence of ionic liquid additives afforded the dronic derivatives in record yields.

Keywords: hydroxy-methylenebisphosphonic derivatives, dronic acid derivatives, synthesis, optimization, ionic liquids.

10.1 Structure, History and Utilization of Bisphosphonates

The bisphosphonic acid derivatives have two pentavalent tetracoordinate P-atoms, more precisely two phosphonate units and a P–C–P moiety, which gives them a metabolic stability. Their structure is analogous to that of pyrophosphoric acid that contains a P–O–P unit instead of the P–C–P moiety (Figure 10.1). The elimination of calcium from the bones is inhibited by bisphosphonic acids [1].

The bisphosphonic acid derivatives differ from each other in the substitution pattern of the central carbon atom. Special derivatives are tiludronic acid and clodronic acid shown in Figure 10.2 [2, 3].

The real dronic acid derivatives contain a hydroxy group on their central carbon atom. Three groups may be distinguished according to the other substituent. The first generation of dronic derivatives does not have a nitrogen atom in the C-substituent. The representatives of the second and third generations have aminoalkyl or N-heterocyclic substituents, respectively (Table 10.1) [2, 3].

Bisphosphonates are biologically active compounds, and their activity is highly dependent on the substituents. The members of the first generation are not efficient enough, and have an unfavorable side-effect profile, thus the second- and third-generation derivatives have appeared soon. Among the first-generation derivatives, only tiludronic and clodronic acids are used for therapeutic purposes even today. The second-generation pamidronic acid, alendronic acid and ibandronic acid are still in market. The most effective derivatives are the third-generation risedronic and zoledronic acid [2, 3].

https://doi.org/10.1515/9783110535839-010

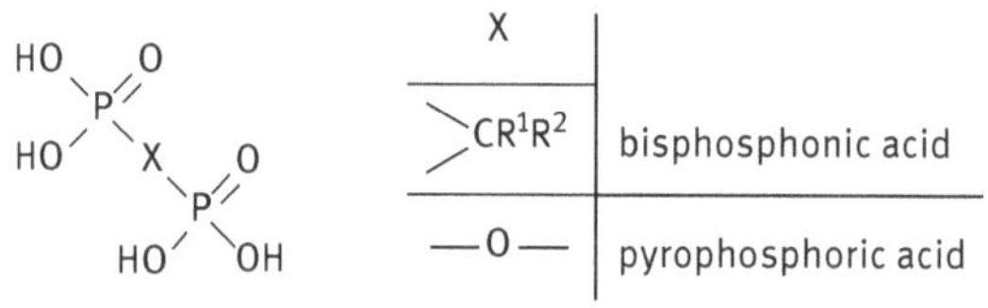

Figure 10.1: Structural similarity of bisphosphonic acids and pyrophosphoric acid.

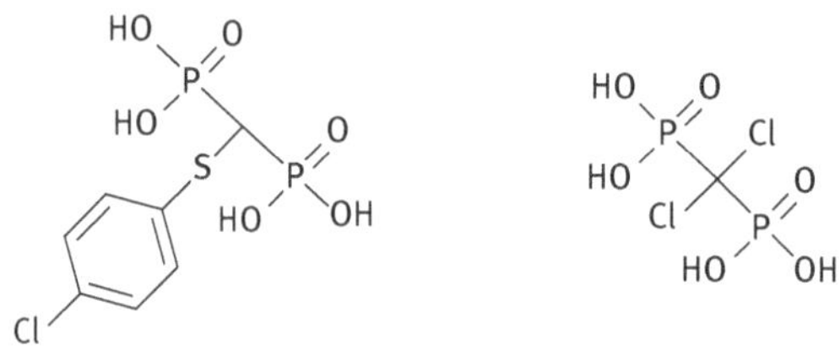

Tiludronic acid

Clodronic acid

Figure 10.2: Structure of special dronic acids.

The synthesis of hydroxy-bisphosphonates was already investigated in the second half of the nineteenth century. The first compound synthesized was etidronic acid in 1865 [4, 5]. In the early decades, the bisphosphonates were used as corrosion inhibitor, complexing agent or water softener. Their biological activity was revealed in the middle of the 1960s, and etidronic acid was the first that got in market at the early 1970s (Figure 10.3) [2, 3, 6–9].

In the beginning, the hydroxy-bisphosphonates were used to treat tumor diseases affecting bone tissue. Nowadays, they are considered the best drugs in the treatment of osteoporosis (Figure 10.4), Paget-disease and tumor-induced hypercalcemia, but they also have direct anticancer (breast, prostate and kidney) and antiparasitic effect [3, 6, 10].

Hydroxy-bisphosphonates are still in the focus, thus the investigation of their syntheses hides challenges.

10.2 Structure–activity Relationship

The structure of bones is not invariant; they are constantly changing, more precisely dissolving and re-forming. Osteoblasts (bone-building cells), osteocyte (the inactive form of osteoblasts) and osteoclasts (bone-degrading cells) are involved in the process [11].

1-Hydroxy-bisphosphonic acids have a tridentate functionality to enable the binding of Ca^{2+} ions, and to promote the affinity for species responsible for the accumulation of phosphates in the bone tissues. The presence of the P–C–P moiety in the molecule is necessary to inhibit bone resorption. Monophosphates and molecules containing a P–C–C–P or a P–N–P moiety do not have an inhibitory effect on bone

Table 10.1: A few important dronic acid derivatives.

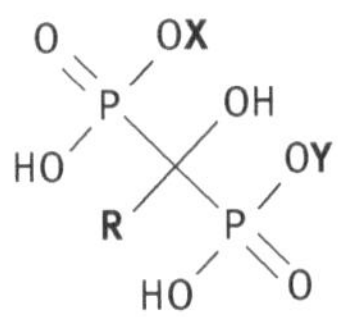

X, Y *e.g.*: H, Na

Generation	R	Commercial name
First	Me	Etidronic acid
		Fenidronic acid
	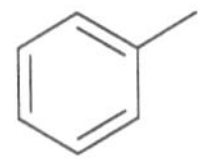	Benzidronic acid
	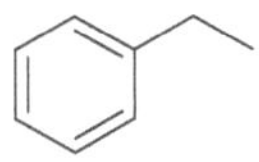	3-Phenylpropidronic acid
Second	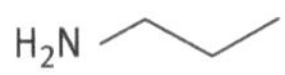	Pamidronic acid
	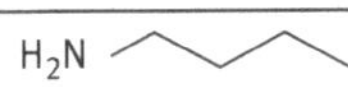	Alendronic acid
	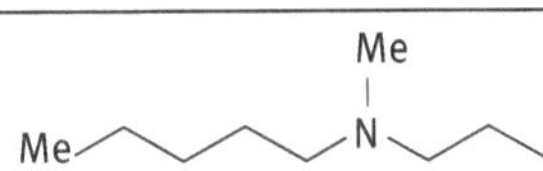	Ibandronic acid
Third	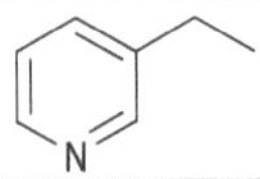	Risedronic acid
	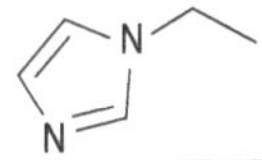	Zoledronic acid

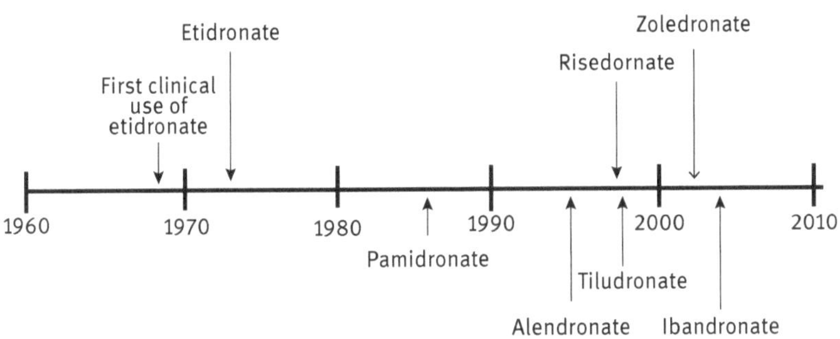

Figure 10.3: The discovery of different dronates [6].

resorption [12]. Beyond the attachment of hydroxy-bisphosphonic acids to the bone surface (physical effect), the bone resorption is also inhibited by their action on osteoclasts: they may induce cell death (apoptosis) of the osteoclasts (Figure 10.5) [12–15].

Depending on the side chains of bisphosphonates, they can cause apoptosis in two ways. In the case of the first-generation derivatives, nonhydrolyzable ATP analogues may be formed [shown on the example of etidronate (Figure 10.6)], which eventually inhibit cell metabolism [16].

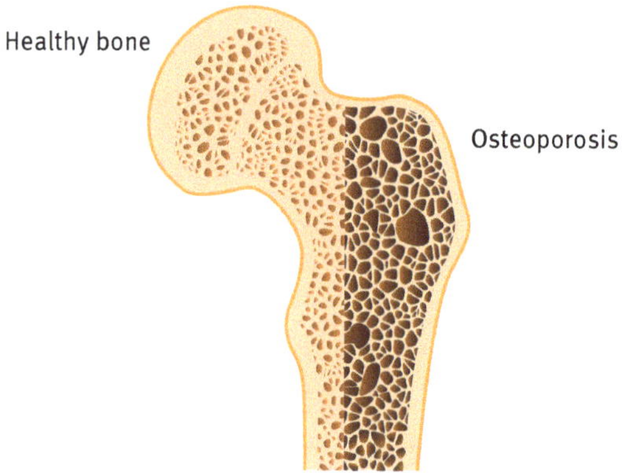

Figure 10.4: Differences between a healthy and osteoporotic bone tissue.

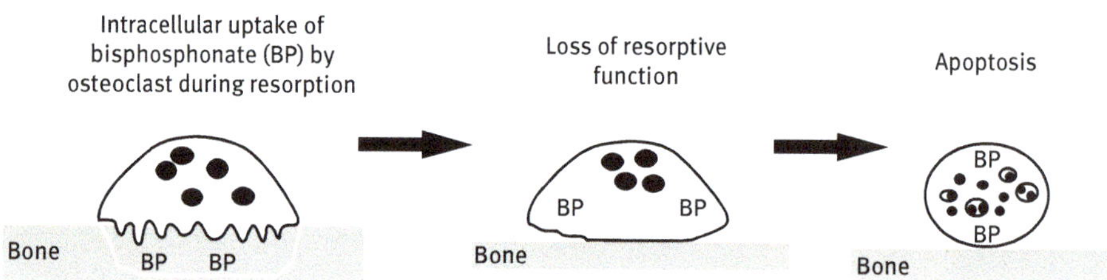

Figure 10.5: The effect of bisphosphonates on osteoclasts [15].

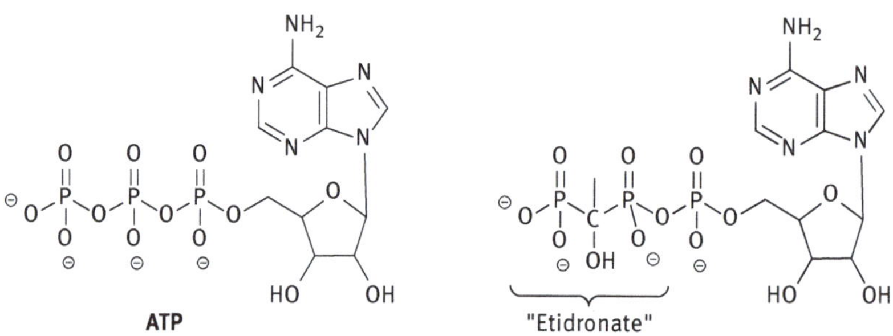

Figure 10.6: Nonhydrolyzable ATP analogs [16].

The members of second- and third-generation dronic acids inhibit the enzymes of mevalonate pathways (mainly the farnesyl pyrophosphate synthase), and thus prevent the synthesis of terpene and steroidal compounds, causing again apoptosis [16].

10.3 Possibilities for the Synthesis of Bisphosphonic Acid Derivatives

There are a number of possibilities for the synthesis of bisphosphonic acids. Three major synthetic routes are known for common bisphosphonates without a hydroxy

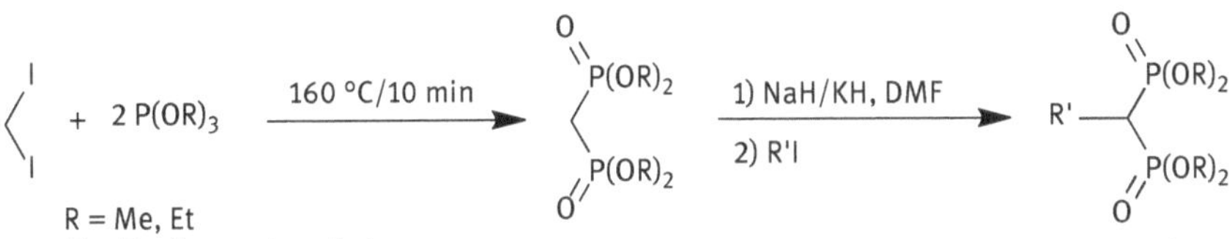

Figure 10.7: The simplest preparation of substituted methylenebisphosphonates (**2**).

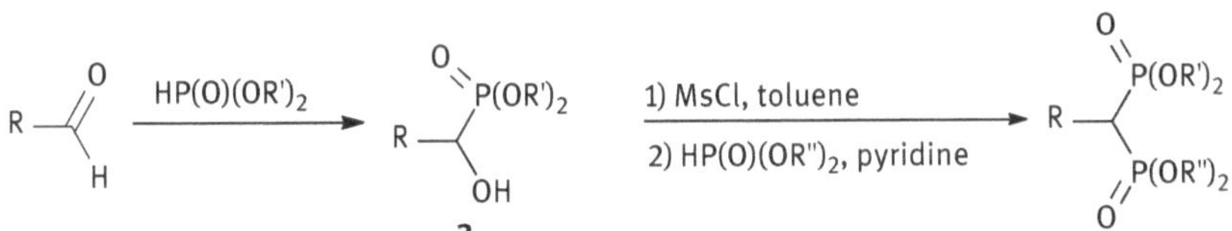

Figure 10.8: Another simple way to methylenebisphosphonates (**2**).

group at the central carbon atom (**2**). According to the first possibility, diiodomethane is reacted with 2 equivalents of trialkyl phosphite (in two consecutive Arbuzov reactions), then the CH-acidic methylenebisphosphonate (**1**) formed may be alkylated on its central carbon atom (Figure 10.7) [17, 18].

Starting from aldehydes, bisphosphonates (**2**) may be obtained in an addition–substitution reaction sequence. In the first step, dialkyl phosphite is added to the C=O group of aldehyde to give α-hydroxyphosphonate (**3**). Then, the hydroxy group of adduct **3** is mesylated with methanesulfonyl chloride to establish a good leaving group, and the intermediate so formed is reacted with a second molecule of dialkyl phosphite to furnish bisphosphonic acid (**2**) as the product (Figure 10.8) [19].

A special synthetic method of α-substituted methylenebisphosphonates (**2**) is based on the generation of an anion (**5**) from a phosphonate (**4**) by lithium isopropylamide followed by its reaction with dialkyl chlorophosphate (Figure 10.9) [20].

The free acidic form of methylenebisphosphonic acids may be obtained by the acidic hydrolysis of the corresponding esters (**2**).

The syntheses of α-hydroxy-methylenebisphosphonic acid derivatives (**8**) may be divided into two main groups, called as "direct" and "indirect" syntheses [2, 20–23].

During the "indirect" synthesis, the corresponding carboxylic acid chloride is reacted with trialkyl phosphite (in Arbuzov reaction), then dialkyl phosphite is added to the carbonyl group of the resulting α-ketophosphonate (**6**). The hydrolysis of tetraalkyl bisphosphonate (**7**) so obtained leads to the target α-hydroxy-methylenebisphosphonic acid (**8**) (Figure 10.10) [24].

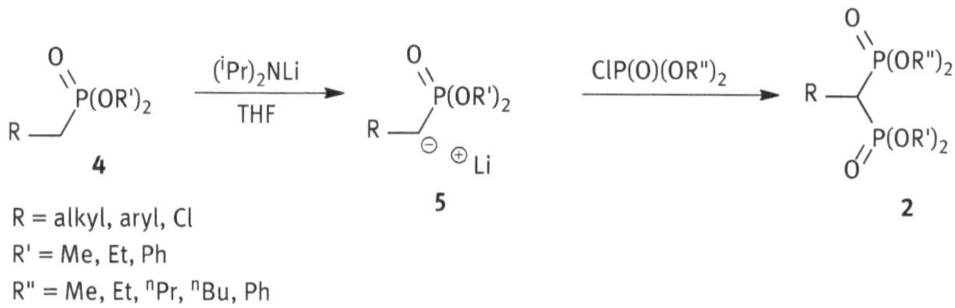

Figure 10.9: The third basic approach to methylenebisphosphonates (2).

The synthesis of the esters of etidronic, fenidronic and benzidronic acids (**8**, R = Me, Ph, $PhCH_2$) was performed under microwave conditions [25, 26].

The most common "direct" preparation involves the reaction of the corresponding carboxylic acid or its derivatives (acid chloride, anhydride or ester) with phosphorus trichloride and/or phosphorous acid (or in a few instances with phosphoric acid, phosphoryl chloride, phosphorus pentachloride or phosphorus trioxide). Upon the completion of the reaction, the target dronic acid (**8**) was obtained after hydrolysis and a suitable purification (Figure 10.11) [2, 20–23].

The "direct" synthetic route may be attractive as a consequence of the low price and the easy availability of the reagents, but the reaction has drawbacks as well. The reaction mixtures have a high degree of heterogeneity and often unstirrable. During the reaction, and the subsequent hydrolysis, intensive gas evolvement may occur. The data published are misleading and unreliable, as the required type and amount of the P-reagents have not been clarified for long. The role of the P-reactants and the solvents has not been investigated either, and the mechanism has not been explored.

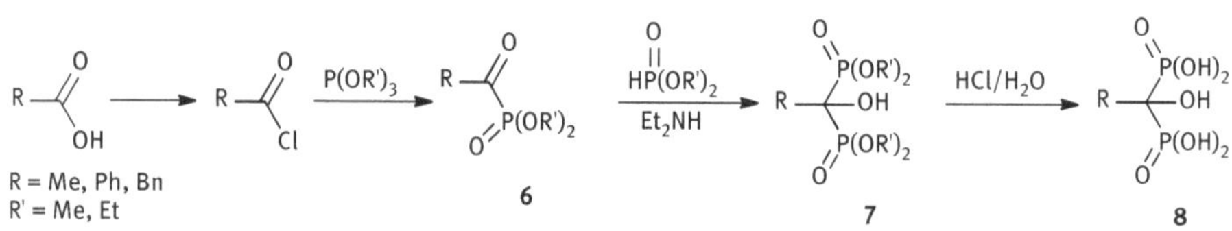

Figure 10.10: "Indirect" synthesis of dronic acids (8).

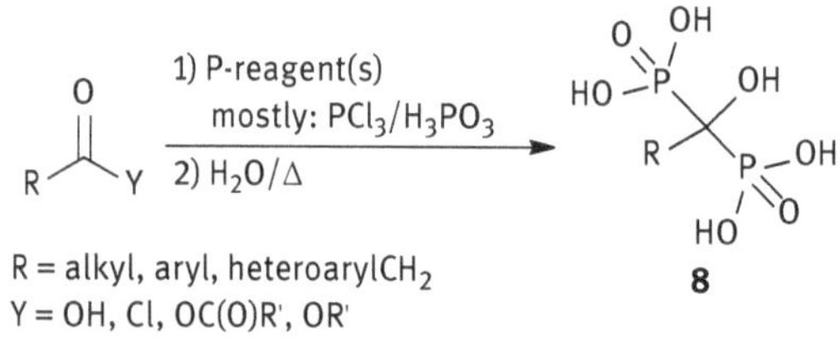

Figure 10.11: "Direct" synthesis of dronic acids (8).

Moreover, the products prepared were not supplied with adequate criteria for purity. The most commonly used solvents are methanesulfonic acid (MSA), chlorobenzene, sulfolane and toluene, but in many cases, the synthesis of dronic acid derivatives was performed in the absence of any solvent. However, according to Keglevich et al., the major representatives mentioned above cannot be synthesized in good yields under solvent-free conditions. A yield of 30–40% can be considered a good result if it relates to a pure product [2, 22, 27]. Remarkable and important results have been achieved by the authors of this chapter in the field of the syntheses of α-hydroxy-methylenebisphosphonic acids. The most important results are summarized in the next part [22, 23, 28–38].

10.4 Pioneering Results in the Synthesis of Hydroxy-methylenebisphosphonic Acids

The research was started with the investigation of the preparation of zoledronate (**9**) and risedronate (**10**) in MSA. The corresponding carboxylic acids were reacted with phosphorus trichloride and phosphorous acid in different ratios (Figure 10.12) [28]. It was proved that in MSA, the real P-reagent is phosphorus trichloride and its optimum amount is 3 equivalents. Phosphorous acid does not take part in the reaction because of its low nucleophilicity, it is only an unnecessary ballast. The pure dronic acids were obtained in yields of 53% (zoledronic acid (**9**)) and 74% (risedronic acid (**10**)) [28].

A putative reaction mechanism is proposed in Figure 10.13. In the first step, the heteroarylacetic acid is converted into the corresponding acid chloride, then it may be transformed to double phosphonium salt (**12-1**) by the action of two molecules of phosphorus trichloride via intermediate **11**. The bisphosphonium intermediate **12-1** may also exist as resonant structures **12-2** and **12-3**, but form **12-3** it is not too probable due to the distortion around the pentavalent and pentacoordinated P-atom. Hydrolysis of species **12** may lead to the target dronic acids (**9** and **10**) [28].

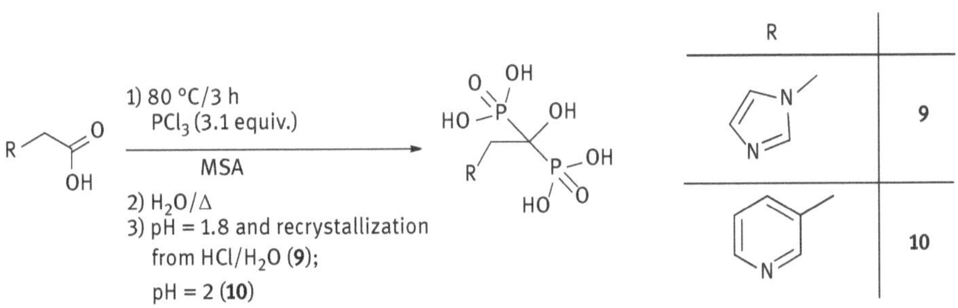

Figure 10.12: Preparation of zoledronic (**9**) and risedronic acid (**10**) in MSA by an optimized method [28].

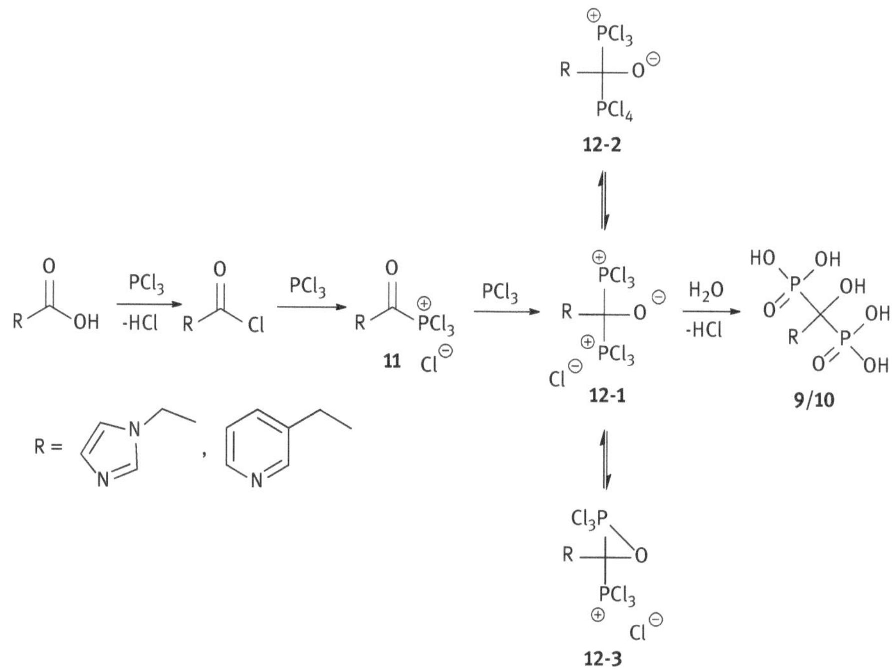

Figure 10.13: Formation of bisphosphonic acids by an initial concept [28].

The method was extended to the preparation of pamidronate (**13-Na**) [30], alendronate (**14**) [29], ibandronate (**15**) [29], etidronate (**16**) [31] and fenidronate (**17**) [32]. Phosphorus trichloride was the only P-reactant applied in a quantity of 3.1 equivalents using MSA as the solvent (Figure 10.14). Involvement of MSA in the reaction was also proved (see later).

In the cases of second-generation bisphosphonates (**13-Na**)-**15**, the yields were 46–58%, while in the synthesis of etidronate (**16**) and fenidronate (**17**), yields of 36% and 46%, respectively, were obtained [31, 32].

The mechanism shown in Figure 10.13 was refined, as it was also taken into account that MSA used as the solvent may react with the components of the reaction mixture. It can activate the carboxylic acid via the formation of a carboxylic

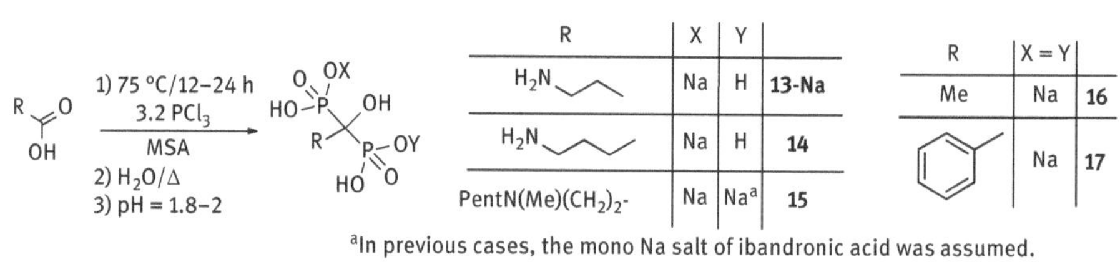

Figure 10.14: Extending the scope of the rationalized procedure.

acid–MSA mixed anhydride (**18**), which may be formed in two ways. The mixed anhydride (**18**) may be formed either by the reaction of acid chloride with MSA, or by the reaction of the starting carboxylic acid with methanesulfonyl chloride formed by the interaction of MSA and phosphorus trichloride. The reaction of phosphorus trichloride and MSA may also lead to Cl_2P–O–SO_2Me (**19**) that is also an anhydride-type intermediate, and may attack the carbonyl group of the acid chloride or mixed anhydride (**18**) to afford adducts **20a** and **20b**, respectively, that may be stabilized by the loss of the leaving group, resulting in α-keto intermediates **21a** and **21b**, respectively. The latter species (**21a** and **21b**) may react with another molecule of Cl_2P–O–SO_2Me (**19**), then hydrolysis of the intermediates (**22a** and **22b**) so formed furnishes the target dronic acid derivatives (**8**) (Figure 10.15) [30]. It was justified by quantum chemical calculations that the anhydride (**18**) is more reactive than the acid chloride [33, 34].

In the synthesis of dronic acid derivatives, the other frequently used solvent is sulfolane. The preparation of pamidronic acid (**13**), alendronate (**14**) and ibandronate (**15**) was studied in this solvent. Surprisingly, when only phosphorous trichloride was the P-reactants, no dronic acid was formed. There was need to use both P-reagents (PCl_3 and H_3PO_3) at the same time in a 2:2 or 3:2 molar ratio (Figure 10.16) [30, 36, 37].

In the case of pamidronic acid (**13**), alendronate (**14**) and ibandronate (**15**), the best yields were 63%, 52% and 83%, respectively. The experiments revealed that the solvent used has an important role, as determines, which P-reactant participates in

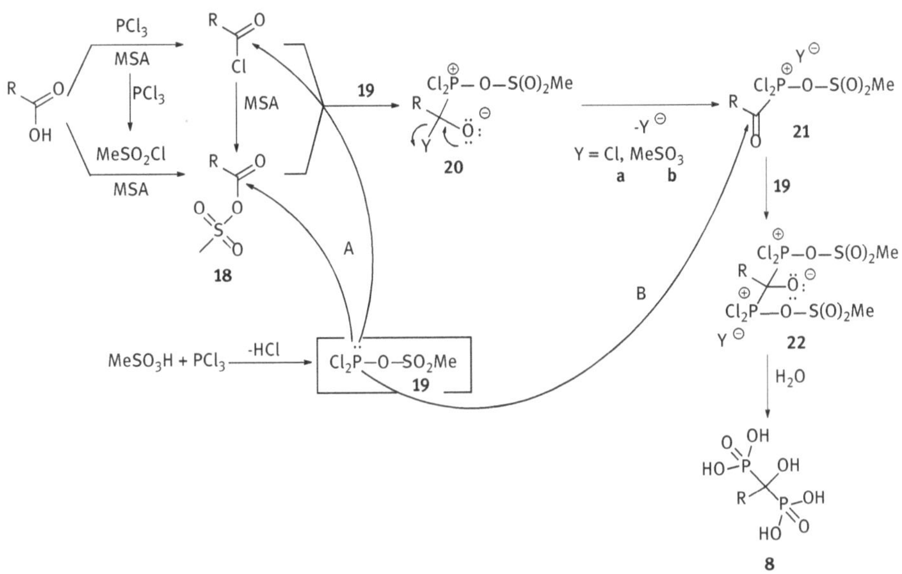

Figure 10.15: A refined mechanism for the formation of dronic acid derivatives (**8**) starting from the corresponding carboxylic acid and phosphorus trichloride in MSA as the solvent [30].

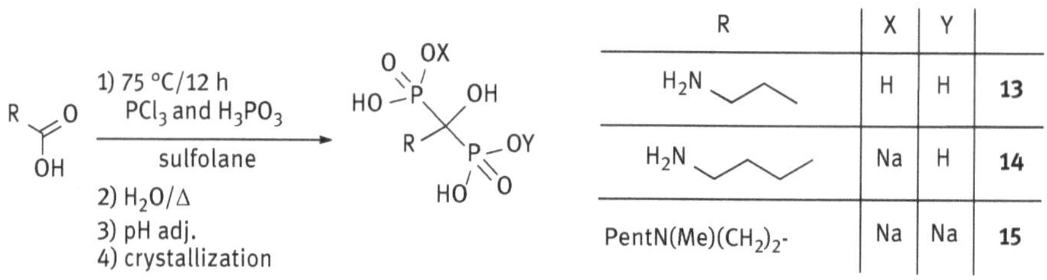

Figure 10.16: Preparation of pamidronic acid (**13**), alendronate (**14**) and ibandronate (**15**) in sulfolane.

the synthetic sequence [30, 36, 37]. It was suggested that in the first step, the corresponding carboxylic acid may react with 1 equivalent of phosphorus trichloride, resulting in acid chloride as a more active starting material. In the case under discussion, the interaction of phosphorus trichloride and phosphorous acid may result in the formation of intermediates $(HO)_2P{-}O{-}PCl_2$ (**23**) and $(HO)_2P{-}O{-}PCl{-}O{-}P(OH)_2$ (**24**). The reaction mechanism is similar to that shown in Figure 10.15. Instead of $Cl_2P{-}O{-}SO_2Me$ (**19**), species $(HO)_2P{-}O{-}PCl_2$ (**23**) and $(HO)_2P{-}O{-}PCl{-}O{-}P(OH)_2$ (**24**) are the active P-reagents which may react with the starting carboxylic acid or acid chloride (Figure 10.17) [30, 36, 37].

During the preparation of benzidronate (**25**) [33] and 3-phenylpropidronate (**26**) [34], unexpected results were observed, when MSA was the solvent (Figure 10.18). In contrast to our earlier experiences, lower yields were achieved, when only 3 equivalents of phosphorus trichloride were used in MSA [46% for benzidronate (**25**) and 2% for 3-phenylpropidronate (**26**)], as compared to the cases, when phosphorous acid was also applied beside phosphorus trichloride. Using the P-reagents in quantities of 3:1/3:2 or 2:3/2:4, benzidronate (**25**) and 3-phenylpropidronate (**26**) were obtained in yields of ca. 80% and 64%, respectively [33, 34].

As an explanation, again species $(HO)_2P{-}O{-}PCl_2$ (**23**) and $(HO)_2P{-}O{-}PCl{-}O{-}P(OH)_2$ (**24**) were assumed as the nucleophiles, and not $Cl_2P{-}O{-}SO_2Me$ (**19**) that would be reasonable in MSA. The anhydride-type intermediate **19** may formed, but it may have a lower nucleophilicity than species **23** and **24** [33, 34].

Studying the synthesis of 3-phenylpropidronate (**26**) by B3LYP/6–31G(d,p)//PCM(ACN) calculations, it was found that $(HO)_2P{-}$ may be the attacking nucleophilic moiety of P-reactants **23** and **24** that reacts with the carbonyl group of the starting 3-phenylpropionic acid (**27**), 3-phenylpropionyl chloride (**28**) or 3-phenylpropionic acid–MSA mixed anhydride (**29**), which may be present as intermediates. The activation barrier is the lowest, when activated acid derivatives **28** and **29** react with the $(HO)_2P{-}$ moiety of the P-nucleophiles (**23** and **24**). Reaction of carboxylic acid (**27**) may be less favorable (Figure 10.19) [34].

Nowadays, the use of ionic liquids (ILs) is spreading in synthetic organic chemistry, as they are considered "green" solvents, but in many cases, the ILs were used as only additives or catalysts [38–41]. The synthesis of α-hydroxy-methylene-bisphosphonic acid derivatives has also been investigated in ILs. We found that

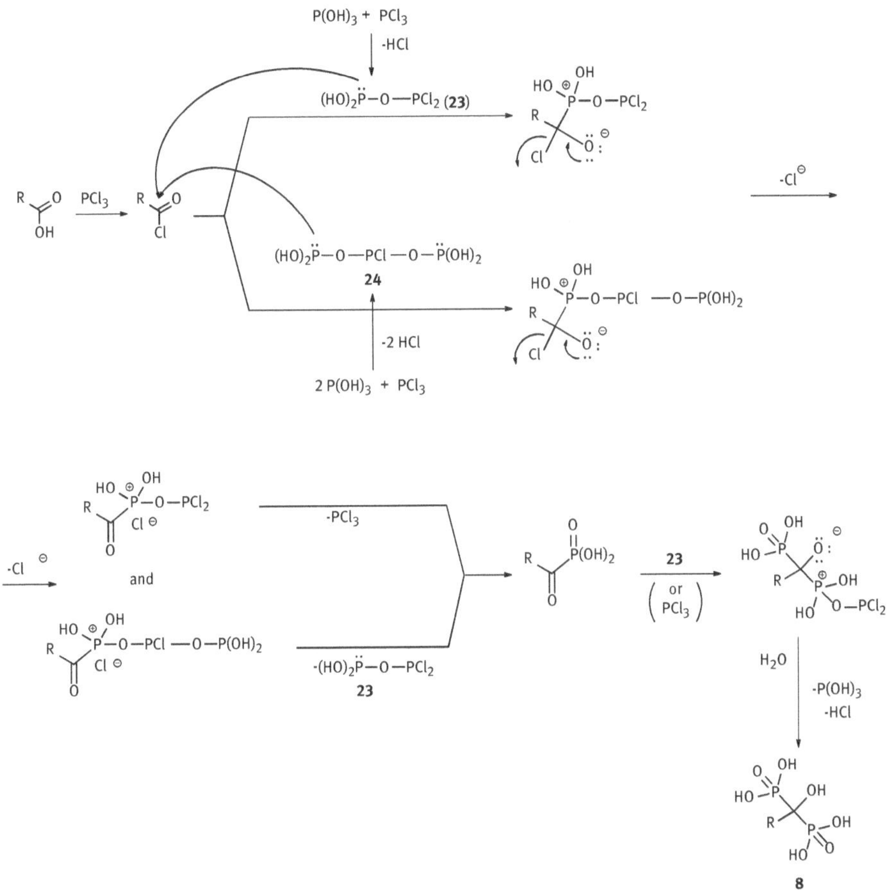

Figure 10.17: The mechanism in sulfolane – a novel protocol.

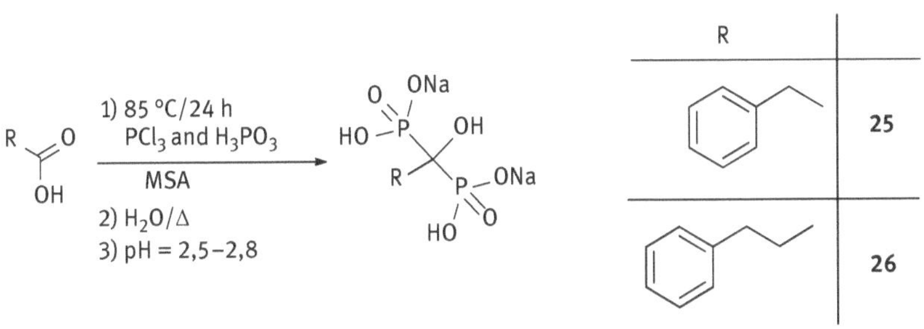

Figure 10.18: Synthesis of benzidronate (**25**) and 3-phenylpropidronate (**26**) in MSA.

there was no need to use the ILs (e.g. [bmim][BF_4] and [bmim][PF_6]) as a solvent, already smaller, catalytic amounts were also efficient. Their optimal amounts fell in a range of 0.1–0.6 equivalents. It was found that it is necessary to measure in

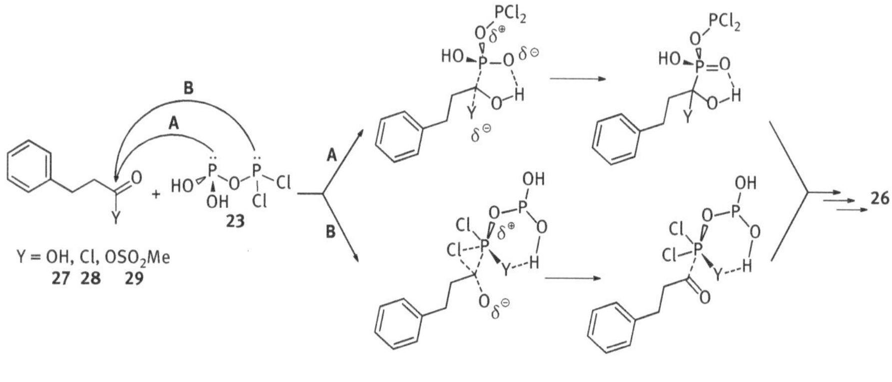

Figure 10.19: Initial steps during the formation of 3-phenylpropidronate (**26**) in MSA [34].

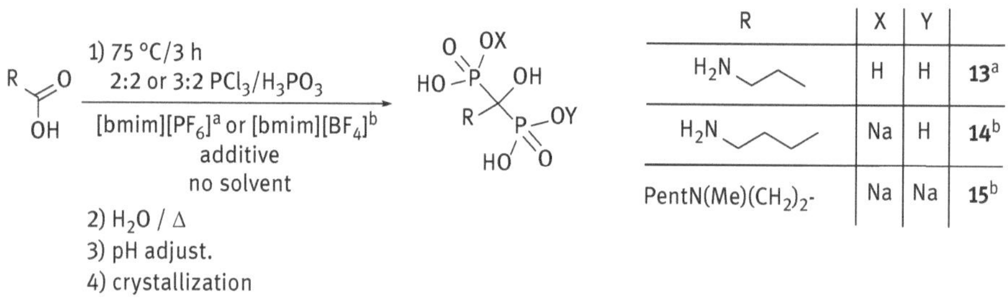

Figure 10.20: Synthesis of dronic acid derivatives (**13–15**) in the presence of ILs [35–38].

both phosphorus trichloride and phosphorous acid as the P-reagents (Figure 10.20) [35–38].

Pamidronic acid (**13**) and alendronate (**14**) were obtained in a yield of 72% and 66%, respectively, in the presence of 0.3 equivalents of [bmim][PF_6], or [bmim][BF_4], respectively. The yields were almost independent of the ratio (2:2 or 3:2) of the P-reactants. In the case of ibandronate (**15**), the catalytic effect of IL was even more pronounced. When phosphorus trichloride and phosphorous acid were applied in amounts of 3:2 equivalents, an outstandingly high yield (90%) was reached in the presence of only 0.1 equivalents of [bmim][BF_4]. The ILs caused the most significant increase in the yield of alendronate (**14**), as it was increased from 0 to 66% (Figure 10.21) [35–38].

Differing from the case of other bisphosphonates, ibandronate (**15**) could be prepared in high yields also in the absence of any solvent or IL (~75%) [37]. It may be concluded that ILs may be advantageous in the synthesis of dronic derivatives. However, there is no general rule, which IL, and in what quantity should be measured in.

It was assumed that the presence of an IL may increase the electrophilic character of the carbonyl group of the carboxylic acid derivatives (Figure 10.22) [35–38].

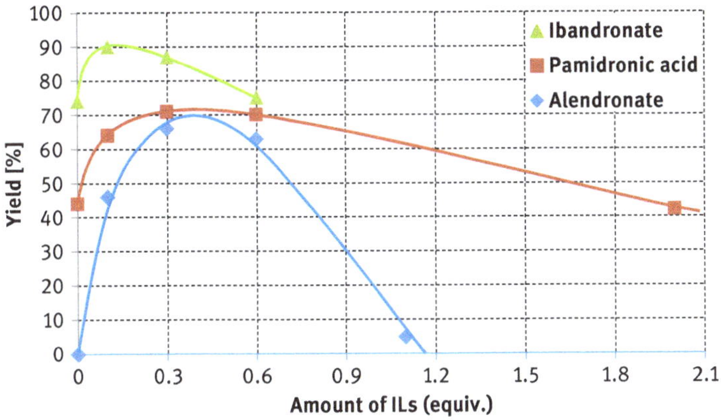

Figure 10.21: The effect of the amount of ILs on the yield of ibandronate (**15**) ([bmim][BF_4]), pamidronic acid (**13**) ([bmim][BF_4]) and alendronate (**14**) ([bmim][PF_6]) [35–38].

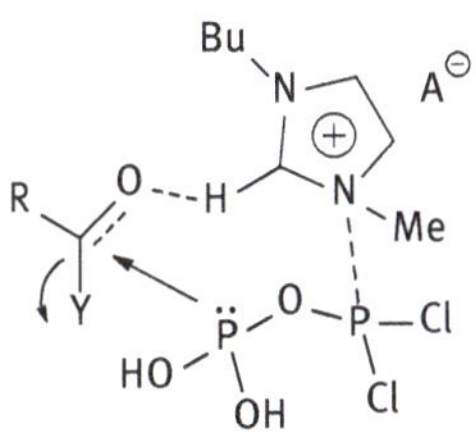

Y = OH, Cl

Figure 10.22: Explanation of the catalytic effect of the IL additive [35–38].

10.5 Summary

In summary, the recent results achieved in the synthesis of hydroxy-bisphosphonic acids are stimulating. Many contradictions have been clarified, and the syntheses of a series of dronic derivatives were rationalized and optimized. The role of the P-reactants and solvents was explored, and novel reaction mechanisms were proposed. The catalytic effect of ILs has been demonstrated in the syntheses of a few α-hydroxy-methylenebisphosphonic derivatives.

Acknowledgment: This work was partially supported by the National Research, Development and Innovation Office (K119202).

References

[1] H. Fleisch, R. G. G. Russell, F. Straumann, *Nature*, 1966, **212**, 901–903.
[2] H. R. Hudson, N. J. Wardle, S. W. A. Bligh, I. Greiner, A. Grün, G. Keglevich, *Mini Rev. Med. Chem.*, 2012, **12**, 313–325.
[3] E. Breuer, E. *In Analogue-based drug discovery*, 2006, 371–384.

[4] N. Menschutkin, *Eur. J. Org. Chem.*, 1865, **133**, 317–320.
[5] H. V. Baeyer, K. A. Hofmann, *Chem. Ber.*, 1897, **30**, 1975.
[6] R. G. G. Russell, *Bone*, 2011, **49**, 2–19.
[7] W. S. Rolf, M. Irene, *Pharm. World Sci.*, 1998, **20**, 206–213.
[8] R. G. G. Russell, *Pediatrics*, 2007, **119**, 150–162.
[9] L. Widler, K. A. Jaeggi, M. Glatt, K. Müller, R. Bahmann, M. Bisping, A. R. Born, R. Cortesi, G. Guiglia, H. Jeker, R. Klein, U. Ramseier, J. Schmid, G. Schreiber, Y. Seltenmeyer, J. R. Green, *J. Med. Chem.*, 2002, **45**, 3721–3738.
[10] J. B. Catterall, T. E. Cawston, *Arthritis Res. Ther.*, 2002, **5**, 12–24.
[11] R. Marcus, D. Feldman, D. Nelson, C. J. Rosen, *Osteoporosis;* Elsevier Academic Press: Burlington, 2007, 14–16.
[12] R. G. G. Russell, P. I. Croucher, M. J. Rogers, *Osteoporosis Int.*, 1999, **9**, 66–80.
[13] A. J. Roelofs, K. Thompson, S. Gordon, M. J. Rogers, *Clin. Cancer Res.*, 2006, **12**, 6222–6230.
[14] S. P. Luckman, F. P. Coxon, F. H. Ebetino, R. Graham, R. G. G. Russell, M. J. Rogers, *J. Bone Miner. Res.* 1998, **13**, 1668–1678.
[15] R. G. G. Russell, M. J. Rogers, *Bone,* 1999, **25**, 97–106.
[16] M. J. Rogers, J. C. Crockett, F. P. Coxon, J. Mönkökken, *Bone*, 2011, **49**, 34–41.
[17] G. M. Kosolapoff, *J. Am. Chem. Soc.*, 1953, **75**, 1500–1501.
[18] D. R. Magnin, J. K. Dickson, J. V. Logan, R. M. Lawrence, Y. Chen, R. B. Sulsky, C. P. Ciosek, S. A. Biller, T. W. Harrity, K. G. Jolibois, L. K. Kunselman, L. C. Rich, D. A. Slusarchyk, *J. Med. Chem.*, 1995, **38**, 2596–2605.
[19] L. I. Kruse, V. P. Shah, *US 5204336*, 1993.
[20] M.-P. Teulade, P. Savignac, E. E. Aboujaoude, S. Lietge, N. J. Collignon, *Organomet. Chem.*, 1986, **304**, 283–300.
[21] M. Lecouvey, Y. Leroux, *Heteroat. Chem.*, 2000, **11**, 556–561.
[22] D. I. Nagy, A. Grün, S. Garadnay, I. Greiner, G. Keglevich, *Molecules*, 2016, **21**, e1046.
[23] D. I. Nagy, A. Grün, I. Greiner, G. Keglevich, *Curr. Org. Chem.*, 2017, **21**, 1567–1578.
[24] L. M. Nguyen, E. Niesor, C. L. Bentzen, *J. Med. Chem.*, 1987, **30**, 1426–1433.
[25] A. Grün, G. I. Molnár, B. Bertók, I. Greiner, G. Keglevich, *Heteroat. Chem.*, 2009, **20**, 350–354.
[26] G. Keglevich, A. Grün, G. .I. Molnár, I. Greiner *Heteroat. Chem.*, 2011, **22**, 640–648.
[27] G. R. Kieczykowski, R. B. Jobson, D. G. Melillo, D. F. Reinhold, V. J. Grenda, I. Shinkai, *J. Org. Chem.*, 1995, **60**, 8310–8312.
[28] G. Keglevich, A. Grün, K. Aradi, S. Garadnay, I. Greiner, *Tetrahedron Lett.*, 2011, **52**, 2744–2746.
[29] R. Kovács, A. Grün, S. Garadnay, I. Greiner, G. Keglevich, *Curr. Org. Synth.*, 2013, **10**, 640–644.
[30] R. Kovács, A. Grün, O. Németh, S. Garadnay, I. Greiner, G. Keglevich, *Heteroat. Chem.*, 2014, **25**, 186–193.
[31] R. Kovács, D. I. Nagy, A. Grün, G. T. Balogh, S. Garadnay, I. Greiner, G. Keglevich, *Lett. Drug Des. Discovery*, 2013, **10**, 733–737.
[32] A. Grün, R. Kovács, D. I. Nagy, S. Garadnay, I. Greiner, G. Keglevich, *Lett. Org. Chem.*, 2014, **11**, 368–373.
[33] R. Kovács, D. I. Nagy, A. Grün, S. Garadnay, I. Greiner, G. Keglevich, *Lett. Drug. Des. Discovery*, 2015, **12**, 78–84.
[34] A. Grün, D. I. Nagy, O. Németh, Z. Mucsi, S. Garadnay, I. Greiner, G. Keglevich, *Curr. Org. Chem.*, 2016, **20**, 1745–1752.
[35] A. Grün, D. I. Nagy, S. Garadnay, I. Greiner, G. Keglevich, *Lett. Drug. Des. Discovery,* 2016, **13**, 475–478.
[36] D. I. Nagy, A. Grün, S. Garadnay, I. Greiner, G. Keglevich, *Heteroat. Chem.*, 2017, **28**, e21370.

[37] D. I. Nagy, A. Grün, O. Pavela, S. Garadnay, I. Greiner, G. Keglevich, *Lett. Drug. Des. Discovery,* 2018, **15** (in press).
[38] N. Z. Kiss, D. I. Nagy, G. Keglevich, in *Advances in Chemistry Research*; J. C. Taylor, Ed.; Nova Science Publishers: New York, 2017, **37**, 121–140.
[39] N. Z. Kiss, G. Keglevich, *Tetrahedron Lett.*, 2016, **57**, 971–974.
[40] H. Olivier-Bourbigou, L. Magna, D. Morvan, *Appl. Catal. A.*, 2010, **373**, 1–56.
[41] Q. Zhang, S. Zhang, Y. Deng, *Green Chem.*, 2011, **13**, 2619–2637.

Goutam Brahmachari

11 P-Chemistry at ambient conditions: A recent update

Abstract: This chapter offers an update on the recent developments in green synthetic protocols in organophosphorus chemistry at ambient conditions, and covers the literature reported during the period of 2012–mid-2017. The role of solid-supported heterogeneous catalysts, nano-catalysts and low-cost homogeneous catalysts, chiral ligands, catalyst-free conditions, and also the ball-milling technique in accessing a diverse range of organophosphorus compounds at ambient conditions are discussed in detail.

Keywords: organophosphorous reactions, green chemistry, ambient conditions, catalysts, mechanochemistry.

11.1 Introduction

Organophosphorus compounds form an important class of chemicals with extensive commercial applications [1, 2]. As a result, the study and synthesis of organophosphorus compounds have undergone notable advances in recent decades [3–10]. These promising compounds can be used in organic syntheses as ligands for transition-metal catalysts [11–16], as organocatalysts [17–23], and also as building blocks for pharmaceutical and agricultural chemistry [24–30]. Thus, phosphorus–carbon and phosphorus–heteroatom bond formation remains a valid and active exercise in chemical research as a result of which new reactions are developed for the preparation of new families of organophosphorus compounds with potential multifaceted interest.

Since its first proposition in the 1990s until now, research on green and sustainable chemistry has undergone considerable growth over the past 25 years, and become very popular among the researchers working in all branches of chemical sciences. Dedicated research endeavors have yielded innumerable green processes and products so far, and such well-documented quarter century of advances in the fields of green chemistry and green engineering have framed a solid platform for further explorations [31–33]. During the recent past, various greener pathways as alternatives to traditional chemical syntheses and transformations have been invented in diverse fields of chemistry to attain sustainability through newer concepts, such as step- and atom-economy and E-factor. Greener features in designing an alternative protocol for developing useful organic molecules and/or materials include the use of bio-renewable resources, benign reaction media (use of greener solvents

https://doi.org/10.1515/9783110535839-011

or no solvent), recyclable (magnetic) heterogeneous catalysts, or no catalyst at all; minimization of by-products or waste generation, with efficient isolation of the target product; and improving energy efficiency while performing a chemical reaction by judicious screening of reaction conditions capable of carrying out the transformation at ambient conditions or using microwaves, ultrasound, visible light and mechanical mixing (ball-milling) as *green* alternatives for heating and activation of the reactants. As part of such ongoing developments on green synthetic strategies, designing reactions under ambient conditions coupled with other green aspects has now been regarded as an emerging field of research in organic chemistry, and recent literature reports indicate the progress [32]. The present chapter is aimed at offering an update on the recent developments of synthetic protocols in organophosphorus chemistry (P-chemistry) occurring at ambient conditions reported during the period of 2012–mid-2017.

11.2 Recent Advances in the Development of Synthetic Protocols in Organophosphorus Chemistry at Ambient Conditions

11.2.1 Solvent-free Synthesis of α-hydroxyphosphonates using Solid-supported Heterogeneous Catalysts

Solid-supported heterogeneous catalysts have received considerable attention in organic synthesis during the recent past due to their manifold benefits in reducing and/or eliminating the use and generation of hazardous substances [34–44]. In 2013, there were two reports on the application of such heterogeneous catalytic systems in synthesizing α-hydroxyphosphonates via modified Pudovik reaction under solvent-free conditions at room temperature. Vaccaro and group [45] made successful use of recyclable PS-BEMP (2-*tert*-butylimino-2-diethylamino-1,3-dimethylperhydro-1,3,2-diazaphosphorine supported on polystyrene), a polymer-supported organocatalyst, in accomplishing hydrophosphonation of aromatic and aliphatic aldehydes (**1**) with diethyl phosphite (**2**) under solvent-free conditions at 30 °C to give the corresponding α-hydroxyphosphonates (**3**) in almost quantitative yields without the need for further purification (Figure 11.1). Calculated E-factors for the transformations were found to be within the range of 11.6–14.6 (based on a 0.5 mmol scale), which is quite satisfactory. This is to mention herein that even at this ambient temperature PS-BEMP, being strongly basic in nature, was found to facilitate the phospha-Brook rearrangement of the aldehydes with electron-withdrawing groups, such as 2-pyridylcarboxyaldehyde and 4-cyanobenzaldehyde, yielding the rearranged products as major components;

however, complete conversion into α-hydroxyphosphonates for these substrates was achieved by reducing the temperature to 4 °C and using acetonitrile as the reaction medium. In addition, the investigators applied their protocol to a large-scale reaction (50 mmol scale), that is a continuous-flow procedure that enables the reduction of the E-factor of the process by 92.7–96.3% to further improve the sustainability of the entire method [45].

Fluorapatite (FAP, $Ca_{10}(PO_4)_6(F)_2$) has recently attracted much attention for its potential application as an efficient catalyst in organic syntheses [47–51]. Interestingly, FAP has an unusual property in that it contains both acidic and basic sites in a single crystal lattice. Ramananarivo et al. [46] prepared sodium-modified FAP (Na@FAP) and developed another straightforward and eco-friendly method for the synthesis of a series of α-hydroxyphosphonates (**5**) via the Pudovik reaction between carbonyl compounds (**4**) and dialkyl phosphites (**2**) using this new version of FAP, which they reported to be a highly efficient solid catalyst under solvent-free conditions at room temperature (Figure 11.2). The products were purified without the need for column chromatography. .

Both of these solvent-free protocols are advantageous over the available methods in terms of operational simplicity, efficiency, yields, ease of work-up, and catalyst reusability, making them much greener alternatives as a whole. Among organophosphorus compounds, α-hydroxyphosphonates are of significant bio-

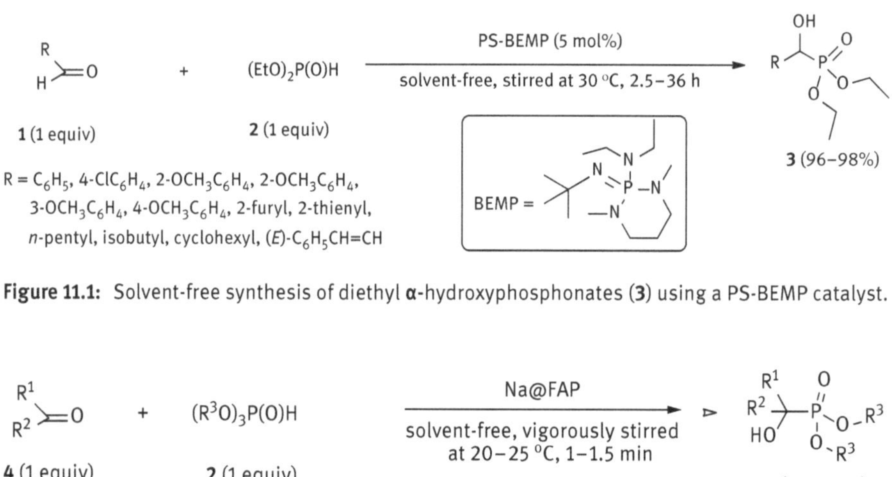

Figure 11.1: Solvent-free synthesis of diethyl **α**-hydroxyphosphonates (**3**) using a PS-BEMP catalyst.

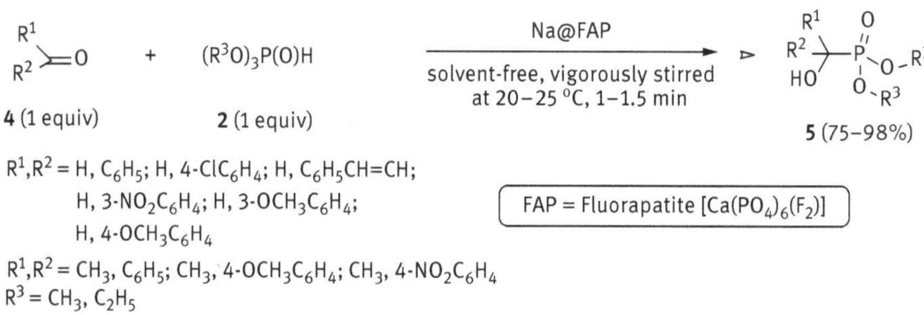

Figure 11.2: Solvent-free synthesis of α-hydroxyphosphonates (**5**) using Na@FAP as the catalyst.

logical and pharmaceutical interest, and are used in the development of antiviral, antibacterial and anticancer agents; pesticides; and inhibitors of various enzymes, such as renin, HIV protease, and polymerase [52–59]. Moreover, such compounds also find applications as useful precursors in the synthesis of various α-substituted phosphonates and phosphonic acids [60, 61].

11.2.2 Synthesis of Alkyl/Aryl-phosphonates/Phosphinates/Phosphine Oxides

Organo-phosphonates/phosphinates/phosphine oxides are useful organic compounds finding immense applications in synthetic, biological, pharmaceutical, and material chemistry [62–70]. Moreover, arylphosphonate scaffolds are used in designing fuel cell membranes [71] and also materials with special optical properties [72, 73].

Dhokale and Mhaske [74] demonstrated a metal-free, mild, and convenient aryl-C-P bond forming reaction protocol for the straightforward P-arylation of diversely substituted aryl silyl triflates, (**6**) leading to the formation of arylphosphonates (**7**), arylphosphinates (**9**), and arylphosphine oxides (**8**) using cesium fluoride as a catalyst under mild reaction conditions at room temperature in acetonitrile (Figure 11.3). In their proposed mechanistic approach (Figure 11.4), the investigators assumed two possible pathways: **Path A,** which followed a concerted type of mechanism involving the participation of a solvent molecule, while **Path B** followed a mechanism of Michaelis–Arbuzov type, both involving an aryne intermediate (**10**). However, the authors were of the opinion that **Path B,** involving fluoride ion participation, appears to be more appropriate, although a detailed study on the mechanism is warranted to clarify the situation.

In another report, Geng et al. [75] accomplished a catalyst-free synthesis of a series of functionalized pyrazole derivatives (**15**) linked with phenylphosphine oxide

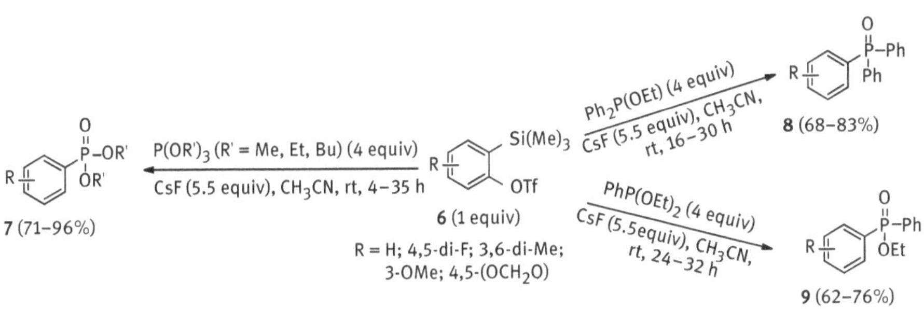

Figure 11.3: Synthesis of aryl-phosphonates/phosphinates/phosphine oxides made possible by facile aryne reactivity.

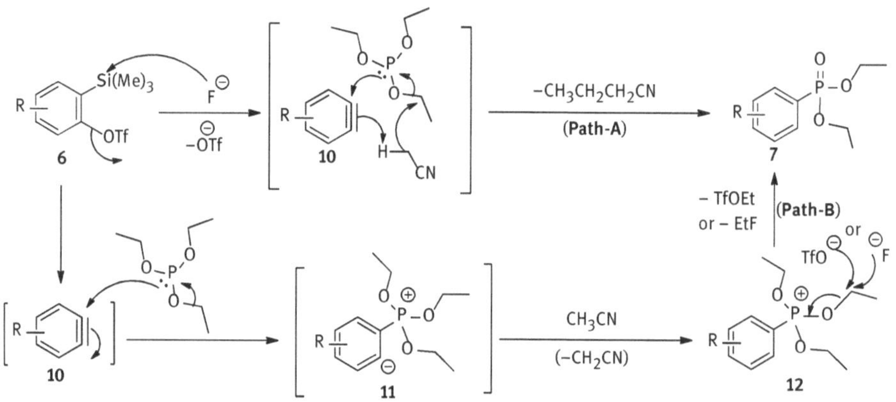

Figure 11.4: Proposed mechanisms for the transformation of aryl silyl triflates into arylphosphonates.

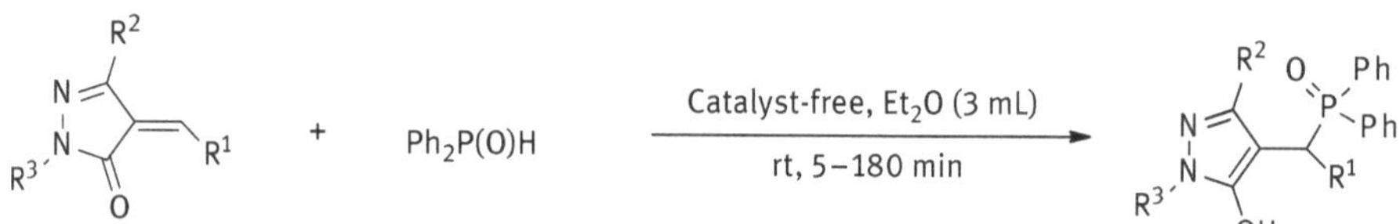

13 (0.60 mmol) **14** (0.03 mmol) **15** (93–99%)

$R^1 = C_6H_5$, 4-ClC_6H_4, 3-ClC_6H_4, 4-BrC_6H_4, 2,4-di-Cl_2-C_6H_3, 4-$OCH_3C_6H_4$, 4-$CH_3C_6H_4$, 4-$SCH_3C_6H_4$, 3,5-di-$(OCH_3)_2C_6H_3$, 3,5-$(OCH_2O)C_6H_3$, 2-naphtyl, 2-furyl

$R^2 = CH_3$, CF_3; $R^3 = C_6H_5$, 4-BrC_6H_4, 4-ClC_6H_4

Figure 11.5: Catalyst-free synthesis of phosphinoylated pyrazole derivatives (**15**).

via a Michael addition reaction between α,β-unsaturated pyrazolones (**13**) and the secondary phosphine oxide (**14**) in diethyl ether at room temperature in high yields (93–99%) (Figure 11.5). This protocol appears to be a promising and versatile green alternative for the synthesis of such biologically significant heterocyclic scaffolds.

11.2.3 Synthesis of α-aminophosphonates via the Kabachnik-Fields Reaction

Kabachnik–Fields reaction (also known as phospha-Mannich condensation) has been enormously studied, and appears to be a powerful reaction to obtain α-aminophosphonates by the one-pot three-component condensation of primary or secondary amines, carbonyl compounds (aldehydes and ketones), and >P(O)H species, especially dialkyl phosphites [76–80]. α-Aminophosphonates and related derivatives are evergreen targets in the bio- and medicinal chemistry due to their versatile bioactivity [81–90].

A series of α-aminophosphonate derivatives (**18**) were prepared by Fang et al. [91] through the Kabachnik–Fields reaction of aromatic aldehydes (**16**), aromatic amines (**17**), and triethyl phosphite/diethyl phosphite at room temperature, using a biodegradable SO_3H-functionalized ionic liquid (IL), 3-(*N*,*N*-dimethyldodecylammonium) propanesulfonic acid ([DDPA][HSO_4]), under solvent-free conditions, or in aqueous media (Figure 11.6).

Later on, Shaterian et al. [92] also reported an eco-friendly and novel method for the synthesis of another series of such compounds (**18**) by the one-pot Kabachnik–Fields reaction employing nano-TiO_2 as a reusable heterogeneous catalyst under ambient and solvent-free conditions (Figure 11.7). The method is efficient, and provides excellent yields in short reaction times, which makes the protocol an environmentally friendly and economically valuable process for the synthesis of this class of compounds.

In another report, Abdel-Megeed et al. [93] demonstrated the synthesis of a series of antimicrobial diphenyl 1-(arylamino)(pyridin-3-yl)methylphosphonates (**20**) with good yields by the reacting nicotinaldehyde (**19**) with aromatic amines (**17**) and triphenyl phosphite in the presence of titanium tetrachloride as a catalyst in dichlorometane at room temperature (Figure 11.8).

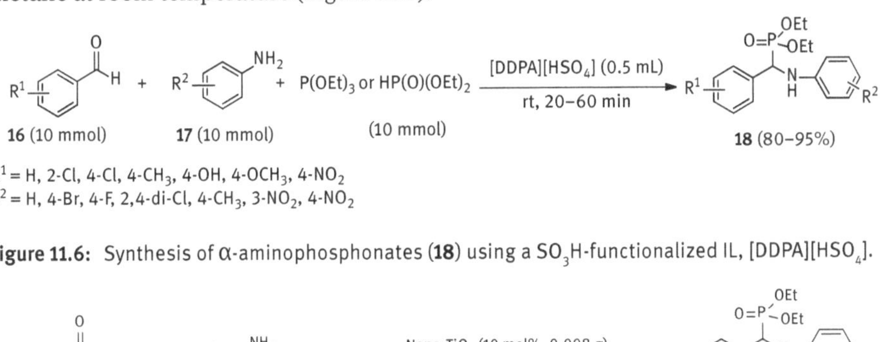

16 (10 mmol) **17** (10 mmol) (10 mmol) **18** (80–95%)

R^1 = H, 2-Cl, 4-Cl, 4-CH_3, 4-OH, 4-OCH_3, 4-NO_2
R^2 = H, 4-Br, 4-F, 2,4-di-Cl, 4-CH_3, 3-NO_2, 4-NO_2

Figure 11.6: Synthesis of α-aminophosphonates (**18**) using a SO_3H-functionalized IL, [DDPA][HSO_4].

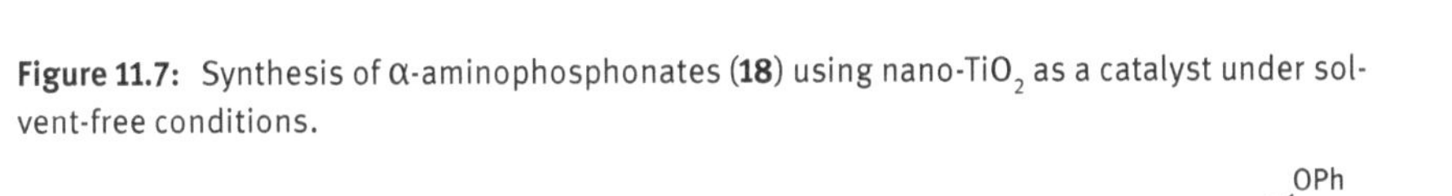

16 (1 mmol) **17** (1 mmol) (1 mmol) **18** (85–94%)

R^1 = H, 4-Cl, 4-CH_3, 2-OH, 4-OH, 4-OCH_3, 3, 4-di-OCH_3, 3-NO_2, 4-NO_2
R^2 = H, 4-Br, 4-CH_3, 3-NO_2, 4-NO_2, (4-Cl,2-NO_2)

Figure 11.7: Synthesis of α-aminophosphonates (**18**) using nano-TiO_2 as a catalyst under solvent-free conditions.

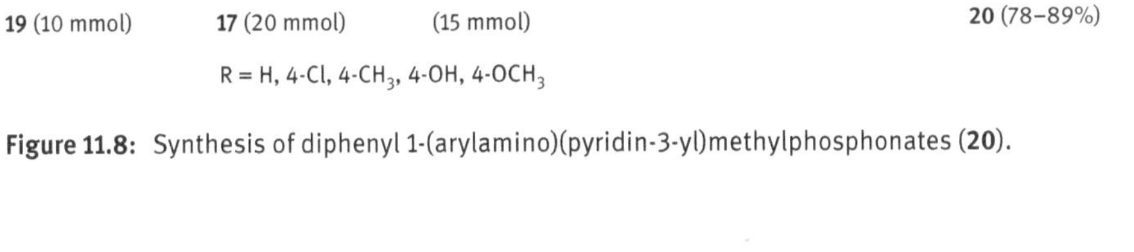

19 (10 mmol) **17** (20 mmol) (15 mmol) **20** (78–89%)

R = H, 4-Cl, 4-CH_3, 4-OH, 4-OCH_3

Figure 11.8: Synthesis of diphenyl 1-(arylamino)(pyridin-3-yl)methylphosphonates (**20**).

11.2.4 Synthesis of (2-amino-3-cyano-4*H*-chromen-4-yl)Pposphonic Acid Diethyl Esters

2-Amino-4*H*-chromene scaffolds represent a *privileged* structural motif, well distributed in naturally occurring compounds [94–96] with a broad spectrum of pharmacological efficacies that include antimicrobial [97], antiviral [98, 99], anti-inflammatory [100], antimalarial [101, 102], sex-hormonal [103], antiproliferative [104], antitumor [105], and anticancer activities [106], and are also widely employed in making cosmetics and pigments [107], and potent biodegradable agrochemicals [108, 109]. As mentioned in earlier sections, phosphonates and their derivatives also find immense applications in synthetic and medicinal chemistry. The wide range of promising biological activities and pharmacological properties of both phosphonate derivatives and 2-amino-4*H*-chromenes has led to attempts at synthesizing a variety of phosphonate derivatives coupled with 2-aminochromenyl rings.

Rao Kolla and Lee [110] reported a room temperature method for the rapid and efficient synthesis of biologically relevant 2-amino-4*H*-chromen-4-ylphosphonate derivatives (**23**) by the one-pot three-component condensation of substituted salicylaldehydes (**21**), malononitrile (or ethylcyanoacetate) (**22**), and triethyl phosphite in the presence of ethylenediamine diacetate as a catalyst in alcohol at room temperature (Figure 11.9). The research group of the author of this chapter [111] also developed an energy-efficient multicomponent one-pot procedure for the facile synthesis of a wide range of diverse (2-amino-3-cyano-4*H*-chromen-4-yl)phosphonic acid diethyl esters (**23′**) at room temperature by the reaction of salicylaldehydes, malononitrile (or ethyl cyanoacetate), and triethyl phosphite on the surface of a reusable heterogeneous MgO-nanocatalyst in aqueous ethanolic medium (Figure 11.10). The significant features of this eco-friendly green protocol include operational simplicity, easy reusability of the catalyst, room temperature condition, energy efficiency, clean reaction profiles, and good yields.

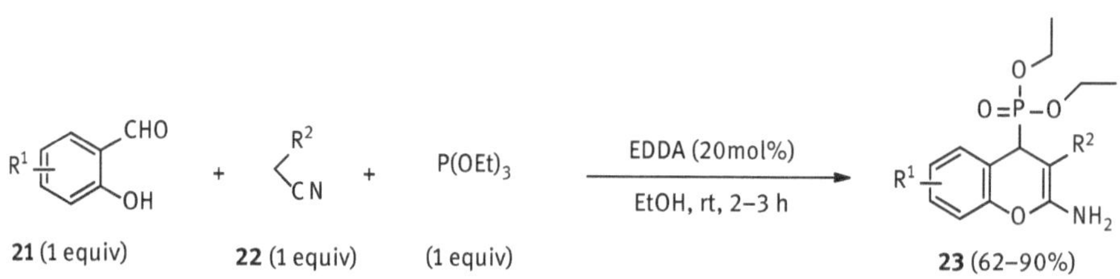

R^1 = H, 3-CH_3, 5-CH_3, 4-OCH_3, 5-OCH_3, 5-Br, 3, 5-di-Br, 5-NO_2
R^2 = CN, COOEt

Figure 11.9: Synthesis of (2-amino-4*H*-chromen-4-yl)phosphonate derivatives (**23**) using ethylenediamine diacetate as the catalyst.

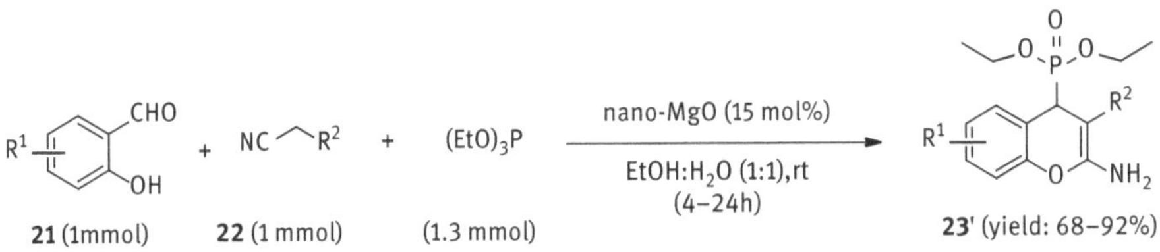

R^1 = H, 5-Cl, 5-Br, 3,5-di-Cl, 3,5-di-Br, 5-Me, 5-OMe, 5-NO_2
R^2 = CN, COOEt

Figure 11.10: One-pot synthesis of diethyl (2-amino-3-cyano-4*H*-chromen-4-yl)phosphonates (**23′**) using a nanocrystalline MgO catalyst.

11.2.5 Synthesis of Vicinal Amphiphilic Phosphine Oxide Boronates

Phosphine-substituted borane and boronate esters have received enhanced interest recently for their promising applications in the emerging field of frustrated Lewis pair chemistry [112–115], and as organocatalysts or ligands in metal-catalyzed transformations [116–118], where the boron atom also has the ability to bind transition metals by acting as σ-acceptor ligand, thereby offering new fascinating possibilities in terms of controlling reactivity [119]. The enantioselective synthesis of optically active vicinal amphiphilic phosphine oxide boronates at room temperature was reported as a significant research in organophosphorus chemistry [120]. The investigators elaborated an efficient asymmetric β-boration of α,β-unsaturated phosphine oxides (**24**) using bis(pinacolato)diboron (**25**) to synthesize a series of boronate esters **26** in good yields (62–95%), and with high enantioselectivities (up to 96%) under ambient conditions using a chiral catalytic system based on copper(I) and ligand (R_C,S_P)-Josiphos (Figure 11.11). Broad structural scope, functional group tolerance, mild reaction conditions, and high enantiomeric excess are the major advantages of this elegant method.

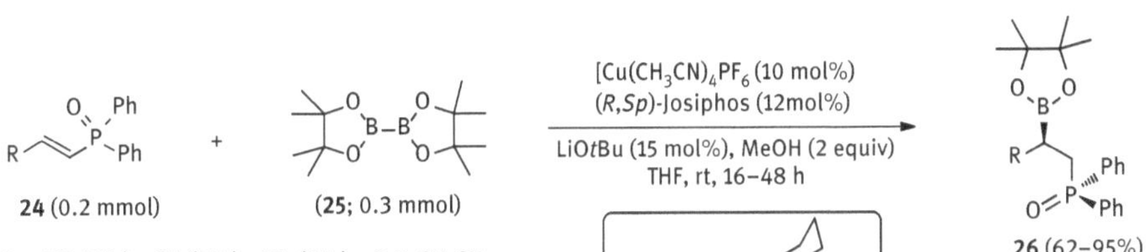

R = $CH_3(CH_2)_3$, $CH_3(CH_2)_5$, $CH_3(CH_2)_9$, $C_6H_5CH_2CH_2$, C_6H_5, *c*-propyl, *c*-pentylmethyl, *c*-hexylmethyl, $C_2H_5COOCH_2CH_2$, $TBSOCH_2CH_2$, $(CH_3)_3Si$ (TBS = *tert*-butyldimethylsilyl)

Figure 11.11: Chiral Cu(I)-complex-catalyzed enantioselective synthesis of phosphine oxide boronates (**26**).

11.2.6 Synthesis of Organophosphorus Compounds using Mechanochemistry under Ambient Conditions

Recently, mechanochemistry, i.e. chemical synthesis conducted by grinding the reactants together under neat conditions, has attracted much attention as an alternative and green chemical approach [121–123]. This is a solvent-free technique [124–128]. Although the term *mechanochemistry* has been adopted recently, it was known to all early chemists as a hand-grinding technique. Today, this hand-grinding technique has been automated through a suitable equipment, called *ball mill*. Nowadays, ball milling has emerged as a powerful tool to realize chemical reactions by mechanical energy. Allowing a variety of reactions to occur at ambient temperatures and under solvent-free conditions, ball milling offers a greener route for initiating many chemical reactions; this method has been summarized in many review papers and book chapters [129]. Hence, organic reactions taking place under neat conditions by means of ball milling activation demonstrate significant advantages, including substantial improvements in energy efficiency as compared to conventional solution-based or microwave-assisted syntheses, short reaction times, straightforward work-up, quantitative yields, higher safety, and the potential for scale-up over other solvent-free techniques [130–134].

Marandi et al. [135] used the mechanochemical technique, and developed a catalyst- and solvent-free protocol for the one-pot synthesis of a variety of functionalized phosphorus ylides (**31**) and 1,4-diionic organophosphorus compounds (**32**) by the reaction of dialkyl acetylenedicarboxylates (**27**), triphenylphosphine (**28**), and diverse NH- and C-H activated acids (**29** and **30**) by grinding the reactants at room temperature (Figure 11.12). The products were obtained in excellent yields and with high stereoselectivity. The present protocol provides notable advantages, particularly the avoidance of the catalyst and toxic organic solvent, and hence ensuring safe and mild reaction conditions [135].

Recently, another study on the use of mechanochemistry in organophosphorus chemistry has been published. Wang and his group [136] elaborated the first solvent-free synthesis of a series of C_2-phosphonylated thiazoles/benzothiazoles (**35**) via the manganese(III) acetate-promoted C(sp^2)-P cross-coupling reaction between thiazole/benzothiazole derivatives (**33**) and different organophosphorus compounds (**34**) including phosphine oxides, phosphinates, and phosphonates under ball-milling conditions (Figure 11.13). Previously reported methods for such phosphonylated heterocycles required palladium acetate and silver nitrate as catalysts, proline or 2,2′-bipyridine as a ligand, $K_2S_2O_8$ or di-*tert*-butyl peroxide as the oxidant, and acetonitrile as the solvent [137–140]. As compared to the earlier methods, the present ball-milling protocol offers several advantages, such as the solvent-free condition, the use of manganese(III) acetate as a low-cost and eco-friendly catalyst, the use of air instead of any oxidant, and shorter reaction times with moderate to good yields up to 97%.

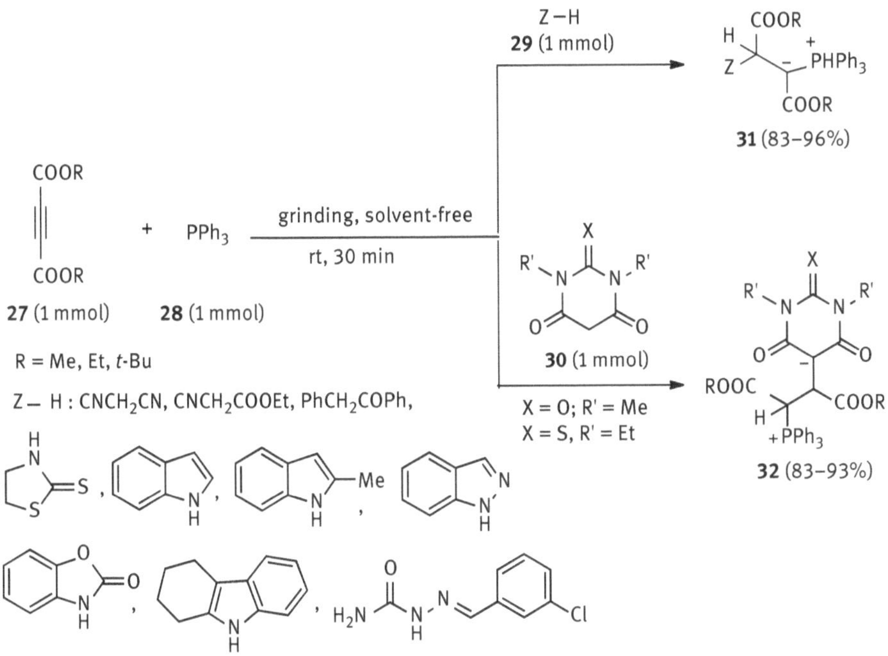

Figure 11.12: Catalyst- and solvent-free synthesis of phosphorus ylides (**31**) and 1,4-diionic organophosphorus compounds (**32**).

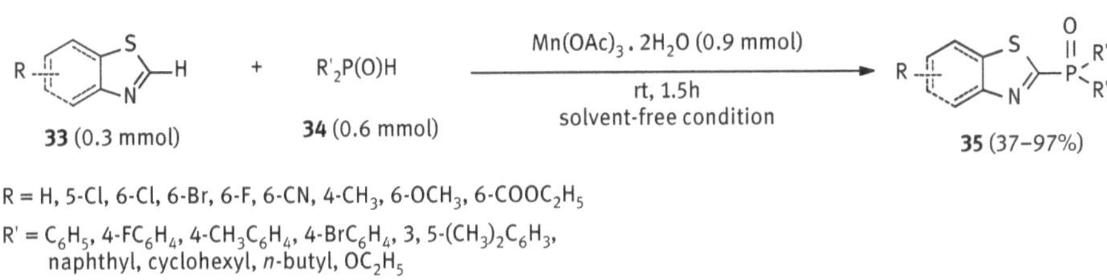

Figure 11.13: Manganese(III) acetate-catalyzed synthesis of C_2-phosphonylated thiazoles/benzothiazoles (**35**) via a C(sp^2)-P cross-coupling reaction under ball milling.

As a scale-up, benzothiazole (6 mmol) and diphenylphosphine oxide (12 mmol) were reacted under the optimal conditions to furnish benzo[*d*]thiazol-2-yldiphenylphosphine oxide (**35a**) in a yield of 95% on a gram-scale (Figure 11.14) [136].

11.3 Conclusions

Organophosphorous compounds have gained spectacular interest recently among researchers, including synthetic chemists, medicinal chemists, pharmacologists, biologists, and others working in interdisciplinary areas owing to their inherent biologi-

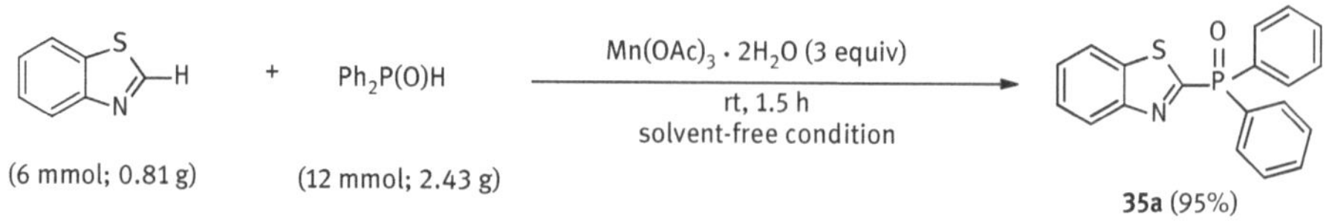

Figure 11.14: Synthesis of a benzo[*d*]thiazol-2-yldiphenylphosphine oxide (**35a**) using manganese(III) acetate as the catalyst under ball milling.

cal and physical properties. Synthetic organic chemists are, thus, deeply involved in exploring newer methodologies for the generation of organophosphorus compounds of both known and unknown skeletons and their analogs that may offer different kinds of physical and biological properties. With the advent of the concept of *green chemistry approaches*, chemists currently pay their attention to fulfill the expectations and to satisfy certain green chemical parameters in developing a new synthetic protocol, or modifying existing ones. As a result, a considerable development in the green synthetic approaches of organophosphorus compounds has taken place, highlighting various aspects of green chemistry during the last decade. The present chapter offers an update on such developments of synthetic protocols in organophosphorus chemistry at ambient conditions, reported during the period of 2012–mid-2017. Room temperature synthetic protocols for a variety of organophosphorus compounds, including α-hydroxyphosphonates, alkyl/aryl-phosphonates/phosphinates/phosphine oxides, α-aminophosphonates, vicinal amphiphilic phosphine oxide boronates, phosphorus ylides, and 1,4-diionic organophosphorus compounds, are discussed in detail. These green synthetic protocols offer new information to readers on the use of solid-supported heterogeneous catalysts, nano-catalysts, low-cost catalysts, chiral ligands, catalyst-free conditions, and also the application of ball mill as a part of mechanochemical reactions. Mild reaction protocols for the synthesis of a wide range of biologically relevant phosphorus-containing organic compounds are documented, and the author hopes that this overview boosts the ongoing research in organophosphorus chemistry coupled with green chemistry objectives.

Acknowledgment: The author thanks the Science and Engineering Research Board (SERB), Department of Science and Technology (DST), New Delhi (Grant No. EMR/2014/001220) is for their financial support for conducting this study.

References

[1] D. W. Allen, J. C. Tebby, D. Loakes (eds.), *Organophosphorus Chemistry* (Specialist periodical reports), Vol. 40–46, Royal Society of Chemistry; Cambridge, England, 2011–2017.

[2] D. E. C. Corbridge, *Phosphorus: Chemistry, Biochemistry and Technology*, 6th ed., CRC Press: Boca Raton, 2013.

[3] H. Guo, A. Yoshimura, T. Chen, Y. Saga, L.-B. Han, air-induced double addition of P(O)–H bonds to alkynes: A clean and practical method for the preparation of 1,2-bisphosphorylethane, *Green Chem.*, 2017, **19**, 1502–1506.
[4] D. S. Flett, Solvent extraction in hydrometallurgy: The role of organophosphorus extractant, *J. Organomet. Chem.*, 2005, **690**, 2426–2438.
[5] O. Delacroix, A. C. Gaumont, Hydrophosphination of unactivated alkenes, dienes and alkynes: A versatile and valuable approach for the synthesis of phosphines, *Curr. Org. Chem.* 2005, **9**, 1851–1882.
[6] A. L. Schwan, Palladium catalyzed cross-coupling reactions for phosphorus-carbon bond formation, *Chem. Soc. Rev.*, 2004, **33**, 218–224.
[7] F. Alonso, I. P. Beletskaya, M. Yus, Transition-metal-catalyzed addition of heteroatom-hydrogen bonds to alkynes, *Chem. Rev.*, 2004, **104**, 3079–3159.
[8] P. Savignac, B. Iorga, *Modern Phosphonate Chemistry;* CRC Press: Boca Raton, 2003.
[9] R. Engel, J. I. Cohen, *Synthesis of Carbon-phosphorus Bonds*, 2nd ed., CRC Press, Boca Raton, 2003.
[10] L. D. Quin, *A Guide to Organophosphorus Chemistry;* Wiley–Interscience: Hoboken, 2000.
[11] X.-T. Ma, Y. Wang, R.-H. Dai, C.-R. Liu, S.-K. Tian, Catalytic allylation of stabilized phosphonium ylides with primary allylic amines, *J. Org. Chem.*, 2013, **78**, 11071–11075.
[12] D. T. Kolio, *Chemistry and Application of H-phosphonates;* Elsevier Science: Oxford, 2006.
[13] J.-L. Montchamp, Recent advances in phosphorus–carbon bond formation: synthesis of *H*-phosphinic acid derivatives from hypophosphorous compounds, *J. Org. Chem.*, 2005, **690**, 2388–2406.
[14] G. Chelucci, G. Orrù, G. A. Pinna, Chiral *P,N*-ligands with pyridine-nitrogen and phosphorus donor atoms. Syntheses and applications in asymmetric catalysis, *Tetrahedron*, 2003, **59**, 9471–9515.
[15] O. I. Kolodiazhnyi, *Phosphorus ylides;* Wiley–VCH: Weinheim, 1999.
[16] B. E. Maryanoff, A. B. Reitz, The Wittig olefination reaction and modifications involving phosphoryl-stabilized carbanions. Stereochemistry, mechanism, and selected synthetic aspects, *Chem. Rev.*, 1989, **89**, 863–927.
[17] W. Tang, X. Zhang, New chiral phosphorus ligands for enantioselective hydrogenation. *Chem. Rev.*, 2003, **103**, 3029–3070.
[18] P. C. J. Kamer, P. W. N. M. van Leeuwen. (eds.), *Phosphorus(III) Ligands in homogeneous Catalysis: Design and Synthesis;* Wiley–VCH: Chichester, 2012.
[19] H. Fernández-Pérez, P. Etayo, A. Panossian, A. Vidal-Ferran, Phosphine–phosphinite and phosphine-phosphite ligands: preparation and applications in asymmetric catalysis, *Chem. Rev.*, 2011, **111**, 2119–2176.
[20] M.-N. Birkholz, Z. Freixa, P. W. N. M. van Leeuwen, Bite angle effects of diphosphines in C–C and C–X bond forming cross coupling reactions, *Chem. Soc. Rev.*, 2009, **38**, 1099–1118.
[21] D. J. Gorin, B. D. Sherry, F. D. Toste, Ligand effects in homogeneous Au catalysis, *Chem. Rev.*, 2008, **108**, 3351–3378.
[22] A. Borner. (ed), *Phosphorus Ligands in Asymmetric Catalysis – Synthesis and Application*; Wiley–VCH: Weinheim, 2008.
[23] H. A. McManus, P. J. Guiry, Recent developments in the application of oxazoline-containing ligands in asymmetric catalysis, *Chem. Rev.*, 2004, **104**, 4151–4202.
[24] S. O. Jeon, J. Y. Lee, Comparison of symmetric and asymmetric bipolar type high triplet energy host materials for deep blue phosphorescent organic light-emitting diodes, *J. Mater. Chem.*, 2012, 22, 7239–7244.
[25] X. Chen, D. J. Kopecky, J. Mihalic et al., Structure-guided design, synthesis, and evaluation of guanine-derived inhibitors of the eIF4E *m*RNA-cap interaction, *J. Med. Chem.*, 2012, **55**, 3837–3851.

[26] Q. Dang, Y. Liu, D. K. Cashion, S. R. Kasibhatla et al., Discovery of a series of phosphonic acid-containing thiazoles and orally bioavailable diamide prodrugs that lower glucose in diabetic animals through inhibition of fructose-1,6-bisphosphatase, *J. Med. Chem.*, 2011, **54**, 153–165.
[27] Y. J. Cho, J. Y. Lee, Modified *n,n′*-dicarbazolyl-3,5-benzene as a high triplet energy host material for deep-blue phosphorescent organic light-emitting diodes, *Chem. Eur. J.*, 2011, **17**, 11415–11418.
[28] V. Spampinato, N. Tuccitto, S. Quici, V. Calabrese, G. Marletta, A. Torrisi, A. Licciardello, Functionalization of oxide surfaces by terpyridine phosphonate ligands: surface reactions and anchoring geometry, *Langmuir*, 2010, **26**, 8400–8406.
[29] T. S. Kumar, S.-Y. Zhou, B. V. Joshi et al., Structure–activity relationship of (*n*)-methanocarba phosphonate analogues of 5′-AMP as cardioprotective agents acting through a cardiac P2X receptor, *J. Med. Chem.*, 2010, **53**, 2562–2576.
[30] T. Baumgartner, R. Réau, Organophosphorus π-conjugated materials, *Chem. Rev.*, 2006, **106**, 4681–4727.
[31] G. Brahmachari, *Green Synthetic Approaches for Biologically Relevant Heterocycles;* 1st ed., Elsevier: Amsterdam, The Netherland, 2015.
[32] G. Brahmachari, *Room Temperature Organic Synthesis;* 1st ed., Elsevier: Amsterdam, The Netherland, 2015.
[33] G. Brahmachari, *Catalyst-free Organic Synthesis;* 1st ed., Royal Society of Chemistry: Cambridge, 2018.
[34] Q. Wang, W. Ma, Q. Tong et al., Graphene oxide foam supported Titanium(IV): recoverable heterogeneous catalyst for efficient, selective oxidation of arylalkyl sulfides to sulfoxides under mild conditions, *Sci. Rep.*, 2017, **7**, 7209 (DOI:10.1038/s41598-017-07590-1).
[35] Y. Zhou, Y. Li, W. Shen, Shape engineering of oxide nanoparticles for heterogeneous catalysis, *Chem. Asian. J.*, 2016, **11**, 1470–1488.
[36] R. Schlögl, Heterogeneous catalysis, *Angew. Chem. Int. Ed.*, 2015, **54**, 3465–3520.
[37] G. Strappaveccia, D. Lanari, D. Gelman et al., Efficient synthesis of cyanohydrin trimethylsilyl ethers *via* 1,2-chemoselective cyanosilylation of carbonyls, *Green Chem.*, 2013, **15**, 199–204.
[38] S. Liu, S.-Q. Bai, Y. Zheng, K. W. Shah, M.-Y. Han, Composite metal–oxide nanocatalysts, *ChemCatChem*, 2012, **4**, 1462–1484.
[39] S. Bonollo, D. Lanari, J. M. Longo, L. Vaccaro, E-factor minimized protocols for the polystyryl-BEMP catalyzed conjugate additions of various nucleophiles to α,β-unsaturated carbonyl compounds, *Green Chem.*, 2012, **14**, 164–169.
[40] D. Lanari, R. Ballini, A. Palmieri, F. Pizzo, L. Vaccaro, Diastereoselective three-step route to *o*-(6-nitrocyclohex-3-en-1-yl)phenol and tetrahydro-6*H*-benzo[*c*]chromen-6-ol derivatives from salicylaldehydes, *Eur. J. Org. Chem.*, 2011, 2874–2884 (DOI: 10.1002/ejoc.201001476).
[41] A. Zvagulis, S. Bonollo, D. Lanari, F. Pizzo, L. Vaccaro, 2-*tert*-Butylimino-2-diethylamino-1,3-dimethylperhydro-1,3,2- diazaphosphorine supported on polystyrene (PS-BEMP) as an efficient recoverable and reusable catalyst for the phenolysis of epoxides under solvent-free conditions, *Adv. Synth. Catal.*, 2010, **352**, 2489–2496.
[42] P. T. Anastas, The transformative innovations needed by green chemistry for sustainability, *ChemSusChem*, 2009, **2**, 391–392.
[43] J. H. Clark, Catalysis for green chemistry, *Pure. Appl. Chem.*, 2001, **73**, 103–111.
[44] N. Mizuno, M. Misono, Heterogeneous catalysis, *Chem. Rev.*, 1998, **98**, 199–218.
[45] T. Angelini, S. Bonollo, D. Lanari, F. Pizzo, L. Vaccaro, E-Factor minimized hydrophosphonylation of aldehydes catalyzed by polystyryl-BEMP under solvent-free conditions, *Org. Biomol. Chem.*, 2013, **11**, 5042–5046.
[46] H. R. Ramananarivo, A. Solhy, J. Sebti et al., An eco-friendly paradigm for the synthesis of α-hydroxyphosphonates using sodium-modified fluorapatite under solvent-less conditions, *ACS Sustainable Chem. Eng.*, 2013, **1**, 403–409.

[47] M. L. Kantam, G. T. Venkanna, K. B. Shiva Kumar et al., Mild and efficient allylation of aldehydes by using copper fluorapatite as catalyst, *Adv. Synth. Catal.*, 2008, **350**, 1497–1502.
[48] M. L. Kantam, K. B. Shiva Kumar, P. Srinivas, B. Sreedhar, Fluorapatite-supported palladium catalyst for Suzuki and Heck coupling reactions of haloarenes, *Adv. Synth. Catal.* 2007, **349**, 1141–9.
[49] A. Solhy, J. H. Clark, R. Tahir, S. Sebti, M. Larzek, Transesterifications catalysed by solid, reusable apatite–zinc chloride catalysts, *Green Chem.*, 2006, **8**, 871–4.
[50] R. Tahir, K. Banert, S. Sebti. Natural and synthetic phosphates: New and clean heterogeneous catalysts for the synthesis of 5-arylhydantoins, *Appl. Catal. A.*, 2006, 298, 261–4.
[51] H. El Badaoui, F. Bazi, R. Tahir, H. B. Lazrek, S. Sebti, Synthesis of 3,4-dihydropyrimidin-2-ones catalysed by fluorapatite doped with metal halides, *Catal. Commun.*, 2005, **6**, 455–458.
[52] A. H. Kategaonkar, R. U. Pokalwar, S. S. Sonar, V. U. Gawali, B. B. Shingate, M. S. Shingare, Synthesis, in vitro antibacterial andantifungal evaluations of new α-hydroxyphosphonate and new α-acetoxyphosphonate derivatives of tetrazolo[1,5-*a*]quinolone, *Eur. J. Med. Chem.*, 2010, **45**, 1128–1132.
[53] P. Merino, E. Marqués-Lopez, R. P. Herrera, Catalytic enantioselective hydrophosphonylation of aldehydes and imines, *Adv. Synth. Catal.*, 2008, **350**, 1195–1208.
[54] R. Snoeck, A. Holý, C. Dewolf-Peeters, J. Van Den Oord, E. De Clercq, G. Andrei, Antivaccinia activities of acyclic nucleoside phosphonate derivatives in epithelial cells and organotypic cultures, *Antimicrob. Agents Chemother.*, 2002, **46**, 3356–3361.
[55] M. L. Peters, M. Leonard, A. A. Licata, Role of alendronate and risedronate in preventing and treating osteoporosis, *Clev. Clin. J. Med.*, 2001, **68**, 945–951.
[56] M. V. Lee, E. M. Fong, F. R. Singere, R. S. Guenette, Bisphosphonate treatment inhibits the growth of prostate cancer cells, *Cancer Res.*, 2001, **61**, 2602–2608.
[57] B. Z. Leder, H. M. Kronenberg, Gastroenterologists and choosing the right bisphosphonate, *Gastroenterology*, 2000, **119**, 866–869.
[58] M. Tao, R. Bihovsky, G. J. Wells, J. P. Mallamo, Novel peptidyl phosphorus derivatives as inhibitors of human calpain I, *J. Med. Chem.*, 1998, **41**, 3912–3916.
[59] D. V. Patel, K. Rielly-Gauvin, D. E. Ryono et al. Alpha-hydroxy phosphinyl-based inhibitors of human rennin, *J. Med. Chem.*, 1995, **38**, 4557–4569.
[60] S. Sobhani, Z. Tashrifi, Synthesis of α-functionalized phosphonates from α-hydroxyphosphonate, *Tetrahedron*, 2010, **66**, 14291439.
[61] B. Kaboudin, A convenient synthesis of 1-aminophosphonates from 1-hydroxyphosphonates, *Tetrahedron Lett.*, 2003, **44**, 1051–3.
[62] W. W. Metcalf, W. A. van der Donk, Biosynthesis of phosphonic and phosphinic acid natural products, *Annu. Rev. Biochem.*, 2009, **78**, 65–94.
[63] K. Panigrahi, M. Eggen, J. H. Maeng, Q. R. Shen, D. B. Berkowitz, The α,α-difluorinated phosphonate L-pSer-analogue: An accessible chemical tool for studying kinase- dependent signal transduction, *Chem. Biol.* 2009, **16**, 928–936.
[64] A. K. White, W. W. Metcalf, Microbial metabolism of reduced phosphorus compounds, *Ann. Rev. Microbiol.*, 2007, **61**, 379–400.
[65] D. M. Rothman, E. J. Petersson, M. E. Vazquez, G. S. Brandt, D. A. Dougherty, B. Imperiali. Caged phosphoproteins, *J. Am. Chem. Soc.*, 2005, **127**, 846–847.
[66] E. De Clercq, A. Holy, Acyclic nucleoside phosphonates: A key class of antiviral drugs, *Nat. Rev. Drug Discovery*, 2005, **4**, 928–940.
[67] W. P. Zheng, D. Schwarzer, A. LeBeau, J. L. Weller, D. C. Klein, P. A. Cole, Cellular stability of serotonin *N*-Acetyltransferase conferred by phosphonodifluoromethylene alanine (Pfa) substitution for Ser-205, *J. Biol. Chem.*, 2005, **280**, 10462–10467.
[68] W. P. Zheng, Z. S. Zhang, S. Ganguly, J. L. Weller, D. C. Klein, P. A. Cole, Cellular stabilization of the melatonin rhythm enzyme induced by nonhydrolyzable phosphonate incorporation, *Nat. Struct. Biol.*, 2003, **10**, 1054–1057.

[69] R. Engel, J. I. Cohen, *Synthesis of Carbon-phosphorus Bonds*; CRC Press: Boca Raton, 2004.
[70] P. Savignac, B. Iorga, *Modern Phosphonate Chemistry;* CRC Press: Boca Raton, 2003.
[71] T. Bock, H. Moehwald, R. Muelhaupt, Arylphosphonic acid-functionalized polyelectrolytes as fuel cell membrane material, *Macromol Chem Phys.,* 2007, **208**, 1324–1340.
[72] T. Ogawa, N. Usuki, N. Ono. A new synthesis of π-electron conjugated phosphonates and phosphonic bis(diethylamides) and their SHG activities, *J Chem Soc Perkin Trans 1*, 1998, 2953–2958 (DOI:10.1039/A709082J).
[73] S. M. Zakeeruddin, M. K. Nazeeruddin, P. Pechy et al., Molecular engineering of photosensitizers for nanocrystalline solar cells: synthesis and characterization of ru dyes based on phosphonated terpyridines, *Inorg. Chem.*, 1997, **36**, 5937–5946.
[74] R. A. Dhokale, S. B. Mhaske. P-arylation: Arynes to aryl-phosphonates, -phosphinates, and -phosphine oxides, *Org. Lett.*, 2013, **15**, 2218–2221.
[75] Z.-C. Geng, J.-X. Zhang, N. Li et al., Construction of highly substituted pyrazole derivatives with P-C bond: Access to racemic and enantioselective forms by conjugate addition of diarylphosphane oxides to α,β-unsaturated pyrazolones, *Tetrahedron*, 2014, **70**, 417–426.
[76] M. I. Kabachnik, T. Y. Medved, New synthesis of aminophosphonic acids, *Dokl. Akad. Nauk. SSSR,* 1952, **83**, 689–692.
[77] E. K. Fields, The synthesis of esters of substituted amino phosphonic acids, *J. Am. Chem. Soc.*, 1952, **74**, 1528–1531.
[78] R. A. Cherkasov, V. I. Galkin, The Kabachnik–Fields reaction: synthetic potential and the problem of the mechanism, *Russ. Chem. Rev.*, 1998, **67**, 857–882.
[79] G. Keglevich, A. Szekrényi, Eco-friendly accomplishment of the extended Kabachnik-Fields reaction: A solvent- and catalyst-free microwave-assisted synthesis of aminophosphonates and α-aminophosphine oxides, *Lett. Org. Chem.*, 2008, **5**, 616–622.
[80] N. S. Zefirov, E. D. Matveeva, Catalytic Kabachnik-Fields reaction: New horizons for old reaction. *ARKIVOC,* 2008, **(i)**, 1–17.
[81] V. Koteswara Rao, S. Subba Reddy, B. Sathish Krishna et al., Design, synthesis and anti colon cancer activity evaluation of phosphorylated derivatives of lamivudine (3TC), *Lett. Drug Des. Discovery,* 2011, **8**, 59–64.
[82] S. S. Sonar, S. A. Sadaphal, V. B. Labade, B. B. Shingate, M. S. Shingare, An efficient synthesis and antibacterial screening of novel oxazepine α-aminophosphonates by ultrasound approach. Phosphorus, *Sulfur Silicon Relat. Elem.*, 2010, **185**, 65–73.
[83] F. Orsini, G. Sello, M. Sisti, Aminophosphonic acids and derivatives. Synthesis and biological applications, *Curr. Med. Chem.*, 2010, **17**, 264–289.
[84] E. D. Naydenova, P. Todorov, K. Troev, Recent synthesis of aminophosphonic acids as potential biological importance, *Amino Acids,* 2010, **38**, 23–30.
[85] A. Janardhan Rao, P. Visweswara Rao, V. Koteswara Rao, C. M. C. Naga Raju, C. Suresh Reddy, Microwave assisted one-pot synthesis of novel α-aminophosphonates and their biological activity, *Bull Korean Chem Soc.*, 2010, **31**, 1863–1868.
[86] M. Siénczyk, J. Oleksyszyn, Irreversible inhibition of serine proteases-design and *in vivo* activity of diaryl alpha-aminophosphonate derivatives, *Curr. Med. Chem.*, 2009, **16**, 1673–1687.
[87] Z. Razaei, H. Friouzabadi, N. Iranpoor et al., Design and one-pot synthesis of α-aminophosphonates and bis(α-aminophosphonates) by iron(III) chloride and cytotoxic activity, *Eur. J. Med. Chem.*, 2009, **44**, 4266–4275.
[88] J. Grembecka, A. Mucha, T. Cierpicki, P. Kafarski, The most potent organophosphorus inhibitors of leucine amino peptidase. Structure-based design, chemistry, and activity, *J. Med. Chem.*, 2003, **46**, 2641–2655.
[89] W. Liu, C. J. Royers, A. J. Fisher, M. Toney, Aminophosphonate inhibitors of dialkylglycine decarboxylase: structural basis for slow binding inhibition, *Biochem.*, 2002, **41**, 12320–12328.

[90] V. P. Kukhar, H. R. Hudson, *Aminophosphonic and Aminophosphinic Acids: Chemistry and Biological Activity;* Wiley-VCH Verlag GmbH & Co: Weinhaeim, Germany, 2000.
[91] D. Fang, Y. Cao, J. Yang, Clean procedure for the synthesis of α-aminophosphonates catalyzed by biodegradable ionic liquid, *Phosphorus Sulfur Silicon Relat. Elem.*, 2013, **188**, 826–832.
[92] H. R. Shaterian, J. Farbodeh, M. Mohammadnia, Nano-TiO_2: An eco-friendly and clean reusable heterogeneous catalyst for preparation of α-aminophosphonates under ambientand solvent-free conditions, *Phosphorus Sulfur Silicon Relat, Elem.*, 2013, **188**, 850–854.
[93] M. F. Abdel-Megeed, G. A. El-Hiti, B. E. Badr, M. M. Azaam, Synthesis of a series of diphenyl (arylamino)(pyridin-3-yl)methylphosphonates as potential antimicrobial agents, *Phosphorus Sulfur Silicon Relat, Elem.*, 2013, **188**, 879–885.
[94] G. Brahmachari, *Handbook of Pharmaceutical Natural Products;* Wiley-VCH: Weinheim, 2010.
[95] S. Hatakeyama, N. Ochi, H. Numata, S. Takano, A new route to substituted 3-methoxycarbonyldihydropyrans; enantioselective synthesis of (–)-methyl elenolate, *J. Chem. Soc. Chem. Commun.*, 1988, 1202–1204.
[96] E. A. Hafez, M. H. Elnagdi, A. A. Elagamey, F. A. M. El-Taweel, Nitriles in heterocyclic synthesis: novel synthesis of benzo[*c*]coumarin and of benzo[*c*]pyrano[3,2-*c*]quinoline derivatives, *Heterocycles*, 1987, **26**, 903–907.
[97] M. M. Khafagy, A. H. F. A. El-Wahas, F. A. Eid, A. M. El-Agrody, Synthesis of halogen derivatives of benzo[*h*]chromene and benzo[*a*]anthracene with promising antimicrobial activities, *Farmaco*, 2002, **57**, 715–722.
[98] P. W. Smith, S. L. Sollis, P. D. Howes et al., Dihydropyrancarboxamides related to Zanamivir: A new series of inhibitors of influenza virus *Sialidases*. 1. Discovery, synthesis, biological activity, and structure–activity relationships of 4-guanidino- and 4-amino-4*H*-pyran-6-carboxamides, *J. Med. Chem.*, 1998, **41**, 787–797.
[99] A. G. Martinez, L. J. Marck, Friedländer reaction on 2-amino-3-cyano-4*H*-pyrans: Synthesis of derivatives of 4*H*-pyran [2,3-*b*] quinoline, new tacrine analogues, *Bioorg. Med. Chem. Lett.*, 1997, **7**, 3165–3170.
[100] D. O. Moon, Y. H. Choi, N. D. Kim, Y. M. Park, G. Y. Kim, Anti-inflammatory effects of β-lapachone in lipopolysaccharide-stimulated BV2 microglia, *Int. Immunopharmacol.*, 2007, **7**, 506–514.
[101] P. S. Elisa, E. B. Ana, A. G. Ravelo, D. J. Yapu, A. G. Turba, Antiplasmodial activity of naphthoquinones related to lapachol and *β*-lapachone, *Chem. Biodivers.*, 2005, **2**, 264–274.
[102] V. F. De Andrade-Neto, M. O. F. Goulart, J. F. Da Silva Filho et al., Antimalarial activity of phenazines from lapachol, β-lapachone and its derivatives against *Plasmodium falciparum* in *vitro* and *Plasmodium berghei* in vivo, *Bioorg. Med. Chem. Lett.*, 2004, **14**, 1145–1149.
[103] S. J. Mohr, M. A. Chirigios, F. S. Fuhrman, J. W. Pryor, Pyran copolymer as an effective adjuvant to chemotherapy against a murine leukemia and solid tumor, *Cancer Res.*, 1975, **35**, 3750–3754.
[104] G. Bianchi, Tava, Synthesis of (2*R*)-(+)-2,3-dihydro-2,6-dimethyl-4*H*-pyran-4-one, a homologue of Pheromones of a species in the Hepialidae family, *Agric. Biol. Chem.*, 1987, **51**, 2001–2002.
[105] D. R. Anderson, S. Hegde, E. Reinhard et al., Aminocyanopyridine inhibitors of mitogen activated protein kinase-activated protein kinase 2 (MK-2), *Bioorg. Med. Chem. Lett.*, 2005, **15**, 1587–1590.
[106] M. Ough, A. Lewis, E. A. Bey et al., Efficacy of beta-lapachone in pancreatic cancer treatment: Exploiting the novel, therapeutic target NQQ1, *J. Cancer. Biol. Therapy*, 2005, **4**, 95–102.
[107] G. P. Ellis (ed), *The Chemistry of Heterocyclic Compounds: Chromenes, Chromanones, and Chromones*; Vol 31, John Wiley & Sons: New York, 2008.
[108] M. A. Sofan, F. M. A. El-Taweel, A. G. A. Elagamey, M. H. Elnagdi, Studies on cinnamonitriles: The reaction of cinnamonitriles with cyclopentanone, *Eur. J. Org. Chem.*, 1989, 935–936 (DOI: 10.1002/jlac.198919890246).

[109] F. M. Abdel Galil, B. Y. Riad, S. M. Sherif, M. H. Elnagdi, Activated nitriles in heterocyclic synthesis: a novel synthesis of 4-azoloyl-2-aminoquinolines, *Chem. Lett.*, 1982, 1123–1126 (https://doi.org/10.1246/cl.1982.1123).

[110] S. Rao Kolla, Y. R. Lee, Efficient one-pot synthesis of β-phosphono malonates and 2-amino-3-cyano-4*H*-chromen-4yl phosphonate derivatives by ethylene diacetate-catalyzed three-component reactions, *Tetrahedron*, 2012, **68**, 226–237.

[111] G. Brahmachari, S. Laskar, Nano-mgo-catalyzed one-pot synthesis of phosphonate ester functionalized 2-amino-3-cyano-4*H*-chromene scaffolds at room temperature, *Phosphorus Sulfur Silicon Relat. Elem.*, 2014, **189**, 873–888.

[112] D. W. Stephan, Frustrated Lewis pairs: From concept to catalysis, *Acc. Chem. Res.*, 2015, **48**, 306–316.

[113] R. Liedtke, F. Scheidt, J. Ren et al., Frustrated Lewis pair modification by 1,1-carboboration: disclosure of a phosphine oxide triggered nitrogen monoxide addition to an intramolecular P/B frustrated Lewis pair, *J. Am. Chem. Soc.*, 2014, **136**, 9014–9027.

[114] G. Kehr, S. Schwendemann, G. Erker, Intramolecular frustrated Lewis pairs: Formation and chemical features, *Top. Curr. Chem.*, 2013, **332**, 45–83.

[115] B.-H. Xu, G. Kehr, R. Frçhlich et al., Reaction of frustrated Lewis pairs with conjugated ynones-selective hydrogenation of the carbon–carbon triple bond, *Angew. Chem. Int. Ed.*, 2011, **50**, 7183–7186.

[116] M.-A. Courtemanche, M.-A. Légaré, L. Maron, F.-G. Fontaine, A highly active phosphine–borane organocatalyst for the reduction of CO_2 to methanol using hydroboranes, *J. Am. Chem. Soc.*, 2013, **135**, 9326–9329.

[117] O. Baslé, S. Porcel, S. Ladeira, G. Bouhadir, D. Bourissou, Phosphine-boronates: efficient bifunctional organocatalysts for Michael addition, *Chem. Commun.*, 2012, **48**, 4495–4497.

[118] J. Vergnaud, M. Grellier, G. Bouhadir, L. Vendier, S. Sabo-Etienne, D. Bourissou, Synthesis and reactivity of ruthenium arene complexes incorporating novel $Ph_2PCH_2CH_2BR_2$ ligands. Easy access to the four-membered ruthenacycle [(*p*-cymene)RuCl($K^{C,P}$-$CH_2CH_2PPh_2$)], *Organometallics* 2008, **27**, 1140–1146.

[119] M. Devillard, G. Bouhadir, D. Bourissou, Cooperation between transition metals and lewis acids: A way to activate H_2 and H-E bonds, *Angew. Chem. Int. Ed.*, 2015, **54**, 730–732.

[120] V. Hornillos, C. Vila, E. Otten, B. L. Feringa, Catalytic asymmetric synthesis of phosphine boronates, *Angew. Chem. Int. Ed.*, 2015, **54**, 7867–7871.

[121] S. L. James, C. J. Adams, C. Bolm et al., Mechanochemistry: Opportunities for new and cleaner synthesis, *Chem. Soc. Rev.*, 2012, **41**, 413–447.

[122] T. Friščić, Supramolecular concepts and new techniques in mechanochemistry: Cocrystals, cages, rotaxanes, open metal-organic frameworks, *Chem. Soc. Rev.*, 2012, **41**, 3493–3510.

[123] A. Stolle, T. Szuppa, S. E. S. Leonhardt, B. Ondruschka, Ball milling in organic synthesis: solutions and challenges, *Chem. Soc. Rev.*, 2011, **40**, 2317–2329.

[124] C. Li, X.-Y. Mu, Y.-L. Li, Y. Liu, X.-S. Wang, Combinatorial synthesis of fused tetracyclic heterocycles containing [1,6]naphthyridine derivatives under catalyst free conditions, *ACS Comb. Sci.*, 2013, **15**, 267–272.

[125] T. Szuppa, A. Stolle, B. Ondruschka, W. Hopfe, Solvent-free dehydrogenation of γ-terpinene in a ball mill: investigation of reaction parameters, *Green Chem.*, 2010, **12**, 1288–1294.

[126] M. R. Naimi-Jamal, J. Mokhtari, M. G. Dekamin, G. Kaupp, Sodium tetraalkoxyborates: intermediates for the quantitative reduction of aldehydes and ketones to alcohols through ball milling with NaBH4, *Eur. J. Org. Chem.*, 2009, 3567–3572 (DOI: 10.1002/ejoc.200900352).

[127] D. A. Fulmer, W. C. Shearouse, S. T. Medonza, J. Mack, Solvent-free Sonogashira coupling reaction *via* high speed ball milling, *Green Chem.*, 2009, **11**, 1821–1825.

[128] J. Mack, D. Fulmer, S. Sofel, N. Santos, The first solvent-free method for the reduction of esters, *Green Chem.*, 2007, **9**, 1041–1043.

[129] A. Stolle, B. C. Ranu (eds.), *Ball Milling Towards Green Synthesis: Applications, Projects, Challenges;* Royal Society of Chemistry: Cambridge, 2014.
[130] M. G. Dekamin, M. Eslami, Highly efficient organocatalytic synthesis of diverse and densely functionalized 2-amino-3-cyano-4*H*-pyrans under mechanochemical ball milling, *Green Chem.*, 2014, **16**, 4914–4921.
[131] O. H. Qareaghaj, S. Mashkouri, M. R. Naimi-Jamal, G. Kaupp, Ball milling for the quantitative and specific solvent-free Knoevenagel condensation + Michael addition cascade in the synthesis of various 2-amino-4-aryl-3-cyano-4*H*-chromenes without heating, *RSC. Adv.*, 2014, **4**, 48191–48201.
[132] V. Kumar, N. Taxak, R. Jangir, P. V. Bharatam, K. P. Ravindranathan Kartha, In(III) triflate-mediated solvent-free synthesis and activation of thioglycosides by ball milling and structural analysis of long chain alkyl thioglycosides by TEM and quantum chemical methods, *J. Org. Chem.*, 2014, **79**, 3427–3439.
[133] G. Brahmachari, S. Das. L-Proline catalyzed multicomponent one-pot synthesis of *gem*-diheteroarylmethane derivatives using facile grinding operation under solvent-free conditions at room temperature, *RSC. Adv.*, 2014, **4**, 7380–7388.
[134] J. Mokhtari, M. R. Naimi-Jamal, H. Hamzeali, M. G. Dekamin, G. Kaupp, Kneading ball-milling and stoichiometric melts for the quantitative derivatization of carbonyl compounds with gas–solid recovery, *ChemSusChem*, 2009, **2**, 248–254.
[135] G. Marandi, N. A. Torbati, R. Heydari et al., Solvent-free conditions as an eco-friendly strategy for synthesis of organophosphorus compounds, *Phosphorus Sulfur Silicon Relat. Elem.*, 2012, **187**, 1450–1461.
[136] L. Li, J.-J. Wang, G.-W. Wang, Manganese(III) acetate-promoted cross-coupling reaction of benzothiazole/thiazole derivatives with organophosphorus compounds under ball-milling conditions, *J. Org. Chem.*, 2016, **81**, 5433–5439.
[137] H.-J. Zhang, W.-D. Lin, Z.-J. Wu, W.-Q. Ruan, T.-B. Wen, Silver-mediated direct phosphorylation of benzothiazoles and thiazoles with diarylphosphine oxides, *Chem. Commun.*, 2015, **51**, 3450–3453.
[138] X.-L. Chen, X. Li, L.-B. Qu et al., Peroxides as “switches” of dialkyl H-Phosphonate: Two mild and metal-free methods for preparation of 2-acylbenzothiazoles and dialkyl benzothiazol-2-ylphosphonates, *J. Org. Chem.*, 2014, **79**, 8407–8416.
[139] C.-B. Xiang, Y.-J. Bian, X.-R. Mao, Z.-Z. Huang, Coupling reactions of heteroarenes with phosphites under silver catalysis, *J. Org. Chem.*, 2012, **77**, 7706–7710.
[140] C. Hou, Y. Ren, R. Lang, X. Hu, C. Xia, F. Li, Palladium-catalyzed direct phosphonation of azoles with dialkyl phosphites, *Chem. Commun.*, 2012, **48**, 5181–5183.

Péter Ábrányi-Balogh

12 Recent developments in the synthesis of new P-heterocycles

Abstract: A number of new phosphorus-containing heterocycles have been isolated and characterized in the recent years. The novel organophosphorus species contain single rings and ring systems from four-membered up to nine-membered rings. Among the methods reported for their syntheses, a wide range of organic reactions, from nucleophilic additions through cycloadditions to the nowadays fashionable ring-closing metathesis, can be found. In most of the cases, the heterocycles could be derivatized during their syntheses, but post-synthetic derivatizations were also found to occur. Beside the scientific impact associated with the isolation of new molecules, the significance of these new P-heterocycles is demonstrated by the biological applications reported, and by the transition metal (Au, Pt, Mo) complex-forming abilities.

Keywords: synthesis, heterocycles, organophosphorus scaffolds, derivatization.

12.1 Introduction

Organophosphorus chemistry is a dynamically developing discipline expanding the available chemical space year by year [1–3]. The analogy between the classical heterocycles and their phosphorus analogs is undeniable. Moreover, the diverse coordination states of phosphorus give rise to various geometrical architectures, and apparently lend variety to the libraries constructed from P-containing molecules [4]. Nowadays, the progress in chemical synthesis involves the application of new up-to-date synthetic methods in the construction of phosphorus-containing compounds, particularly P-heterocycles [5, 6]. This chapter includes the synthesis of novel P-heterocycles developed in the past five years (2012–2017). The importance of P-heterocycles cannot be overstated. This class of molecules, among others, is part of an organic chemist's toolkit for transition metal catalysis [7], and finds medical [8], agricultural [9], analytical, and biochemical applications [4].

https://doi.org/10.1515/9783110535839-0012

12.2 New four-membered P-heterocycles

Phosphaalkenes (**1**) have been used as building blocks for four-membered heterocyclic cations (Figure 12.1) [10]. Although P_2C_2 rings have been observed within polycyclic structures before [11], the molecule synthesized represented the first example of a structurally characterized 1,3-diphosphetanium cation and dication. The reaction of phosphaalkene **1** with methyl triflate led to a methylenephosphonium intermediate followed by cyclization with a second phosphaalkene (**1**) to give the 1,3-diphosphetanium salt (**2**). The latter could be alkylated further with a second molecule of MeOTf to form a diphosphetanium product (**3**) (Figure 12.1).

Four-membered phosphorus-chalcogen (Ch) rings with a P_2Ch_2 core (**5**) were synthesized as the first representatives of such cyclic structures with trivalent phosphorus atoms (Figure 12.2) [12]. The chalcogens incorporated were sulfur and selenium. The products formed were prepared by the 1:1 addition of $S(TMS)_2$ [bis(trimethylsilyl)sulfide] to $Ar\text{-}PCl_2$ (**4**) in THF at room temperature. After a reaction time of 16 h the corresponding heterocycles (**5**) were obtained in good yields. An analogous synthetic procedure was used to get the seleno derivative, and the reaction of $(ArPSe)_2$ with 2,3-dimethylbutadiene led to a six-membered Diels–Alder product (**6**). Besides, the success in preparing

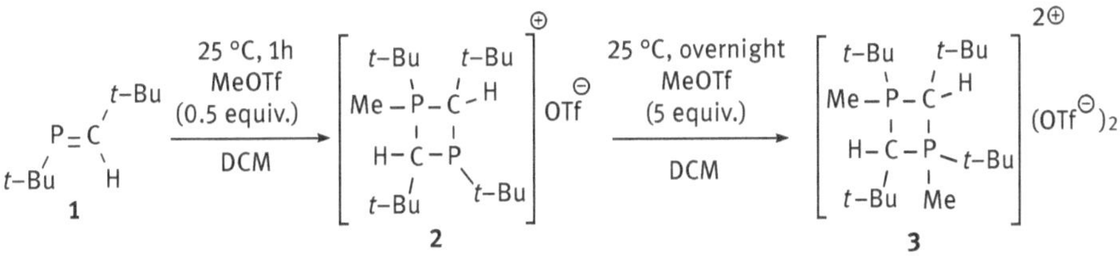

Figure 12.1: Synthesis of diphosphetanium cationic (**2**) and dicationic (**3**) species.

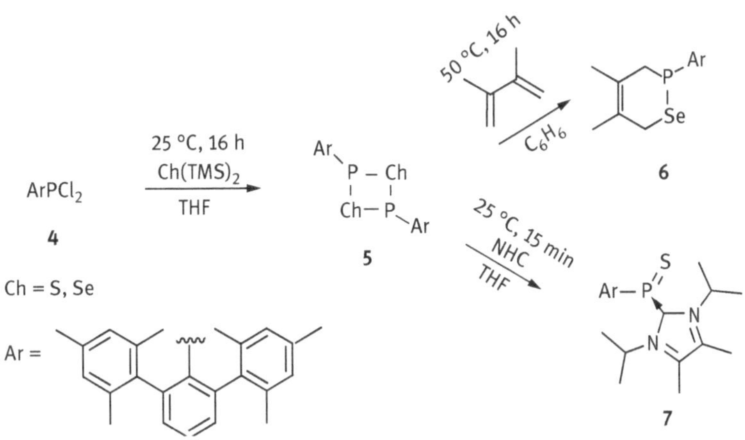

Figure 12.2: Synthesis and reactions of new $(ArP)_2Ch_2$ cores (**5**).

the $(ArPS)_2$ ring provided access to an *N*-heterocyclic carbene (NHC)-stabilized phosphinidene sulfide (**7**) by reacting with 1,3-diisopropyl-4,5-dimethylimidazole-2-ylidene.

12.3 New five-membered P-heterocycles

The development of thieno[3,4-*c*]phosphole-4,6-diones (**9**, **10**) as versatile building blocks was reported by Takeda et al. (Figure 12.3) [13]. The synthesis was carried out by the reaction of thiophene-3,4-dicarboxylic acid chlorides (**7**) or (**8**) with *in situ* generated $[(Tipp)P(SiMe_3)_2]$ in diethyl ether to furnish the products (**9**, **10**) in moderate yields. The usefulness of the new compounds was demonstrated by converting them to new D-A-D (donor-acceptor-donor) functional molecules and low-bandgap conjugated polymers. In the latter case, the bandgap control was achieved through post-polymerization modification.

The Diels–Alder cycloadducts (**12**, **13**) of 1,3-benzothiaphosphole (**11**) with butadienes and cyclopentadiene were synthesized (Figure 12.4) [14]. The starting heterocycle reacted with the dienes within 1–6 days at 60–65 °C to give dihydro-1*H*-

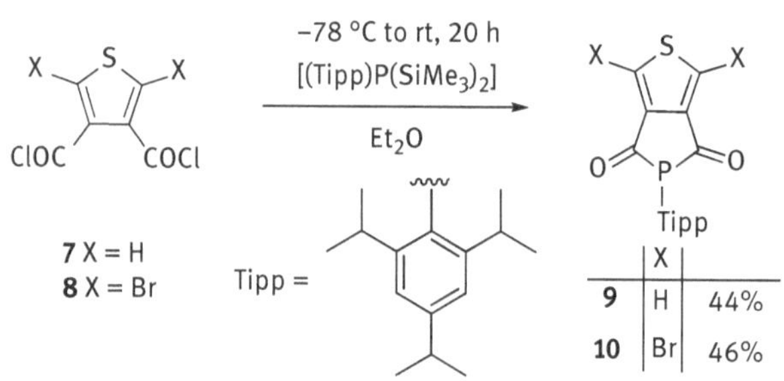

Figure 12.3: Synthesis of two thieno[3,4-c]phosphole-4,6-diones (**9** and **10**).

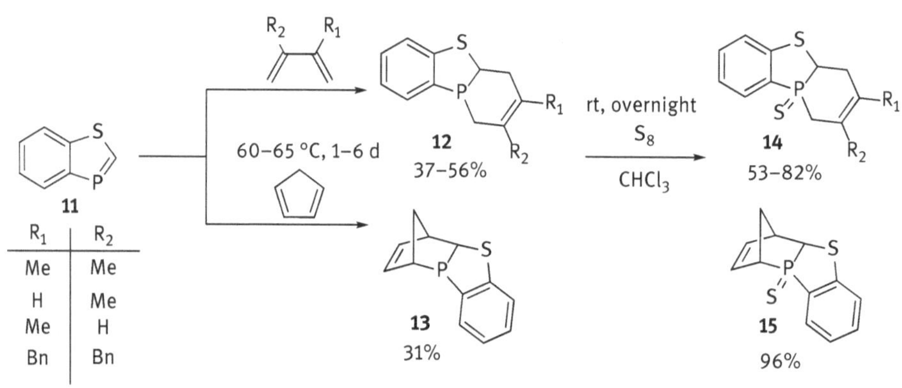

Figure 12.4: Synthesis and modification of cycloadducts (**12** and **13**) of 1,3-benzothiaphosphole (**11**) and dienes.

benzo[d]-phosphinino[2,1-*b*][1,3]thiaphospholes (**12**) and endo-/exo-1,4,4a-trihydro-1,4-methanobenzo[d]phosphinino[2,1-b][1,3]thiaphosphole (**13**), respectively, which were transformed to air-stable sulfides (**14, 15**) by reacting with elemental sulfur.

Novel fluorinated spiro[oxindole-thiazolidinone] derivatives fused with phosphorus heterocycles were synthesized starting from 3'-(4-fluorophenyl)-4'*H*-spiro [indole-3,2'-thiazolidine]-2,4'(1*H*)-dione (**16**) (Figure 12.5) [15]. First, thiazolidine **16** was reacted with 2-chloro-6-fluorobenzaldehyde, creating compound **17** containing an activated double bond. A Michael-type addition took place with diethyl phosphite in the presence of boron trifluoride diethyl etherate, followed by intramolecular nucleophilic attack of the oxygen at the phosphorus atom, causing an O-P ring closure with the elimination of EtOH. The reaction led to spiro{indole-3,5'-[1,2]oxaphospholo[5,4-*d*] thiazol}-2-one (**19**) in a good yield. Moreover, when the starting thiazolidinone (**16**) was reacted with hydrazine, hydrazone **20** was formed and this intermediate (**20**) was reacted with $POCl_3$ in the presence of TEA, and the corresponding spiro{indole-3,5'-thiazolo[5,4-*d*][1,2,3]diazaphosphol}-2-one derivative (**21**) was obtained after hydrolysis in a good yield.

The synthesized compounds were tested for antioxidant activity by the 1,1-diphenyl-2-picrylhydrazyl method, and species **19** and **20** showed excellent activities in the range of 63–71%.

The synergistic 1,1-addition of unsaturated vicinal frustrated phosphorus-boron (P/B) Lewis pairs (FLP) with isocyanides afforded new heterocycles in moderate to good yields (Figure 12.6) [16]. In the first reaction step, the isonitrile ligands coordinated conventionally at the boron atom that is the Lewis acidic site of compound **22**, and this coordination was followed by a ring closure under thermodynamic control. The products (**23**) were isolated as mixtures of E/Z isomers.

The first example of an isolable diazadiphosphapentalene derivative (**28**) was synthesized and derivatized recently (Figure 12.7) [17, 18]. The heterocycle (**28**) was obtained in a four-step synthesis starting from a methyl *tert*-butyl imine (**24**) that

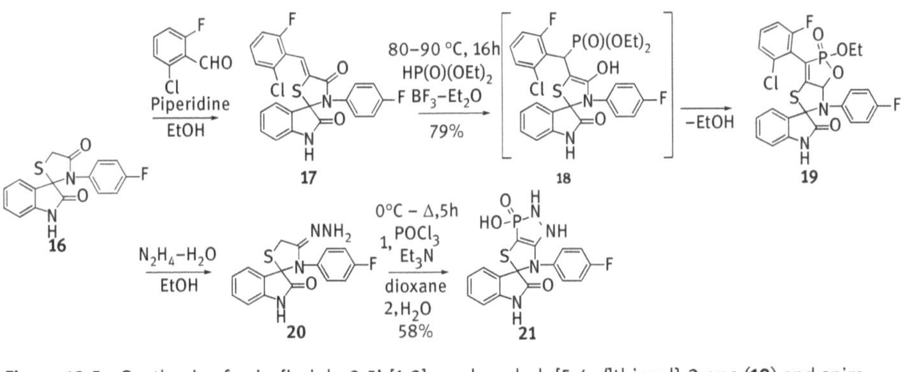

Figure 12.5: Synthesis of spiro{indole-3,5'-[1,2]oxaphospholo[5,4-*d*]thiazol}-2-one (**19**) and spiro {indole-3,5'-thiazolo[5,4-*d*][1,2,3]diazaphosphol}-2-one (**21**).

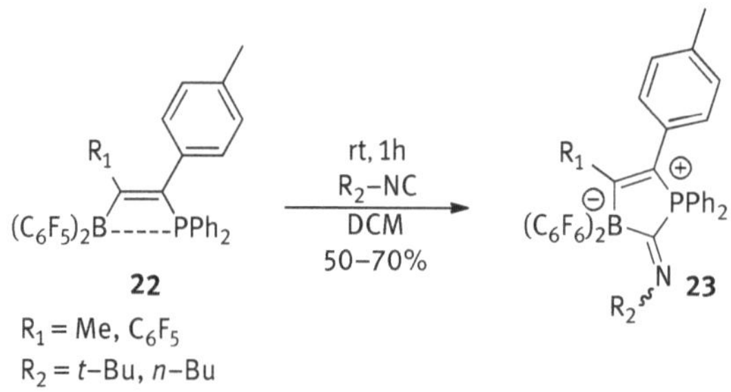

Figure 12.6: Reaction of isocyanides with P/B FLPs.

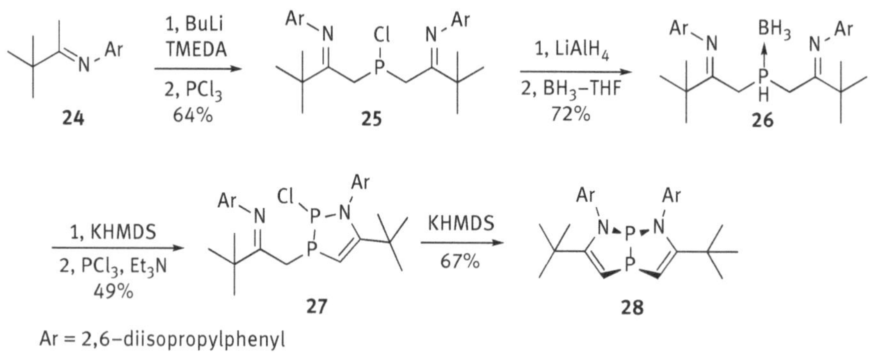

Figure 12.7: Synthesis of diazadiphosphapentalene **28**.

was deprotonated on the methyl group, and reacted with phosphorus trichloride in a nucleophilic substitution affording intermediate **25**, which was successfully reduced by LAH. The formed *P*(III) moiety was protected by reaction with borane. The borane complex (**26**) was then converted to **27** azadiphosphole by reacting with KHMDS and PCl_3, and further treatment of the intermediate (**27**) with KHMDS resulted in the formation of diazadiphosphapentalene **28**, which was stable in an inert atmosphere.

To investigate the chemical properties of the new heterocycle, various reactions were performed (Figure 12.8). The diazadiphosphapentalene (**28**) could be oxidized (**29**) in the presence of TEMPO, fluorinated (**30**) and methylated (**31**) on the phosphorus atom, and converted to the complexes with borane (**32**) or $PtCl_2$. Moreover, a photoinduced isomerization was also observed leading to a new diazadiphosphapentalene isomer (**33**) that was proven to be more stable and also could be transformed to borane complex **34**.

A novel aromatic 6π electron ring compound, 2,4-diphospha-3,5-diaza-thiole (cyclo-SNPNP, **38**), was obtained from $SP(N_3)_3$ (**35**) in gas phase at 1,000 °C under an Ar atmosphere by flash pyrolysis (Figure 12.9) [19]. The starting compound **35** was converted to the reactive 1,3-dipolar SPN intermediate (**36**) that underwent an intermolecular cycloaddition/dimerization leading to molecule **37**, followed by a ring contraction with sulfur elimination, resulting in heterocycle **38**. The expelled sulfur atom dimerized to disulfur (S_2), which was detected by UV spectroscopy. The compound **38**

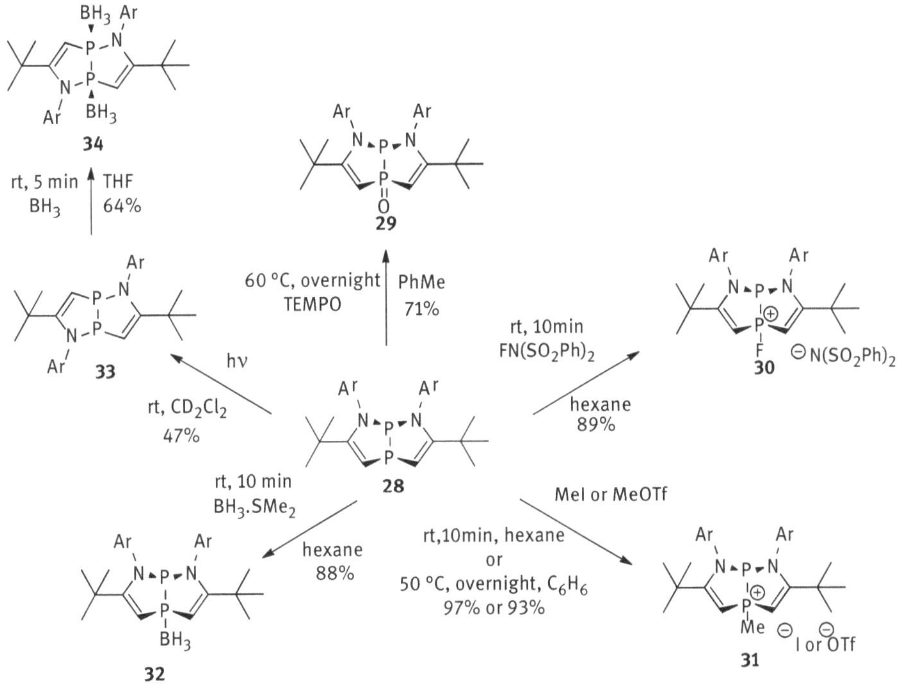

Figure 12.8: Functionalization of diazadiphosphapentalene **28**.

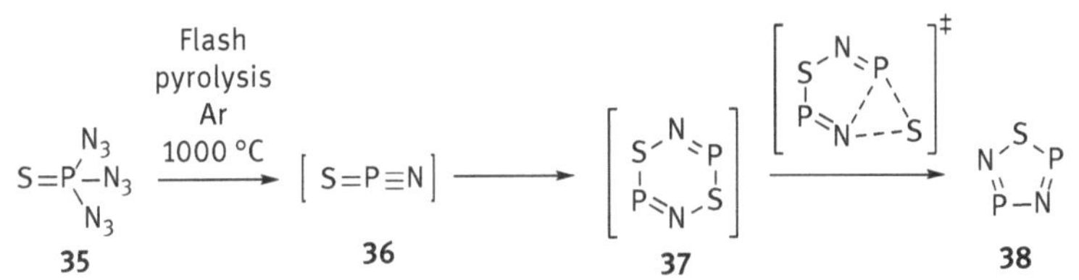

Figure 12.9: Synthesis of cyclo-SNPNP (**38**).

was isolated and deposited in a matrix at −257 °C. The product (**38**) was identified and characterized by IR spectroscopy and ^{15}N isotope labeling.

12.4 New six-membered P-heterocycles

Novel phosphaphenalenes (**40**) fused with various carbocycles and heterocycles were synthesized by a non-catalyzed reaction of lithiated arylnaphthalenes (**39**) with dichlorophenylphosphane at 0 °C followed by quenching with hydrogen peroxide in order to obtain the air-stable phosphorus heterocycles (**40**) (Figure 12.10) [20]. The 3-thiophene derivative could be iodinated, brominated, and reduced to P(III) to form a gold complex.

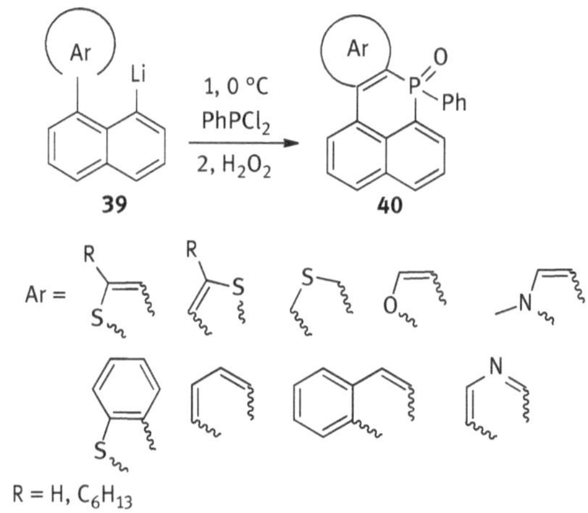

Figure 12.10: Synthesis of **40** phosphaphenalenes fused with carbo- and heterocycles.

The first syntheses of 1-phenyl-1,4-dihydrophosphinoline-1-oxides (**42**) and 1,4-dihydrophosphinolines (**43**) have been described [21–23]. In the acid-promoted reaction, allenes (**41**) were initially transformed to (3-hydroxyalk-1-en-1-yl)diphenylphosphine oxides that were converted to the thermodynamically more stable fused compounds (Figure 12.11). The promotion could be achieved best with TfOH at 120 °C or with $AlCl_3$ even at 20 °C, and the products were isolated in moderate to good yields. The mechanism of the reaction was studied by NMR and DFT calculations. The P(V) derivatives (**42**) were reduced with trichlorosilanes to form trivalent phosphines (**43**). The latter formed Pt and Pd complexes. Moreover, the dihydrophosphinoline-1-oxides (**42**) could be converted to triflate salts (**44**).

Phosphininofurans (**46**), representing a novel heterocyclic system, were synthesized in moderate yields by photochemical reaction of furanyl phosphonium-iodonium ylides (**45**) with acetylenes (Figure 12.12) [24]. A few years later the same approach was used for the synthesis of the isomeric compound (**48**) starting from 3-furyl analogs [25].

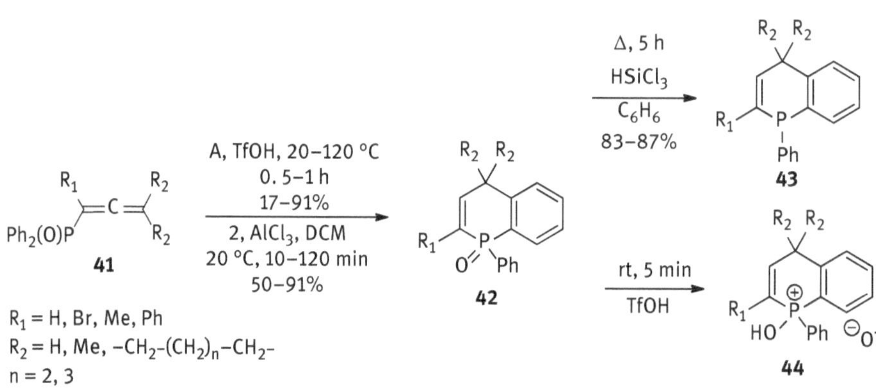

Figure 12.11: Synthesis of 1-phenyl- 1,4-dihydrophosphinolines (**42**, **43**, **44**).

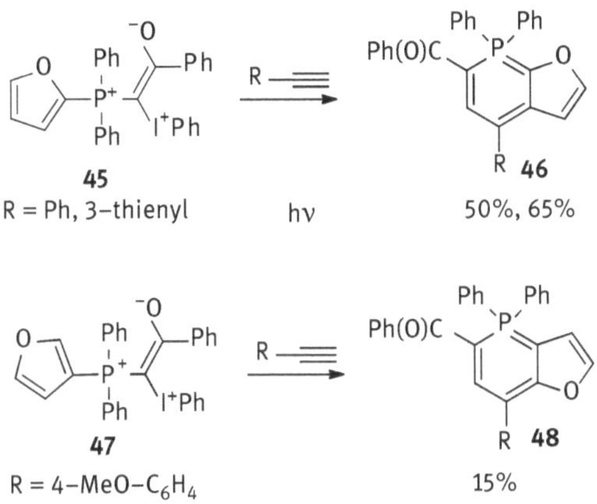

Figure 12.12: Synthesis of phosphininofurans (**45, 48**).

Two novel polyfused *P*-centered heterocycles (**53** and **54**) were synthesized from 2-iodophenyl-methylthioether (**49**) and diacetylenephenylphosphine oxide (**50**) [26]. Initially, a double-Sonogashira reaction afforded phosphine oxide **51**. The latter was transformed to the key intermediate bis(3-iodobenzo[b]thiophen-2-yl)-phenylphosphine oxide (**52**) which was converted with moderate yields to **53** or **54** by a double-Heck reaction with styrene or double-Ullmann reaction with *N*,*N*-dimethylethylenediamine (Figure 12.13).

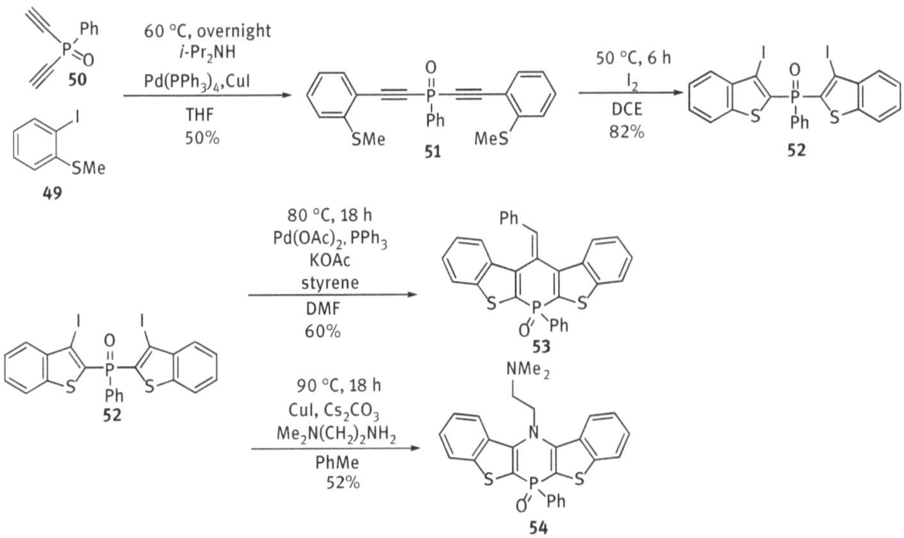

Figure 12.13: Synthesis of bis(benzo[b]thiophen-2-yl)-1-phosphinine-oxide (**53**) and benzo[4,5]-thieno[3,2-*b*]benzo[4,5]thieno[2,3-*e*][1,4]azaphosphinine-oxide (**54**).

An iterative iodocyclization led to benzothiopheno-fused oxaphosphinines (**57**) [26]. Phosphonate (**55**) was at first partially hydrolyzed, followed by an electrophilic cyclization promoted by methanesulfonic acid or *N*-iodosuccinimide to give benzo[4,5] thieno-[2,3-*c*][1,2]oxaphosphinines (**57**) in good yields (Figure 12.14).

The complexation properties of *ortho*-(diphenylphosphino)phenyl isocyanide (**58**) with gold(I), palladium(II) and nickel(II) chlorides was studied [27]. The application of $NiCl_2$ promoted unexpectedly a head-to-tail dimerization of isocyanide **58**, leading to an organic dimer (**59**) as a new heterocyclic core in a low yield (Figure 12.15).

N-Benzyl-4-phenyldeoxyestrone-*P*-phenylphosphamide (**60**) was modified by a radical oxidative C-H amination with *N*-iodosuccinimide in dichloromethane, leading to the chiral *P*-stereogenic atropoisomeric derivative **61** in a high yield and diastereoselectivity (Figure 12.16) [28, 29]. The starting compound was *N*-benzyl-4-phenyldeoxyestrone-*P*-phenylphosphamide (**60**) and the ring closure went smoothly leading to the chiral *P*-stereogenic atropoisomeric **61** (Figure 12.16). The product was reported to have been used as a drug and diagnostic reagent in cancer therapy. Moreover, on changing the coupling reagent to elemental iodine in the presence of $PhI(OAc)_2$, the iodine atom could be inserted into the A phenyl ring (**62**). This could open the way to further modify the hexacycle by various coupling reactions.

The reaction of 5-fluoro-3'-(4-fluorophenyl)-5'-(trifluoroacetyl)-spiro[indole-3,2'-thiazolidine]-2,4'(1*H*)-dione **63** with methyl phosphonic diamide in refluxing ethanol containing a catalytic amount of piperidine affords a novel phosphorus heterocycle (**64**) fused with the spiro[oxindole-thiazolidinone] ring system [15]. The ^{1}HNMR data indicated an equilibrium between the amino-imino tautomers (Figure 12.17).

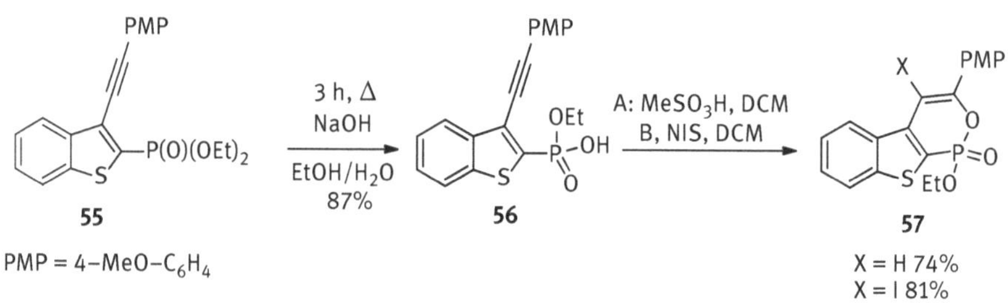

Figure 12.14: Synthesis of benzo[4,5]thieno-[2,3-*c*][1,2]oxaphosphinines (**57**).

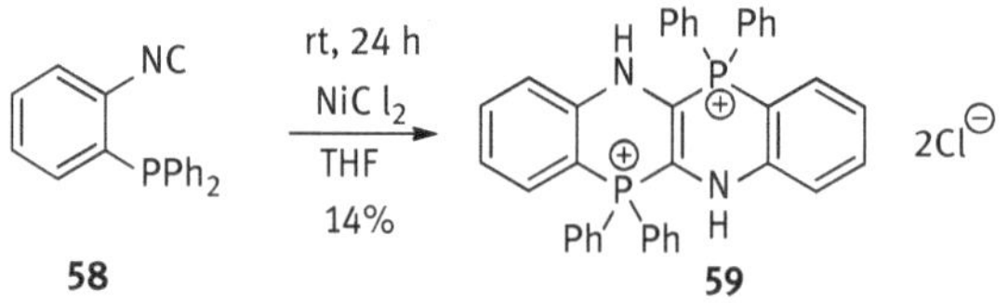

Figure 12.15: $NiCl_2$ promoted dimerization of *ortho*-(diphenylphosphino)phenyl isocyanide (**58**).

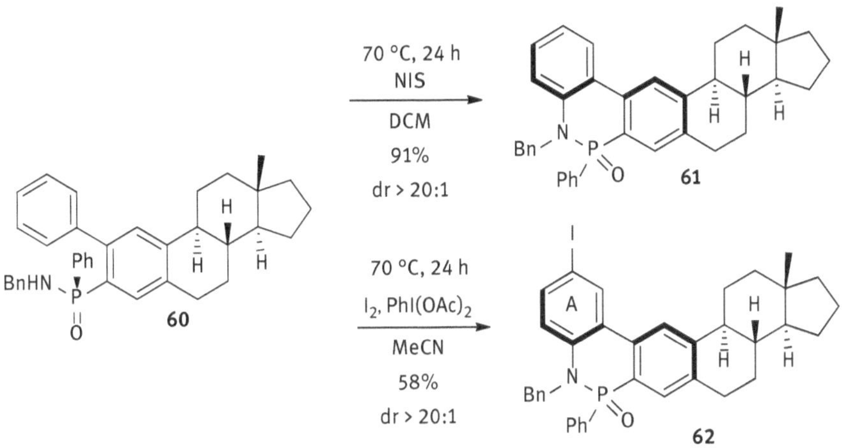

Figure 12.16: Synthesis of estrone-derived atropoisomeric *P*-stereogenic phosphinamides (**61**, **62**).

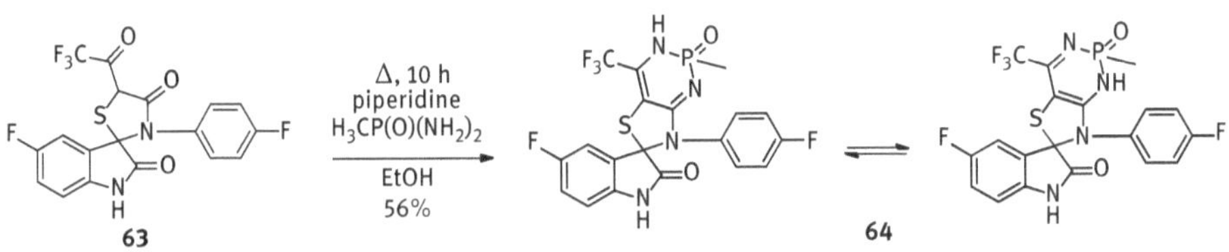

Figure 12.17: Synthesis of spiro{indole-3,6'-thiazolo[4',5'-*d*][1,3,2]diazaphosphinin}-2-one (**64**).

Cyanoacetohydrazide was reacted with 1,3-diphenyl-1*H*-pyrazole-4-carbaldehyde (**65**) to yield 2-cyano-*N*'-[1,3-diphenyl-1*H*-pyrazol-4-ylmethylidene] acetohydrazide (**66**) [30]. This intermediate reacted smoothly with diethyl phosphite at 70–80 °C in the presence of boron trifluoride diethyl etherate, resulting in the formation of the corresponding novel oxadiazaphosphininyl-acetonitrile (**67**) (Figure 12.18).

The formation of **67** could be explained by a P-H addition to C=N bond and subsequent cyclization by the enol tautomer eliminating EtOH. The final step was an auto-oxidation affording pyrazolyloxadiazaphosphorine **67**.

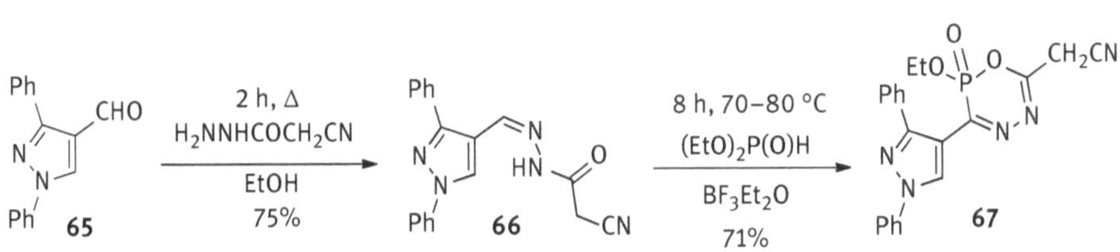

Figure 12.18: Synthesis of pyrazolyloxadiazaphosphorine (**67**).

The coupling of carbon monoxide and nitrogen monoxide at **68** FLP template was studied [31]. These experiments led to new phosphorus-boron heterocycles (**69**, **70**), when the "Piers' borane carbonyl" (**68**) was reacted with nitrogen monoxide at room temperature (Figure 12.19). The two major products were isolated in moderate yields.

The synthesis of new phosphine-boronates was achieved through a unique rearrangement [32]. Initially, phosphoranyl dihydroborate (**71**) was generated and by an extraction with diethyl ether the new heterocycle was formed (Figure 12.20). The formation of phosphine-boronate **72** was imagined by a double migration of the alkoxy groups from the phosphorus to the boron atom, as well as the migration of a hydrogen from the hydroborane to the phosphorus. This mechanism was proven by an experiment with deuterated derivatives. This double ring expansion was also studied by the treatment of the dihydroborane **71** with but-2-yne. The formation of the butenylated derivative (**73**) showed the same hydroboration reactivity as the starting compound. The mechanism of this reaction is presumably analogous, as one hydrogen and two

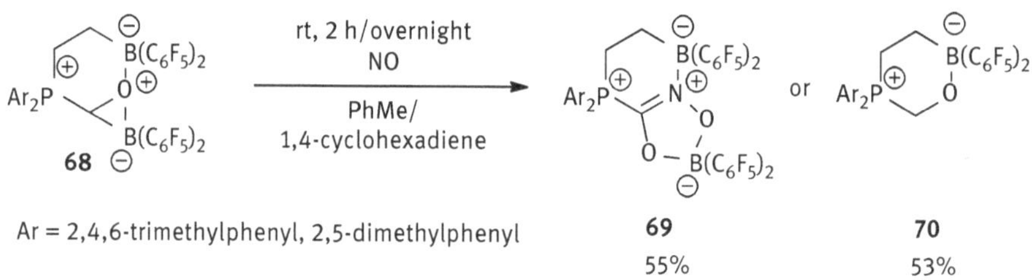

Figure 12.19: Synthesis of novel phosphorus heterocycles (**69** and **70**) from **68** FLP template.

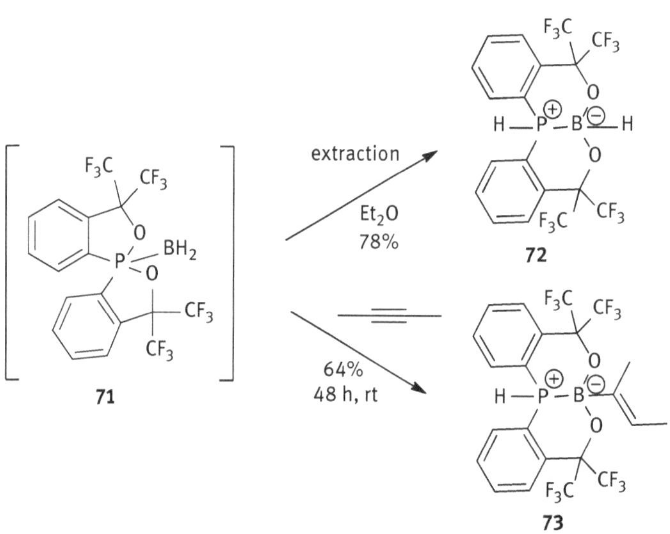

Figure 12.20: Synthesis of new phosphine-boronates (**72**, **73**).

oxygen atoms of the corresponding phosphoranyl-substituted 2-butenyl(hydro)-borane quickly rearrange. The rearrangement was so rapid that the second hydroboration did not occur. This property of the heterocycles could be advantageous for a selective single hydroboration reaction.

The hitherto unknown phosphorus analog of cyanuric acid was synthesized by Suter et al. [33]. 2,4,6-Tri(hydroxy)-1,3,5-triphosphinine was obtained in a straightforward synthesis involving the reaction of NaOCP and chlorodiisopinocampheylborane ($(ipc)_2BCl$, **74**) at −40 °C, which initially led to the formation of the $(ipc)_2B$-O-C---P phosphaacetylene and the phosphaketene $(ipc)_2B$-O=C=P. Both compounds dimerized at room temperature either to phosphabutadiene or to an 1,3-diketone, respectively, followed by a ring expansion by the reaction with a third OCP molecule forming the desired $P_3C_3[OB(ipc)_2]_3$ (**75**). The boronic protecting group could be removed with *tert*-butanol leading to heterocycle **76**, and the hydroxy groups were silylated, forming compound **77** (Figure 12.21). The borylated and the silylated derivative could be transformed to a $Mo(CO)_3$ complex.

Recently, organophosphorus(III)-tellurium heterocycles were also isolated and their crystal structures were elucidated [34]. Applying new synthetic methods, the first examples of solid-state structures have been reported and the first representatives of ring systems containing equal numbers of alternating phosphorus and tellurium atoms were synthesized. The reaction of supermesitylphosphorus dichloride (**78a**) or tritylphosphorus dichloride (**78b**) with Na_2Te_3 led to the new tritelluratriphosphorinanes (**79a,b**) in low to acceptable yields (Figure 12.22). In both reactions the new heterocycle was only a minor product, but isolable as a solid that was proven to be highly air sensitive and slightly light sensitive. XRD analysis of heterocycle **79b** shows that the P_3Te_3 ring crystallizes in the chair conformation, in which the bulky trityl ligands are able to take the sterically advantageous equatorial positions.

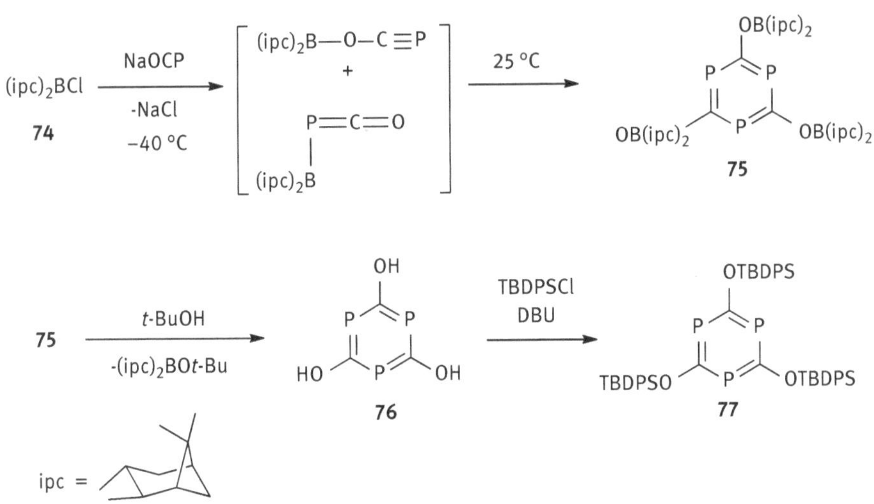

Figure 12.21: Synthesis of 2,4,6-tri(hydroxy)-1,3,5-triphosphinine (**76**).

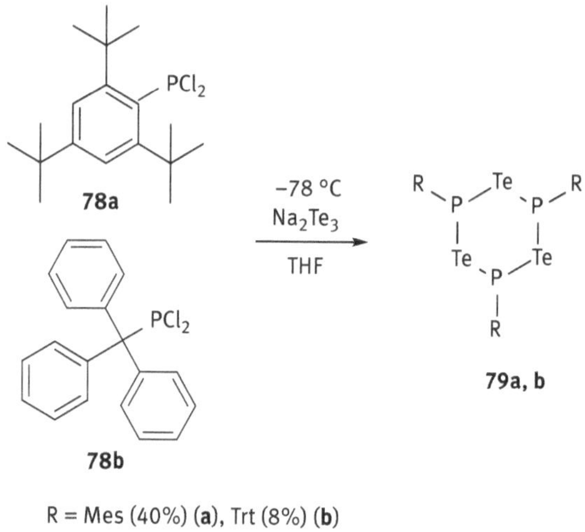

Figure 12.22: Synthesis of tritelluratriphosphorinanes (**79a** and **b**).

12.5 New seven-, eight-, and nine-membered P-heterocycles

Seven- and eight-membered phosphorus heterocycles were synthesized by ring-closing metathesis [35]. Hitherto unreported benzo-fused oxaphosphocine and oxaphosphopine derivatives (**82**) were obtained from substrates with unsymmetrical alkenyl groups linked directly to a phosphorus atom (**80**, **81**). Grubbs' first-generation catalyst was applied to the reagents under a nitrogen atmosphere at room temperature in dichloromethane for 5–8 h (Figure 12.23). The corresponding seven- and eight-membered products were isolated in excellent yields. The reaction was extended to different substituents in the benzene ring. Moreover, naphthalene or *N,N'*-diethyl-pyrimidine-dione derivatives could also serve as reactants. A few of the compounds could be hydrogenated quantitatively, leading to their saturated analogs (**83**).

The first synthesis of a diazaphosphocane (**85**) and a diazaphosphonane (**86**) was reported by a nucleophilic substitution reaction – an intramolecular aza-Michael sequence with *N,N'*-dimethyl secondary amines and a phosphonochloridate precursor (**84**) in *t*-BuOH (Figure 12.24) [36]. The eight-membered ring (**85**) was stable, while the nine-membered diazaphosphonane (**86**) showed limited stability over a few days. The products were isolated as diastereomeric mixtures, as suggested by the ^{13}C and ^{31}P NMR data, and in the case of diazaphosphocane **85** those could be separated by flash chromatography.

The Woolins' reagent (**87**) was applied for the synthesis of novel organophosphorus-selenium heteroatom compounds [37]. *Trans*-1,2-cyclohexanediamine was reacted with compound **87** forming a zwitterionic intermediate (**88**), followed by the addition of *ortho*-xylylenedibromide (Figure 12.25). The alkylating agent was attacked by the

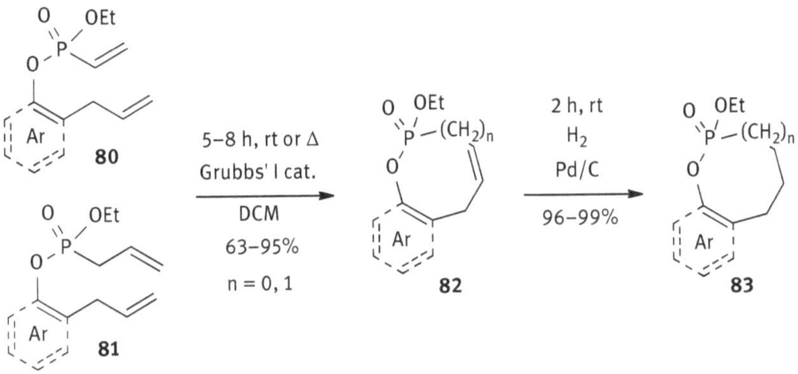

Ar = aryl, *N,N*'-diethylpyrimidine-dione

Figure 12.23: Synthesis of oxaphosphocine and oxaphosphopine derivatives (**82** and **83**).

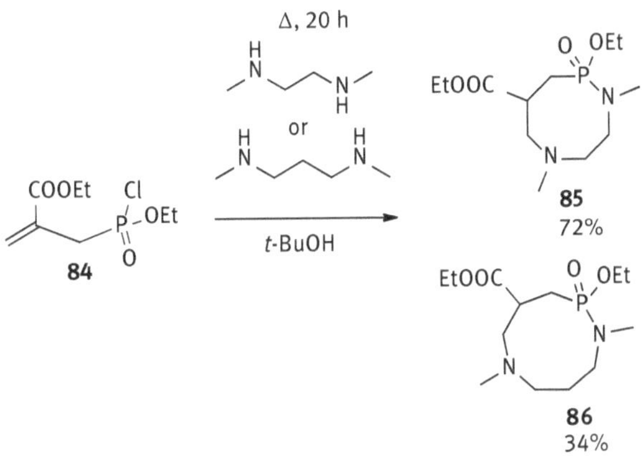

Figure 12.24: Synthesis of diazaphosphocane (**85**) and a diazaphosphonane (**86**) derivatives.

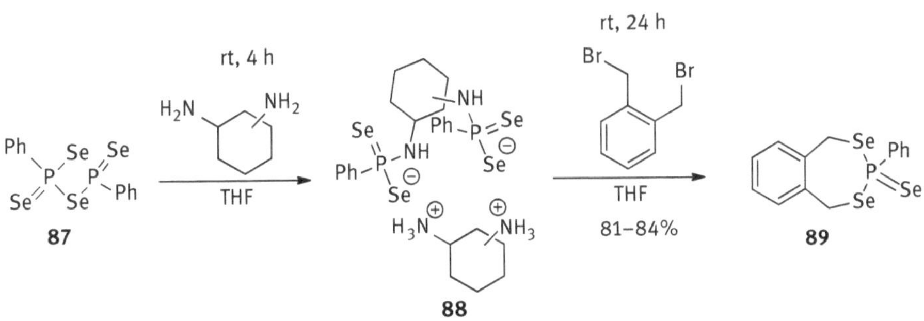

Figure 12.25: Synthesis of diselenaphosphepine (**89**).

selenide anions, leading to 3-phenyl-1,5-dihydrobenzo[e][1,3,2]diselenaphosphepine 3-selenide (**89**). The same product (**89**) was obtained when 1,3-cyclohexanediamine was applied in the first step of the reaction sequence.

References

[1] Louis D.Quin, *A Guide to Organophosphorus Chemistry*; Wiley: New York, 2000, Ch. 8, 233.
[2] Francois Mathey, *Phosphorus-carbon Heterocyclic Chemistry: The Rise of a New Domain;* Elsevier Science Ltd.: Oxford, 2001.
[3] G. Keglevich, 6-Membered P-heterocycles: 1,2-Dihydro-, 1,2,3,6-tetrahydro- and 1,2,3,4,5,6-hexahydrophosphinine 1-oxides. *Curr. Org. Chem.*, 2006, **10**, 93.
[4] *Topics in Heterocyclic Chemistry, Vol. 20, Phosphorus Heterocycles I*, Ed., R. R. Gupta, Vol. Ed.: R. K. Bansal; Springer: Berlin, 2009.
[5] G. Keglevich, Application of Microwave Irradiation in the Synthesis of P-Heterocycles. In: *Green Synthetic Approaches for Biologically Relevant Heterocycles*, Ed. G. Brahmachari, Elsevier: Amsterdam, 2015, Ch. 20, 559–570.
[6] *Topics in Heterocyclic Chemistry, Vol. 21, Phosphorus Heterocycles II*, Ed., R. R. Gupta, Vol. Ed., R. K. Bansal; Springer: Heidelberg, 2010.
[7] L. Kollár, G. Keglevich, *Chem. Rev.*, 2010, **110**, 4257.
[8] A. R. Ahmed, S. M. Hombal, *J. Am. Acad. Dermatol.*, 1984, **11**, 115.
[9] A. Hirashima, I. Ishaaya, R. Ueno, Y. Ichiyama, S.-Y. Wu, M. Eto, *Agr. Biol. Chem.*, 1989, **53**, 175.
[10] J. I. Bates, D. P. Gates, *Chem. Eur. J.*, 2012, **18**, 1674.
[11] K. K. Laali, B. Geissler, M. Regitz, J. J. Houser, *J. Org. Chem.*, 1995, **60**, 47.
[12] C. M. E. Graham, T. E. Pritchard, P. D. Boyle, J. Valjus, H. M. Tuononen, P. J. Ragogna, *Angew. Chem. Int. Ed.*, 2017, **56**, 6236.
[13] Y. Takeda, K. Hatanaka, T. Nishida, S. Minakata, *Chem. Eur. J.*, 2016, **22**, 10360.
[14] J. C. Worch, E. Hellemann, G. Pros, C. Gayathri, T. Pintauer, R. R. Gil, K. J. T. Noonan, *Organometallics*, 2015, **34**, 5366.
[15] T. E. Ali, R. M. Abdel-Rahman, *J. Sulfur Chem.*, 2014, **35**, 399.
[16] O. Ekkert, G. G. Miera, T. Wiegand, H. Eckert, B. Schirmer, J. L. Petersen, C. G. Daniliuc, R. Fröhlich, S. Grimme, G. Kehr, G. Erker, *Chem. Sci.*, 2013, **4**, 2657.
[17] J. Cui, Y. Li, R. Ganguly, A. Inthirarajah, H. Hirao, R. Kinjo, *J. Am. Chem. Soc.*, 2014, **136**, 16764.
[18] J. Cui, Y. Li, R. Ganguly, R. Kinjo, *Chem. Eur. J.*, 2016, **22**, 9976.
[19] X. Zeng, H. Li, H. Sun, H. Beckers, H. Willner, H. F. Schaefer, *Angew. Chem. Int. Ed.*, 2015, **54**, 1327.
[20] C. Romero-Nieto, A. López-Andarias, C. Egler-Lucas, F. Gerbert, J.-P. Neus, O. Pilgram, *Angew. Chem. Int. Ed.*, 2015, **54**, 15872.
[21] M. A. Klingenberg, A. S. Bogachenkov, M. A. Kinzhalov, A. V. Vasilyev, V. P. Boyarskiy, *New J. Chem.*, 2016, **40**, 3336.
[22] A. S. Bogachenkov, A. V. Dogadina, V. O. Boyarskiy, A. V. Vasilyev, *Org. Biomol. Chem.*, 2015, **13**, 1333.
[23] A. S. Bogachenkov, A. V. Dogadina, I. A. Boyarskaya, V. M. Boyarskiy, A. V. Vasilyev, *Org. Biomol. Chem.*, 2016, **14**, 1370.
[24] E. D. Matveeva, T. A. Podrugina, M. A. Taranova, A. M. Ivanova, R. Gleiter, N. S. Zefirov, *J. Org. Chem.*, 2012, **77**, 7324.
[25] E. D. Matveeva, D. S. Vinogradov, T. A. Podrugina, T. D. Nekipelova, A. V. Mironov, R. Gleiter, N. S. Zefirov, *Eur. J. Org. Chem.*, 2015, 7324.

[26] A. Gupta, B. L. Flynn, *J. Org. Chem.*, 2016, **81**, 4012.
[27] L. Zhang, W. Yu, C. Liu, Y. Xu, Z. Duan, F. Mathey, *Organometallics*, 2015, **34**, 5697.
[28] Y.-N. Ma, M.-X. Cheng, S.-D. Yang, *Org. Lett.*, 2017, **19**, 600.
[29] Y.-N. Ma, X. Zhang, S.-D. Yang, *Chem. Eur. J.*, 2017, **23**, 3007.
[30] T. E. Ali, *Heteroat. Chem.*, 2013, **24**, 426.
[31] K.-Y. Ye, G. Kehr, C. G. Daniliuc, L. Liu, S. Grimme, G. Erker, *Angew. Chem. Int. Ed.*, 2016, **55**, 9216.
[32] N. Kano, N. J. O'Brien, R. Uematsu, R. Ramozzi, K. Morokuma, *Angew. Chem. Int. Ed.*, 2017, **56**, 5882.
[33] R. Suter, Y. Mei, M. Baker, Z. Benkő, Z. Li, H. Grützmacher, *Angew. Chem. Int. Ed.*, 2017, **56**, 1356.
[34] A. Nordheier, T. Chivers, O. Schön, K. Karaghiosoff, K. S. A. Arachchige, A. M. Z. Slawin, J. D. Woolins, *Chem. Eur. J.*, 2014, **20**, 704.
[35] K. C. Majumdar, R. K. Nandi, S. Ganai, *Tetrahedron Lett.*, 2014, **55**, 1247.
[36] H. Boufoura, A. Abdelli, F. Bourdreux, A. Gaucher, G. Clavier, M. L. Efrit, H. M'rabet, D. Prim, *RSC Adv.*, 2017, **7**, 18211.
[37] G. Hua, R. A. M. Randall, A. M. Z. Slawin, J. D. Woolins, *Tetrahedron*, 2013, **69**, 5299.

Bubun Banerjee, Ádám Tajti and György Keglevich

13 Ultrasound-assisted synthesis of organophosphorus compounds

Abstract: Compared to conventional methods, the application of ultrasonic irradiation has many advantages in organic syntheses, such as faster and more selective reactions, as well as higher yields and cleaner products. Most of the ultrasound-mediated approaches satisfy the goals of green chemistry. In this chapter, the ultrasound-assisted synthesis of various organophosphorus compounds including α-aminophosphonates, α-hydroxyphosphonates, and their derivatives using the Kabachnik–Fields condensation, the Pudovik, click, and other reactions is discussed.

Keywords: ultrasonic irradiation, α-aminophosphonates, α-hydroxyphosphonates, Kabachnik–Fields condensation, Pudovik reaction, click reaction.

13.1 Introduction

The use of ultrasonic irradiation in organic syntheses has led to significant developments in the last decades. Ultrasound-mediated reactions may be more advantageous as compared to their traditional thermal analogues in respect of reaction rate and selectivity, as well as the yield and the purity of the products [1–4]. Organophosphorus compounds have been receiving continuous attention due to their potential biological activity (see also Chapter 3) [5, 6]. This chapter summarizes recent developments in the field of ultrasound-assisted synthesis of α-aminophosphonates, α-hydroxyphosphonates, and related derivatives.

13.2 Ultrasound-assisted Kabachnik–Fields Condensation

The Kabachnik–Fields condensation is a one-pot three-component condensation of an amine, an oxo compound and a P-reagent [7]. This reaction is generally applied for the synthesis of α-aminophosphonates [8] receiving continuous attention due to their potential pharmacological applications as anticancer [9], antithrombotic [10], antihypertensive [11], antiinflammatory [12], antiviral [13], antimicrobial [14–16] agents, and as antioxidants [17], or enzyme inhibitors [18]. A number of approaches have been reported for the synthesis of these important organophosphorus

https://doi.org/10.1515/9783110535839-013

compounds (see also Chapter 6) [19–22]. Many of these processes require long reaction times and/or harsh conditions. As a possible development, the ultrasonic irradiation was tried out in a series of Kabachnik–Fields reactions.

13.2.1 Kabachnik–Fields Reactions Starting from Dialkyl Phosphites

A number of α-aminophosphonates were synthesized by the one-pot three-component reaction of various benzaldehyde derivatives, substituted anilines and dialkyl phosphites at room temperature (Table 13.1). It should be noted that there were no temperature data in the references, and due to this, the results cannot be precisely compared with each other. Condensations were performed using nano-SiO_2 as an efficient and reusable catalyst in aqueous media under ultrasonic irradiation (Table 13.1, entry 1). Another simple and efficient protocol for the rapid synthesis of α-aminophosphonates was carried out in the presence of a catalytic amount of sulfonic acid-functionalized Santa Barbara Amorphous-15 (SBA-15/SO_3H) nano-composite under ultrasonic irradiation in ethanol (Table 13.1, entry 2). Graphene oxide was found to be an

Table 13.1: Ultrasound-assisted Kabachnik–Fields reactions starting from benzaldehyde derivatives and substituted anilines at room temperature.

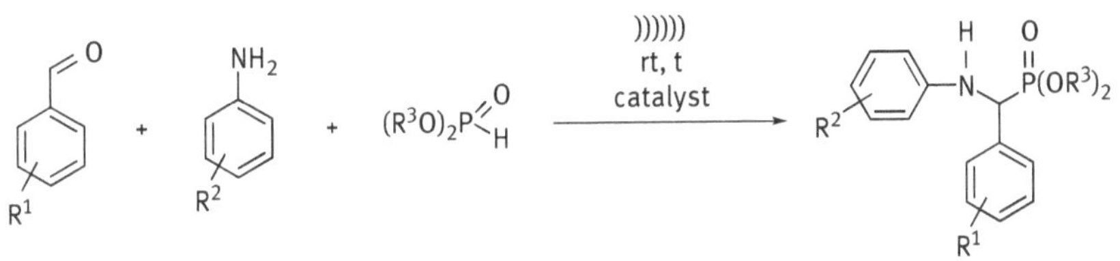

Entry	R^1	R^2	R^3	Catalyst	Solvent	t (min)	Yield (%)	Ref.
1	H, 4-MeO, 2-NO_2, 4-NO_2, 4-Cl, 4-Me	H, 2-Cl, 3-Cl, 4-Cl, 4-F, 4-Me, 4-MeO, 4-NO_2	Et	Nano-SiO_2	H_2O	10–20	91–97	[23]
2	H, 3-NO_2, 4-NO_2, 2-Cl, 4-Cl, 2-MeO	H, 4-Me	Me	Nano-SBA-15-SO_3H	EtOH	4–5	94–95	[24]
3	H, 4-MeO, 4-Cl, 2-Cl, 4-Br, 2-HO, 4-CH_3, 2,5-diMeO	H, 4-I, 2-NO_2, 3-NO_2, 4-NO_2, 4-C(O)Me, 4-CN	Et	graphene oxide	–	5–30	76–98	[25]
4	H, 4-Cl, 2-Cl, 3-NO_2, 4-NO_2, 2-MeO	H, 4-Me	Me	[MPIm]Cl@ SBA-15	EtOH	4–7	91–96	[26]
5	H, 4-Cl, 2-Cl, 3-NO_2, 4-NO_2, 2-MeO	H, 4-Me	Me	[bmim] $AlCl_4$	EtOH	4–7	89–93	[27]

efficient catalyst for the rapid synthesis of α-aminophosphonates under the influence of ultrasonic irradiation (Table 13.1, entry 3). Methyl propylimidazolium chloride-immobilized Santa Barbara Amorphous-15 ([MPIm]Cl@SBA-15) was also found to be an efficient alternative catalytic system under ultrasonic irradiation for the three-component reactions in ethanol (Table 13.1, entry 4). Simultaneous application of ultrasonic irradiation and [bmim]$AlCl_4$ also proved to be efficient in accelerating the reaction of aromatic aldehydes, amines, and dimethyl phosphite (Table 13.1, entry 5).

In the presence of ultrasonic irradiation, 10 mol% $BF_3.Et_2O$ was sufficient to catalyze the reaction of 2-fluorobenzoaldehyde, 2-trifluoromethyl-4-bromoaniline and dialkyl phosphites at 75–80 °C under solvent-free conditions to afford the corresponding α-aminophosphonates in good yields (Figure 13.1) [28].

Silica-supported zinc bromide (SiO_2-$ZnBr_2$) was also found to be an efficient catalyst for the synthesis of α-aminophosphonates starting from various aromatic aldehydes, 4-(4-chlorophenoxy)aniline and diethyl phosphite at 50 °C (Figure 13.2) [29]. The comparative reactions performed in the absence of ultrasonic irradiation required 2–3 h for full conversion.

Another simple ultrasound-assisted protocol was developed for the synthesis of α-aminophosphonates containing an isoxazole moiety by the reaction of fluorobenzaldehydes, 3-amino-5-methylisoxazole and dialkyl phosphites (Figure 13.3) [30]. The reaction took place at 70–80 °C in the absence of any catalyst under solvent-free conditions.

α-Aminophosphonates were also prepared by the condensation of substituted salicylic aldehydes, secondary amines and dialkyl phosphites in water (Figure 13.4) [31]. The reactions were carried out in the absence of catalysts under ultrasonic irradiation at room temperature.

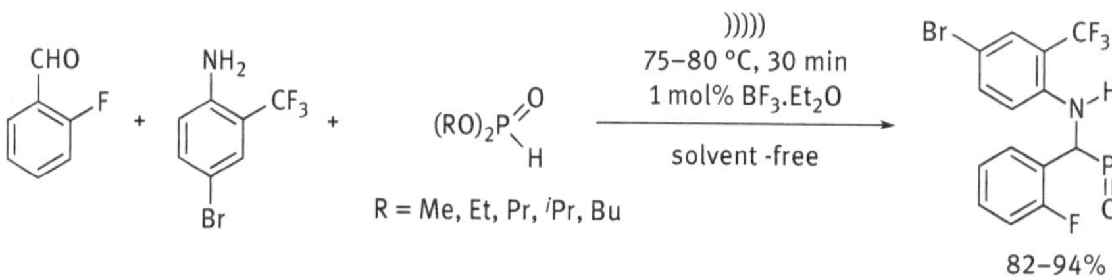

Figure 13.1: $BF_3.Et_2O$-catalyzed ultrasound-assisted synthesis of α-aminophosphonates.

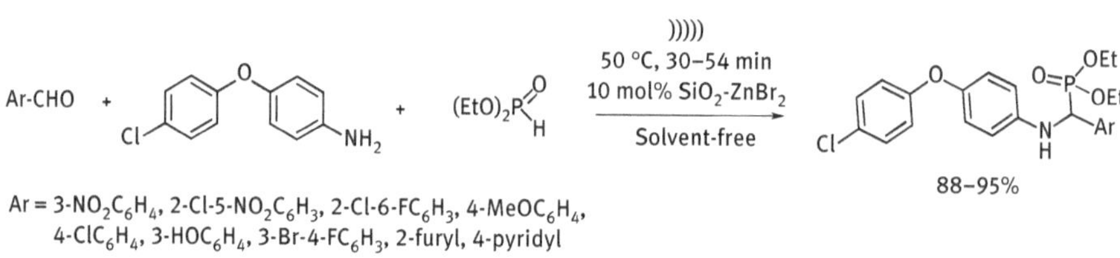

Figure 13.2: SiO_2-$ZnBr_2$-catalyzed ultrasound-assisted synthesis of α-aminophosphonates.

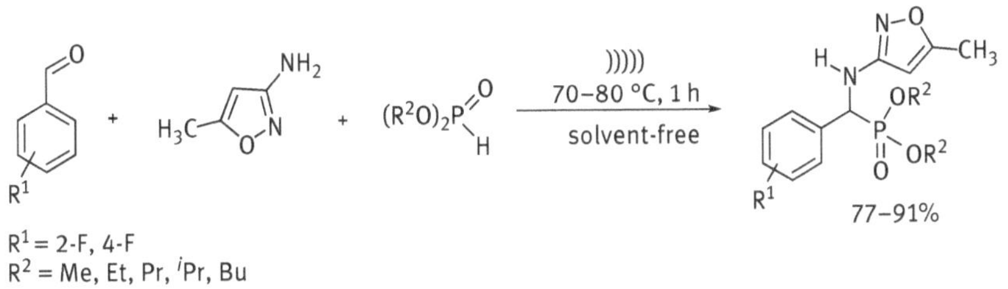

Figure 13.3: Ultrasound-assisted synthesis of α-aminophosphonates containing an isoxazole moiety.

A series of bis(α-aminophosphonate) derivatives was synthesized by the Kabachnik–Fields reaction at 45 °C in polyethylene glycol–water mixture as the solvent under ultrasonic irradiation (Figure 13.5) [32]. The compounds synthesized were screened for cytotoxic activity.

The Kabachnik–Fields reaction of ketones had to be carried out at a temperature of 70 °C for 2–3.5 h under ultrasonic irradiation to prepare the target products that were obtained in variable yields (Figure 13.6) [33].

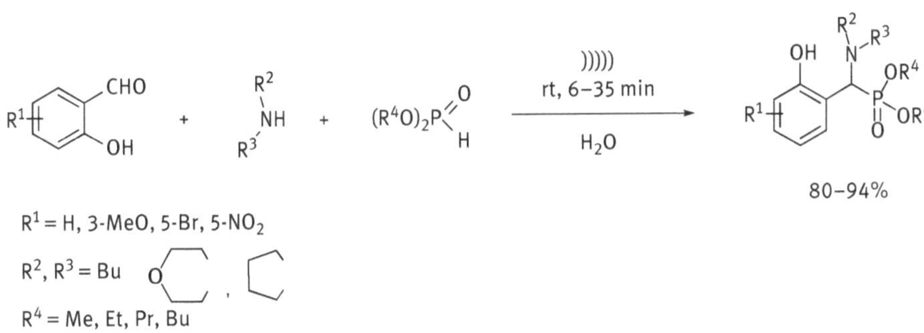

Figure 13.4: Ultrasound-assisted catalyst-free synthesis of α-aminophosphonates from salicylic aldehyde derivatives and secondary amines.

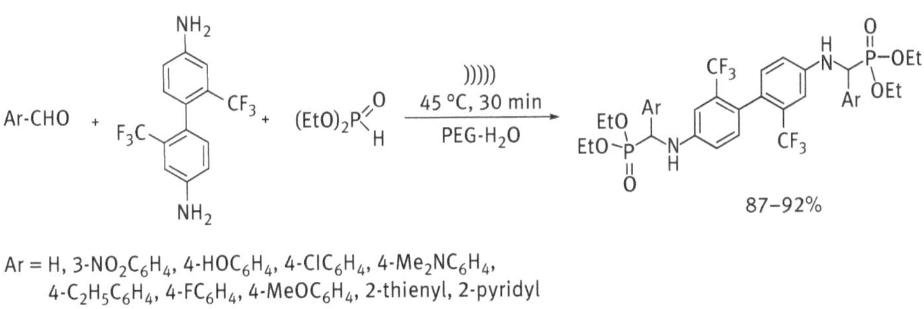

Figure 13.5: Ultrasound-assisted synthesis of bis(α-aminophosphonates) in PEG-H_2O.

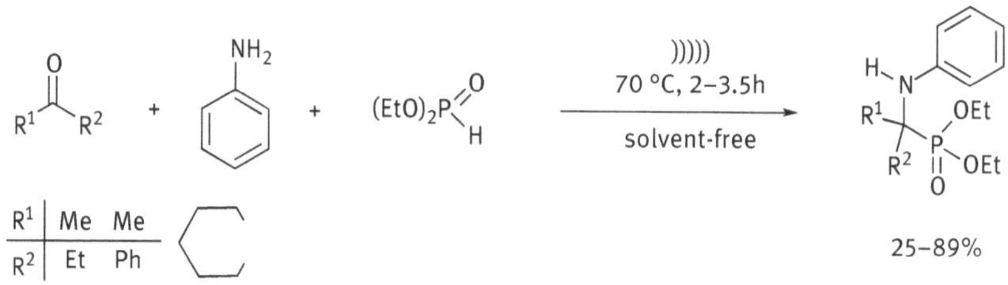

Figure 13.6: Ultrasound-assisted synthesis of α-aminophosphonates from ketones.

13.2.2 Kabachnik–Fields Reactions Starting from Trialkyl Phosphites

A number of methods have been reported for the synthesis of α-aminophosphonates using trialkyl phosphites as the P-reagent. Most of the methods involved substituted benzaldehydes, aniline derivatives and triethyl phosphite (Table 13.2). A simple and efficient ultrasound-assisted method was developed *via* the one-pot three-component condensation of various aryl aldehydes, amines and triethyl phosphite using CeO_2 nanoparticles as the catalyst under solvent-free conditions at ambient temperature (Table 13.2, entry 1). α-Aminophosphonates were also synthesized in the presence of a catalytic amount of thiamine hydrochloride under ultrasonic irradiation in water (Table 13.2, entry 2). Sodium salt of 1-hexanesulfonic acid also proved to be an efficient catalyst under ultrasonic irradiation for the rapid synthesis of α-aminophosphonates (Table 13.2, entry 3). Another simple protocol was also developed for the one-pot three-component condensation of benzaldehyde derivatives, anilines and triethyl phosphite using silica-supported Fe_3O_4 nanoparticles containing phosphomolybdic acid and imidazole moieties (Fe_3O_4@SiO_2-imid-PMA^n NPs) as a magnetically recoverable heterogeneous catalyst. (Table 13.2, entry 4). A series of α-aminophosphonates was synthesized from aromatic aldehydes, aryl amines and triethyl phosphite at 30 °C in the absence of any catalyst and solvent (Table 13.2, entry 5). Camphorsulfonic acid was also able to catalyze the reaction of aromatic aldehydes, aryl amines and triethyl phosphite under the influence of ultrasonic irradiation (Table 13.2, entry 6).

The synthesis of oxazaphosphinane derivatives was accomplished by the ultrasound-assisted three-component reaction of benzaldehyde, amino alcohols and triethyl phosphite. The condensation was carried out at room temperature in the absence of any catalyst and solvent (Figure 13.7) [40].

Starting from various aldehydes, urea or thiourea and triethyl phosphite, a series of α-ureidophosphonate derivatives was synthesized at 75 °C under catalyst- and solvent-free conditions (Figure 13.8) [41].

α-Sulfamidophosphonates were obtained by a simple ultrasound-assisted one-pot condensation among benzaldehyde, sulfonamides and trialkyl phosphites at room temperature (Figure 13.9) [42].

Table 13.2: Ultrasound-assisted synthesis of α-aminophosphonates using triethyl phosphite as the P-reagent.

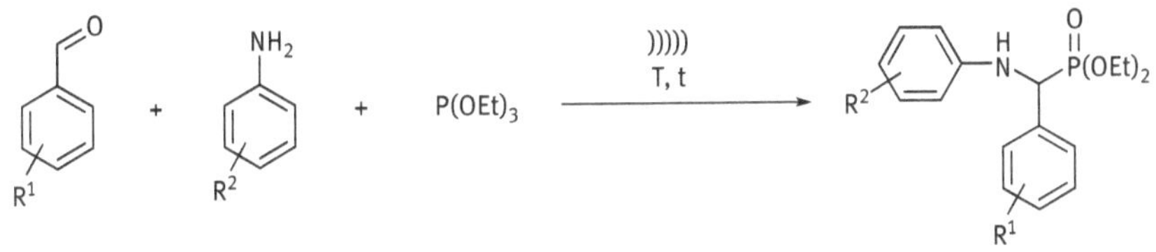

Entry	R¹	R²	Catalyst	Solvent	T (°C)	t (min)	Yield (%)	Ref.
1	H, 4-Cl, 4-HO, 4-NO_2, 4-Me, 4-MeO	H, 4-Me, 4-MeO, 4-HO, 4-NO_2, 4-Cl	Nano-CeO_2	–	rt[a]	5–30	84–99	[34]
2	H, 2-Me, 4-MeO, 4-NO_2, 2-Cl, 4-HO, 2,6-diCl	H, 4-Me, 4-Cl, 4-MeO	thiamine hydrochloride	H_2O	rt[a]	3–9	85–95	[35]
3	H, 4-Me, 4-MeO, 4-NO_2, 2,6-diCl, 3-Cl, 4-HO	H, 4-Me, 4-MeO, 4-Cl	1-hexanesulfonic acid sodium salt	–	rt[a]	12–20	86–94	[36]
4	H, 4-HO, 4-MeO, 2-Me, 4-Me, 2-Cl, 4-Cl, 4-Br, 4NO_2, 3-NO_2	H, 4-NO_2, 4-MeO, 4-Me	Fe_3O_4@SiO_2-imid-PMA[n] NPs	–	rt[a]	2–9	91–98	[37]
5	4-Cl, 2,4,5-triMeO, 4-Me_2N, 2-MeO-5-Br, 2-HO, 2,6-diCl, 3,4-OCH_2O-, 4-F, 3,5-diF, 4-MeO	4-Cl, 4-Me, 4-CF_3O, 4-F	–	–	30	0.33–0.75	80–99	[38]
6	H, 4-HO, 4-Cl, 4-MeO, 4-NO_2, 3,4-OCH_2O-	H, 4-Me	camphor sulfonic acid	–	rt[a]	8–20	83–93	[39]

[a]No exact temperature was provided.
rt: Room temperature.

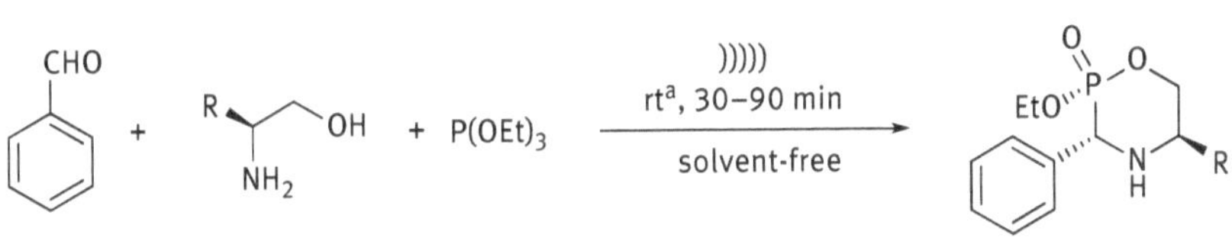

79–93%

R = H, Me, Pr, ^{i}Bu, Bn, CH_2(indol-3-yl)
[a]no exact temperature was provided

Figure 13.7: Ultrasound-assisted catalyst-free synthesis of novel oxazaphosphinanes.

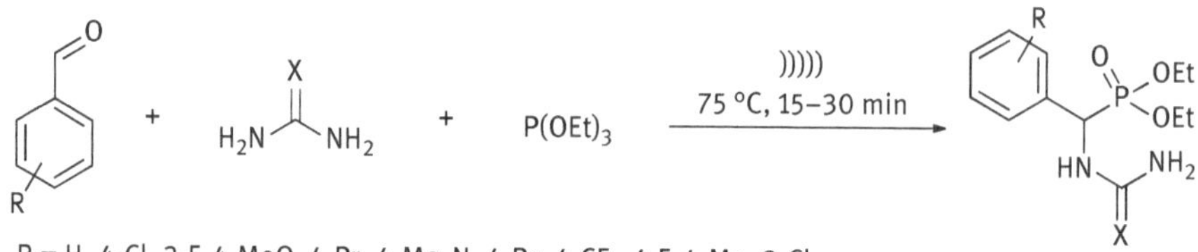

$R = H, 4\text{-}Cl, 3\text{-}F, 4\text{-}MeO, 4\text{-}Pr, 4\text{-}Me_2N, 4\text{-}Br, 4\text{-}CF_3, 4\text{-}F, 4\text{-}Me, 2\text{-}Cl$

X = O, S

75–85%

Figure 13.8: Ultrasound-assisted synthesis of α-ureidophosphonates.

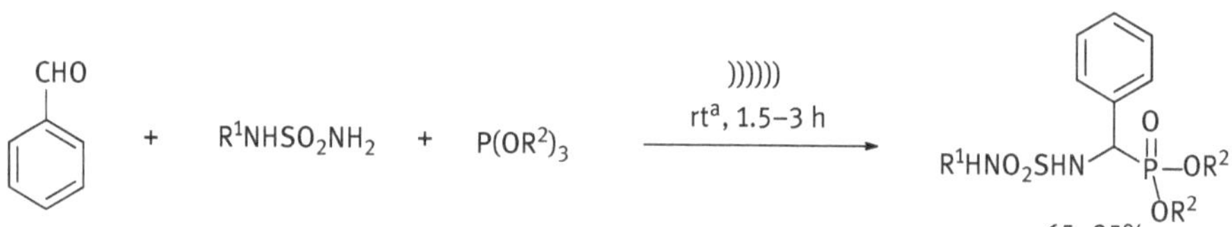

65–95%

$R^1 = Et, Pr, Bu, Bn, Ph, 4\text{-}MeC_6H_4, 4\text{-}ClC_6H_4,$
$3\text{-}FC_6H_4, 2\text{-}MeOC_6H_4, 4\text{-}MeOC_6H_4, CHMeC_6H_5$
$R^2 = Me, Et$
[a]no exact temperature was provided

Figure 13.9: Ultrasound-assisted synthesis of α-sulfamidophosphonates.

13.3 Ultrasound-assisted Pudovik Reactions

13.3.1 Synthesis of α-hydroxyphosphonates

α-Hydroxyphosphonates have potential pharmacological properties, such as antiviral [43], antibacterial [44], anticancer [45], rennin inhibiting [46], pesticidal [47] and anti-HIV activities [48]. They are also used as important precursors in the synthesis of α-halophosphonates, α-ketophosphonates, α-aminophosphonates, α-thiocyanatophosphonates, α-azidophosphonates, and α-carboxyphosphonates [49]. Further data on the synthesis and reactions of α-hydroxyphosphonates can be found in Chapter 5. A number of ultrasound-assisted methods are also available to carry out the synthesis of this important scaffold under various conditions.

α-Hydroxyphosphonate derivatives were synthesized by the reaction of aldehydes with dialkyl phosphites or alkyl phenyl-*H*-phosphinates at 60 °C using 1,4-dimethylpiperazine (DMP) as a basic catalyst in tetrahydrofuran under ultrasonic irradiation (Figure 13.10) [50]. All compounds synthesized were screened for their *in vitro* and *in vivo* antioxidant activity.

Another simple and ultrasound-assisted protocol was developed for the synthesis of α-hydroxyphosphonates by the reaction of aryl or hetaryl aldehydes and dialkyl phosphites at 50 °C using zinc oxide nano-sheets as the catalyst in ethanol (Figure 13.11) [51].

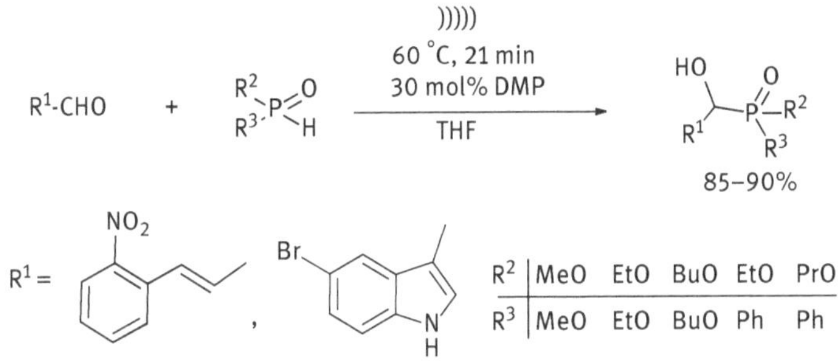

Figure 13.10: Ultrasound-assisted DMP-catalyzed synthesis of α-hydroxyphosphonates.

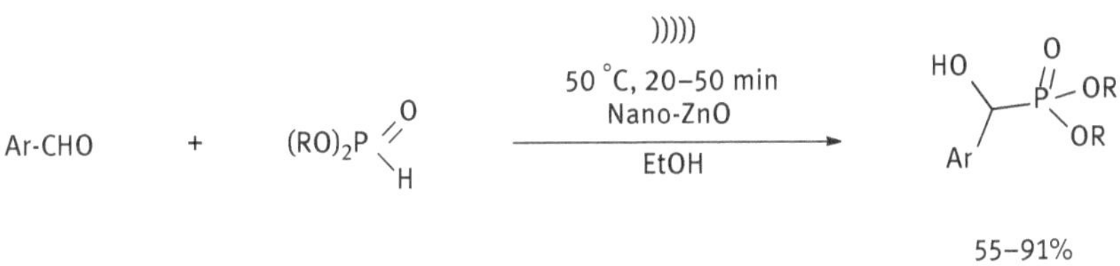

Ar = 2-Cl-6-FC_6H_3, 3-NO_2-4-HOC_6H_3, 3-NO_2-4-ClC_6H_3,
2-Br-4-FC_6H_3, 4-$CF_3OC_6H_4$, 2-HO-3,5-diClC_6H_2,
2-HO-3-Br-5-ClC_6H_2, 2-HO-5-ClC_6H_3, 2-HO-4-$Me_2NC_6H_3$,
4-pyridyl, 5-Br-2-thienyl, 5-NO_2-thienyl

R = Me, Et, ^{i}Pr, ^{t}Bu

Figure 13.11: Ultrasound-assisted nano-ZnO-catalyzed synthesis of α-hydroxyphosphonates.

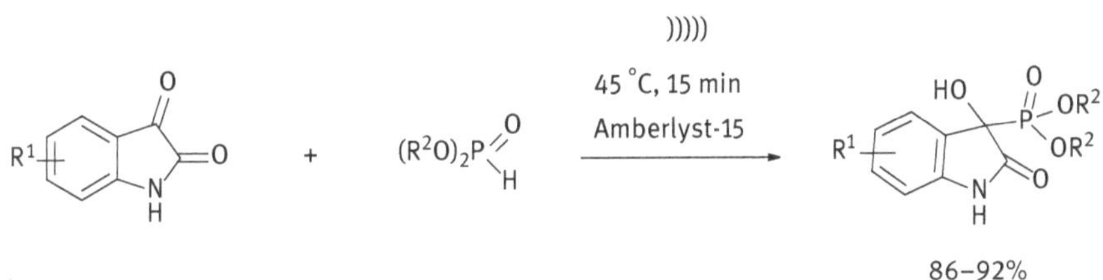

R^1 = H, 5-CH_3O, 5-NO_2, 5,7-diCl, 5-Br, 5-F
R^2 = Me, Et

Figure 13.12: The use of Amberlyst-15 in the synthesis of α-hydroxyphosphonates.

Amberlyst-15 efficiently catalyzed the reaction of substituted isatins and dialkyl phosphites to afford the corresponding α-oxindole-α-hydroxyphosphonates at 45 °C under solvent-free ultrasonic conditions (Figure 13.12) [52].

Sulfamic acid was employed as an efficient catalyst in the reaction of substituted benzaldehydes with triethyl phosphite at room temperature under ultrasonic irradiation for the synthesis of α- hydroxyphosphonates (Table 13.3, entry 1). With the influence of ultrasonic irradiation, potassium dihydrogen phosphite (KH_2PO_4) also proved to be a suitable catalyst in the reaction of benzaldehydes and triethyl phosphite under solvent-free conditions (Table 13.3, entry 2). Camphorsulfonic acid was also used in

Table 13.3: Ultrasound-assisted synthesis of α-hydroxyphosphonates from benzaldehyde derivatives and triethyl phosphite.

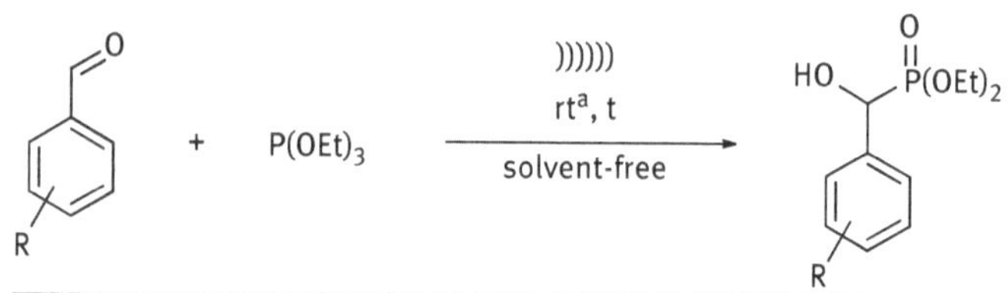

Entry	R	Catalyst	t (min)	Yield (%)	Ref.
1	H, 4-Cl, 4-HO, 4-Me, 4-MeO	NH_2SO_3H	20–35	89–98	[53]
2	H, 4-Me, 4-MeO, 4-Cl, 2-Cl, 3-HO	KH_2PO_4	5–14	80–92	[54]
3	H, 4-Me, 4-MeO, 4-NO_2, 3,4-OCH_2O-, 4-Cl, 4-HO	camphor sulfonic acid	8–12	85–93	[39]

[a]No exact temperature was provided.
rt: Room temperature.

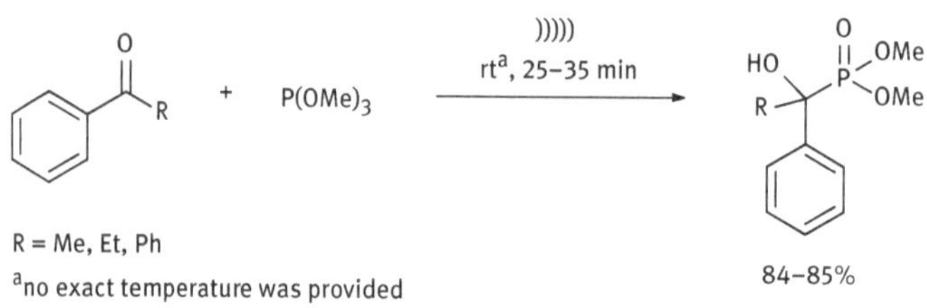

Figure 13.13: Ultrasound-assisted catalyst-free synthesis of α-hydroxyphosphonates from ketones and trimethyl phosphite.

the synthesis of α-hydroxyphosphonates under solvent-free conditions at room temperature (Table 13.3, entry 3).

The use of ultrasonic irradiation was also efficient in the synthesis of α-hydroxyphosponates by the reaction of ketones with trimethyl phosphite at room temperature without any catalyst or solvent (Figure 13.13) [55].

13.3.2 Synthesis of α-aminophosphonates

A number of α-aminophosphonate derivatives were synthesized by the reaction of imines derived from amlodipine. The Shiff-base was reacted with diethyl phosphite or dibutyl phosphite in the presence of a catalytic amount of $SnCl_2.2H_2O$ in boiling ethanol under ultrasonic irradiation (Figure 13.14) [56]. All compounds synthesized were screened for antimicrobial activities, and a few of them showed antibacterial, antifungal or antiviral activities.

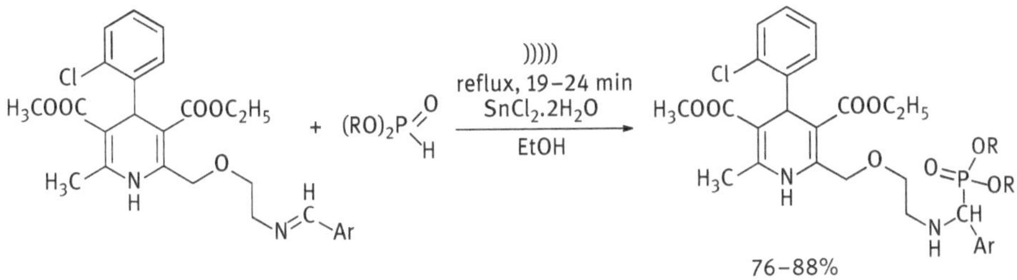

Ar = 2-Cl-6-FC_6H_3, 3-Br-4-FC_6H_3, 3-NO_2-4-HOC_6H_3, 2,4-diClC_6H_3, 2-thienyl, 3-pyridyl
R = Et, Bu

Figure 13.14: Ultrasound-assisted $SnCl_2.H_2O$-catalyzed synthesis of α-aminophosphonates from imines.

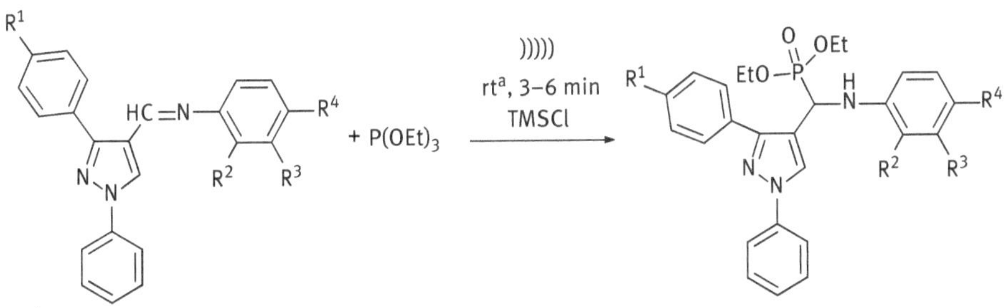

R^1 = H, Me, NO_2
R^2, R^3, R^4 = H, F
[a]no exact temperature was provided

77–94%

Figure 13.15: Ultrasound-assisted TMSCl-mediated synthesis of novel α-aminophosphonates from imines by the reaction with triethyl phosphite.

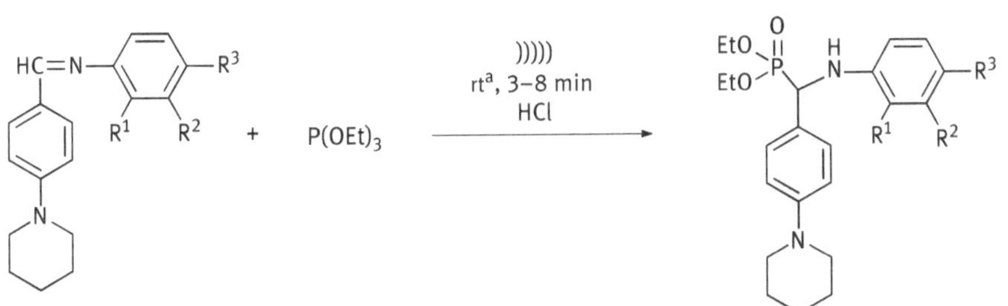

R^1 = H, F, Me, MeO, EtO, Cl, NO_2
R^2 = H, F, MeO, NO_2
R^3 = H, F
[a]no exact temperature was provided

73–88%

Figure 13.16: Ultrasound-assisted HCl-catalyzed synthesis of α-aminophosphonates by the reaction of imines with triethyl phosphite.

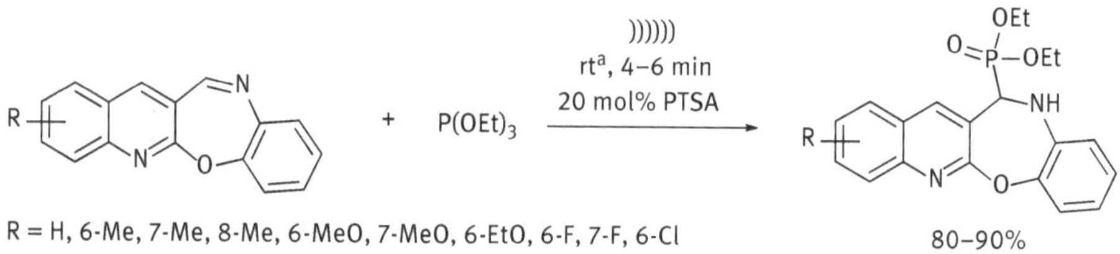

R = H, 6-Me, 7-Me, 8-Me, 6-MeO, 7-MeO, 6-EtO, 6-F, 7-F, 6-Cl
[a]no exact temperature was provided

Figure 13.17: Ultrasound-assisted catalyst-free synthesis of oxazepinephosphonates.

A simple and ultrasound-assisted TMSCl-mediated protocol was reported for the rapid synthesis of α-aminophosphonates by the reaction of triethyl phosphite and imines containing a pyrazole ring at room temperature under solvent-free conditions (Figure 13.15) [57].

Reactions between triethyl phosphite and imines bearing a piperazine ring were catalyzed by HCl to afford α-aminophosphonate derivatives under the influence of ultrasonic irradiation (Figure 13.16) [58].

Oxazepinephosphonates were synthesized by the ultrasound-assisted reaction of quino[2,3-*b*][1,5]benzoxazepines and triethyl phosphite in the presence of a catalytic amount of *p*-toluenesulfonic acid (Figure 13.17) [59]. The products were found to be active as antibiotics.

13.4 Ultrasound-assisted Click Reactions

Aminobisphosphonate derivatives containing a 1,2,3-triazole moiety were prepared by ultrasound-assisted click reactions of tetraethyl (2-(prop-2-yn-1-ylamino)-ethane-1,1-diyl)bisphosphonate and azides (Figure 13.18) [60]. The reactions were carried out at 65 °C in the presence of sodium ascorbate and copper sulfate as the catalysts in aqueous tert-butanol.

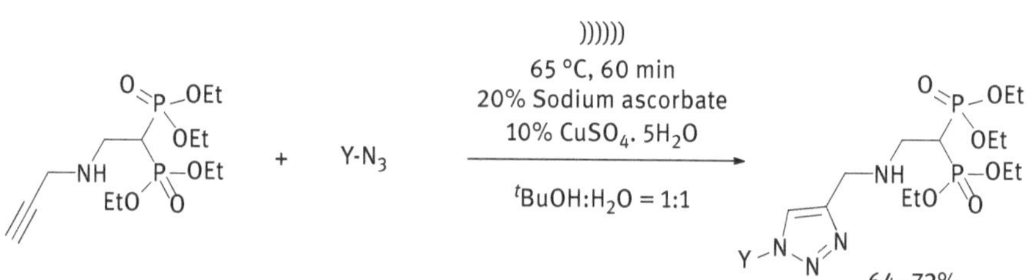

Y = Bn, Ph, 2-MeC$_6$H$_4$, 4-MeC$_6$H$_4$, 3-MeC$_6$H$_4$, 2-ClC$_6$H$_4$, 4-ClC$_6$H$_4$, 2-NO$_2$C$_6$H$_4$

Figure 13.18: Ultrasound-assisted synthesis of 1,2,3-triazole-aminobisphosphonates by a click reaction.

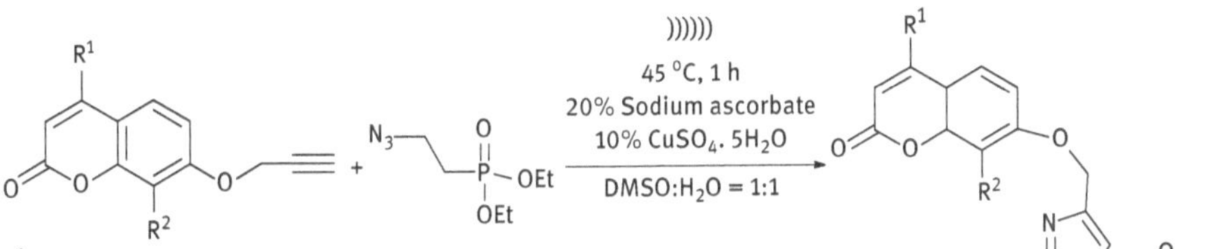

Figure 13.19: Ultrasound-assisted synthesis of phosphonate-functionalized coumarin derivatives by a click reaction.

The same research group synthesized 1,2,3-triazole-funtionalized phosphonate derivatives of a series of coumarines by click reactions at 45 °C in the mixture of dimethyl-sulfoxide–water as the solvent (Figure 13.19) [61].

13.5 Miscellaneous Ultrasound-assisted Organophosphorus Transformations

Diethyl phenylcarbamoylphosphonates and phenylthiocarbamoylphosphonates were prepared by the ultrasound-assisted reaction of isocyanates or isothiocyanates and diethyl phosphite applying silica-supported cesium chloride (SiO_2-$CeCl_3.7H_2O$) as the catalyst under solvent-free conditions (Figure 13.20) [62]. A few compounds synthesized were found to be of antibacterial and antifungal activity.

One-pot three-component reactions of various aldehydes, diethyl phosphite and isocyanates were carried out applying ultrasonic irradiation in the presence of a catalytic amount of magnesium oxide at room temperature to afford carbamoyloxyphosphonates (Figure 13.21) [63].

An allylphosphonate was synthesized *via* the ultrasound-assisted intermolecular olefin cross-metathesis of dimethyl allylphosphonate and 1,3-diacetoxy-2-methylenepropane using ruthenium complexes as the catalyst at 55 °C in dichloromethane (Figure 13.22) [64].

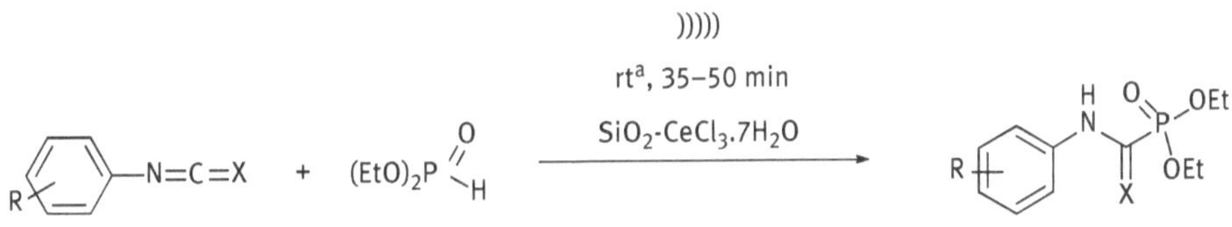

X = O, S
R = H, 3,4-diCl, 4-F, 2,4-diF, 3,5-diF, 2,6-diF, 2-F-5-NO_2, 3-CF_3, 4-NO_2, 3-Cl-4-NO_2
[a]no exact temperature was provided

Figure 13.20: Ultrasound-assisted synthesis of carbamoyl and thiocarbamoyl phosphonates.

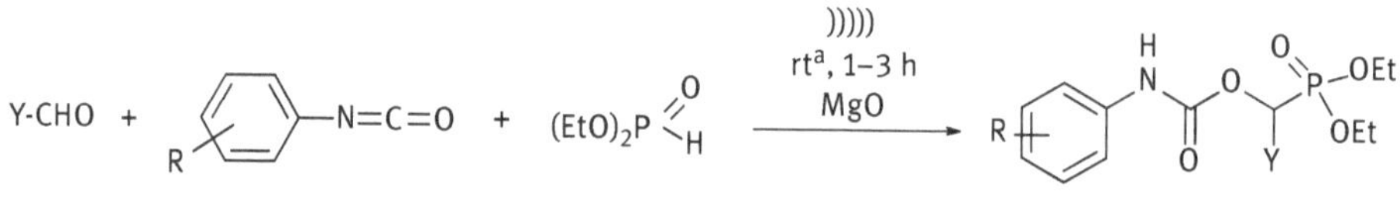

50–80%

Y = Ph, 4-FC_6H_4, 4-MeC_6H_4, 4-$MeSC_6H_4$, 4-$MeOC_6H_4$, 2-NO_2-C_6H_4CH=CH-, 2-thienyl
R = ^{c}Hex, Bn, Ph, 3-Cl-4-MeC_6H_3
[a]no exact temperature was provided

Figure 13.21: MgO-catalyzed ultrasound-assisted synthesis of α-carbamoyloxy phosphonates.

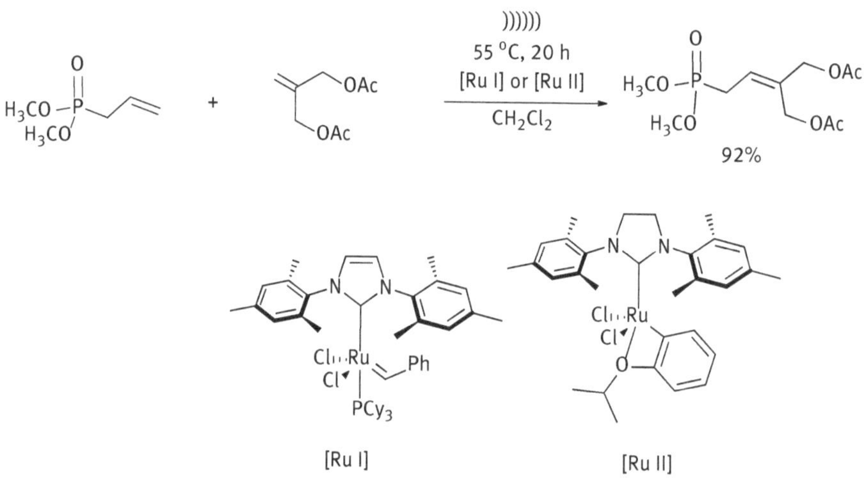

Figure 13.22: Ultrasound-assisted synthesis of an allylphosphonate.

Imidazole-immobilized Santa Barbara Amorphous-15 (SBA-IM/SO_3H) nano-composite with sulfonic acid functions was applied in the synthesis of 4*H*-chromen-4-yl-phosphonates. The target scaffold was prepared by the ultrasound-assisted three-component reaction of salicylic aldehydes, CH-acid derivatives and triethyl phosphite under solvent-free conditions (Figure 13.23) [65].

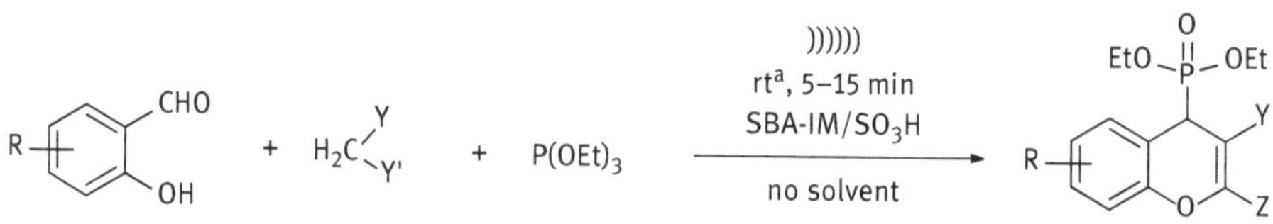

R = H, 5-Cl, 4-MeO
Y, Y' = CN, C(O)OEt, C(O)Et
Z = H_2N, EtO
[a]no exact temperature was provided

50–95%

Figure 13.23: Ultrasound-assisted synthesis of 4*H*-chromen-4-yl-phosphonates.

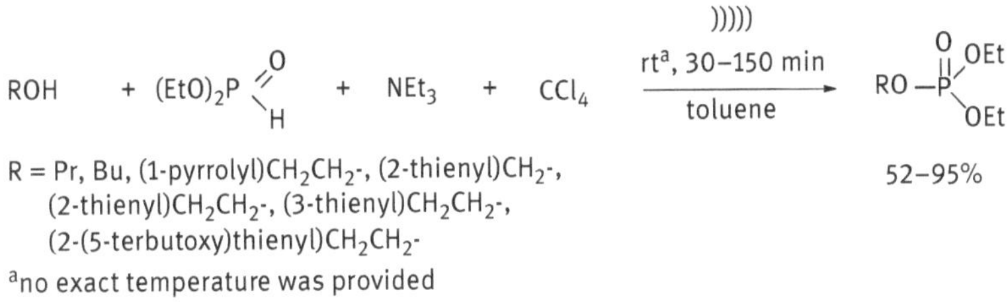

Figure 13.24: Ultrasound-assisted phosphorylation of alcohols.

The phosphorylation of alcohols was carried out under the influence of ultrasonic irradiation. The Atherton–Todd reaction of alcohols, diethyl phosphite, carbon tetrachloride and triethylamine at room temperature afforded the mixed trialkyl phosphates (Figure 13.24) [66].

13.6 Conclusions

The aim of this chapter was to provide an overview about the recent developments in the synthesis of organophosphorus compounds under ultrasonic irradiation. Various reactions including the Kabachnik–Fields condensation, the Pudovik reaction, the click and miscellaneous reactions were discussed. Ultrasonic irradiation can be applied efficiently to afford important organophosphorus compounds, such as α-aminophosphonates, α-hydroxyphosphonates, and related derivatives.

Acknowledgments: Bubun Banerjee thanks the Indus International University (Una, Himachal Pradesh, India) for their assistance, and the Kartha Education Society (Mumbai, India) for their financial support. Ádám Tajti thanks Gedeon Richter Talentum Foundation and Pro Progressio Foundation for financial support.

References

[1] B. Banerjee, *Ultrason. Sonochem.*, 2017, **35**, 1.
[2] B. Banerjee, *Ultrason. Sonochem.*, 2017, **35**, 15.
[3] B. Banerjee, *Aust. J. Chem.*, 2017, **70**, 872.
[4] B. Banerjee, *J. Serb. Chem. Soc.*, 2017, **82**, 755.
[5] D. E. C. Corbridge, *Phosphorus: Chemistry,Biochemistry and Technology*; CRC Press: Boca Raton, 2013.
[6] V. P. Kukhar, H. R. Hudson, *Aminophosphonic and Aminophosphinic Acids: Chemistry and Biological Activity*; Wiley: Chichester, 2000.
[7] G. Keglevich, E. Bálint, *Molecules*, 2012, **17**, 12821.
[8] P. Kafarski, M. Gorny vel Gorniak, I. Andrasiak. *Curr. Green Chem.*, 2015, **2**, 218.
[9] F. R. Atherton, C. H. Hassall, R. W. Lambert, *J. Med. Chem.*, 1987, **30**, 1603.

[10] F. R. Atherton, C. H. Hassall, R. W. Lambert, *J. Med. Chem.*, 1986, **29**, 29.
[11] R. Rozenfeld, X. Iturrioz, M. Okada, B. Maigret, C. Llorens-Cortes, *Biochemistry*, 2003, **42**, 14785.
[12] R. D. Dillard, D. E. Pavey, D. N. Benslay, *J. Med. Chem.*, 1973, **16**, 251.
[13] A. Peyman, W. Stahl, K. Wagner, D. Ruppert, K. H. Budt, *Bioorg. Med. Chem. Lett.*, 1994, **4**, 2601.
[14] L. Maier, H. Spoerri, *Phosphorus Sulfur Silicon*, 1991, **61**, 69.
[15] H. Kleszczynska, D. Bonarska, K. Bielecki, J. Sarapuk, *Cell. Mol. Biol. Lett.*, 2002, **7**, 929.
[16] J. Emsley, D. Hall, *The Chemistry of Phosphorus;* Harper & Row: London, 1976.
[17] H. Kleszczynska, J. Sarapuk, *Cell. Biol. Mol. Lett.*, 2001, **6**, 83.
[18] S. D. Lombaert, L. Blanchard, J. Tan, Y. Sakane, C. Berry, R. D. Ghai, *Bioorg. Med. Chem. Lett.*, 1995, **5**, 145.
[19] R. A. Cherkasov, V. I. Galkin, *Russ. Chem. Rev.*, 1998, **67**, 857.
[20] H. Ghafuri, A. Rashidizadeh, H. R. E. Zand, *RSC Adv.*, 2016, **6**, 16046.
[21] B. R. P. Reddy, P. V. G. Reddy, B. N. Reddy, *New J. Chem.*, 2015, **39**, 9605.
[22] E. D. Matveeva, T. A. Podrugina, E. V. Tishkovskaya, L. G. Tomilova, N. S. Zefirov, *Synlett*, 2003, 2321.
[23] M. A. Nasseri, M. Sadeghzadeh, *J. Chem. Sci.*, 2013, **125**, 537.
[24] S. Rostamnia, H. Xin, X. Liu, K. Lamei, *J. Mol. Catal. A: Chem.*, 2013, **374–375**, 85.
[25] T. Bhattacharya, B. Majumdar, D. Dey, T. K. Sarma, *RSC Adv.*, 2014, **4**, 45831.
[26] S. Rostamnia, H. Xin, *J. Mol. Liq.*, 2014, **195**, 30.
[27] S. Rostamnia, M. Amini, *Chem. Papers*, 2014, **68**, 834.
[28] B.-A. Song, G.-P. Zhang, S. Yang, D.-Y. Hu, L.-H. Jin, *Ultrason. Sonochem.*, 2006, **13**, 139.
[29] M. R. Sivala, S. R. Devineni, M. Golla, V. Medarametla, G. K. Pothuru, N. R. Chamarthi, *J. Chem. Sci.*, 2016, **128**, 1303.
[30] B. Song, S. Yang, Y. Hong, G. Zhang, L. Jin, D. Hu, *J. Fluorine Chem.*, 2005, **126**, 1419.
[31] R. M. N. Kalla, J. Bae, I. Kim, *New J. Chem.*, 2012, **41**, 6653.
[32] K. U. M. Rao, S. Namkoong, H.-C. Yu, J. Park, C.-M. Chung, S. Y. Oh, *Arch. Pharm. Chem. Life Sci.*, 2013, **346**, 851.
[33] M. Xia, Y.-D. Lu, *Ultrason. Sonochem.*, 2007, **14**, 235.
[34] S. M. Agawane, J. M. Nagarkar, *Tetrahedron Lett.*, 2011, **52**, 3499.
[35] P. G. Mandhane, R. S. Joshi, D. R. Nagargoje, C. H. Gill, *Chin. Chem. Lett.*, 2011, **22**, 563.
[36] K. S. Niralwad, B. B. Shingate, M. S. Shingare, *Ultrason. Sonochem.*, 2010, **17**, 760.
[37] M. Esmaeilpour, S. Zahmatkesh, J. Javidi, E. Rezaei, *Phosphorus Sulfur Silicon Relat. Elem.*, 2016, **192**, 530.
[38] B. Dar, A. Singh, A. Sahu, P. Patidar, A. Chakraborty, M. Sharma, B. Singh, *Tetrahedron Lett.*, 2012, **53**, 5497.
[39] P. V. Shinde, A. H. Kategaonkar, B. B. Shingate, M. S. Shingare, *Tetrahedron Lett.*, 2011, **52**, 2889.
[40] B. Belhani, A. Bouzina, M. Berredjem, N.-E. Aouf, *Monatsh Chem.*, 2015, **146**, 1871.
[41] A. Bouzina, M. Berredjem, S. Bouacida, H. Merazig, N.-E. Aouf, *RSC Adv.*, 2015, **5**, 99775.
[42] B. Belhani, M. Berredjem, M. L. Borgne, Z. Bouaziz, J. Lebreton, N.-E. Aouf, *RSC Adv.*, 2015, **5**, 39324.
[43] J. Neyts, E. De Clercq, *Antimicrob. Agents Chemother.*, 1997, **41**, 2754.
[44] H. Fleisch, *Endocr. Rev.*, 1998, **19**, 80.
[45] M. V. Lee, E. M. Fong, F. R. Singer, R. S. Guenett, *Cancer Res.*, 2001, **61**, 2602.
[46] D. V. Patel, K. R. Gauvin, D. E. Ryono, C. A. Free, W. L. Rogers, S. A. Smith, J. M. De Forrest, R. S. Oehl, E. W. Jr. Petrillo, *J. Med. Chem.*, 1995, **38**, 4557.
[47] P. Kafarski, B. Lejczak, *J. Mol. Catal. B: Enzym.*, 2004, **29**, 99.
[48] X. Zheng, V. Nair, *Tetrahedron*, 1999, **55**, 11803.

[49] D. V. Patel, K. R. Gauvin, D. E. Ryono, *Tetrahedron Lett.*, 1990, **31**, 5591.

[50] S. Rasheed, K. Venkata Ramana, G. Madhava, D. Subba Rao, C. Naga Raju, *Phosphorus Sulfur Silicon Relat. Elem.*, 2014, **189**, 606.

[51] S. K. T. Basha, S. Rasheed, K. C Sekhar, C. N. Raju, Md. S. Ali, C. A. Rao, *Phosphorus Sulfur Silicon Relat. Elem.*, 2014, **189**, 1.

[52] K. U. M. Rao, G. D. Reddy, C.-M. Chung, *Phosphorus Sulfur Silicon Relat. Elem.*, 2013, **188**, 1104.

[53] S. A. Sadaphal, S. S. Sonar, R. U. Pokalwar, N. V. Shitole, M. S. Shingare, *J. Korean Chem. Soc.*, 2009, **53**, 536.

[54] P. G. Mandhane, R. S. Joshi, D. R. Nagargoje, C. H. Gill, *Tetrahedron Lett.*, 2010, **51**, 1490.

[55] A. Bouzina, N.-E. Aouf, Ma Berredjem, *Res. Chem. Intermed.*, 2016, **42**, 5993.

[56] S. Rasheed, K. Venkataramana, K. Chandrasekhar, G. Fareeda, C. N. Raju, *Arch. Pharm. Chem. Life Sci.*, 2012, **345**, 294.

[57] D. R. Nagargoje, P. G. Mandhane, S. B. Bhagat, N. M. Chavanc, P. V. Badadhed, C. H. Gill, *Der Pharma. Chemica*, 2011, **3**, 1.

[58] P. G. Mandhane, R. S. Joshi, D. R. Nagargoje, A. V. Chate, C. H. Gill, *Phosphorus Sulfur Silicon Relat. Elem.*, 2011, **186**,149.

[59] S. S. Sonar, S. A. Sadaphal, V. B. Labade, B. B. Shingate, M. S. Shingare, *Phosphorus Sulfur Silicon Relat. Elem.*, 2010, **185**, 65.

[60] S. Liu, W. Bi, X. Li, X. Chen, L. Qu, Y. Zhao, *Phosphorus Sulfur Silicon Relat. Elem.*, 2015, **190**, 1735.

[61] X. Li, X. Chen, J. Yuan, Y. Liu, P. Li, L. Qu, Y. Zhao, *Phosphorus Sulfur Silicon Relat. Elem.*, 2015, **190**, 961.

[62] D. S. Rao, G. Madhava, S. Rasheed, S. K. T. Basha, M. N. L. Devamma, C. N. Raju, *Phosphorus Sulfur Silicon Relat. Elem.*, 2015, **190**, 574.

[63] B. Kaboudin, M. Fallahi, *Tetrahedron Lett.*, 2011, **52**, 4346.

[64] O. Sari, M. Hamada, V. Roy, S. P. Nolan, L. A. Agrofoglio, *Org. Lett.*, 2013, **15**, 4390.

[65] M. J. Nasab, A. R. Kiasat, *Microporous Mesoporous Mater.*, 2016, **223**, 10.

[66] B. Oussaid, M. Soufiaoui, B. Garrigues, *Synth. Commun.*, 1995, **25**, 871.

Zoltán Mucsi, Imre G. Csizmadia, Nóra Zs. Kiss
and György Keglevich

14 Computational study of the aromaticity and antiaromaticity of cyclic organophosphorus compounds containing a single phosphorus atom

Abstract: Organophosphorus chemistry has become one of the most rapidly developing research areas in organic and organometallic chemistry. Studies conducted in the last half century have revealed many interesting chemical structures, important intermediates, useful metal ligands, and biologically important pharmacophores. In this chapter, we have collected and analysed selected four- and five-membered antiaromatic, as well as aromatic organic compounds containing a single phosphorus atom in the ring, which were considered, studied and prepared during the last decades.

Keywords: P-heterocycles, aromaticity, antiaromaticity, theoretical calculations, aromaticity indexes.

14.1 Aromaticity and Antiaromaticity

Aromaticity and antiaromaticity are characteristic properties of cyclic unsaturated hydrocarbons. Typical examples for aromatic 6π-electron containing unsaturated cyclic hydrocarbons are benzene and the cyclopentadiene anion. Similar examples for antiaromaticity are simple 4π-electron systems, such as cyclobutadiene and the cyclopentadiene cation. Heterocyclic analogues can also be constructed by the same way, involving N, or in certain cases B atom from the first period. Analogous heterocyclic derivatives from the second period may include a P and seldom an Al atom. These typical cases are summarized in Figure 14.1. The aromaticity ($AR\% > 0$) and antiaromaticity ($AR\% < 0$) together with the ring size show the pattern illustrated in Figure 14.2.

The three families represent different patterns, and thus, different chemical behaviours. It is clear that the area of rhomb is reduced as we go from homo to heteroatom-substituted rings. It is also interesting that at the "left corner" of 4-atomic rings, the antiaromaticity values are increasing on going from C, N to P, that is the antiaromatic character decreases on moving towards the aromatic region. Similarly, at the "right-hand corner," the aromaticity decreases on going from C and N to P, that is the aromatic character decreases on moving towards the antiaromatic region.

https://doi.org/10.1515/9783110535839-014

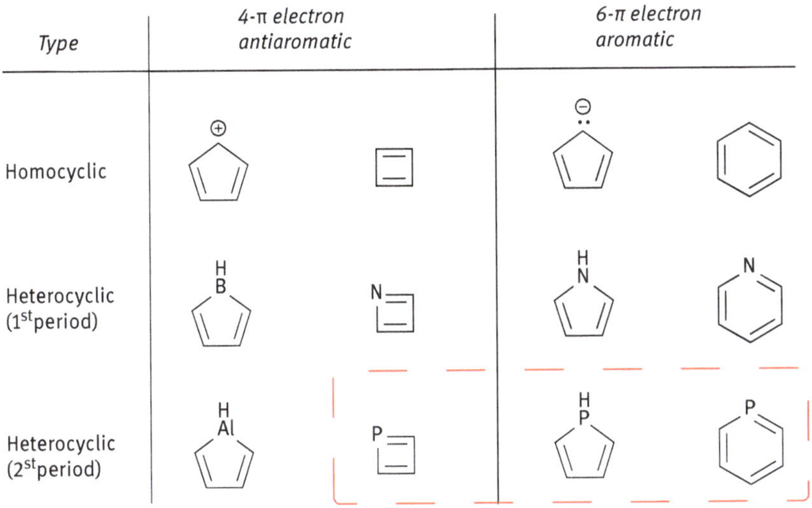

Figure 14.1: Homoatom and heteroatom containing unsaturated ring structures exhibiting aromaticity and antiaromaticity.

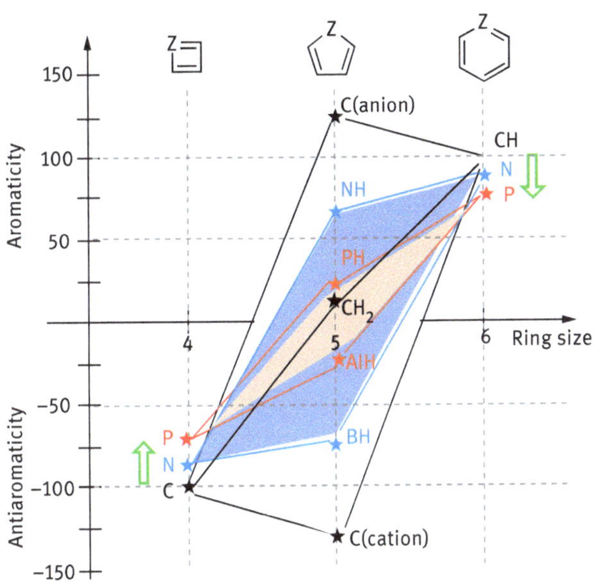

Figure 14.2: Aromaticity (AR% > 0) and antiaromaticity (AR% < 0) patterns of four-, five- and six-membered rings of homo- and heteroatomic compounds.

In contrast to benzene, the synthesis of cyclobutadiene has remained unsuccessful [1–8] in spite of nearly a century's effort. The failure is explained by the new concept of antiaromaticity [6–8]. The synthesis of other compounds having antiaromatic character have been a challenge for preparative organic chemists. The concept of π-aromaticity of benzene (1920s) [9–12] and heterocycles [13] had been known for long, but the antiaromaticity of cyclobutadiene (1965) was introduced only later [1–5]. The π-aromaticity (1979) [14, 15] was followed by the concept of σ-aromaticity (2004) [16] of metal clusters [9–16]. Beyond planar aromaticity, the concept of spherical aromaticity (1978) [17] was also introduced with the discovery of the fullerene family.

Aromatic stabilizing or antiaromatic destabilizing effects determine, whether a particular molecule is isolable or not. According to the well-known 4n+2 and 4n rules, it may seem trivial to distinguish between aromaticity and antiaromaticity. However, sophisticated and accurate theoretical and experimental investigations have often led to opposite conclusions. The precise quantification of aromatic characteristics of ring structures is not trivial. Quantitative methods are required to replace the initial, inexact descriptions. In most cases, these theoretical methods may be classified into four categories based on fundamentally different concepts:

(1) On energetic grounds, several methods are favoured. First of all, aromaticity (AM%) may be a useful parameter for classifying these methods [18]. Aromatic Stabilization Enthalpy (H_{ASE}) or Energy (E_{ASE}) values are also indicative [19–24], and homodesmic [20, 23, 25] hydrogenation [18] or isomerization (H_{ISE}) reactions [26] may also be suitable for the definition of antiaromatic destabilization.

(2) On geometrical grounds: The indices HOMA [27, 28], Bird [29]; and BDSHRT [30] are based on the analysis of geometrical data obtained experimentally or theoretically; however, this method is not suitable for the determination of the antiaromatic character.

(3) On NMR characteristics: Magnetic shielding properties [31], such as nucleus independent chemical shift (NICS) [21, 32], may also be good indicators of aromaticity/antiaromaticity.

(4) Using reactivity criteria [23] to study the transition states (TS) of various reactions of aromatic and antiaromatic compounds may also provide useful information.

Previously, only aromatic phosphorus compounds (phosphinines and phospholes) appeared in the literature, which exhibited aromatic rings [13, 33–41]. Here we focus on the degree of aromaticity and antiaromaticity (Figure 14.1) of single P-heterocyclic structures with or without an additional (N or O) heteroatom, and we also discuss their synthetic consequences. Surprisingly, detailed investigations have shown many four-membered phosphorus compounds to be non-aromatic rather than antiaromatic. The three- [38, 39, 42, 43] and seven-membered rings containing a P atom [44], as well as the four-membered phosphazenes [45], are not covered in this study. Although phosphinine derivatives play a prominent role in organic chemistry, and exhibit high, benzene-like aromaticity, six-membered aromatic P-heterocycles have also been left out of this chapter. It is surprising that, in most of the relevant articles on four- and five-membered P-heterocycles, their antiaromatic property was not usually mentioned and analysed; only in a few cases the problem was treated qualitatively (Figue 14.3) [34].

Third-row elements, such as P have different numbers of valence states (trivalent: λ^3 or pentavalent: λ^5), which results in an increased number of possible structures.

All 4π-electron systems try to minimize their unfavourable antiaromatic electronic effect to somehow decrease the significant destabilizing energy. A certain extent of stabilization may be attained in five different ways that promote the synthesis of these structures.

The possibilities are the following:

(1) In transition metal complexes, where antiaromatic P-species are involved as ligands, the overall electronic structure may be modified to lead to aromatic species.
(2) The electronic structure may be affected by exocyclic substituents resulting in stabilization.
(3) In a few reaction sequences, antiaromatic species may be involved as TSs or reactive intermediates that may be transformed to non-aromatic structures. Ceasing the antiaromaticity means the driving force for such transformations.
(4) The high instability and reactivity of antiaromatic compounds may be decreased dramatically if bulky substituents (e.g. ^{t}Bu or Me_3Si) are attached to the ring. Such systems could be isolated as stable structures.

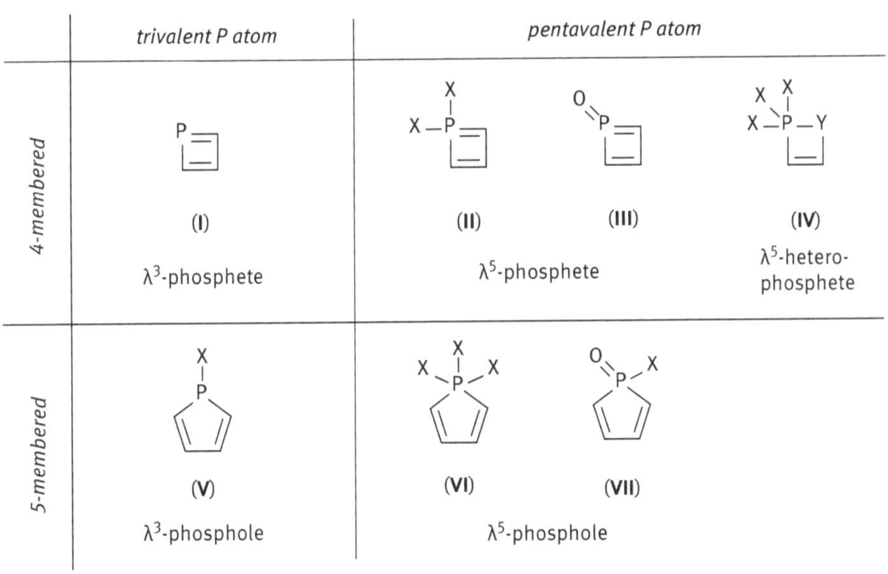

Figure 14.3: P-heterocycles discussed in the present review. The roman numbers refer to the chemical scaffold neglecting the substituents.

(5) In the absence of any external influence, the geometry of unstable cyclic scaffolds may be distorted, minimizing the non-beneficial overlap between the atomic orbitals (AOs) causing antiaromaticity.

For all these fundamental phosphorous-containing structures, the aromaticity and antiaromaticity parameters [18, 41, 46] {calculated at B3LYP/6-31G(d,p) [47]}, as well as nucleus independent chemical shifts {NICS; calculated at B3LYP/6-31G(d,p) [47, 48]} were calculated using Gaussian09 [49].

14.2 Four-membered Rings (Phosphetes I, II, III and IV)

14.2.1 λ³-Phosphete Structure (I)

In the case of the four-membered unsaturated rings with a single P atom (called λ³-phosphetes, **I**), the recognition of the antiaromatic character may be trivial due to the two genuine double bonds in a conjugated 4π-electron system. This ring is analogous to that of cyclobutadiene, where there is a "homogenic" antiaromaticity due to the four p_z AOs involved. The aromaticity parameters for these compounds were measured by the linear scale of aromaticity methodology [18, 24, 41, 46].

Compound λ³-PC_3H_3 (**I**) has a planar structure with two unequal C–P bonds, and exhibits a stretched cyclobutadiene-like structure, as showed by theoretical calculations [50, 51]. The aromaticity parameter (AR%) calculated for phosphete **I** is –77.4%, while the standard value for cyclobutadiene is –100% (Table 14.1) [18, 41, 46]. One may conclude that the degree of antiaromaticity is significantly influenced by the introduction of a P atom into the ring.

Interestingly, the valence isomerization of cyclobutadiene and its P-analogue (**I**) showed structurally similar (square *vs* rectangle) TSs, implying that their antiaromatic character is comparable. In respect of phosphete **I**, the enthalpy of activation is somewhat lower ($\Delta H^{\#}_{iso}$ = 30.1 kJ mol^{-1}) than that for cyclobutadiene (where $\Delta H^{\#}_{iso}$ = 35.1 kJ mol^{-1}). It may mean that the P=C double bond is "weaker" than the corresponding C=C unsaturation. This also implies that the delocalization of the C=C π electrons is more extensive than that of the P=C double bond [50].

Another theoretical study of the λ³-PC_3H_3 structure (**I**) led to the conclusion that species **I** is almost as antiaromatic as cyclobutadiene. The NICS value pairs for species **I** and cyclobutadiene are very close to each other. These values suggest that the incorporation of the λ³-P unit has only a slight impact on the antiaromaticity of the cyclobutadiene scaffold [52, 53].

These antiaromatic molecules are expected to be isolable synthetically if their antiaromaticity is reduced. This can be achieved either by association with a transition

Table 14.1: Various computed aromaticity parameters for compounds **I–IV** in comparison to cyclobutadiene.

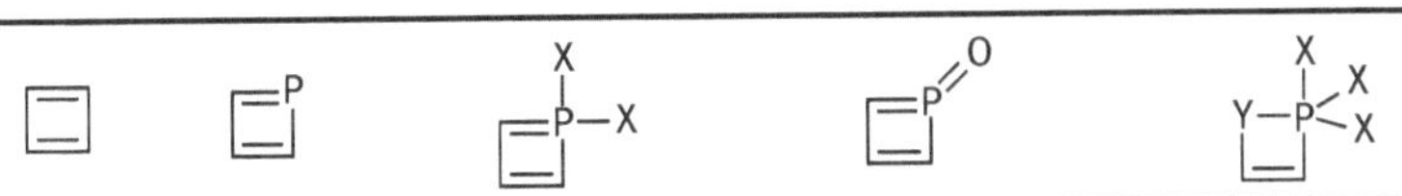

	$(CH)_4$	I	II		III	IV	
			X = Me	X = F		Y = S X = F	Y = N X = F
NICS(0)[a]	25.7	27.8	1.69	5.3	8.1	X	X
NICS(1)[a]	17.5	19.9	5.1	10.7	13.7	–	–
AR% (%)[b]	−100.0	−77.4	−1.2	−41.1	−74.2	−26.8	−49.6

[a]The NICS values were computed at B3LYP/6-31G(d,p) level of theory [46].
[b]The aromaticity percentages were calculated according to reference [46].

metal ligand sphere, or by structural modifications, which may lead to unique chemical scaffolds (involving a λ^5 P atom). Not surprisingly, the λ^3-P=C and λ^5-P=C double bonds (as in **I** and **II**, respectively) differ considerably from each other. Interestingly, the poetic statement that the λ^3-P atom is a "carbon copy" is quite correct [36], since the energies of isodesmic reactions and π-ionizations exhibited by the C=C and λ^3-P=C double bond are rather close [54, 55]. In contrast, other studies have reported that the λ^5-P=C scaffold may be characterized as an intermediate state between the zwitter-ionic and the genuine double bond system [56–58].

Regitz and co-workers [59] reported the first preparation of a λ^3-phosphete scaffold. They synthetized Co-complex **3** from bis(trimethylsilyl)acetylene (**2**) and phosphacetylene **1** (that is a "phospha-alkyne") (Figure 14.4) [52, 53]. In this case, the antiaromatic character was stabilized by the Co(I) ion, and the phosphete moiety was present as a metal ligand. Complex **3** was characterized by NMR spectroscopy. Simplified calculations for the complex without $SiMe_3$ substituents suggested a high aromatic character of 77.4% [41].

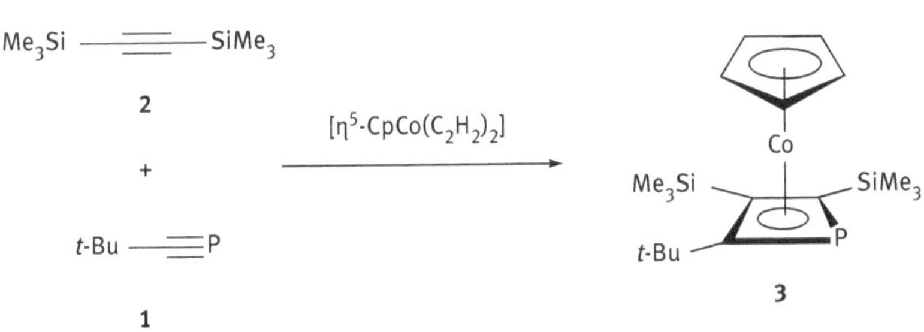

Figure 14.4: First synthesis of a λ^3-phosphete scaffold stabilized as a Co-complex.

However, a subsequent theoretical investigation concluded that structure **I** cannot be the intermediate for the conversion of **1** ⟶ **3** due to the forbidden dissociation of phosphete **I** to acetylene and phosphacetylene [53].

14.2.2 λ⁵-Phosphetes (II and III)

Computed structures of the λ^5-$PX_2C_3H_3$ molecules (**II**; X = Me or F) exhibit almost a C_{2V} symmetry with a planar four-membered ring [51]. However, the species under discussion show different electronic structures, as compared to that of λ^3-phosphete **I** [51]. Depending on the P-substituents, species **II** either shows no aromaticity (X=Me) or possesses moderate antiaromaticity (X=F) [60]. The aromaticity character and the NICS values are listed in Table 14.1.

The practically non-aromatic character of **II**, when X = Me, allowed the isolation of an analogous derivative (**7**) that was characterized experimentally (see later) [61]. Cyclobutadiene, and other sterically hindered cyclobutadiene derivatives were observed typically as transient species [62]. The difference discussed above may be the consequence of the different electronic nature of the X substituents F and Me. In case of X = F, a contribution of pseudo π electrons from the two P–F bonds to the 4π-electron rings may occur. The resonance structures of species **II** include allyl-type and ylide-type substructures (Figure 14.5) [34, 63]. The electronic interaction on the ylide structure may be significant, causing a higher antiaromaticity parameter (–41.1%), as compared to the case of the non-aromatic dimethyl derivative (–1.2%), where no electronic stabilization is possible. It should be mentioned that the MO associated with the P–Me bond is closer to π-orbital energies, than that associated with the P–F bond [51].

A stable λ^5-$R_2PC_3H_3$ derivative (**7**) could be isolated [64] by carrying out the following reaction sequence (Figure 14.6). The reaction of a chloro-containing phosphaylide (**4**) with $AlCl_3$ gave dihydrophosphetium salt **5** as an intermediate that rearranged to isomer **6**. Cyclic phosphonium salt (**6**) yielded λ^5-phosphete **7** by deprotonation with sodium bis(trimethylsilyl)amide [61, 64]. In this case, the presence of the fused

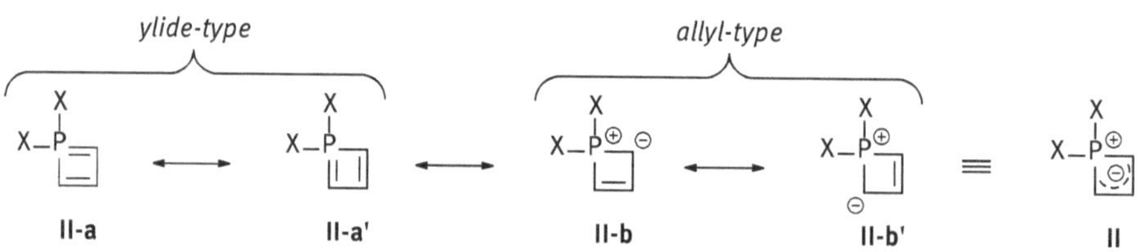

Figure 14.5: Possible resonance structures of phosphate **II**.

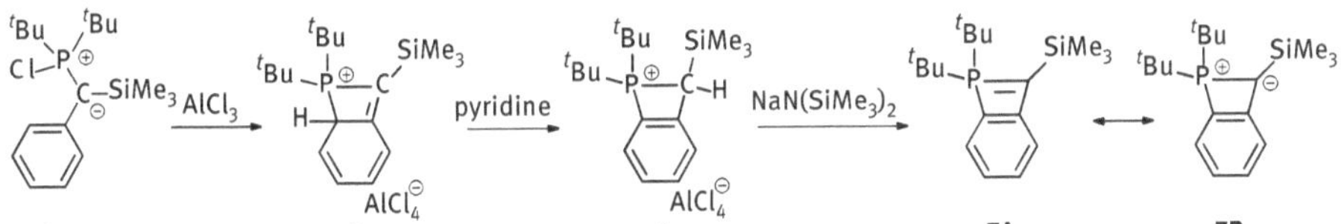

Figure 14.6: Synthesis of a stabilized λ5-phosphete.

aromatic benzene ring stabilizes the four-membered ring, decreasing further the lower antiaromatic character of the parent λ5-phosphete ring.

It was shown experimentally that, despite all attempts at utilizing heat or UV irradiation, λ5-phosphete **9** was not obtainable from λ5-diazaphosphinine (**8**) by the elimination of N_2 (Figure 14.7). Heterocycle **8** featuring the >P=N-N=C< structural motif (as in the Staudinger–Meyer adducts formed from phosphines and diazo derivatives) is of high stability, and cannot be the precursor of phosphetes [64–66].

A close λ5-phosphete analogue, where the four-membered ring contains an additional nitrogen heteroatom (**13**), which is the subject of the next subchapter, is shown in Figure 14.8. Refluxing cyclic adduct **12** obtained from azide **10** and acetylene **11**, the elimination of dinitrogen took place yielding azaphosphete **13** (Figure 14.8). This product exhibits four π-electrons, however, it can be regarded as a non-antiaromatic species, rather than an antiaromatic one. This is in accord with the fact that this azaphosphete could be prepared [67].

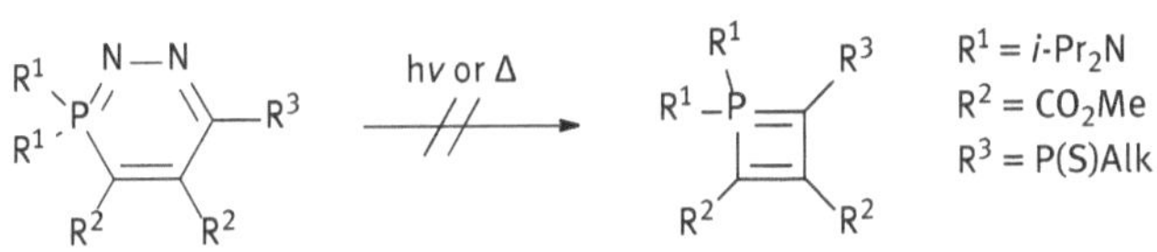

Figure 14.7: Failure of the **8**→**9** conversion.

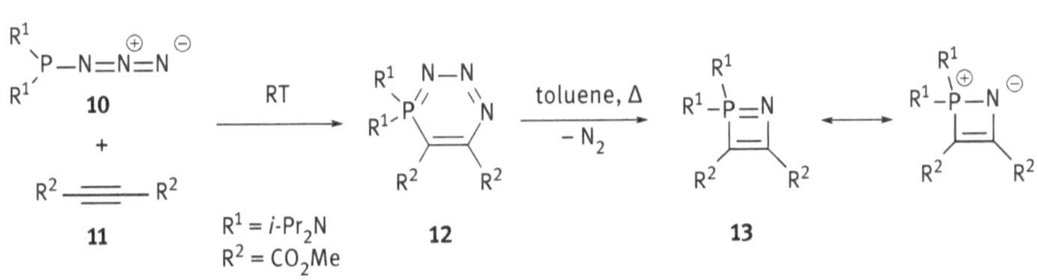

Figure 14.8: Synthesis of azaphosphetes **13**.

There is no experimental or computational support for the existence of the P-oxide of the phosphete (**III**). This species has a high antiaromatic character of –74.2, suggested by the calculations, which is comparable with that (–77.4) of phosphete **I**. Eventually, the large instability of the P-oxide of phosphete **III** obstructs its synthesis. A related theoretical study involves only the thermodynamic parameters of the P-oxide of phosphete from the aspect of ring strain and hydrogen bond strength [67].

14.2.3 λ^5-heterophosphetes (IV)

Heterophosphetes (oxaphosphetes, aminophosphetes and thiophosphetes) with a pentacoordinated P atom (**IV**) may have two distinctly different conformations (**IV-A** and **IV-B**). In one of them (**IV-A**), the phosphorus–heteroatom bond (P–Y) is equatorial, while in the other one (**IV-B**), the P–Y bond is apical. The interconversion connecting them is called pseudorotation (Figure 14.9) [68–72]. In addition to the pseudorotation, heterophosphetes **IV** may undergo ring-opening reactions to form *β*- oxo-, amino- and thioxophosphorane **14** [68]. This ring opening is the reason that earlier synthetic attempts to prepare heterophosphetes of type **IV** failed, and always led to phosphoranes of type **14** formed by ring opening [68, 73–78].

The reversible transformation of heterophosphete **IV-B** to **14** may be regarded as a ring-chain valence tautomerism. The electronic effect in **IV,** i.e. the strong conjugation between the lone electron pair of Y and the pentavalent P results in the antiaromatic character of species **IV**. The bond order of P–Y increases, and the double bond may interact with the C=C double bond, causing an unfavourable cyclobutadiene-like resonance (Figure 14.9) [64]. It was shown that in spite of their structural similarities, form **IV-A** exhibits a higher antiaromaticity (depending on the nature of R, –35% to –20%) than structures **IV-B** (–5% to +5%).

With certain substituents, the **IV-B** conformer of the heterophosphete turned out to be non-aromatic rather than antiaromatic. One may fine-tune the antiaromaticity parameters of these compounds by varying the exocyclic substituents X and the ring

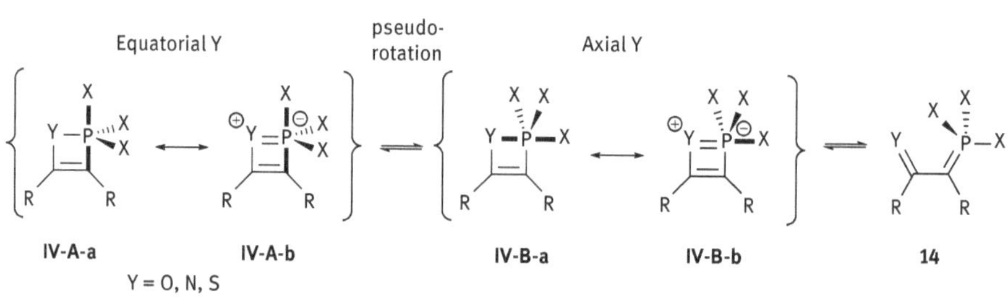

Figure 14.9: Interconversion of conformers **IV-A** and **IV-B** by pseudorotation together with the ring-opened form.

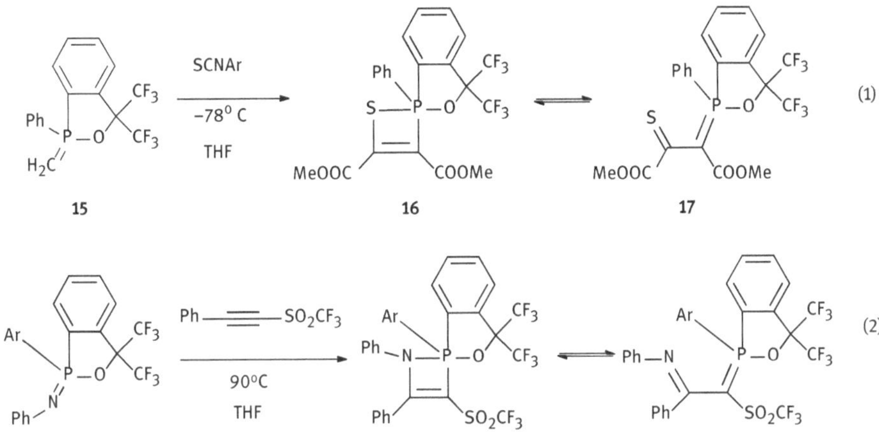

Figure 14.10: Synthesis of a spirocyclic thiophosphete (**15**) and azaphosphete (**18**).

heteroatom Y [69]. The difference in the internal molecular symmetry of **IV-A** and **IV-B** may also contribute to the antiaromaticity [69]. During the decade of preparative attempts, only two related structures could be isolated [79–81]. Thiaphosphete **16** having a spirocyclic form was prepared from heterocyclic phosphorane **15** by reaction with aryl thioisocianates (Figure 14.10-(1)) [79]. In solution, a ring chain equilibrium involving cyclic **16** and open chain form **17** may occur.

The other stable compound, azaphosphete **19,** was prepared by the reaction of imino derivative **18** and an acetylene derivative (Figure 14.10-(2)) [80, 81]. The structure of spirocyclic **19** was established by X-ray crystallography, indicating a **IV-B**-type structure. The N···P distance in species **19** is considerably longer (2.17 Å) [80–82] than the sum of the covalent radii (1.06 Å + 0.75 Å = 1.81 Å). The long N–P bond length was in accord with an ionic-type connection between the P and N atoms [69], suggesting an equilibrium of azaphosphete **19** and phosphorane **20,** which was substantiated by ^{31}P NMR. At a higher temperature, the ring-chain equilibrium was shifted to the ring-opened form (**20**) [80]. Compound **19** revealed a non-aromatic character (−4.3%) [69].

14.3 Five-membered Rings (V, VI and VII)

The aromaticity of phosphole derivatives (**V**) has been studied for long [83–92], and the generally accepted view is that on the basis of their bond length and spectral data (e.g. NMR) [83–85] they exhibit a rather low level of aromatic character [29, 93, 94]. In this respect, the structural data are more reliable than the spectral ones. Their chemical properties (e.g. their reactivity in aromatic electrophilic substitution) also suggested a low degree of aromaticity [41–43]. The phosphole ring lacks the extensive electron delocalization, as compared to the well-known *S*- and *N*-heterocyclic counterparts (thiophene and furane), where there may be a certain degree of delocalization. It is due to the pyramidality of the P atom that its lone pair is prevented from an efficient overlap with the p_z orbitals of the sp^2 carbon atoms. X-ray diffraction analysis showed that the P-substituent is by 66.9° out of the plane of the hetero ring in 1-benzyl-phosphole. The aromaticity problem of phosphole derivatives (**V**) was studied in detail by Quin and Keglevich [95–108].

The aromatic stabilization energies of phosphole derivatives (related on skeleton **V**) were found to be as high as 30–60 kJ mol^{-1}, which is considerably lower [4, 29, 93, 94] than that of pyrrole derivatives (70–130 kJ mol^{-1}) [18, 29, 93, 94]. The level of NICS values are on the edge of the aromaticity and non-aromaticity. The weak aromaticity of phosphole **V** (characterized by a Bird index of 35.5 and an AR% of 18.4%) (Table 14.2) may be attributed to the pyramidal geometry around the P atom preventing the participation of its lone pair in delocalization [109–115]. These molecules exhibit a high inversion barrier [86–92].

The best synthetic method for phospholes (whose skeleton is marked in general by **V**, and exemplified by e.g. **23**) was described by Mathey [116, 117]. In this method, various butadiene derivatives (**21**) were reacted with alkyl or aryl dihalophosphines, and a subsequent treatment of the phosphonium salts (**22**) so obtained by a strong base (DBU or α-picoline) furnished the target phospholes (**23**) (Figure 14.11/Method A). An alternative method utilizes the cycloaddition reaction of bis-alkyne **24** and $PhPH_2$ to afford phosphole **25** (Figure 14.11/Method B) [118, 119].

Table 14.2: Various aromaticity parameters for phosphole derivatives **V**, **VI** and **VII** in comparison with pyrrole and borole.

			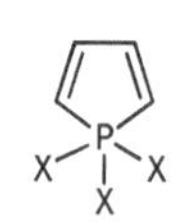		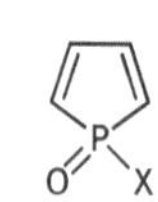		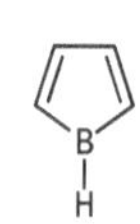
	pyrrole	**V**	**VI**		**VII**		**borole**
			X = Me	X = F	X = H	X = F	
NICS(0) (ppm)[a]	−15.9	−8.2	−3.2	1.8	−0.5	1.2	20.3
NICS(1) (ppm)[a]	−11.5	−5.8	–	1.0	−2.3	0.0	12.1
BIRD index[b]	59	35.5	–	–	–	–	–
AR%[c]	56.6	18.4	−30.4	−16.3	−13.5	−19.4	−59.3

[a]The NICS values were computed at B3LYP/6-31G(d,p) level of theory [46].
[b]The BIRD indexes were taken from reference [29].
[c]The aromaticity percentages were calculated according to reference [46].

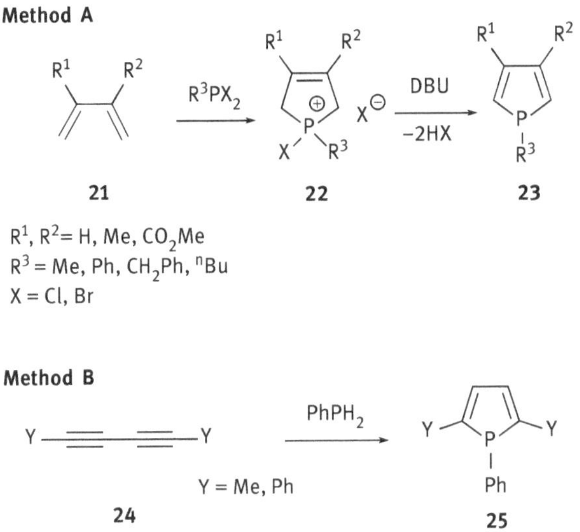

Figure 14.11: Simple methods for the synthesis of the phosphole scaffold.

The synthesis of sterically hindered phospholes (**28a–f**) was elaborated by Quin and Keglevich as shown in Figure 14.12. 1-Chloro-3-phospholene **26** obtained from the corresponding 1-chlorophospholene oxide by deoxygenation was converted to aryl phospholenes **27** by reaction with the suitable Grignard reagents. Then, oxidation, the addition of bromine, deoxygenation and spontaneous aromatization (Method **A**), or a quaternization reaction with bromine and HBr elimination (Method **B**) led to the arylphosphole (**28**) [95–108].

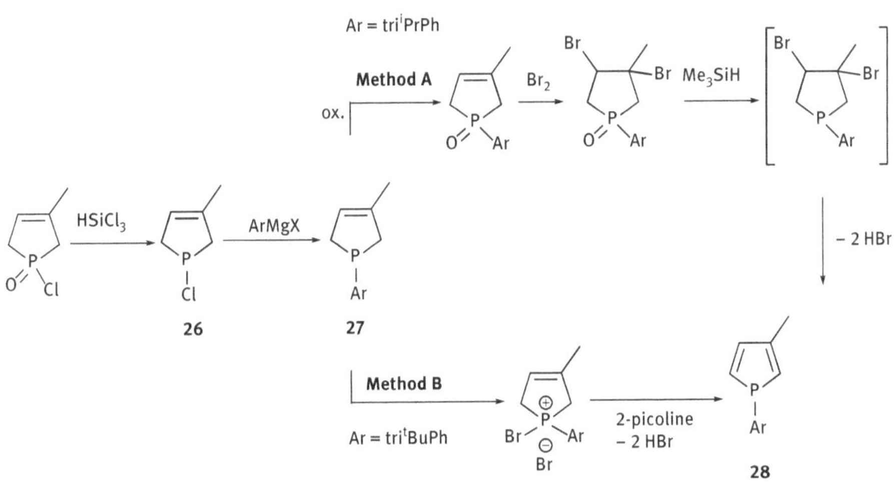

Figure 14.12: Syntheses of phospholes with sterically demanding substituent on the P atom.

Theoretical and experimental studies revealed that a bulky substituent on the P atom has a significant planarizing effect on the P-pyramid, resulting in an electron delocalization (Table 14.3). Hence, 2,4,6-trialkylphenylphospholes [40, 95–98] may possess a certain degree of aromaticity. The Bird index of aromaticity for 1-(2,4,6-triisopropylphenyl)phosphole (**28f**) was found to be 40.4 [103], while that of the 1-(2,4,6-tri-tert-butylphenyl derivative (**28f**) was as high as 56.5 [103]. The aromaticity values calculated for arylphospholes **28a–e** are summarized in Table 14.3. The percentage values are in good agreement with expectation and the previously reported data [29, 93, 94]. As can be seen, phenylphosphole (**28a**) is of a very weak aromatic character, but in accordance with earlier experiences, 2,6-ditert-butylphenyl phosphole **28e** exhibits a significant degree of aromaticity, and species **28e** may be considered as an aromatic molecule [96–108].

The sterically hindered phospholes, e.g. 1-(2,6-di-tertierbutylphenyl)phosphole (**28e**) and 1-(2,4,6-triisopropylphenyl)phosphole (**28c**), were involved in a sigmatropic Ar[1,5]-rearrangement at a higher temperature yielding reactive intermediates **26**/**27**. The Diels–Alder cycloaddition of 2*H*-phosphole **26** with diphenylacetylene afforded phosphanorbornadiene **29e** (Figure 14.13-(1)), while the "dimerization" of isomers **26** and **27** led to adduct **30c** (Figure 14.13-(2)) [97], which adopts an *endo* geometry with relatively long and weak P–P and C–C bonds at the junction of the dimer [101, 102].

In spite of the stability of phospholes (**V**), phosphole oxide derivatives (in general **VII**, specifically **31** and **33**) are unstable [41, 120–123], and undergo a regio- and stereospecific Diels–Alder dimerization to afford phosphole oxide dimers with a

Table 14.3: Dependence of the aromaticity (AR, %) of arylphospholes **28** on the out of plane angle (α, °°).

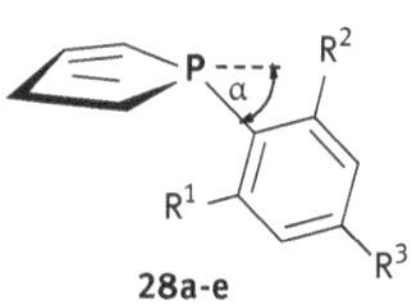

28a-e

Compound	R^1	R^2	R^3	α^a	AR%^{b}
28a	H	H	H	68.5°	8.0
28b	Me	Me	Me	65.2°	16.2
28c	*i*-Pr	*i*-Pr	*i*-Pr	62.6°^{c}	16.6^c
28d	*t*-Bu	Me	*t*-Bu	58.0°	24.1
28e	*t*-Bu	*t*-Bu	H	56.0°	26.2

[a]Taken from reference [91].
[b]Taken from reference [40].
[c]Computed at B3LYP/6–31(d,p) [56].

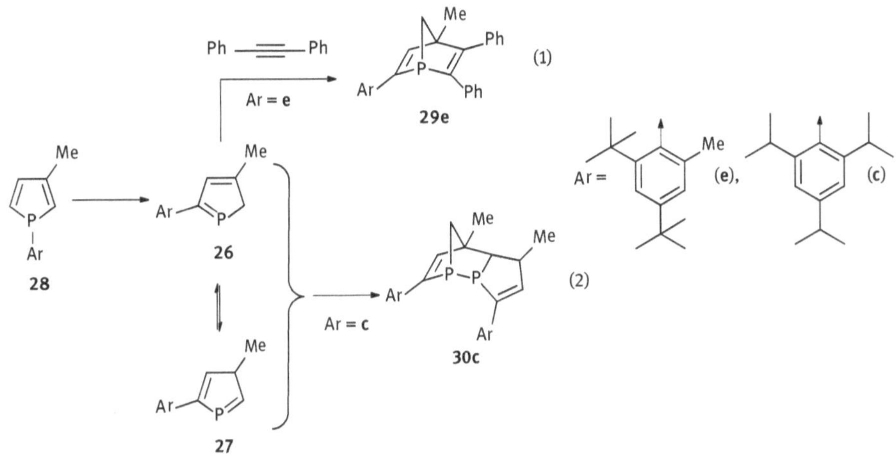

Figure 14.13: Sigmatropic rearrangement of sterically demanding phospholes **28** followed by Diels–Alder cycloaddition.

7-phosphanorbornene ring system (**32**) (Figure 14.14) [120–131], or are isomerized to the corresponding cyclohexene-2-phospholene oxides **34** [41, 107].

Theoretical investigations revealed that the ceasing of antiaromaticity may be the driving force for different transformations and reaction sequences starting from phosphole oxides. Depending on the substituents, the antiaromatic character of phosphole oxide **31** may vary between –11 and –20%. It means that the antiaromatic character of phosphole oxides (in general **VII**) may be fine-tuned by the

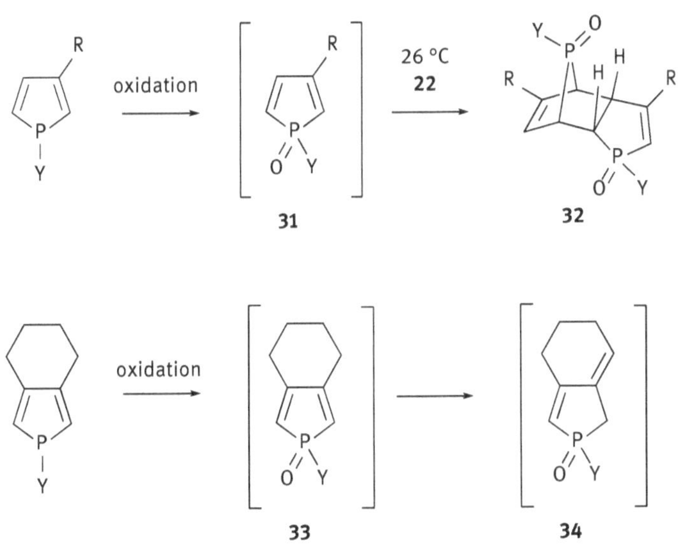

Figure 14.14: Two examples for the instability of phosphole oxides (**31** and **33**).

Figure 14.15: Possible orientations of the phenyl ring in phosphole oxides.

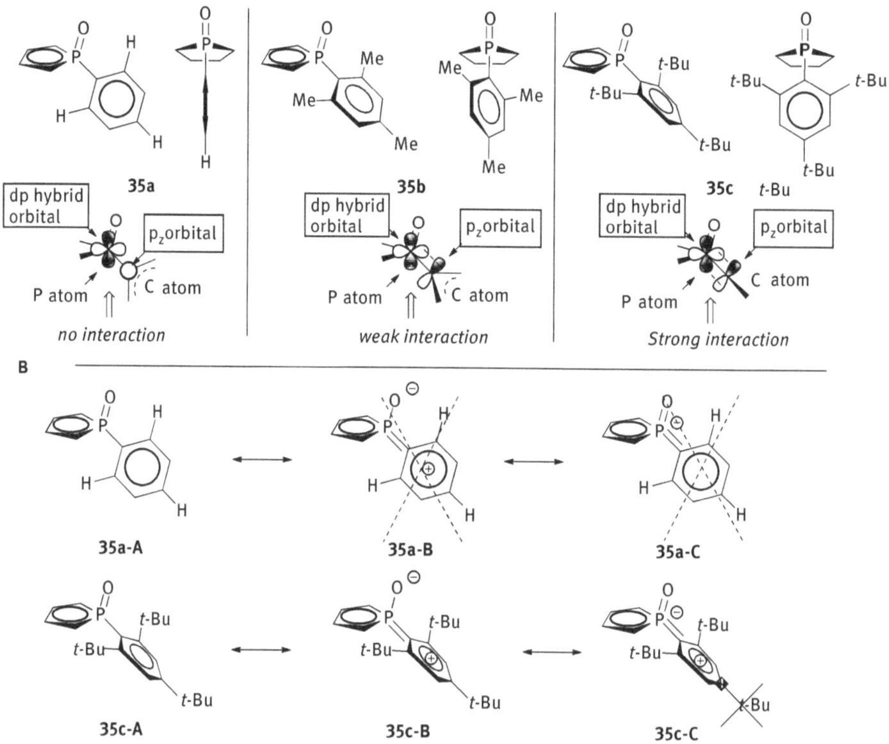

Figure 14.16: "A": Electronic effects depending on the different orientation of the aryl groups in **35a–c**. "B": resonance structures (**A**, **B** and **C**) of **35a** and **35c**.

P-substituents. The higher the antiaromatic character of the phosphole oxides is, the lower the activation enthalpy barrier is for the dimerization. Hence, the elimination of the antiaromaticity of the starting molecule may be the driving force for certain transformations. It was shown that the electronic properties of the phenyl group depend on the O=P–C_α–C_β torsion angle. When the O=P–C–C torsion angle is parallel with the P=O unit (as shown on Figure 14.15-(A)), the phenyl group acts as an electron-withdrawing group, while, when the torsion angle is orthogonal with the P=O unit (as shown on Figure 14.15-(B)), the phenyl group behaves as an electron-releasing substituent. In the latter case, a charge transfer may take place from the Ph ring towards the phosphole oxide ring.

The substituent of the P-aryl ring may influence the antiaromatic character of phosphole oxides (**35a–c**) [40]. In general, an electron-withdrawing substituent, e.g. the Ph group, as in phosphole **35a**, increases, while electron-donating groups, e.g. the 2,4,6-triMeC_6H_2 and 2,4,6-tri$^tBuC_6H_2$ substituents, as in phospholes **35b** and **35c**, respectively, decrease the unbeneficial property under discussion. Hence, the antiaromatic character of trialkylphenylphospholes **35b** and **35c** is lower than that in the phenyl case (**35a**). As Figure 14.16-(A) shows, these lower antiaromaticity values for phospholes **35b** and **35c** may also be attributed to the different position of the aryl group, as compared to species **35a**.

In the distorted arrangement of the aryl group (as in **35b** and **35c**), the p_z atomic orbital on the C1 atom of the phenyl group is able to overlap with the dp hybrid orbital of the P atom. The phenyl group in phosphole oxide **35a** behaves as an electron-withdrawing group due to the absence of the resonance structures **B** and **C** (Figure 14.16-(B)), while the bulky electron-donating aryl substituent (as in phosphole oxides **35b** and **35c**) may also be represented by resonance structures **B** and **C** (Figure 14.16-(B)). The aromatic stabilization energy for sterically congested phosphole oxide **35c** is 7 kJ mol^{-1} [40].

14.4 Summary

The concept of aromaticity had been known for long, but the opposite phenomenon of antiaromaticity was explored only much later, only 15 years ago, in the context of P-heterocycles. In this chapter, the four- and five-membered P-heterocycles were systematically surveyed from the point of view of aromaticity and antiaromaticity, as compared to other species including cyclobutadiene, the cyclopentadiene cation and anion based on experimental data and quantum chemical calculations. On the one hand, the antiaromatic character may serve as a driving force for chemical transformations, while on the other hand, the antiaromaticity of a few species may be influenced in different ways. The concept of antiaromaticity represents a new discipline in the study of P-heterocycles.

References

[1] R. Breslow, *Pure Appl. Chem.*, 1971, **28**, 111.
[2] R. Breslow, *Acc. Chem. Res.*, 1973, **6**, 393.
[3] R. Breslow, *Chem. Br.*, 1968, **4**, 100.
[4] G. Maier, *Angew. Chem. Int. Ed.*, 1974, **13**, 425.
[5] A. Fattahi, L. Lis, Z. Tian, S. R. Kass, *Angew. Chem. Int. Ed.*, 2006, **45**, 4984.
[6] A. Fattahi, L. Lis, S. R. Kass, *J. Am. Chem. Soc.*, 2005, **127**, 3089.

[7] V. I. Minkin, M. N. Glukhotsev, B. Y. Simkin, *Aromaticy and Antiaromaticy: Electronic and Structural Aspects*; Wiley: New York, 1994.
[8] G. Maier, *Angew. Chem. Int. Ed.*, 1988, **27**, 309.
[9] P. v. R. Schleyer, *Chem. Rev.*, 2001, **101**, 1115.
[10] A. R. Katritzky, K. Jug, D. C. Oniciu, *Chem. Rev.*, 2001, **101**, 1421.
[11] P. v. R. Schleyer, *Chem. Rev.*, 2004, **105**, 3433.
[12] A. T. Balaban, P. v R. Schleyer, H. S. Rzepa, *Chem. Rev.*, 2005, **105**, 3436.
[13] A. T. Balaban, D.C. Oniciu, A. R. Katritzky, *Chem. Rev.*, 2004, **104**, 2777.
[14] S. Shaik, P. C. Hilberty, *J. Am. Chem. Soc.*, 1985, **107**, 3089.
[15] P. C. Hilberty, D. Danovich, A. Shurki, S. Shaik, *J. Am. Chem. Soc.*, 1995, **117**, 7760.
[16] C. S. Wannere, C. Corminboeuf, Z-X. Wang, M. D. Wodrich, R. B. King, P. v. R. Schleyer, *J. Am. Chem. Soc.*, 2005, **127**, 5701.
[17] Z. Chen, R. B. King, *Chem. Rev.*, 2005, **105**, 3613.
[18] Z. Mucsi, B. Viskolcz, I. G. Csizmadia, *J. Phys. Chem. A*, 2007, **111**, 1123.
[19] M. K. Cyrański, P. v. R. Schleyer, T. M. Krygowski, H. J. Jiao, G. Hohlneicher, *Tetrahedron*, 2003, **59**, 1657.
[20] P. George, M. Trachtman, C. W. Bock, A. M. Brett, *J. Chem. Soc. Perkin 2*, 1976, 1222.
[21] P. v. R. Schleyer, H. Jiao, *Pure Appl. Chem.*, 1996, **68**, 209.
[22] K. B. Wiberg, *Chem. Rev.*, 2001, **101**, 1317.
[23] M. K. Cyrański, *Chem. Rev.*, 2005, **105**, 3773.
[24] Y. Mo, P. v. R. Schleyer, *Chem. Eur. J.*, 2006, **12**, 2009.
[25] Z. Chen, C. S. Wannere, C. Corminboeuf, R. Puchta, P. v. R. Schleyer, *Chem. Rev.*, 2005, **105**, 3842.
[26] P. v. R. Schleyer, F. Puhlhofer, *Org. Lett.*, 2002, **4**, 2873.
[27] J. Poater, M. Duran, M. Solà, B. Silvi, *Chem. Rev.*, 2005, **105**, 3911.
[28] T. M. Krygowski, B. T. Stêpiñ, *Chem. Rev.*, 2005, **105**, 3482.
[29] C. W. Bird, *Tetrahedron*, 1985, **41**, 1409.
[30] A. Dransfeld, L. Nyulászi, P. v. R. Schleyer, *Inorg. Chem.*, 1998, **37**, 4413.
[21] T. Heine, C. Corminboeuf, G. Seifer, *Chem. Rev.*, 2005, **105**, 3889.
[32] P. v. R. Schleyer, C. Maerker, A. Dransfeld, H. Jiao, N. J. R. v. E. Hommes, *J. Am. Chem. Soc.*, 1996, **118**, 6317.
[33] L. D. Quin, *Curr. Org. Chem.*, 2006, **10**, 43.
[34] L. Nyulászi, *Chem. Rev.*, 2001, **101**, 1229.
[35] M. Regitz, *in Multiple Bonds and Low Coordination in Phosphorous Chemistry.* Ed., M. Regitz, O. J. Scherer; Thieme Verlag: Stuttgart, 1990.
[36] K. B. Dillon, F. Mathey, J. F. Nixon, *in Phosphorous: The Carbon Copy;* John Wiley: Chichester, 1988.
[37] F. Mathey, *Angew. Chem. Int. Ed.*, 2003, **42**, 1578.
[38] O. Wagner, G. Maas, M. Regitz, *Angew. Chem. Int. Ed.*, 1987, **26**, 1257.
[39] L. Nyulászi, O. Hollóczki, C. Lescop, M. Hissler, R. Réau, *Org. Biomol. Chem.*, 2006, **4**, 996.
[40] Z. Mucsi, G. Keglevich, *Eur. J. Org. Chem.*, 2007, 4765.
[41] Z. Mucsi, I. G. Csizmadia, *Curr. Org. Chem.* 2008, **12**, 83.
[42] O. Wagner, M. Ehle, M. Regitz, *Angew. Chem. Int. Ed.*, 1989, **28**, 225.
[43] A. Marinetti, F. Mathey, J. Fisher, A. Mitschler, *J. Chem. Soc. Chem. Commun.*, 1984, 45.
[44] M. v. Gottfried, B. Willibald, *Angew. Chem.*, 1984, **96**, 896.
[45] J. Tejeda, R. Reau, F. Dahan, G. Bertrand, *J. Am. Chem. Soc.*, 1993, **115**, 7880.
[46] Z. Mucsi, *Unpublished Results*: Computed at B3LYP/6-31G(d,p) Level of Theory.
[47] A. D. Beke, *J. Chem. Phys.*, 1993, **98**, 5648.
[48] A. G. Baboul, L. A. Curtiss, P. C. Redfern, K. J. Raghavachari, *Chem. Phys.*, 1999, **110**, 7650.

[49] M. J. Frisch, G. W. Trucks, H. B. Schlegel, G. E. Scuseria, M. A. Robb, J. R. Cheeseman, V. G. Zakrzewski, J. A. Montgomery, Jr. R. E. Stratmann, J. C. Burant, S. Dapprich, J. M. Millam, A. D. Daniels, K. N. Kudin, M. C. Strain, O. Farkas, J. Tomasi, V. Barone, M. Cossi, R. Cammi, B. Mennucci, C. Pomelli, C. Adamo, S. Clifford, J. Ochterski, G. A. Petersson, P. Y. Ayala, Q. Cui, K. Morokuma, D. K. Malick, A. D. Rabuck, K. Raghavachari, J. B. Foresman, J. Cioslowski, J. V. Ortiz, A. G. Baboul, B. B. Stefanov, G. Liu, A. Liashenko, P. Piskorz, I. Komaromi, R. Gomperts, R. L. Martin, D. J. Fox, T. Keith, M. A. Al-Laham, C. Y. Peng, A. Nanayakkara, M. Challacombe, P. M. W. Gill, B. Johnson, W. Chen, M. W. Wong, J. L. Andres, C. Gonzalez, M. Head-Gordon, E. S. Replogle, J. A. Pople, *Gaussian 09 A1*, Gaussian, Inc., Pittsburgh PA, 2009.
[50] W. W. Schoeller, T. Busch, *Angew. Chem. Int. Ed.*, 1993, **32**, 617.
[51] Z-X. Wang, P. v. R. Schleyer, *Helv. Chim. Acta*, 2001, **84**, 1578.
[52] P. Le Floch, F. Mathey, *J. Chem. Soc. Chem. Commun.*, 1993, 1295.
[53] F. Mathey, P. Le Floch, *J. Org. Chem.*, 2004, **69**, 5100.
[54] L. Nyulászi, T. Veszprémi, J. Réffy, *J. Phys. Chem.*, 1993, **97**, 4011.
[55] F. Mathey, *Acc. Chem. Res.*, 1992, **25**, 90.
[56] L. Nyulászi, T. Veszprémi, J. Réffy, *J. Phys. Chem.*, 1995, **99**, 10142.
[57] L. Nyulászi, T. Veszprémi, *J. Phys. Chem.*, 1996, **100**, 6456.
[58] L. Nyulászi, T. Veszprémi, *Phosphorous, Sulfur, Silicon*, 1996, **109–110**, 1.
[59] P. Binger, R. Milczarek, R. Mynott, M. Regitz, *J. Organomet. Chem.*, 1987, **323**, C35.
[60] Z.-X. Wang, P. v. R. Schleyer, *Helv. Chim. Acta*, 2001, **84**, 1578.
[61] G. Bertrand, *Angew. Chem. Int. Ed.*, 1998, **37**, 270.
[62] A. A. Deniz, K. Peters, G. J. Snyder, *Science*, 1999, **286**, 1119.
[63] G. Alcaraz, A. Baceiredo, M. Nieger, W. W. Schoeller, G. Bertrand, *Inorg. Chem.*, 1996, **35**, 2458.
[64] U. Heim, H. Pritzkow, U. Fleischer, H. Gruetzmacher, M. Sanchez, R. Reau, G. Bertrand, *Chem. Eur. J.*, 1996, **2**, 68.
[65] M. Regitz, U. Bergstraesser, *Sci. Synth.*, 2002, **9**, 135.
[66] K. Bieger, R. Tejeda, R. Reau, F. Dahan, G. Bertrand, *J. Am. Chem. Soc.*, 1994, **116**, 8087.
[67] C. Trujillo, G. Sanchez-Sanz, I. Alkorta, J. Elguero, *Comput. Theor. Chem.*, 2012, **994**, 81.
[68] Z. Mucsi, T. Körtvélyesi, B. Viskolcz, I. G. Csizmadia, T. Novák, G. Keglevich, *Eur. J. Org. Chem.*, 2007, **11**, 1759.
[69] Z. Mucsi, T. Körtvélyesi, Tibor Novák, G. Keglevich, *Phosphorus, Sulfur, Silicon*, 2008, **183**, 726.
[70] Z. Mucsi, I. Hermecz, B. Viskolcz, I. G. Csizmadia, G. Keglevich, *Tetrahedron*, 2008, **64**, 1868.
[71] J. Altman, K. Yates, I. G. Csizmadia, *J. Am. Chem. Soc.*, 1976, **98**, 1450.
[72] D. J. H. Smith, *Comprehensive Org. Chem.*, Vol. 2, Ed., D. Barton; Pergamon: Oxford, 1979.
[73] G. Keglevich, H. Forintos, T. Körtvélyesi, *Curr. Org. Chem.*, 2004, **8**, 1245.
[74] G. Keglevich, T. Körtvélyesi, H. Forintos, S. Lovas, *J. Chem. Soc. Perkin Trans 2*, 2002, 1645.
[75] G. Keglevich, H. Forintos, T. Körtvélyesi, L. Tőke, *J. Chem. Soc. Perkin Trans. 1*, 2001, 1.
[76] G. Keglevich, T. Körtvélyesi, H. Forintos, Á. G. Vaskó, I. Vladiszlav, L. Tőke, *Tetrahedron*, 2002, **58**, 3721.
[77] G. Keglevich, H. Forintos, H. Szelke, A. Tamás, Á. G. Vaskó, J. Kovács, T. Körtvélyesi, L. Kollár, L. Tőke, *Phoshorus, Sulfur, Silicon*, 2002, **177**, 1681.
[78] G. Keglevich, T. Körtvélyesi, A. Ujvári, E. Dudás, *J. Organomet. Chem.* 2005, **690**, 2497.
[79] T. Kawashima, T. Iijima, A. Kikuchi, R. Okazaki, *Phoshorus, Sulfur, Silicon*, 1999, **144–146**, 149.
[80] N. Kano, J. H. Xing, A. Kikuchi, T. Kawashima, *Heteroat. Chem.*, 2001, **12**, 282.
[81] N. Kano, A. Kikuchi, T. Kawashima, *Chem. Commun.*, 2001, 2096.
[82] N. Kano, J. H. Xing, A. Kikuchi, S. Kawa, T. Kawashima, *Phoshorus, Sulfur, Silicon*, 2002, **177**, 1685.

[83] L. D. Quin, in *Comprehensive Heterocyclic Chemistry II*, Eds., A. R. Katritzky, C. W. Rees, E. F. V. Scriven; vol. Ed., C. W. Bird, vol. 2; Pergamon: Oxford, 1996.
[84] L. D. Quin, in *Phosphorus-Carbon Heterocyclic Chemistry: The Rise of a New Domain* (Ed.: F. Mathey), Pergamon: Amsterdam, 2001.
[85] F. Mathey, *Chem. Rev.*, 1988, **88**, 429.
[86] G. Keglevich, in *Targets in Heterocyclic Systems*, Vol. 6, Eds., O. Attanasi, D. Spinelli; Italian Society of Chemistry: Roma, 2002.
[87] N. Mézailles, P. le Floch, *Curr. Org. Chem.*, 2006, **10**, 43.
[88] F. C. Leavitt, T. A. Manuel, F. Johnson, *J. Am. Chem. Soc.*, 1959, **81**, 3163.
[89] L. D. Quin, J. G. Bryson, *J. Am. Chem. Soc.*, 1967, **89**, 5984.
[90] H. Kaye, *J. Am. Chem. Soc.*, 1970, **92**, 5779.
[91] W. Egan, R. Tang, G. Zon, K. Mislow, *J. Am. Chem. Soc.*, 1971, **93**, 6205.
[92] I. G. Csizmadia, A. H. Cowley, M. W. Taylor, S. Wolfe, *J. Chem. Soc. Chem. Commun.*, 1974, **11**, 432.
[93] M. H. Palmer, R. H. Findlay, *J. Chem. Soc. Perkin Trans. 2*, 1975, 974.
[94] M. H. Palmer, R. H. Findlay, J. A. Gaskell, *J. Chem. Soc. Perkin Trans. 2*, 1974, 420.
[95] F. Mathey, J. Fischer, J. H. Nelson, *Complexing Modes of the Phosphole Moiety, in Structure and Bonding*, Spfinger-Verlag: Berlin Heidelberg, 1983.
[96] L. D. Quin, A. S. Ionkin, R. Kalgutkar, G. Keglevich, *Phosphorus, Sulfur, Silicon*, 1996, **109–110**, 433.
[97] L. D. Quin, G. Keglevich, A. S. Ionkin, R. Kalgutkar, G. Szalontai, *J. Org. Chem.*, 1996, **61**, 7801.
[98] L. Nyulászi, G. Keglevich, L. D. Quin, *J. Org. Chem.*, 1996, **61**, 7808.
[99] G. Keglevich, L. D. Quin, Z. Böcskei, G. M. Keserű, R. Kalgutkar, P. M. Lahti, *J. Organomet. Chem.*, 1997, **532**, 109.
[100] G. Keglevich, Z. Böcskei, G. M. Keserű, K. Újszászy, L. D. Quin, *J. Am. Chem. Soc.*, 1997, **119**, 5095.
[101] G. Keglevich, R. Farkas, T. Imre, K. Ludányi, Á. Szöllősy, L. Tőke, *Heteroat. Chem.* 2003, **14**, 316.
[102] G. Keglevich, R. Farkas, K. Ludányi, V. Kudar, M. Hanusz, K. Simon, *Heteroatom Chem.* 2005, **16**, 104.
[103] G. Keglevich, T. Chuluunbaatar, B. Dajka, A. Dobó, Á. Szöllősy, L. Tőke, *J. Chem. Soc., Perkin Trans.*, 2000, **1**, 2895.
[104] G. Keglevich, L. Nyulászi, T. Chuluunbaatar, B.-A. Namkhainyambuu, K. Ludányi, T. Imre, L. Tőke, *Tetrahedron*, 2002, **58**, 9801.
[105] G. Keglevich, T. Chuluunbaatar, B. Dajka, B.-A. Namkhainyambuu, M. Fekete, L. Kollár. L. Tőke, *Phosphorus, Sulfur, Silicon*, 2002, **177**, 1991.
[106] G. Keglevich, 1-(2,4,6-Trialkylphenyl)-1H-phospholes with a Flattened P-pyramid: Synthesis and Reactivity. In: *Topics in Heterocyclic Chemistry, Vol. 21, Phosphorus Heterocycles II*, Ed., R. R. Gupta, Vol. Ed., R. K. Bansal; Springer: Heidelberg, 2010.
[107] G. M. Keserű, G. Keglevich, *J. Organomet. Chem.*, 1999, **586**, 166.
[108] G. Keglevich, T. Chuluunbaatar, B. Dajka, B.-A. Namkhainyambuu, K. Ludányi, L. Tőke, *Heteroat. Chem.*, 2001, **12**, 633.
[109] L. Nyulászi, *Chem. Rev.*, 2001, **101**, 1229.
[110] W. Schafer, A. Schweig, F. Mathey, *J. Am. Chem. Soc.*, 1976, **98**, 407.
[111] D. Delaere, A. Dransfeld, M. T. Nguyen, L. G. Vanquickenborne, *J. Org. Chem.*, 2000, **65**, 2631.
[112] E. Mattmann, F. Mathey, A. Sevin, G. Frison, *J. Org. Chem.*, 2002, **67**, 1208.
[113] L. Nyulászi, *J. Phys. Chem.*, 1995, **99**, 586.
[114] L. Nyulászi, *Tetrahedron*, 2000, **56**, 79.
[115] F. G. N. Cloke, P. B. Hitchcock, P. Hunnable, J. F. Nixon, L. Nyulászi, E. Niecke, V. Thelen, *Angew. Chem. Int. Ed.*, 1998, **37**, 1083.
[116] A. Breque, F. Mathey, P. Savignac, *Synthesis*, 1981, 983.
[117] F. Mathey, R. Mankowski-Favelier, *Org. Magn. Reson.*, 1972, **4**, 171.

[118] D. Klintuch, K. Krekic, C. Bruhn, Z. Benkő, R. Pietschnig, *Eur. J. Inorg. Chem.*, 2016, **5**, 718.
[119] F. Mathey, *Sci. Synth.*, 2002, **9**, 553.
[120] L. D. Quin, *Rev. Heteroat. Chem.* 1990, **3**, 39.
[121] L. D. Quin, *The Heterocyclic Chemistry of Phosphorus;* Wiley Interscience: New York, 1981.
[122] G. Keglevich, L. Tőke, Z. Böcskei, V. Harmat, *Heteroat. Chem.*, 1997, **8**, 527.
[123] L. D. Quin, *Heteroat. Chem.*, 1991, **2**, 359.
[124] F. Mathey, *Top. Phosphorus Chem.*, 1980, **10**, 43.
[125] D. A. Usher, F. H. Westheimer, *J. Am. Chem. Soc.*, 1964, **86**, 4732.
[126] R. Kluger, F. Kerst, D. G. Lee, E. A. Dennis, F. H. Westheimer, *J. Am. Chem. Soc.*, 1967, **89**, 3919.
[127] G. Keglevich, L. Tőke, K. Újszászi, K. Ludányi, *Heteroat. Chem.*, 1996, **7**, 337.
[128] G. Keglevich, K. Ludányi, L. D. Quin, *Heteroat. Chem.*, 1997, **8**, 135.
[129] L. D. Quin, K. C. Caster, J. C. Kisalus, K. Mesch, *J. Am. Chem. Soc.*, 1984, **106**, 7021.
[130] Y.-Y. H. Chiu, W. N. Lipscomb, *J. Am. Chem. Soc.*, 1969, **91**, 4150.
[131] L. D. Quin, J. Szewczyk, A. T. McPhail, *J. Org. Chem.*, 1986, **51**, 3341.

Hajnalka Szabó-Szentjóbi, Tamás Szabó, Tünde Tóth
and Péter Huszthy

15 Crown ethers containing phosphorus in the macroring

Abstract: Although the first phosphorus-containing macrocycles [$(PNCl_2)_n$, n=5, 6, 7] were prepared by Stokes in as early as 1897 [1], intensive studies in this field appeared only in the 1970s, a few years after Pedersen discovered the crown ethers and their unique complex-forming properties in 1967 [2]. In this chapter, we wish to summarize the developments made in this field, and report on the crown ethers containing phosphorus in the macroring and their various applications since Pedersen's discovery.

There are two types of phosphorous-containing crown ethers, one in which a side chain contains it (so called lariat ethers), and the other, in which it is part of the macroring. We focus only on the latter, which is classified according to the type of the phosphorus compound(s) it contains.

Keywords: crown ethers, phosphates, phosphines, phosphine oxides, phosphinic acids.

15.1 Hydrogen Phosphates, Phosphates, Thiophosphates, and Iminophosphates

A new series of macrocyclic ligands **1a–f** (Figure 15.1), containing a proton-ionizable dialkyl hydrogen phosphate moiety with relatively low pK_a values has been reported by Bradshaw et al. These compounds were prepared by the hydrolysis of the corresponding chlorides. These ligands – containing a proton-ionizable unit – held a lot of promise in the field of anion-free cation transport in liquid membrane systems. Compound **1d** showed a good transport ability in the cases of Pb^{2+}, Zn^{2+}, Ag^{+}, and alkali metal ions [3–5].

Extending this work, seven new enantiopure crown ethers containing a dialkyl hydrogen phosphate moiety (*S*,*S*)-**2a** – (*S*,*S*)-**2g** (Figure 15.2) were prepared by Kovács et al. The last steps of the syntheses were performed using optically active dialkyl-substituted oligoethylene glycols and phosphorus oxychloride followed by mild hydrolysis of the resulting macrocyclic chlorophosphates [6].

Pinchuk and Shtepanek et al. published the synthesis, characterization, and the complexation properties of the crown ethers **3a–c** (Figure 15.3) containing a

https://doi.org/10.1515/9783110535839-015

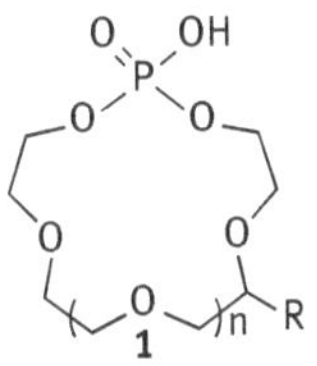

1

	R	n
1a	H	1
1b	n-octyl	1
1c	H	2
1d	n-octyl	2
1e	n-decyl	2
1f	n-hexadecyl	2

Figure 15.1: Racemic crown ethers containing a dialkyl hydrogen phosphate unit.

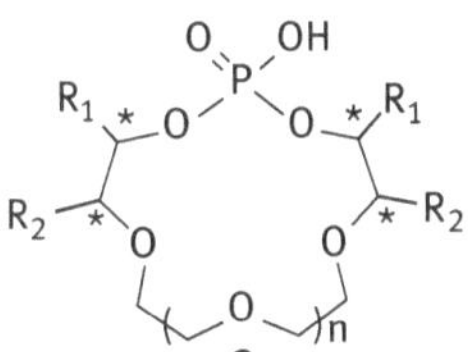

2

	R^1	R^2	n
(*S*,*S*)-**2a**	H	*i*-Bu	1
(*S*,*S*)-**2b**	H	Me	2
(*S*,*S*)-**2c**	Me	H	2
(*S*,*S*)-**2d**	H	*i*-Bu	2
(*S*,*S*)-**2e**	*i*-Bu	H	2
(*S*,*S*)-**2f**	H	n-octyl	2
(*S*,*S*)-**2g**	n-octyl	H	2

Figure 15.2: Chiral crown ethers containing a dialkyl hydrogen phosphate unit.

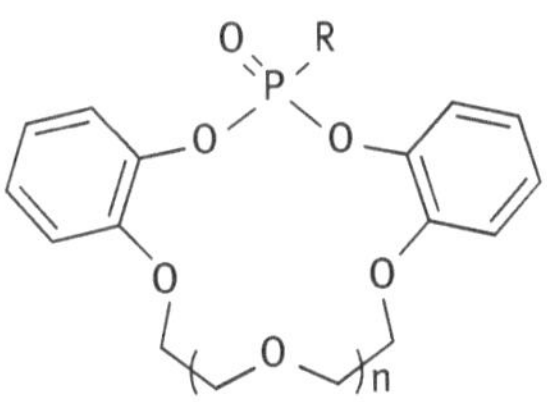

3

	R	n
3a	OPh	1
3b	$N{=}PPh_3$	1
3c	OPh	2

Figure 15.3: Dibenzo-crown ethers containing a phosphate or an iminophosphate unit.

phosphate or an iminophosphate unit. These macrocycles were prepared by treatment of disodium salts of open-chain dibenzo-polyethers with $RP(O)Cl_2$ [7–12].

The synthesis of monobenzo-11-crown-4 ethers **4a–d** (Figure 15.4) containing a phosphorus (V) atom has been described by Podrognyi et al. The researchers also prepared crown ethers containing a phosphonit, as well as a phosphonate, and a thiophosphate unit [13, 14].

Yurchenko et al. described the synthesis of dibenzo-22-crown-8 ether derivatives **5a–c** with two phosphorus atoms (Figure 15.5). Macrocycles **5a** and **5b** containing a phenyl or an adamantyl (Ad) group were prepared by Todd-Atherton reaction and **5c** by oxidation of the corresponding phosphonate (R=H) [15].

The synthesis and characterization of bis(crown ether) **6** containing a diazadiphosphetidine ring (Figure 15.6) were carried out by Dutasta et al., who isolated and characterized both the *cis* and *trans* isomers of the bis-macrocycle **6**. The macrocy-

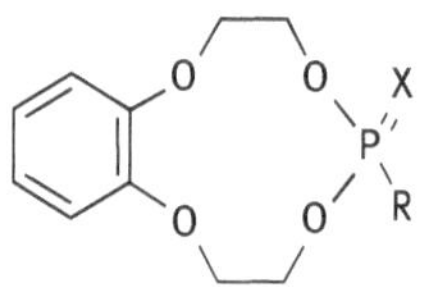

4

	R	X
4a	OEt	O
4b	OPh	O
4c	OEt	S
4d	OPh	S

Figure 15.4: Monobenzo-crown ethers containing a phosphate or a thiophosphate unit.

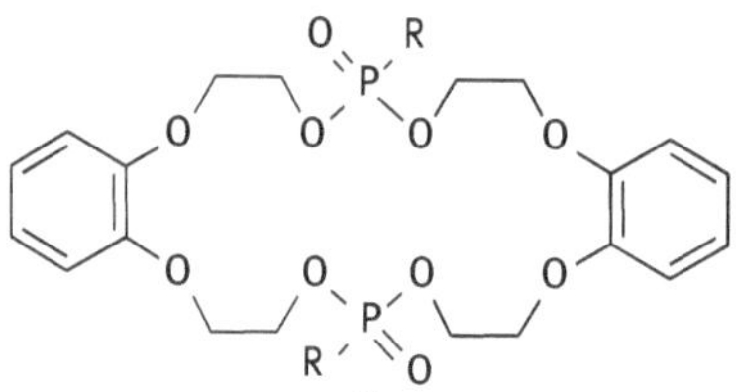

5

	R
5a	NHPh
5b	NHAd
5c	OH

Figure 15.5: Dibenzo-crown ethers containing two phosphorus atoms.

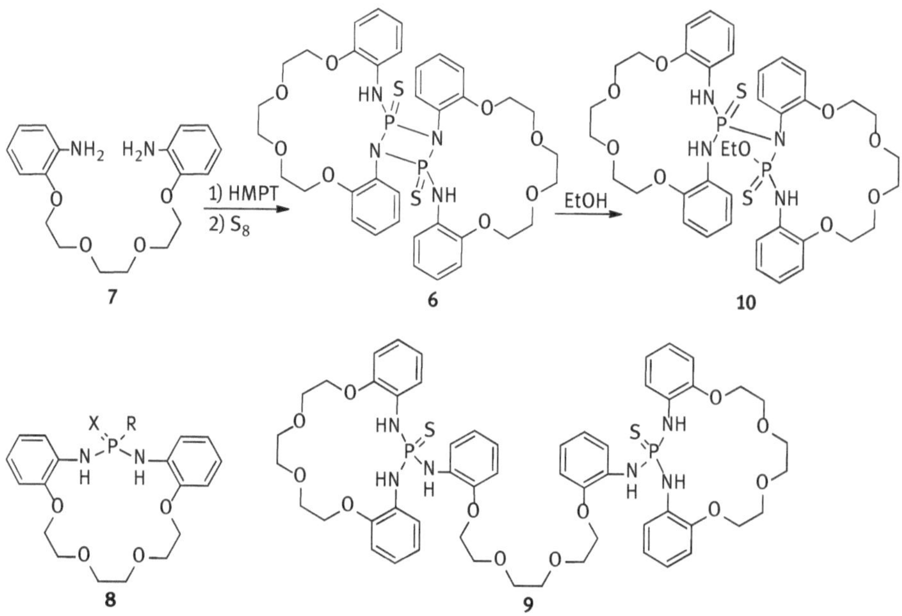

	R	X
8a	NMe_2	S
8b	H	O

Figure 15.6: Synthesis and transformation of crown ether containing a diazadiphosphetidine ring.

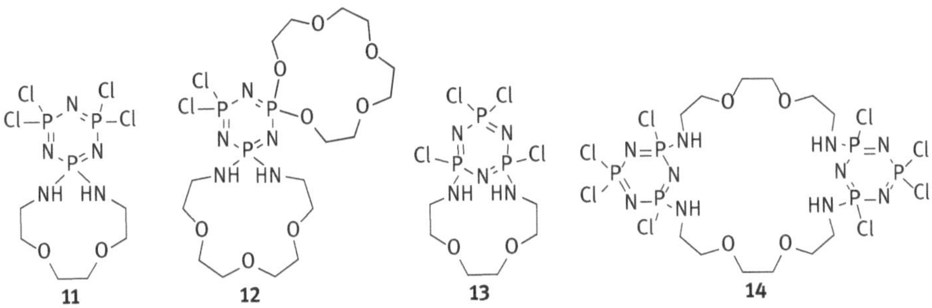

Figure 15.7: Crown ethers containing one or two cyclophosphazene units.

clization started from the diamine **7**, which was treated first with hexamethylphosphorous triamide), and then the resulting bis-macrocycle was subjected to *in situ* sulfurization to obtain **6**. In addition to **6**, two other macrocyclic compounds **8a** and **9** were also isolated and characterized. The researchers concluded that **8a** was directly obtained from the sulfurization of its parent tervalent derivative and **9** might be formed by the condensation reaction of **7** and **8a**. The same group found, that in solution the thiophosphonic acid triamide **8a** was very sensitive to moisture, it hydrolyzed after only a few minutes exposing under open air to macrocyclic phosphonic acid diamide **8b** [16]. A few years later they reported the synthesis of chiral bis-macrocycle **10** (Figure 15.6) having two asymmetric phosphorus centers. The diazadiphosphetidine derivative **6** was reacted with ethanol to form macrocycle **10**, which was isolated both in its chiral and racemic crystalline forms and structurally characterized [17]. The large crown ethers containing two (40- and 52-membered rings) and three (78-membered ring) phosphoramide units were also prepared. The crystal structure analysis showed that voids are minimized in the solid so that the macrocyclic cavity is filled up with part of the molecule itself or with a guest molecule, when the size of the macrocycle does not allow molecular folding for self-filling the cavity [18].

Numerous macrocycles containing cyclophosphazene units were synthesized since 1982, when natural polyamine-linked cyclophosphazenes were developed by Labarre and other researchers [19]. Spiro [20–25] (e.g. **11**), dispiro [20, 26, 27] (e.g. **12**), ansa [20, 28–32] (e.g. **13**), dibino [20, 31–33] (e.g. **14**) (Figure 15.7), and their substituted derivatives were also prepared.

15.2 Phosphonic and Phosphinic Acids and Their Derivatives

The synthesis of numerous monobenzo-11-crown-4 ethers containing a phosphonate, a thiophosphonate, or an iminophosphonate unit **15a–n** (Figure 15.8) has been performed by Podrognyi et al. The macrocycles **15k** and **15l** were prepared from the corresponding macrocyclic phosphonites [13, 14].

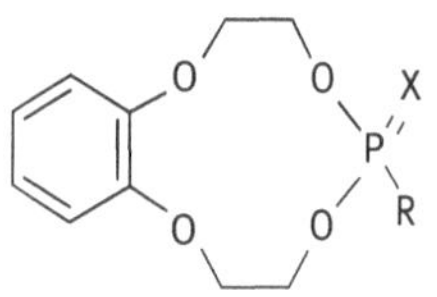

15

	R	X		R	X
15a/b	Me	O/S	**15i/j**	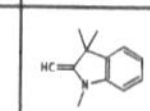	O/S
15c/d	Ph	O/S	**15k/l**	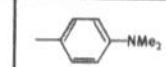	O/S
15e/f	$(CH_2)_2Cl$	O/S	**15m**	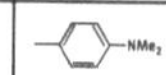	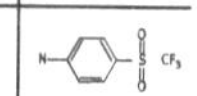
15g/h	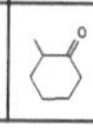	O/S	**15n**	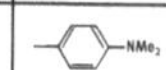	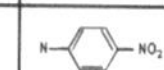

Figure 15.8: Monobenzo-crown ether containing a phosphonate, a thiophosphonate, or an imino-phosphonate unit.

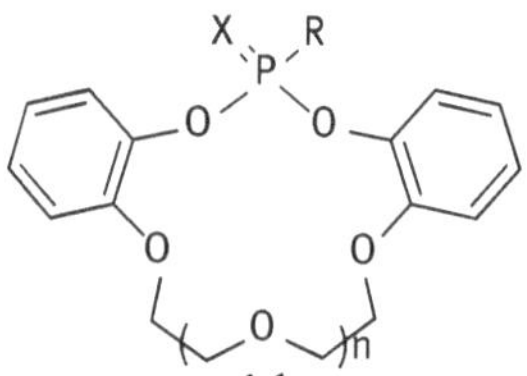

16

	R	X	n		R	X	n
16a	H	O	1	**16g**	Ph	NSO_2Ph	1
16b	Me	O	1	**16h**	Me	O	2
16c	Me	S	1	**16i**	Et	O	2
16d	Ph	O	1	**16j**	Ph	O	2
16e	NMe_2	O	1	**16k**	Me	S	2
16f	Ad	O	1	**16m**	Ph	S	2

Figure 15.9: Dibenzo-crown ethers containing a phosphonate, a thiophosphonate, or an iminophosphonate unit.

The synthesis, characterization, and the complexation studies of numerous dibenzo-14-crown-5 ethers, and dibenzo-17-crown-6 ethers **16a–m** containing a phosphonate, a thiophosphonate, or an iminophosphonate unit (Figure 15.9) have been reported between 1968 and 1992 [5, 7–12, 34–44]. Yurchenko et al. described the synthesis of **16a**, from which many other derivatives can be prepared by Todd-Atherton reaction, e.g. **16d–f** [40]. Yatsimirskii et al. found that the phosphoryl unit in the crown ether significantly influences the complexation and transporting properties of **16b** [45, 46], which forms as stable complexes with Li^+ and Mg^{2+}, as dibenzo-18-crown-6 ether forms with K^+ and Ca^{2+} [41]. Chaikovskaya et al. reported that **16c** generally forms less stable complexes with alkali and alkali earth metal cations than **16b** [9]. The complexation and transporting properties of **16d,f** were studied in the cases of Li^+, Na^+, K^+, Ca^{2+} and Mg^{2+} [36, 47].

Pinchuk et al. reported the synthesis and the crystal structure of crown ether **17** (Figure 15.10), which was found to be structurally similar to macrocycle **16b** with a cavity size of 14-crown-5 ether, although the former is a 20-crown-7 ether [48, 49]. Yatsimirskii et al. reported the bisphosphonate-crown ether **18a** [42] and Rybakova et al. the bisthiophosphonate-crown ether **18b**, which extract Ag^+, Hg^{2+}, and Pd^{2+} into $CHCl_3$ from their aqueous solutions [5, 46].

The syntheses of 17- to 21-membered ring size crown ethers were described by Dutasta et al. The diaminophosphine sulfide derivatives (**19a–e, 20a–d**) (Figure 15.11) were prepared from the corresponding three-coordinated diaminophosphine derivatives by the reaction with sulfur. The diaminophosphine oxide **19c** was obtained either from the three-coordinated parent phosphine by reaction with N_2O_4 or from **19a** by treatment with *m*-chloroperbenzoic acid [50, 51].

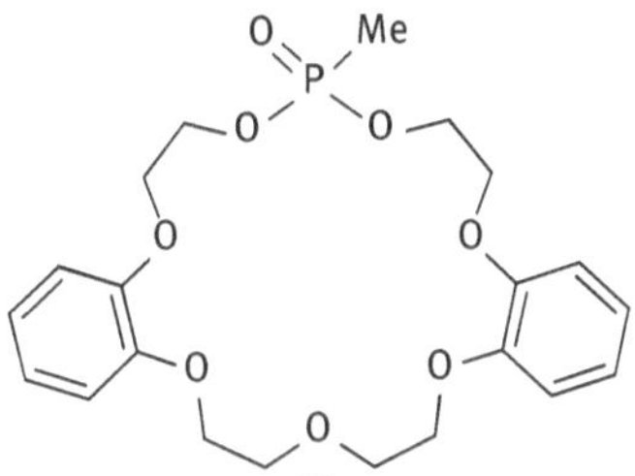

17

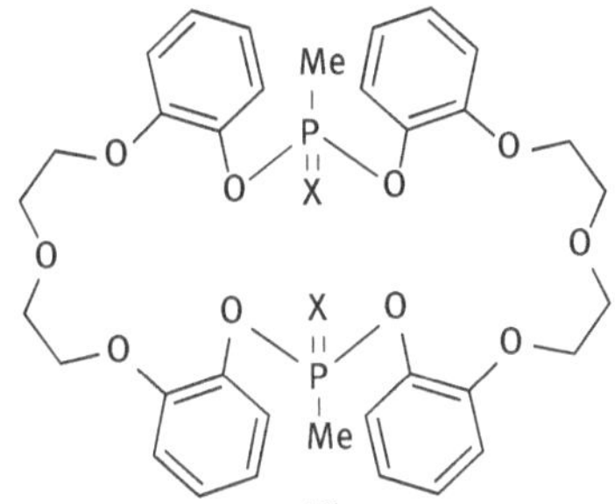

18

	X
18a	O
18b	S

Figure 15.10: Benzo-crown ethers containing phosphonate or thiophosphonate units.

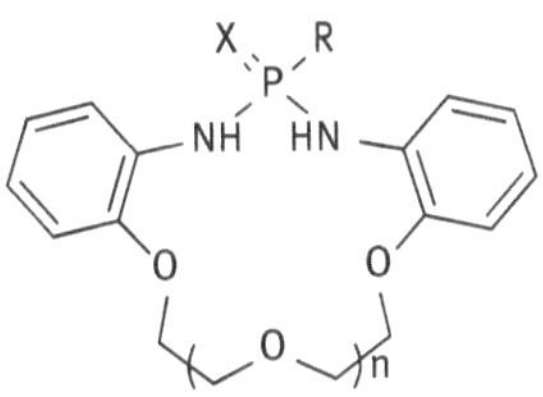

19

	X	n	R
19a	S	1	Ph
19b	S	2	Ph
19c	O	1	Ph
19d	S	1	Me
19e	S	2	Me

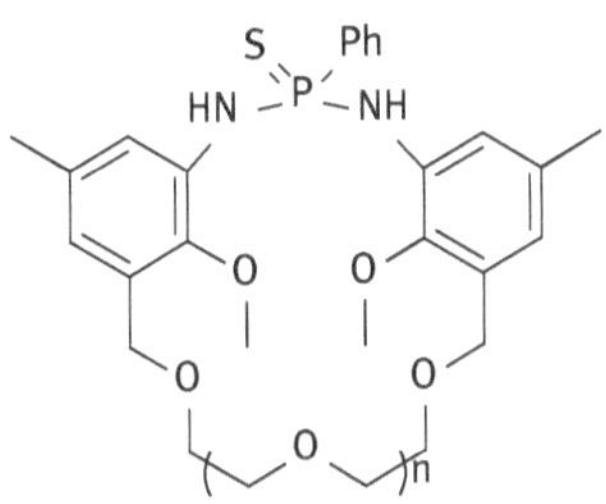

20

	n	R
20a	1	Ph
20b	2	Ph
20c	1	Me
20d	2	Me

Figure 15.11: Crown ethers containing a diaminophosphine sulfide or oxide unit.

The same research group found that macrocycle **21** containing diaminophosphine oxide unit can be made more rigid by substitution at the nitrogen atoms with an ethylene bridge to form compound **22** (Figure 15.12) [52].

Talanova et al. published the synthesis of phosphorus-sulfur-containing crown ethers **23a,b** (Figure 15.13) and their complexes with silver ions. Treatment of these macrocycles with Ag^+ picrate gave complexes in which the encapsulated Ag^+ coordinated mostly with the sulfide S atom [53].

Martin et al. described the synthesis of 16-membered macrocycles **24a–d** (Figure 15.14) containing dithio- or trithiophosphonate units. These macrocycles were obtained by the reaction of the corresponding dithiophosphonites with hydrogen peroxide (**24a,b**), or sulfur (**24c,d**). Macrocycles **24a** and **24b** could complex uranyl nitrate and $SnCl_4$ [54]. The structure of macrocycle *cis* **24d** in solid state was determined by X-ray diffraction [55].

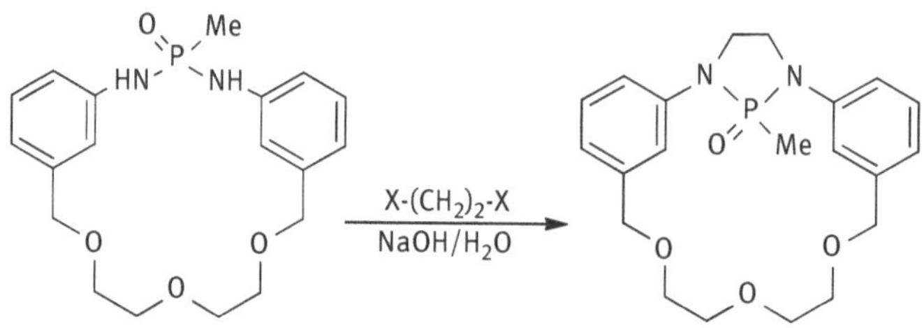

Figure 15.12: Creating more rigid conformation for a crown ether containing a diaminophosphine oxide unit.

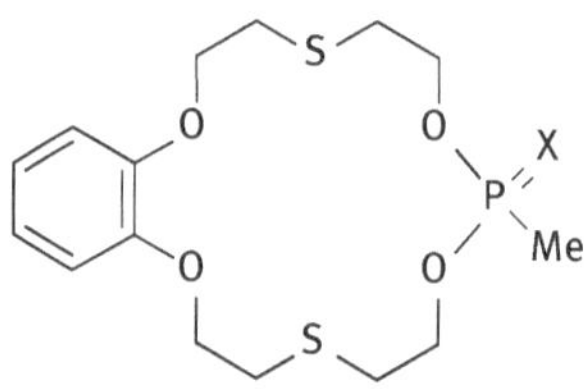

	X
23a	O
23b	S

Figure 15.13: Monobenzo-crown ethers containing a phosphonate or a thiophosphonate unit.

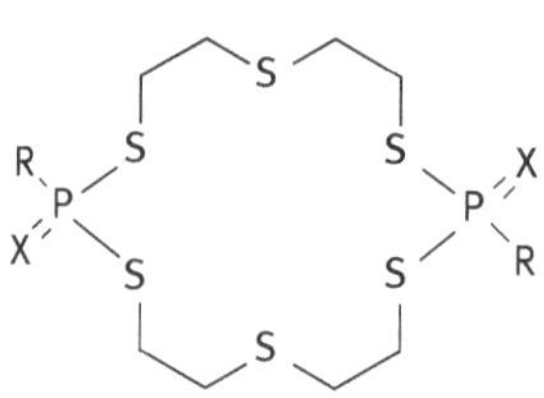

	R	X
24a	Me	O
24b	*t*-Bu	O
24c	Me	S
24d	*t*-Bu	S

Figure 15.14: Macrocycles containing thioether and dithio- or trithiophosphonate units.

Yurchenko et al. described the synthesis of bis(hydrophosphoryl)dibenzo-22-crown-8 ether **25a** (Figure 15.15) [15]. They prepared **25b** and **25c** (Figure 15.15) by Abramov reaction starting from **25a**. Macrocycle **25d** (Figure 15.15) was prepared directly by macrocyclization, and it was the first phosphorus-containing crown ether, which was soluble both in organic solvents and H_2O. Pinchuk et al. also studied the complexation properties of ligand **25d**. The same group found that the stability of alkali metal complexes of **25d** decreased in the order of $Na^+ > Li^+ \gg K^+$ [56].

The ligand **26** containing two macrocyclic and two phosphonate units (Figure 15.16) was prepared by cyclizing the disodium salts of open-chain dibenzo-polyethers with adamantanediphosphonic dichloride [57, 58]. The large macrocycles **27a,b** containing two phosphonate units (Figure 15.16) was obtained in the reaction of a diphenoxy sodium salt with adamantanediphosphonic dichloride [57].

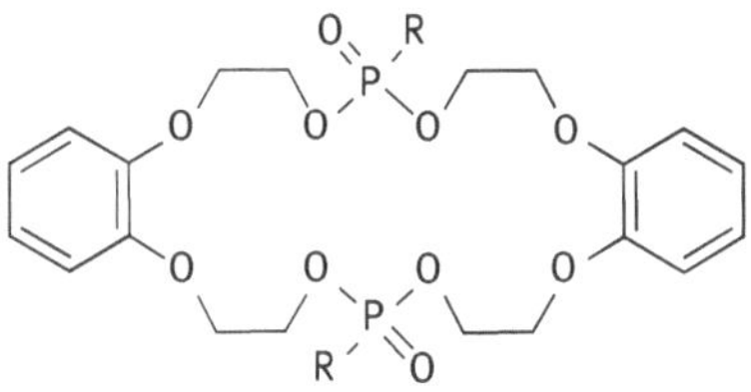

25

	R
25a	H
25b	CH_2OH
25c	$CH(OH)CCl_3$
25d	Me

Figure 15.15: Dibenzo-crown ethers containing two phosphonate units.

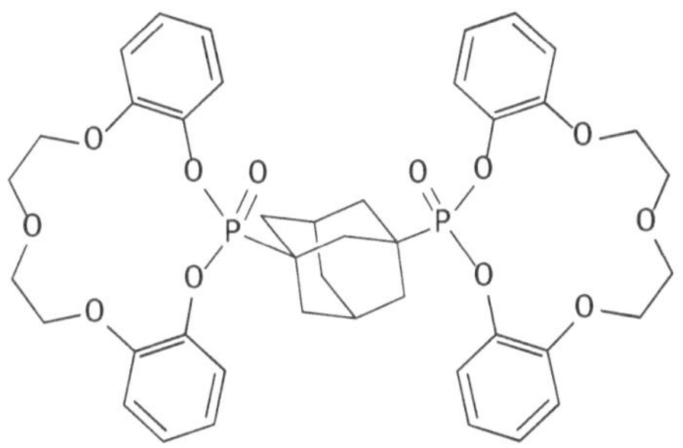

26

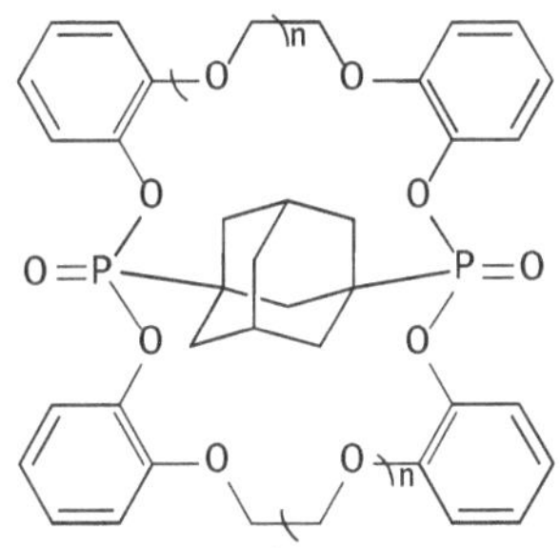

27

	n
27a	1
27b	3

Figure 15.16: Crown ethers containing two phosphonate moiety and an adamantly unit.

Dutasta et al. prepared a few macrocycles containing thiophosphonate units by a ring-opening polymerization method (Figure 15.17). The reaction of open-chain intermediates (e.g. **28**, **29**) with methyl dichlorophosphine and elemental sulfur gave stable thiophosphonates (e.g. **30**, **31**). Considering that the spontaneous ring-opening process of phosphonites was generally very slow, and the yield of the isolated macrocycles was rather low, the method needed to be improved. This was achieved by developing a stepwise controlled synthesis, starting from the ring-opening reaction of phosphonite **32** with diethylene glycol (Figure 15.17) [59, 60]. Treatment of phosphadiols **28** and **29** with methyl dichlorophosphine and sulfur yielded 16- and 24-membered macrocycles **30** and **31**. Both compounds existed in two diastereomeric forms: *cis* and *trans* for **30,** *cis,cis* and *cis,trans* for **31** [61].

Szabó and Székely et al. synthesized numerous enantiopure crown ethers containing a diarylphosphinic acid unit (Figure 15.18), by the hydrolysis of the corresponding esters [62–66]. They proposed that the two aromatic rings made the pseudo-18-crown-6 ether or the 21-crown-7 ether framework more rigid, resulting in higher selectivity for the molecular recognition. The complexation properties were examined by ECD spectroscopy in cases of compounds (*R,R*)-**33a** – (*S,S*)-**33e** [62, 67] (Figure 15.18). The enantioselective transporting ability with chiral amines in a bulk-liquid membrane cell was investigated in the cases of crown ethers (*S,S*)-**34a** – (*S,S*)-**34e** [64, 65] (Figure 15.18). In some cases both enantiomers of these macrocycles were synthesized. Szabó et al. also synthesized differently substituted achiral derivatives **35a–g** (Figure 15.18) and determined their pK_a values, which are very important factors for the transporting abilities [66].

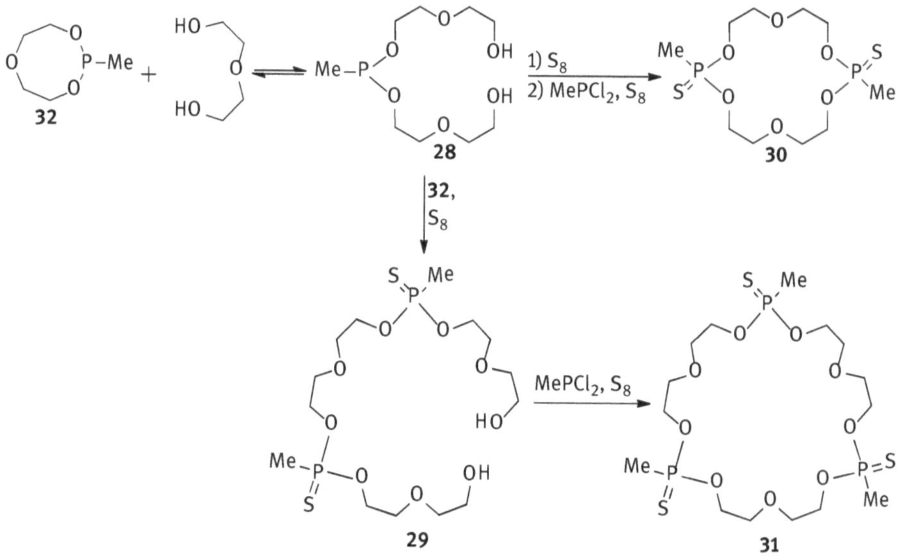

Figure 15.17: Ring-opening polymerization producing macrocycles with thiophosphonate units.

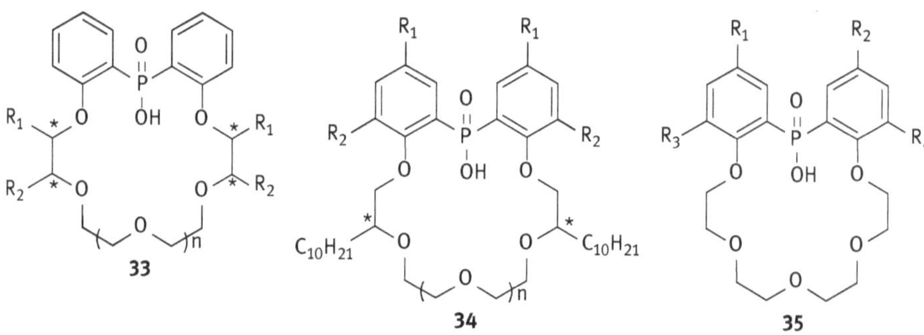

	R^1	R^2
(*R*,*R*)-**33a**	Me	H
(*S*,*S*)-**33b**	H	Me
(*S*,*S*)-**33c**	H	n-octyl
(*S*,*S*)-**33d**	H	Me
(*S*,*S*)-**33e**	H	n-octyl

	R^1	R^2	n
(*S*,*S*)-**34a**	H	H	1
(*S*,*S*)-**34b**	H	H	2
(*S*,*S*)-**34c**	*t*–Bu	H	1
(*S*,*S*)-**34d**	NO_2	H	1
(*S*,*S*)-**34e**	*t*–Bu	NO_2	1

	R^1	R^2	R^3	R^4
35a	H	H	H	H
35b	NO_2	H	H	H
35c	NO_2	NO_2	H	H
35d	NO_2	NO_2	NO_2	H
35e	NO_2	NO_2	NO_2	NO_2
35f	*t*-Bu	*t*-Bu	H	H
35g	*t*-Bu	*t*-Bu	NO_2	NO_2

Figure 15.18: Crown ethers containing a diarylphosphinic acid unit.

15.3 Phosphine Oxides

A 15-crown-5 ether type macrocycle **36** (Figure 15.19) containing two phosphine oxide units and its Co^{2+} and Cu^{2+} complexes was synthesized by Bodrin et al. [68, 69]. The researchers also studied the complexation properties of **36** with alkali metal cations [70].

The synthesis of dibenzo-crown ether **37** (Figure 15.20) containing a phosphine oxide unit was described by Tsvetkov et al. They also reported that this ligand forms a stable complex with Li^+ [71].

Some 18-crown-6 ether type macrocycles (**38a–d**) (Figure 15.21) containing two phosphine oxide units were prepared by Märkl et al. by oxidization of the corresponding crown ethers having phosphine moieties [72].

Cram et al. prepared macrocycle **39** (Figure 15.22) containing two, and macrocycles **40a–d** (Figure 15.22) containing one triaryphosphine oxide unit(s) by

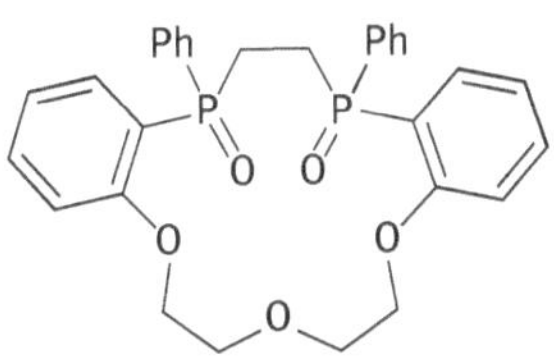

36

Figure 15.19: Macrocycle containing two phosphine oxide units.

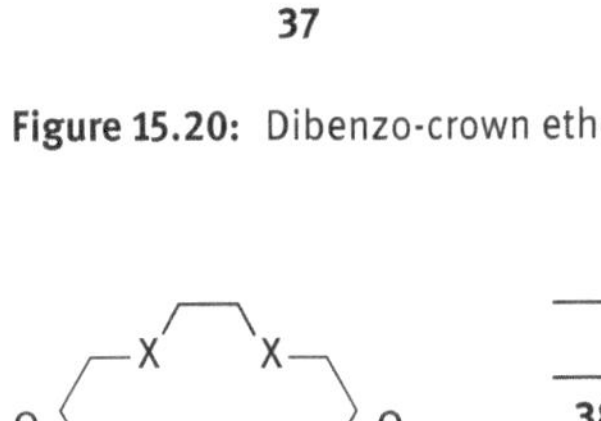

37

Figure 15.20: Dibenzo-crown ether containing a phosphine oxide unit.

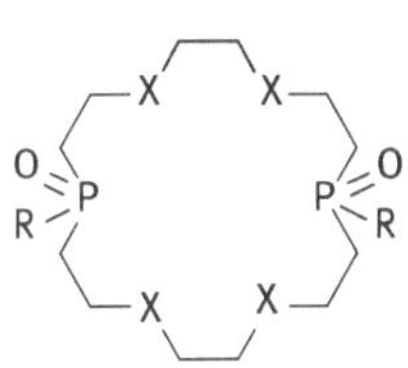

38

	R	X
38a	c-C_6H_{11}	O
38b	Bn	O
38c	Ph	O
38d	Ph	S

Figure 15.21: Macrocycles containing two phosphine oxide units.

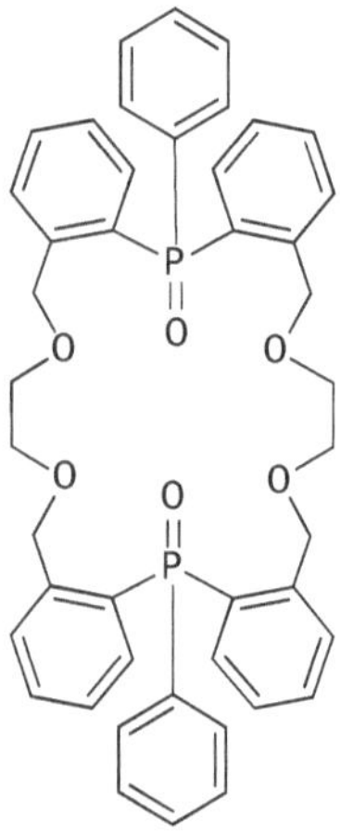

39

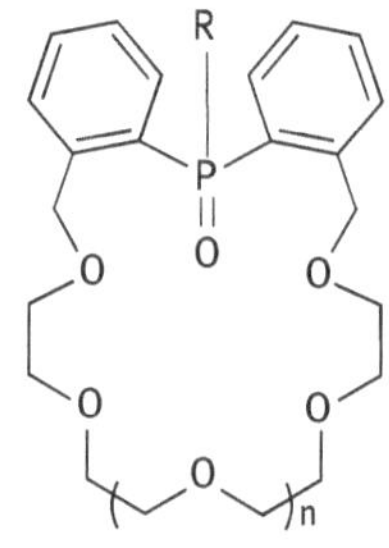

40

	n	R
40a	1	*o*-HO_2C-C_6H_4
40b	1	*o*-H_3CO_2C-C_6H_4
40c	1	Ph
40d	2	Ph

Figure 15.22: Macrocycles containing one or two triarylphosphine oxide units.

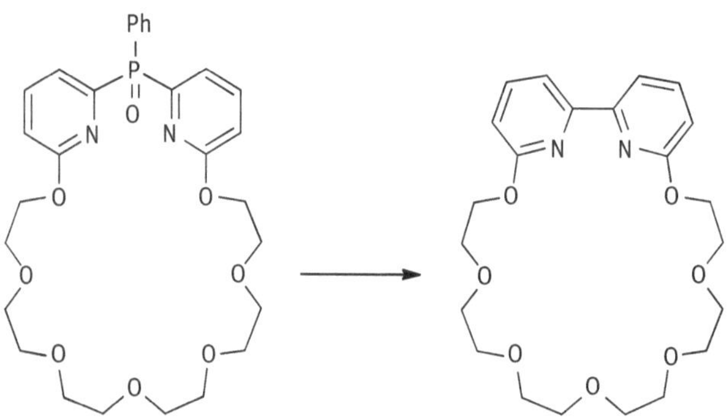

41 42

Figure 15.23: Phosphorus expulsion of a macrocycle containing triarylphosphine oxide unit.

a similar procedure. Depending on the length of the glycol [2+2] (**39**) or [1+1] (**40a–d**) macrocyclization took place. The two isomers of **39** were isolated and characterized, and their complexation properties of with alkali metal and ammonium ions studied by solvent extraction. They found that **40a–c** with smaller cavity sizes were selective for K^+, and **40d** with the larger macroring formed more stable complex with Cs^+. Ligand **39** with two phosphine oxide moieties showed preference for Na^+ [5, 73].

Newkome et al. prepared the macrocycle **41** (Figure 15.23) containing a triarylphosphine oxide unit by oxidation of the corresponding crown ether having a phosphine moiety. The researchers reported a surprising phosphorus expulsion reaction: when **41** was treated with the sodium salt of hexaethylene glycol, it gave the ring-contracted crown ether **42** (Figure 15.23) [74].

15.4 Phosphonium Salts and Phosphoranes

Phosphonium salts **43a** and **43b** (Figure 15.24) were prepared by Märkl et al. [72]. Bisphosphonium salt **43a** was obtained by methylation of the corresponding bisphosphine. The preparation of bisphosphonium salt **43b** was carried out by macrocyclization of triethylene glycol α,ω-dibromide and α,ω-dibenzylphosphine derivative of triethylene glycol. Tetrabenzyl-substituted bisphosphonium salt **43b** could be transformed to the parent crown ether containing two monobenzylphosphine units.

The synthesis, structure, and complexing properties of crown ethers **44a–e** (Figure 15.25) containing a spirophosphorane unit was reported by Markovskii et al. These ligands were found selective for Li^+ and the solubility of LiCl, NaCl, and KCl in the presence of **44a** was comparable to that in the presence of 12-crown-4 ether and it decreased in the above order of the salts [75, 76].

Houalla et al. published the synthesis of the first stable macrocycles **45a** and **45b** (Figure 15.26) containing two bicyclophosphorane units, which were prepared by Atherton-Todd reaction from spirophosphoranes bearing P-H bonds [77, 78]. As a continuation of this work, numerous macrocycles containing one (e.g. **46**), two (e.g. **47a–c**), three (e.g. **48a–c**) (Figure 15.26), and even four bicyclophosphorane units were synthesized. The macrocycles have different cavity sizes, various heteroatoms (S, N*t*-Bu, NPh, NMe), and even d-xylofuranose units [79–86]. Typical examples can be seen in Figure 15.26. This work was driven by the hope that a macrocycle containing both Lewis acid and base centers is capable of recognizing both cations and anions.

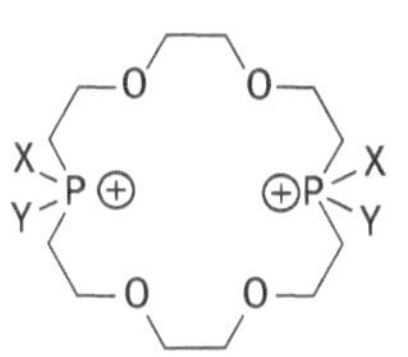

43

	X	Y
43a	Ph	Me
43b	Bn	Bn

Figure 15.24: Crown ethers containing two phosphonium units.

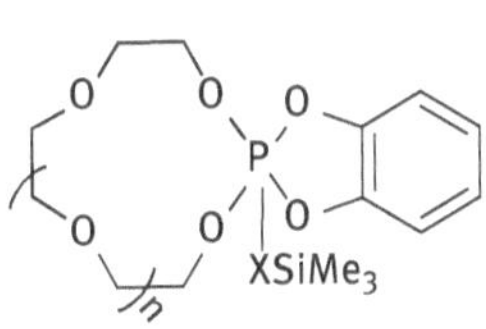

44

	X	n		X	n
44a	O	0	**44c**	S	3
44b	S	0	**44d**	O	5
44c	O	3	**44e**	S	5

Figure 15.25: Crown ethers containing a spirophosphorane unit.

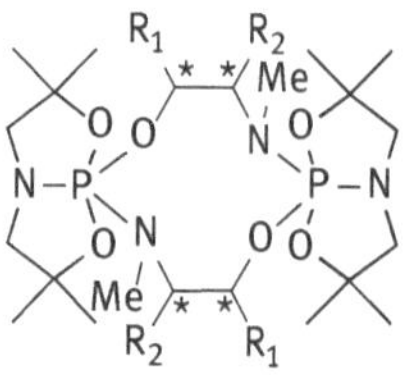

45a: $R^1=R^2=H$
45b: $R^1=Ph$, $R^2=Me$

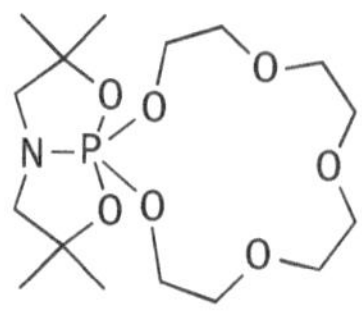

46

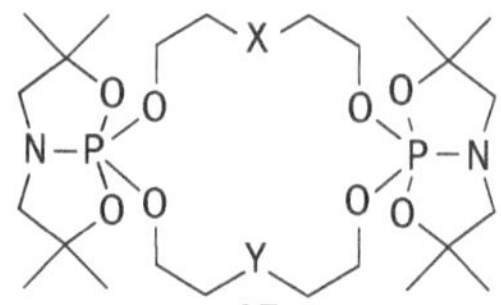

47

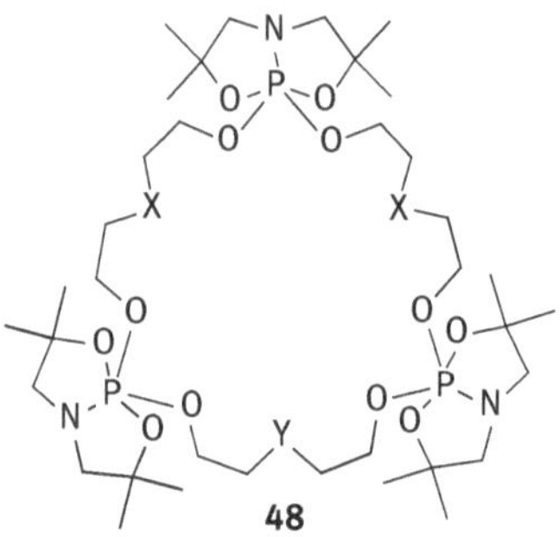

48

	X	Y		X	Y
47a	O	S	**48a**	O	O
47b	NPh	N*t*-Bu	**48b**	O	S
47c	NMe	O	**48c**	S	N*t*-Bu

Figure 15.26: Crown ethers containing bicyclophosphorane units.

15.5 Phosphites, Phosphonites, Phosphinites, and their Derivatives

The synthesis of crown ether **49** (Figure 15.27) containing two phosphinite units was reported by Powell et al. Complex **50** having a "$Mo(CO)_4P_2$" moiety was obtained by reacting **49** with $Mo(CO)_4$(norbornadiene). The structure of complex **50** was determined by single-crystal X-ray diffraction analysis. Addition of different RLi (R = Me, Ph, *n*-Bu, *t*-Bu, NEt_2, N*i*-Pr_2,) reagents to this complex gave products of the type of *fac*-$Mo(CO)_3$(RCOLi)-phosphinites, in which the Li^+ is encapsulated by five oxygens, one of the RCO and four of ligand **49** [87].

Podgornyi et al. reported the synthesis of monobenzo-crown ether **51** (Figure 15.28) containing a phosphonite unit. From the latter macrocycle (**51**) they also prepared the corresponding crown ether having a thiophosphonate moiety, by the reaction with S_8 [13].

Macrocycles **52a–c** and **53a–b** (Figure 15.29) containing a phosphoramidite unit with rigid structure were reported by Dutasta and Simon [50].

Yurchenko et al. reported the synthesis of crown ethers **54** and **55** (Figure 15.30) containing phosphite units [58, 88]. The complexation properties of macrocycle **54** having aryloxy and adamantyloxy moieties were studied with Li^+, Na^+, K^+, Ca^{2+} and Mg^{2+} [36].

An interesting approach to the synthesize various macrocycles with phosphonite units, including crown ethers **56a–d** and **57d** (Figure 15.31) was published by

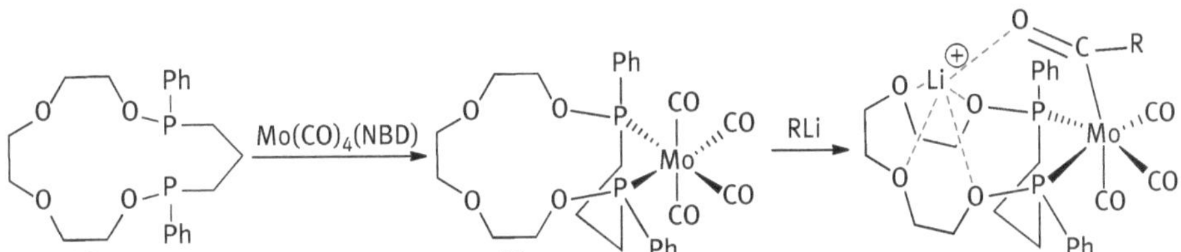

49 **50**

Figure 15.27: Crown ether containing two phosphinite units and its complexes.

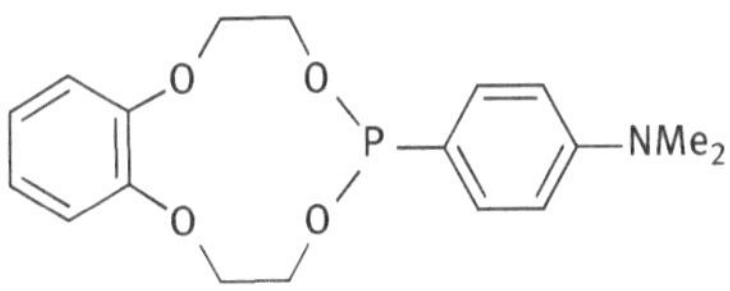

51

Figure 15.28: Crown ether containing a phosphonite unit.

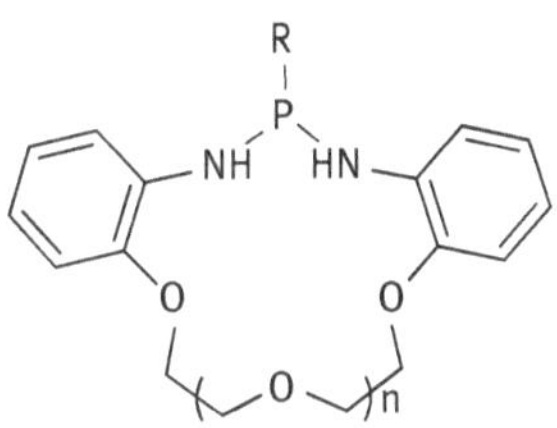

52

	R	n
52a	Ph	1
52b	Ph	2
52c	NMe2	1

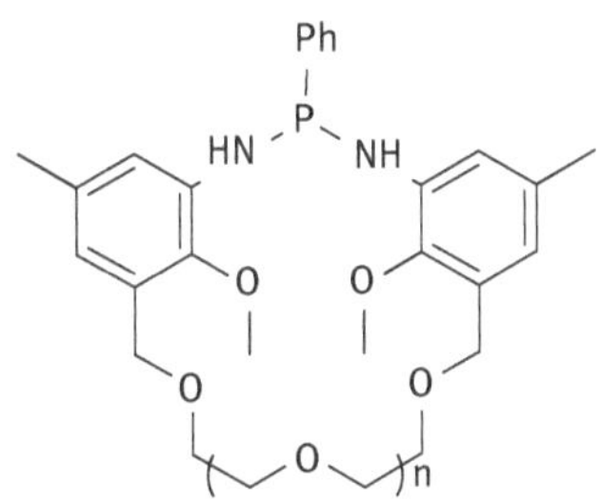

53

	n
53a	1
53b	2

Figure 15.29: Crown ethers containing a phosphoramidite unit.

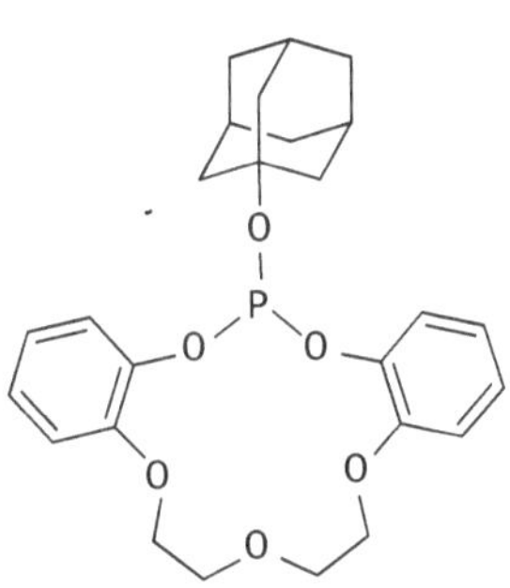

54

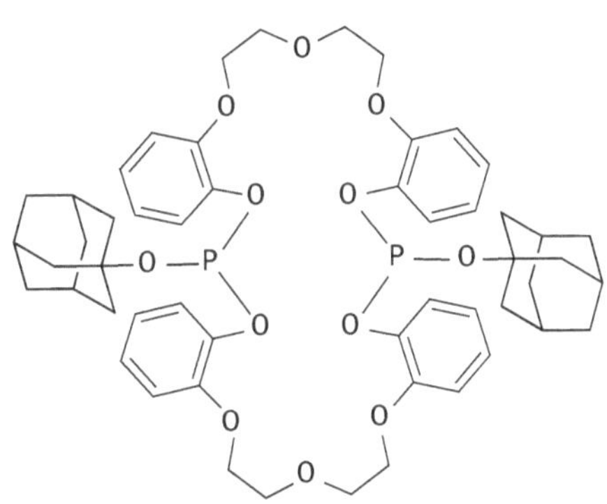

55

Figure 15.30: Crown ethers containing phosphite units.

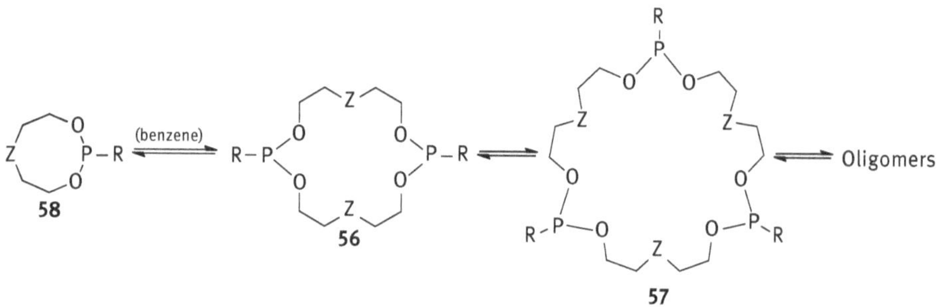

	R	Z
a	Me	O
b	Me	S
c	Me	NMe
d	Ph	O

Figure 15.31: The ring-opening polymerization of phosphonites.

Dutasta et al. The principal step of the synthesis was the ring-opening polymerization of cyclic phosphonites (**58a–d**) in benzene solution. Detailed studies revealed the formation of dimeric and trimeric species along with higher polymeric ones [89]. Several factors influenced the polymerization, e.g. it was accelerated by R substituents in the following order: Ph<<Me<Et<*i*-Pr<*t*-Bu [59, 90]. The size of the ring is also influenced by this reaction [59]. The yields of dimers were very low (<1%) starting from phosphonites **58a–d** (eight-membered rings) [60, 61]. In all cases the formation of two diastereomers *(cis* and *trans)* of **56a–d** were observed. Heat or dilution shifted the equilibrium to the starting monomers [59, 91]. The proneness of these compounds to oxidation made it hard to isolate the corresponding compound containing phosphorus atom in P^{III} oxidation state. On the other hand, the reaction of a mixture of monomers, dimers, and oligomers with elemental sulfur resulted in the formation of stable compounds [59–61, 90, 91].

Martin et al. reported the oligomerization of trithiophosphoctanes **59a,b** (Figure 15.32) in benzene solution. They synthesized the dimers **60a,b** and isolated the *cis* and *trans* isomers of the compounds. The reasearchers found that an equilibrium exists between monomers, dimers and oligomers. For instance, in the case of heating or diluting **60a** the equilibrium was shifted toward the monomer **59a** [54].

In contrast to trithiophosphoctanes, diazaphospholanes do not oligomerize spontaneously [92]. The *ß*-chloroethyloxy substituent at phosphorus in compound **61** (Figure 15.33) induced a quantitative and spontaneous ring-opening reaction leading to the ammonium salt **62**. The reaction of the latter salt with alcohols or amines led to the formation of macrocycles **63a–e**. Carbonyl ($RhCl(CO)(PN_2)$) and ethylene ($RhCl(C_2H_4)(PN_2)$) complexes of **63a–e** with Rh^+ containing alkoxy ligands were

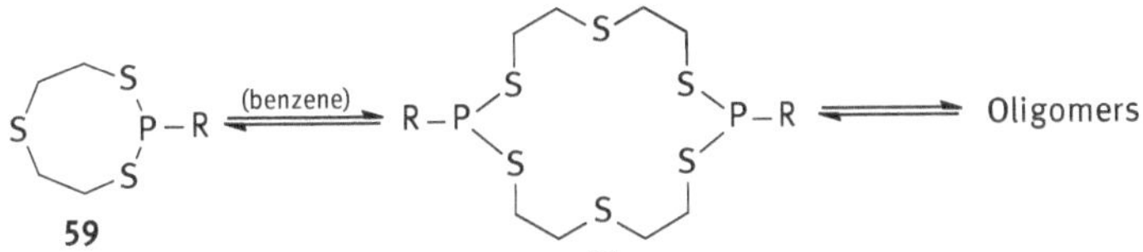

59 60

	R
a	Me
b	*t*-Bu

Figure 15.32: The polymerization of trithiophosphoctanes.

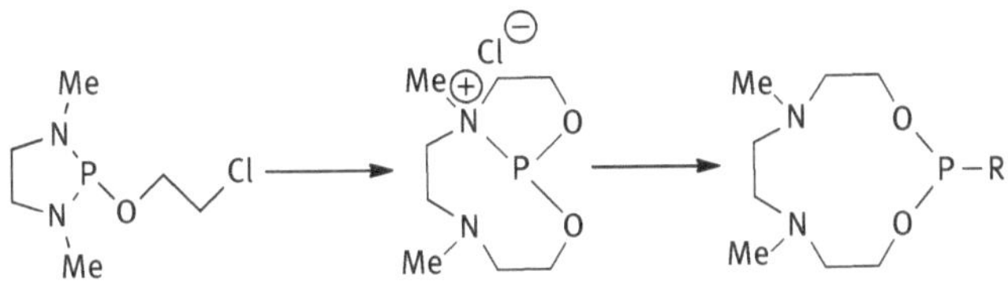

61 62 63

	R		R
63a	O–Me	**63c**	O–CH_2CH(Me)(Et)
63b	O–Et	**63d**	O–1–Menthyl
63c	O–Bu	**63e**	NEt_2

Figure 15.33: Synthesis of crown ethers containing a phosphite unit.

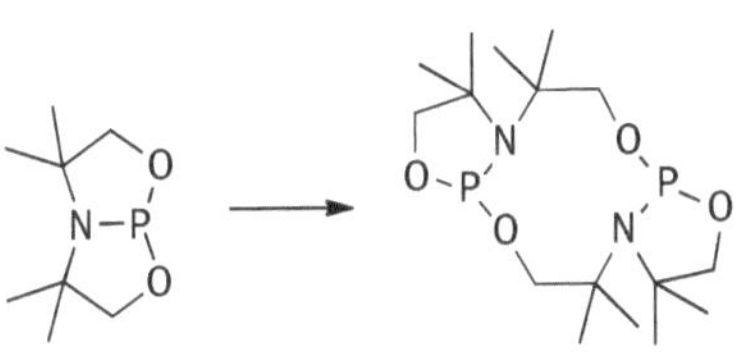

64 65

Figure 15.34: Dimerization of a bicyclic phosphoramidite.

also prepared and shown to be good precursors for the catalytic hydroformylation of olefins [93, 94].

The bicyclic phosphoramidite **64** (Figure 15.34) also undergoes a dimerization at room temperature. In this case only one isomer was obtained. The crystal structure analysis of phosphoramidite **65** (Figure 15.34) showed that this compound was formed by rupture and reformation of two P-O bonds [95, 96]. The orientation of the phosphorus lone electron pairs was *trans* compared to the main plane of the 10-membered ring.

15.6 Phosphines

Van Zon et al. reported the synthesis of two 18-crown-6 ether derivatives (**66a,b**) and a 21-crown-7 ether derivative (**66c**) (Figure 15.35) containing a phosphine unit [97]. The complexation of alkali metal cations and Rh^+ with these macrocycles and their oxide derivatives was studied by NMR methods. It was found that the Rh^+ complexes acted as catalysts for air- or moisture-induced oxidation.

The first examples of optically pure crown ether containing phosphine unit ((*R*,*R*)-**67a**-(*R*,*R*)-**67b**) was prepared by Marinetti et al., by reduction of the corresponding phosphine oxide. The coordinating ability of (*R*,*R*)-**67b** toward transition metal ions has been observed through its reaction with the [(*p*cymene)$RuCl_2]_2$ complex, which quantitatively affords the Ru^{2+} complex of (*R*,*R*)-**67b** (Figure 15.36) [98].

The synthesis and characterization of an achiral (**68a**) and a few enantiopure ((*R*,*R*)-**68b**-(*S*,*S*)-**68d**) crown ethers (Figure 15.37) containing a triarylphosphine unit were described by Szabó et al. The researchers prepared these compounds by the reduction of the corresponding phosphine oxides [99]. The platinum and rhodium complexes of the crown ethers **68b** and **68c** were synthesized and successfully used as catalyst precursors in asymmetric hydroformylation of styrene [100].

Ciampolini et al. reported the synthesis of the first 18-membered sexidentate macrocycles **69a–c** (Figure 15.38) containing four phosphine units [101–103] with sulfide moieties [101, 104–110], or with two tertiary amino groups [111–114]. These macrocycles have five stereoisomers and all of them were isolated and characterized.

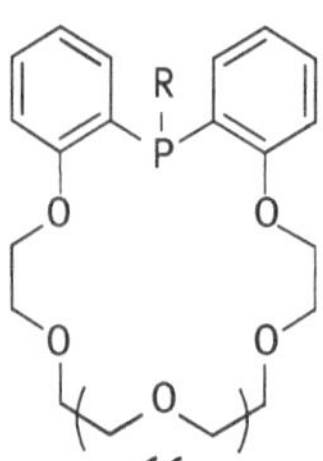

66

	R	n
66a	*t*-Bu	1
66b	Ph	1
66c	Ph	2

Figure 15.35: Crown ethers containing a phosphine unit.

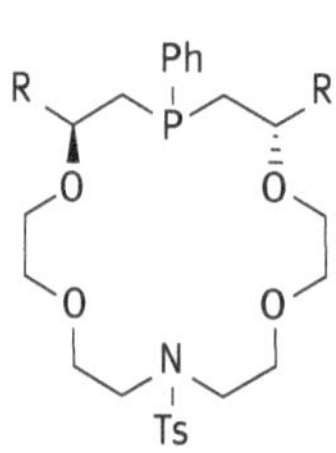

67

	R
(*R*,*R*)-**67a**	Me
(*R*,*R*)-**67b**	Ph

Figure 15.36: Enantiopure crown ethers containing a phosphine unit.

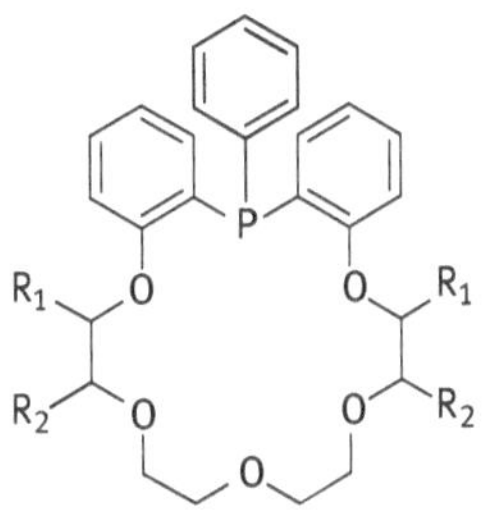

68

	R^1	R^2
68a	H	H
(*R,R*)-**68b**	Me	H
(*S,S*)-**68c**	H	Me
(*S,S*)-**68d**	H	n-octyl

Figure 15.37: Crown ethers containing a triarylphosphine unit.

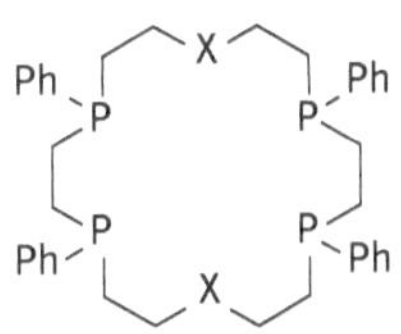

69

	X
69a	O
69b	S
69c	N-Pr

Figure 15.38: Crown ethers containing four phosphine units.

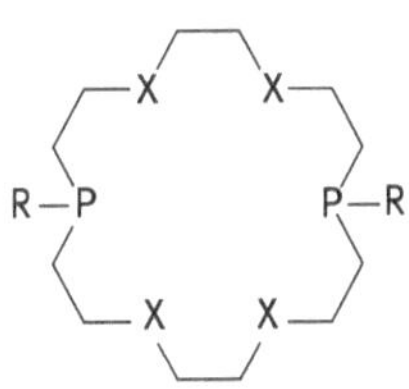

70

	R	X
70a	c-C_6H_{11}	O
70b	Bn	O
70c	Ph	O
70d	Ph	S

Figure 15.39: Crown ethers containing two phosphine units.

The compounds formed stable complexes with Co^{2+} and Ni^{2+}, but they showed significantly different coordinative behaviors according to the stereochemistry at the phosphorus and to the nature of X, as far as the coordination number and geometry of their metal complexes are concerned.

Bisphosphino-18-crown-6 ethers **70a–d** (Figure 15.39) were prepared by Märkl et al. [72]. The researchers oxidized these phosphines to get phosphine oxide derivatives as well. It is interesting that **69c** was reacted with methyl iodide to get the corresponding bis(methylphosphonium) salt, and bisphosphine **70b** was prepared by the reduction of the corresponding bisphosphonium salt having benzyl groups.

Crown ether derivatives containing two chiral phosphine units were synthesized by Morisaki et al. In all cases first the borane-coordinated macrocycles were prepared as precursors, because the free bisphospha-crown ethers are prone to oxidation by air. The borane complexes could be easily decomposed by treatment with organic amines such as 1,4-diazabicyclo[2.2.2]octane. The first two crown ethers of this type were (*R*,*R*)-**71** and (*S*,*S*)-**71** (Figure 15.40). Using these macrocycles Pd^{2+} and Pt^{2+} complexes were synthesized and characterized [115]. Later 12-crown-4 ether type bisphosphino macrocycles (*S*,*S*)-**72a** and 15-crown-5 ether type (*S*,*S*)-**72b** were also prepared. Complex formation of the bisphospha-crown ethers ((*S*,*S*)-**72a** and (*S*,*S*)-**72b**) with Pd^{2+} was carried out and the corresponding palladium(II) complexes were applied for the catalytic asymmetric 1,4-addition of arylboronic acids to α,β-unsaturated ketones [116]. As a continuation of this work the research team synthesized *tert*-butyl-substituted ligands (*S*,*S*)-**72c**-(*S*,*S*)-**72f** (Figure 15.40), (*R*,*R*)-**73a**-(*R*,*R*)-**73c**, (*R*,*R*)-**74a**-(*R*,*R*)-**74c** and (*R*,*R*)-**75** (Figure 15.41) with diverse ring size and different structures. The complexing behavior of (*S*,*S*)-**72a** – (*S*,*S*)-**72f** (Figure 15.40) with alkali metal ions was investigated, comparing the results to the ones of benzo-18-bisphosphacrown-6 and benzo-18-crown-6 ethers [117, 118]. The crown ether (*R*,*R*)-**75** containing pyridine unit exhibited chiral recognition ability for chiral ammonium salts and carboxylic acids. This is the first example of chiral recognition using P-stereogenic macrocycle [119].

This research group also synthesized tetraphospha-crown ethers **76** and **77** (Figure 15.42) containing four P-stereogenic centers in the ring. The tetraphospha-18-crown-6 ether was obtained by a one-pot reaction from the lithiated secondary bisphosphine with diethylene glycol bis(*p*-toluene-sulfonate). On the other hand, P-stereogenic tetraphospha-24-crown-8 ether was synthesized by a stepwise method. The Pt^{2+} complex of 24-tetraphosphacrown-8 ether was also obtained, in which all phosphine units were chelate-coordinated to platinum outside the ring [119].

Oxaphospha-macrocycle **78** and azaphospha-macrocycle **79** (Figure 15.43) containing two stereogenic phosphorus atoms were prepared by Wei et al. The two diastereomers of both bisphosphines were separated and characterized, and several Ni^{2+}, Pd^{2+} and Pt^{2+} complexes were prepared and characterized [120, 121].

Newkome et al. synthesized macrocycle **80** (Figure 15.44) containing a triaryl-phosphine unit, which was easily oxidized to the corresponding phosphine oxide [74].

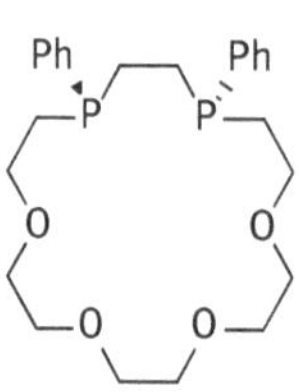

(*R*,*R*)-**71**

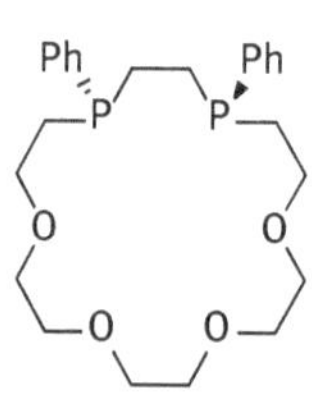

(*S*,*S*)-**71**

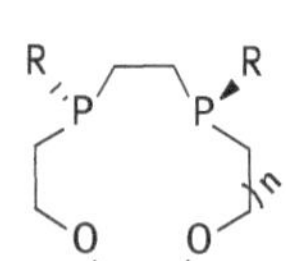

72

	R	n
(*S*,*S*)-**72a**	Ph	1
(*S*,*S*)-**72b**	Ph	2
(*S*,*S*)-**72c**	*t*-Bu	1
(*S*,*S*)-**72d**	*t*-Bu	2
(*S*,*S*)-**72e**	*t*-Bu	3
(*S*,*S*)-**72f**	*t*-Bu	4

Figure 15.40: Chiral crown ethers containing two P-stereogenic phosphine units.

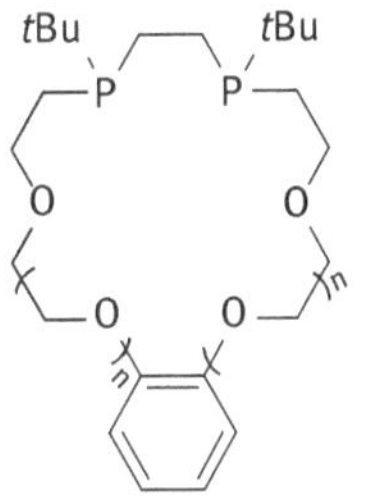

73

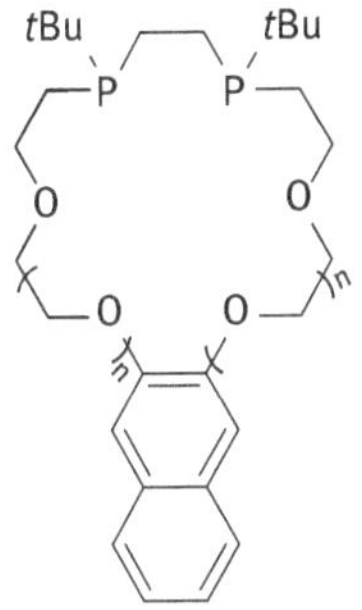

74

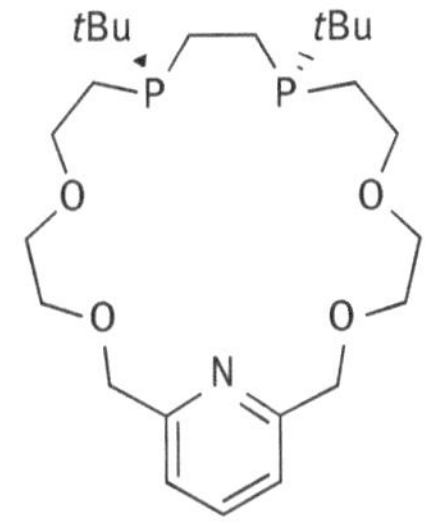

(*R,R*)-**75**

	n
(*R,R*)-**73a**	0
(*R,R*)-**73b**	1
(*R,R*)-**73c**	2

	n
(*R,R*)-**74a**	0
(*R,R*)-**74b**	1
(*R,R*)–**74c**	2

Figure 15.41: Chiral crown ethers containing two P-stereogenic phosphine units and aromatic moieties.

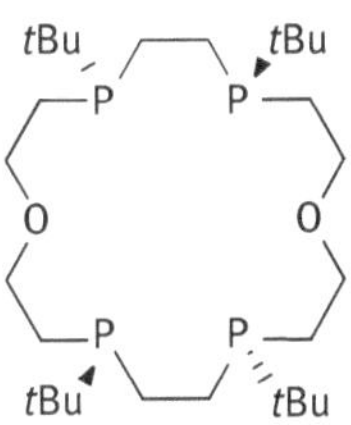

76

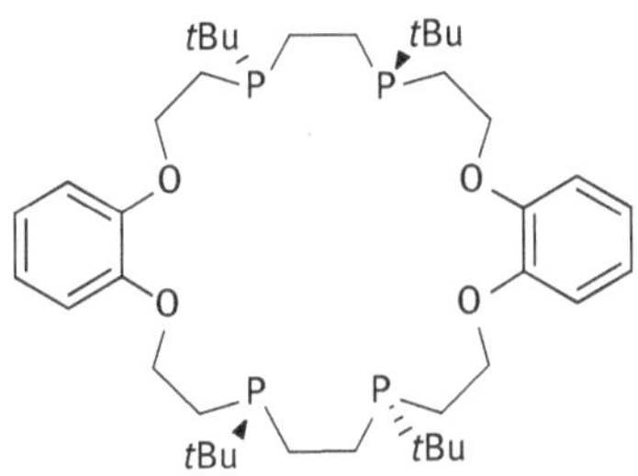

77

Figure 15.42: Crown ethers containing four P-stereogenic phosphine units.

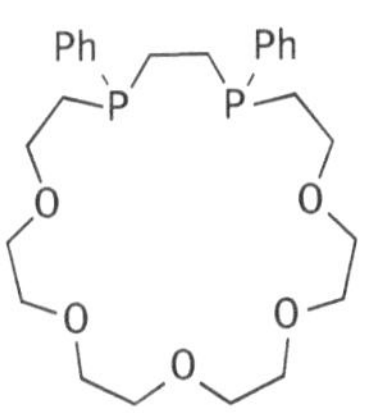

78

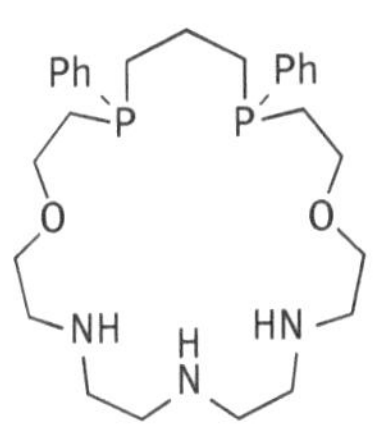

79

Figure 15.43: Oxaphospha and azaphospha-crown ethers.

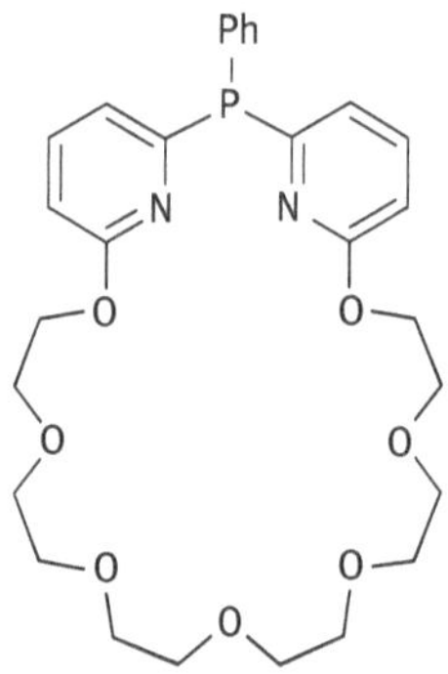

80

Figure 15.44: Crown ether containing a triarylphosphine unit.

15.7 Conclusion

This chapter is intended to summarize the synthesis, complexation studies, and applications of crown ethers containing phosphorus in the macroring, reported from the very beginning to 2017.

Acknowledgment: The authors thank the National Research, Development and Innovation Office (grant number K112289) for the financial support.

References

[1] H. N. Stokes, *Am. Chem. J.*, 1897, **19**, 782.
[2] C. J. Pedersen, *J. Am. Chem. Soc.*, 1967, **89**, 2495.
[3] J. S. Bradshaw, P. Huszthy, R. M. Izatt, *J. Heterocycl. Chem.*, 1986, **23**, 1673.
[4] R. M. Izatt, G. C. Lindh, P. Huszthy, G. A. Clark, R. L. Bruening, J. S. Bradshaw, J. J. Christensen, *J. Inclusion Phenom. Mol. Recog. Chem.*, 1989, **7**, 501.
[5] G. G. Talanova, *Ind. Eng. Chem. Res.*, 2000, **39**, 3550.
[6] I. Kovács, P. Huszthy, F. Bertha, D. Sziebert, *Tetrahedron: Asymmetry*, 2006, **17**, 2538.
[7] A. V. Kirsanov, V. A. Zasorina, A. S. Shtepanek, A. M. Pinchuk, *Dokl. Akad. Nauk SSSR*, 1981, 1112.
[8] V. A. Zasorina, A. S. Shtepanek, A. M. Pinchuk, *Zh. Obshch. Khim.*, 1982, **52**, 1081.
[9] A. A. Chaikovskaya, T. N. Kudrya, A. M. Pinchuk, *Zh. Obshch. Khim.*, 1987, **57**, 671.
[10] T. N. Kudrya, A. A. Chaikovskaya, Z. Z. Rozhkova, A. M. Pinchuk, *Zh. Obshch. Khim.*, 1982, **52**, 1092.
[11] A. V. Kirsanov, T. N. Kudrya, L. V. Balina, A. S. Shtepanek, *Dokl. Akad. Nauk SSSR*, 1979, **247**, 613.
[12] T. N. Kudrya, V. V. Tkachev, L. O. Atovmyan, E. S. Gubnitskaya, A. A. Chaikovskaya, A. M. Pinchuk, *Zh. Strukt. Khim.*, 1991, **32**, 175.

[13] A. V. Podgornyi, A. A. Tolmachev, A. S. Shtepanek, A. A. Kudryavtsev, A. M. Pinchuk, *Zh. Obshch. Khim.*, 1990, **60**, 2440.
[14] A. A. Tolmachev, A. V. Podgornyi, A. N. Kostyuk, I. F. Tsymbal, E. S. Kozlov, A. M. Pinchuk, *Zh. Obshch. Khim.*, 1992, **62**, 1422.
[15] R. I. Yurchenko, V. G. Yurchenko, A. V. Podgornyi, *Zh. Obshch. Khim.*, 1991, **61**, 772.
[16] J.-P. Dutasta, J.-P. Declercq, C. Esteban-Calderon, B. Tinanti, *J. Am. Chem. Soc.* 1989, **111**, 7136.
[17] J.-P. Declercq, B. Tinant, J.-P. Dutasta, J.-C. Mulatier, *Phosphorus, Sulfur, Silicon Relat. Elem.*, 1999, **155**, 1.
[18] B. Tinant, P. Delangle, J.-C. Mulatier, J.-P. Declercq, J. P. Dutasta, *J. Inclusion Phenom.*, 2007, **58**, 139.
[19] J.-F. Labarre, *Top. Curr. Chem.*, 1985, **129**, 173.
[20] J.-F. Labarre, F. Sournies, *Adv. Supramolecular Chemistry*, 1993, **4**, 555.
[21] A. El Bakili, P. Castera, J.-P. Faucher, F. Sournies, J.-F. Labarre, *J. Mol. Struct.*, 1989, **195**, 21.
[22] S. Bilge, B. Özgüç, S. Safran, Ş. Demiriz, H. Isler, M. Hayvalı, Z. Kilic, T. Hökelek, *J. Mol. Struct.*, 2005, **748**, 101.
[23] B. Özgüç, S. Bilge, N. Caylak, Ş. Demiriz, H. Isler, M. Hayvalı, Z. Kilic, T. Hökelek, *J. Mol. Struct.*, 2005, **748**, 39.
[24] Z. Kilic, A. Okumus, Ş. Demiriz, S. Bilge, A. Öztürk, N. Caylak, T. Hökelek, *J. Inclusion Phenom. Macrocyclic Chem.*, 2009, **65**, 269.
[25] A. Okumus, S. Bilge, Z. Kilic, A. Öztürk, T. Hökelek, F. Yilmaz, *Spectrochim. Acta Part A: Mol. Biomol. Spectrosc.*, 2010, **76**, 401.
[26] F. Sournies, A. El Bakili, B. Zanin, J. F. Labarre, J. Jaud, *J. Mol. Struct.*, 1990, **220**, 43.
[27] B. Zanin, J.-P. Faucher, J.-F. Labarre, *Inorg. Chim. Acta*, 1990, **172**, 147.
[28] R. Enjalbert, J. Galy, P. Castera, J. F. Labarre, *Acta Crystallogr. Sect. C Cryst. Struct. Commun.*, 1988, **44**, 1813.
[29] M. Yildiz, Z. Kilic, T. Hökelek, *J. Mol. Struct.*, 1999, **510**, 227.
[30] I. Porwolik-Czomperlik, K. Brandt, T. A. Clayton, D. B. Davies, R. J. Eaton, R. A. Shaw, *Inorg. Chem.*, 2002, **41**, 4944.
[31] K. Brandt, I. Porwolik-Czomperlik, M. Siwy, T. Kupka, R. A. Shaw, D. B. Davies, R. A. Bartsch, *J. Inclusion Phenom.*, 1999, **35**, 281.
[32] K. Brandt, I. Porwolik-Czomperlik, M. Siwy, T. Kupka, B. L. Shaw, S. Ture, A. Clayton, D. B. Davies, M. B. Hursthouse, G. D. Sykara, *J. Org. Chem.*, 1999, **64**, 7299.
[33] F. Sournies, P. Castera, J.-P. Faucher, M. Graffeuil, J.-F. Labarre, *J. Mol. Struct.*, 1990, **221**, 245.
[34] T. N. Kudrya, A. S. Shtepanek, A. V. Kirsanov, *Zh. Obshch. Khim.*, 1978, **48**, 927.
[35] A. M. Pinchuk, V. A. Zasorina, A. S. Shtepanek, Z. Z. Rozhkova, A. F. Solotnov, O. A. Raevskii, *Zh. Obshch. Khim.*, 1983, **53**, 2012.
[36] V. V. Tkachev, A. A. Chaikovskaya, L. O. Atovmyan, G. V. Sazonova, T. N. Kudrya, *Izv. Akad. Nauk SSSR, Seriya Khim.*, 1987, **12**, 2745.
[37] O. A. Raevskii, V. V. Tkachev, L. O. Atovmyan, I. O. Umarova, A. F. Solotnov, T. N. Kudrya, A. A. Chaikovskaya, *Izv. Akad. Nauk SSSR, Seriya Khim.*, 1984, **9**, 2028.
[38] S. T. Malinovskii, T. N. Kudrya, A. A. Chaikovskaya, A. M. Pinchuk, T. I. Malinovskii, *Zh. Strukt. Khim.*, 1986, **27**, 163.
[39] S. T. Malinovskii, Y. A. Simonov, T. N. Kudrya, A. A. Chaikovskaya, T. I. Malinovskii, *Zh. Strukt. Khim.*, 1984, **25**, 130.
[40] R. I. Yurchenko, V. G. Yurchenko, T. I. Klepa, A. M. Pinchuk, *Zh. Obshch. Khim.*, 1991, **61**, 380.
[41] K. B. Yatsimirskii, L. I. Budarin, A. S. Shtepanek, A. I. Telyatnik, V. A. Smirnov, *Teoreticheskaya i Eksperimentalnaya Khim.*, 1976, **312**, 421.
[42] K. B. Yatsimirskii, V. A. Idzilya, L. P. Golovkova, A. S. Shtepanek, *Dokl. Akad. Nauk SSSR*, 1979, **244**, 1142.

[43] K. B. Yatsimirskii, E. N. Korol, V. G. Golovatyo, T. N. Kudrya, G. G. Talanova, *Dokl. Akad. Nauk SSSR*, 1979, **244**, 1359.

[44] O. A. Raevskii, I. O. Umarova, V. V. Tkachev, L. O. Atovmyan, A. S. Shtepanek, T. N. Kudrya, *Izv. Akad. Nauk SSSR, Seriya Khim.*, 1986, **9**, 2003.

[45] K. B. Yatsimirskii, G. G. Talanova, E. V. Rybak-Akimova, A. S. Shtepanek, V. A. Popova, *Theor. Chem. Acc.*, 1986, **22**, 162.

[46] E. V. Rybakova, V. P. Ionov, Y. A. Zolotov, T. N. Kudrya, A. A. Chaikovskaya, *Zh. Neorg. Khim.*, 1987, **32**, 1301.

[47] N. V. Lukoyanov, O. A. Raevskii, A. S. Shtepanek, T. N. Kudrya, *J. Membr. Biol.*, 1984, **1**, 379.

[48] A. V. Kirsanov, T. N. Kudrya, A. S. Shtepanek, *Zh. Obshch. Khim.*, 1980, **50**, 2452.

[49] V. V. Tkachev, L. O. Atovmyan, I. O. Umarova, O. A. Raevskii, T. N. Kudrya, A. M. Pinchuk, *Izv. Akad. Nauk SSSR, Seriya Khim.*, 1985, **8**, 1775.

[50] J.-P. Dutasta, S. Pascal, *Tetrahedron Lett.*, 1987, **28**, 3577.

[51] J.-P. Declercq, P. Delangle, J.-P. Dutasta, L. V. Oostenrick, P. Simon, B. Tinant, *J. Chem. Soc. Perkin Trans.*, **2** 1996, 2471.

[52] J.-P. Dutasta, L. V. Oostenrick, B. Tinant, J.-P. Declercq, *Phosphorus, Sulfur, Silicon Relat. Elem.*, 1993, **75**, 63.

[53] G. G. Talanova, A. V. Podgornyi, V. A. Zasorina, K. B. Yatsimirskii, *Dokl. Akad. Nauk SSSR*, 1991, **318**, 347.

[54] J. Martin, J. B. Robert, *Nouveau J. Chim.*, 1980, **4**, 515.

[55] A. Grand, J. Martin, *Acta Crystallogr.*, 1982, *B***38**, 3052.

[56] A. M. Pinchuk, A. V. Podgornyi, V. A. Zasorina, G. G. Talanova, A. S. Shtepanek, *Dokl. Akad. Nauk Ukrainskoi SSR, Seriya B: Geol., Khim.,Biologich. Nauki* 1987, 53.

[57] A. A. Chaikovskaya, T. N. Kudrya, R. I. Yurchenko, O. M. Voitsekhovskaya, A. M. Pinchuk, *Zh. Obshch. Khim.*, 1988, **58**, 1925.

[58] T. N. Koodrja, A. S. Stepanek, R. I. Yurchenko, A. A. Chaikovskaja, S. E. Lavrona, *Phosphorous, Sulfur Silicone*, 1990, **51**, 384.

[59] J.-P. Dutasta, J. Martin, J. B. Robert, *Heterocycles*, 1980, **14**, 1631.

[60] J.-P. Dutasta, J. B. Robert, *J. Am. Chem. Soc.*, 1978, **100**, 1925.

[61] J.-P. Dutasta, *J. Chem. Res. Synop.*, 1986, **1**, 22.

[62] G. Székely, B. Csordás, V. Farkas, J. Kupai, P. Pogány, Z. Sánta, Z. Szakács, T. Tóth, M. Hollósi, J. Nyitrai, P. Huszthy, *Eur. J. Org. Chem.*, 2012, **18**, 3396.

[63] P. Huszthy, V. Farkas, T. Tóth, G. Székely, M. Hollósi. *Tetrahedron*, 2008, **64**, 10107.

[64] T. Szabó, E. Hirsch, T. Tóth, J. Müller, E. Riethmüller, G. T. Balogh, P. Huszthy, *Tetrahedron: Asymmetry*, 2015, **26**, 650.

[65] T. Szabó, E. Hirsch, T. Tóth, P. Huszthy, *Tetrahedron: Asymmetry*, 2014, **25**, 1443.

[66] T. Szabó, G. Dargó, H. Szentjóbi, T. Tóth, B. Kármos, R. Izrael, J. Oláh, T. Németh, G. T. Balogh, P. Huszthy, *Tetrahedron*, 2016, **72**, 8593.

[67] G. Székely, V. Farkas, L. Párkányi, T. Tóth, M. Hollósi, P. Huszthy, *Struct. Chem.*, 2010, **21**, 277.

[68] G. V. Bodrin, Y. M. Polikarpov, T. Y. Medved, M. I. Kabachnik, *Izv. Akad. Nauk SSSR, Seriya Khim.*, 1978, **8**, 1930.

[69] K. B. Yatsimirskii, M. I. Kabachnik, M. A. Konstantinovskaya, E. I. Sinyavskaya, T. Y. Medved, Y. M. Polikarpov, G. V. Bodrin, *Zh. Neorg. Khim.*, 1980, **25**, 1586.

[70] K. B. Yatsimirskii, M. I. Kabachnik, E. I. Sinyavskaya, T. Y. Medved, Y. M. Polikarpov, G. V. Bodrin, *Zh. Neorg. Khim.*, 1980, **25**, 2355.

[71] E. N. Tsvetkov, T. E. Kron, E. I. Slnyavskaya, *Izv. Akad. Nauk SSSR, Seriya Khim.*, 1986, **11**, 2456.

[72] G. Märkl, M. Hoferer, *Tetrahedron Lett.*, 1992, **33**, 3621.

[73] L. J. Kaplan, G. R. Weisman, D. J. Cram, *J. Organomet. Chem.*, 1979, **44**, 2226.

[74] G. R. Newkome, D. C. Hager, *J. Am. Chem. Soc.*, 1978, **100**, 5567.

[75] V. I. Kal'chenko, L. I. Atamas, Y. A. Serguchev, L. N. Markovskii, *Zh. Obshch. Khim.*, 1984, **54**, 1754.
[76] L. N. Markovskii, V. I. Kal'chenko, A. D. Sinitsa, Y. A. Serguchev, V. V. Negrebetskii, L. Y. Bogel'fer, *Dopovidi Akad. Nauk Ukrains'koi RSR, Seriya B: Geol., Khim. Biol. Nauki*, 1981, **2**, 56.
[77] D. Houalla, Z. Bounja, S. Skouta, L. Riesel, D. Lindemann, *Tetrahedron Lett.*, 1992, **33**, 2817.
[78] D. Houalla, Z. Bounja, S. Skouta, L. Riesel, D. Lindemann, *Phosphorus, Sulfur, Silicon*, 1993, **77**, 216.
[79] D. Houalla, Z. Bounja, S. Skouta, M. Sanchez, R. Wolf, *Phosphorus, Sulfur, Silicon Relat. Elem.*, 1993, **75**, 71.
[80] D. Houalla, Z. Bounja, S. Skouta, R. Wolf, J. Jaud, *Heteroat. Chem.*, 1994, **5**, 175.
[81] M.-r. Mazieres, D. Houalla, M. C. Monje, M. Sanchez, *Phosphorus, Sulfur, Silicon Relat. Elem.*, 1993, **83**, 157.
[82] D. Houalla, L. Moureau, S. Skouta, R. Wolf, *Phosphorus, Sulfur, Silicon Relat. Elem.*, 1994, **88**, 83.
[83] D. Houalla, L. Moureau, S. Skouta, M.-R. Mazieres, *Phosphorus, Sulfur, Silicon Relat. Elem.*, 1995, **103**, 199.
[84] A. Chandrasekaran, R. O. Day, R. R. Holmes, D. Houalla, *Phosphorus, Sulfur, Silicon Relat. Elem.*, 1997, **123**, 219.
[85] D. Houalla, L. Moureau, *Phosphorus, Sulfur, Silicon Relat. Elem.*, 1996, **114**, 51.
[86] D. Houalla, L. Moureau, S. Skouta, R. Wold, *Phosphorus, Sulfur, Silicon Relat. Elem.*, 1994, **90**, 131.
[87] J. Powell, A. Lough, F. Wang, *Organometallics*, 1992, **11**, 2289.
[88] R. I. Yurchenko, T. I. Kelpa, A. M. Pinchuk, *Zh. Obshch. Khim.*, 1990, **10**, 2226.
[89] J. P. Albrand, J. P. Dutasta, J. B. Robert, *J Am. Chem. Soc.*, 1974, **96**, 4584.
[90] J.-P. Dutasta, A. C. Guimaraes, J. B. Robert, *Tetrahedron Lett.*, 1977, **9**, 801.
[91] J. P. Albrand, J.-P. Dutasta, J. B. Robert, *J Am. Chem. Soc.*, 1974, **96**, 4584.
[92] H. Sliwa, J. P. Picavet, *Tetrahedron Letters* 1977, **18**, 1583.
[93] J. Powell, K. S. Ng, W. W. Ng, S. C. Nyburg, *J. Organomet. Chem.*, 1983, **243**, C1.
[94] C. Vaccher, A. Mortreux, F. Petit, J. P. Picavet, H. Sliwa, N. W. Murrall, A. J. Welch, *Inorg. Chem.*, 1984, **23**, 3613.
[95] C. Bonningue, D. Houalla, R. Wolf, J. Jaud, *J. Chem. Soc., Perkin Trans 2*, 1983, **6**, 773.
[96] B. Duthu, K. El Abed, D. Houalla, R. Wolf, J. Jaud, *Can. J. Chem.*, 1992, **70**, 809.
[97] A. Van Zon, G. J. Torny, J. H. G. Frijns, *Recl. Trav. Chim. Pays-Bas*, 1983, **102**, 326.
[98] A. Theil, J. Hitce, P. Retailleau, A. Marinetti, *Eur. J. Org. Chem.*, 2005, **1**, 154.
[99] T. Szabó, L. Petri, S. Gergely, P. Huszthy, *Arkivoc*, 2015, **5**, 20.
[100] P. Pongrácz, H. Szentjóbi, T. Tóth, P. Huszthy, L. Kollár. *Mol. Catal.*, 2017, **439**, 128.
[101] M. Ciampolini, *Pure Appl. Chem.*, 1986, **58**, 1429.
[102] M. Ciampolini, P. Dapporto, N. Nardi, F. Zanobini, *J. Chem. Soc., Chem. Commun.*, 1980, **4**, 177.
[103] M. Ciampolini, P. Dapporto, N. Nardi, F. Zanobini, *Inorg. Chem.*, 1983, 13.
[104] M. Ciampolini, P. Dapporto, A. Dei, N. Nardi, F. Zanobini, *Inorg. Chem.*, 1982, 489.
[105] M. Ciampolini, P. Dapporto, N. Nardi, F. Zanobini, *Inorg. Chim. Acta Lett.*, 1980, L239.
[106] M. Ciampolini, N. Nardi, P. Dapporto, P. Innocenti, F. Zanobini, *J. Chem. Soc., Dalton Trans.: Inorg. Chem.*, 1984, **4**, 575.
[107] M. Ciampolini, N. Nardi, P. Dapporto, F. Zanobini, *J. Chem. Soc., Dalton Trans.: Inorg. Chem.*, 1984, **5**, 995.
[108] M. Ciampolini, N. Nardi, P. L. Orioli, S. Mangani, F. Zanobini, *J. Chem. Soc., Dalton Trans.: Inorg. Chem.*, 1984, 2265.
[109] M. Ciampolini, N. Nardi, P. L. Orioli, S. Mangani, F. Zanobini, *J. Chem. Soc., Dalton Trans.: Inorg. Chem.*, 1985, **6**, 1179.
[110] M. Ciampolini, N. Nardi, P. L. Orioli, S. Mangani, F. Zanobini, *J. Chem. Soc., Dalton Trans.: Inorg. Chem.*, 1985, 1125.

[111] M. Ciampolini, N. Nardi, F. Zanobini, *Inorg. Chim. Acta*, 1983, L17.
[112] M. Ciampolini, N. Nardi, F. Zanobini, R. Cini, P. L. Orioli, *Inorg. Chim. Acta*, 1983, **76**, L17.
[113] C. Mealli, M. Sabat, F. Zanobini, M. Ciampolini, N. Nardi, *J. Chem. Soc., Dalton Trans.: Inorg. Chem.*, 1985, **3**, 479.
[114] S. Mangani, P. Orioli, M. Ciampolini, N. Nardi, F. Zanobini, *Inorg. Chim. Acta*, 1984, **85**, 65.
[115] Y. Morisaki, H. Imoto, K. Tsurui, Y. Chujo, *Org. Lett.*, 2009, **11**, 2241.
[116] Y. Morisaki, H. Imoto, K. Hirano, T. Hayashi, Y. Chujo, *J. Org. Chem.*, 2011, **76**, 1795.
[117] Y. Morisaki, R. Kato, Y. Chujo, *ChemistryOpen*, 2016, **5**, 325.
[118] Y. Morisaki, R. Kato, Y. Chujo, *J. Org. Chem.*, 2013, **78**, 2769.
[119] R. Kato, Y. Morisaki, Y. Chujo, *Asian J. Org. Chem.*, 2015, **4**, 1410.
[120] L. Wei, A. Bell, K.-H. Ahn, M. M. Holl, S. Warner, I. D. Williams, S. J. Lippard, *Inorg. Chem.*, 1990, **29**, 825.
[121] L. Wei, A. Bell, S. Warner, I. D. Williams, S. J. Lippard, *J Am. Chem. Soc.*, 1986, **108**, 8302.

Index

>P(O)H reagent, 45, 93, 96, 108, 123, 126, 131, 136f, 137, 165, 168, 171, 174
1,1,3,3-tetramethyldisiloxane, 189
[2,3]-sigmatropic rearrangement, 4
2,4,6-Tri(hydroxy)-1,3,5-triphosphinine, 243, 243f
2,4-diphospha-3,5-diaza-thiole, 236
2-Amino-4*H*-chromene scaffolds, 220
3-phenylpropidronate, 208, 209f, 210f
4*H*-chromen-4-yl-phosphonates, 260, 260f
4π-electron system(s), 264, 267, 268
6π-electron, 264
7-phosphanorbornene ring system, 277

activating agent, 11, 36–38, 37f, 38f, 156
addition, 1, 15–19, 15f, 16f, 17f, 20f, 21–22, 44, 93, 94, 97–100, 109, 131, 132f, 133, 134, 134f, 136, 136f, 137, 138, 139f, 141, 141f, 148, 159, 164, 168, 172, 182, 184, 187, 189, 203, 216, 218, 233, 235, 241, 244, 272, 275, 287, 296, 302
alcoholysis, 46–47
aldehyde, 91, 93–98, 95t, 98t, 100, 102, 108, 110, 112, 114, 115, 119, 121, 123, 124, 126, 126f, 128, 129, 129f, 139, 152, 154, 154f, 203, 215, 218, 219, 250, 251f, 252, 254, 259, 260
alendronate, 206, 207, 208f, 210, 211f
alkoxyphosphonium salts, 72, 72f
alkyl chloroformates, 39, 39f
alkyl halides, 41–45
allylphosphonate, 259, 260f
α-aliphatic imines, 139–140, 139f
α-aminophosphine oxides, 19, 21, 121, 122, 131, 137
α-aminophosphonate derivatives, 57–58, 115, 123, 124f, 131, 136f, 219, 251, 256, 258
α-aminophosphonate(s), 57, 58, 58f, 108–143, 109f, 113f, 114f, 120f, 121f, 122f, 123f, 132f, 138f, 148, 149, 152–156, 218–219, 219f, 244, 248, 249, 250, 250f, 251f, 252, 252f, 253t, 254, 256–258, 257f
α-aminophosphonic acids, 58f, 108, 109f, 112, 112f
α-hydroxyphosphonate(s), 91–104, 92f, 93f, 94f, 95t, 96f, 98t, 99f, 100f, 103f, 104f, 109, 203, 215, 216, 216f, 224, 248, 254, 255f, 256t, 256f
α-ketophosphonates, 98, 99f, 103, 103f, 203, 254Amberlyst-15, 255, 255f
amidation, 148, 151, 152, 152f
amino acid derivatives, 126, 127, 127f
aminophosphetes, 272
anhydride-type intermediate, 207, 208
anticancer activity of *P*-heterocycles, 59
anticancer drugs, 58, 58f
antimicrobial activities, 256
antioxidant activity, 235, 254
antistroke activity, 60
antiviral organophosphorus compounds, 59
apoptosis, 59, 201, 202
aralkylsilane, 192
Arbuzov reactions, 1, 4f, 5–7, 44–45, 69, 71f, 203
aromatic imines, 132, 135f, 140, 142, 142f
aromatic P-heterocycles, 267
aryl halides, 7, 10, 45, 46, 158, 161, 163, 165, 168
aryl phosphinates, 35, 36, 48
arylboronic acids, 8, 9f, 11, 12f, 163, 163f, 164, 165f, 167f, 168, 169f, 302
arylsilanes, 192
asymmetric, 15, 16f, 17, 18f, 30f, 31f, 66, 73, 84, 100f, 101f, 102, 102f, 131, 140, 141f, 142f, 221, 287, 300, 302
asymmetric phosphorus centers, 287
asymmetric Pudovik reaction, 18f, 101f, 102, 102f, 140
asymmetric β-boration, 221
Atherton–Todd reaction, 47, 261, 261f, 295
atomic orbital(s), 268, 279
atropoisomeric, 240, 241f
axially chiral bis(phosphine oxides), 83, 84
aza-Pudovik reaction, 131

ball-milling, 214, 215, 222
basic or acidic functional groups, 76
benzidronate, 208, 209f
benzo[*d*]thiazol-2-yldiphenylphosphine oxide, 223
benzophosphole oxides, 15, 20f, 21, 26f, 27, 67, 67f
benzothiazole, 14, 113, 120f, 123, 123f, 137, 137f, 222, 223, 223f
bidentate phosphine ligands, 130

https://doi.org/10.1515/9783110535839-016

BINOL, 12, 13f, 19, 79, 79f, 81, 82, 102, 140
bioactivity, 57, 91–93, 218
Bird, 266
Bird index, 274, 276
bis(α-aminophosphonate) derivatives, 251
bis(α-aminophosphonates), 124, 127, 130f, 154, 155, 155f, 251f
bis-macrocycle, 285, 287
bis(phospha-Mannich), 124
bisphosphine, 130, 295, 301, 302
bis(phosphine oxides), 69, 83, 84, 84f, 85f
bisphosphonates, 45, 56–57, 59, 128, 199–203, 202f, 203f, 204f, 206, 210, 258, 288
bis(pinacolato)diboron, 221
[bmim][BF4], 39, 209, 210, 211f
[bmim][PF6], 39, 40, 41t, 209, 210, 211f
borane complexes, 181, 182, 182f, 186, 236, 301
bridged *P*-heterocycle, 48
bridged phosphine oxides, 182, 187
bulky substituent, 267, 276

Ca2+ salts, 81
camphorsulfonic acid, 76, 76f, 77, 79f, 84, 97, 103, 252, 255
carbon copy, 269
catalyst, 7, 8, 9, 9f, 10f, 11, 12, 14, 18, 21, 40, 41t, 42, 44, 45, 91–100, 101f, 102, 103, 110–138, 116t–118t, 122f, 125t, 126f, 130f, 131f, 133f, 138f, 140, 158–161, 163–169, 166f, 170t, 171, 172, 172f, 174, 174f, 175f, 179, 189, 190, 190f, 191, 194, 194f, 195, 208, 214, 215–217, 216f, 218f, 219f, 220, 220f, 221, 221f, 222, 223f, 224, 244, 249, 250, 251f, 252, 253f, 254, 255, 256f, 258, 258f, 259, 299, 300
catalyst-free, 121–124, 125, 129, 130, 135–137, 137f, 174, 175f, 190f, 191, 214, 217, 218f, 224, 251f, 253f, 256f, 258f
catalytic asymmetric 1,4-addition, 302
catalytic effect, 100, 210, 211f
cesium fluoride, 217
chiral aromatic imines, 142, 142f
chiral catalyst, 102, 131
chiral chromatography, 84, 85, 85f
chiral phosphinate, 142, 142f
chiral phosphine, 32, 68, 69f, 76, 76f, 79f, 186, 186f, 187, 188f, 301
chiral phosphine oxides, 69f, 76, 76f, 79f, 186f, 187, 188f
chirality center, 100
chlorophospholene, 186, 275
click reactions, 258–259
clodronic acids, 56, 199
Co-complex, 269, 269f
complexation properties, 240, 284, 290, 291, 293, 294, 296
complexes, 59, 66, 77–84, 91, 95, 97, 102, 102f, 131, 140, 141f, 179, 181, 182f, 186, 193, 194f, 236, 238, 259, 267, 288–290, 293, 297f, 298–302
condensation reactions, 149, 287
condensing agent, 156
coupling agent, 36, 148
covalent diastereomers, 66–77
crown ether(s), 42, 284–304, 286f, 287f, 288f, 289f, 290f, 291f, 292f, 293f, 295f, 296f, 297f, 299f, 300f, 301f, 302f, 303f, 304f
C(sp2)-P cross-coupling, 222, 223f
cyanuric acid, 243
cyclic phosphinates, 46, 47
cyclic phosphinic acids, 37, 37f, 40, 42, 43f, 47, 73, 150, 151f
cyclic platinum complexes, 131
cyclobutadiene, 264, 266, 268, 269f, 270, 272
cyclophosphazene, 287, 287f
cyclo-SNPNP, 236, 237f

D-A-D (donor-acceptor-donor) functional molecules, 234
deoxygenation, 19, 130, 179–196, 181f, 182f, 183f, 184f, 185t, 188f, 275
destabilizing effects, 266
dialkyl phosphite(s), 7, 45, 91, 93–97, 95t, 98, 108–110, 113–115, 122–127, 131, 133, 134, 134f, 136, 137, 139, 139f, 140, 140f, 141f, 158, 159, 161, 163, 164, 168, 172, 203, 216, 218, 249–252, 254, 255
dialkyl phosphites, 7, 45, 91, 93–98, 108–110, 113, 114, 122–127, 131, 133, 134, 134f, 136, 137, 139, 139f, 140, 140f, 141f, 158, 159, 161, 163, 164, 168, 172, 203, 216, 218, 249–252, 254, 255
diarylphosphinic acid, 291, 292f
diarylsilanes, 192
diastereomers, 15, 66–75, 73f, 76, 77, 79–84, 142, 188, 298, 302
diastereoselective synthesis, 123, 123f
diazadiphosphapentalene, 235, 236, 236f, 237f
diazaphosphocane, 244, 245f
diazaphosphonane, 244, 245f

Diels–Alder dimerization, 276
diethyl phenylcarbamoylphosphonates, 259
dihydrophosphinoline-1-oxides, 238
dihydrophosphinolines, 238, 238f
diphenylphosphine oxide, 5, 8, 9–12, 17–19, 21, 24, 136, 163, 165, 167, 169, 171, 172, 174, 189, 223, 238
diphosphetanium, 233, 233f
diselenaphosphepine, 245f, 246
distorted arrangement, 279
double Arbuzov reaction, 5, 6f, 7f
double Kabachnik–Fields, 19, 108, 124–131
driving force, 152, 179, 267, 277, 278
dronic acid derivatives, 199–211, 201t, 207f, 210f
dynamic resolution method, 71

electroreduction, 195, 195f
enantioselective, 12, 15f, 66, 102, 221, 221f, 291
enantioselectivity, 17, 100
esterification, 35–38, 37f, 38f, 39, 39f, 40–44, 40f, 41f, 42f, 43f, 44f, 47, 48, 148–152, 150f, 151t, 151f
estrone, 240, 241f
ethyl phenyl-*H*-phosphinate, 79, 123, 136, 163
ethylenediamine diacetate, 220, 220f
etidronic acid, 56, 200

fenidronate, 206
FLP template, 242, 242f
fluorapatite, 216
fluorescent aminophosphonate, 135
four-membered transition state, 193
frustrated phosphorus-boron (P/B) Lewis pairs (FLP), 235
fungicides, 57, 60, 62–63, 152

green chemical, 94, 110, 124, 158, 168–175, 190, 222, 224
green chemistry, 96, 158, 214, 248
Grignard reagents, 27, 27f, 28, 30, 30f, 31, 32, 99, 275

herbicides, 57, 60, 61, 91, 152
heterocyclic α-aminophosphonates, 123
heterogeneous catalyst, 115, 214–217, 219, 224, 252
hexachlorodisilane, 188, 188f
Hirao reaction, 7–15, 8f, 9f, 13f, 14f, 45–46, 45f, 46f, 158–, 159f, 160f, 161f, 162f, 163f, 164f, 166f, 167f, 168f, 169f, 170f, 171f, 173f
H-phenylphosphinic acid, 38
H-phosphinates, 39, 46–48, 49f, 73, 74, 74f, 75, 75f, 79, 85f, 122, 122f, 123, 126, 131, 136, 141, 142f, 158, 159, 161, 163, 165, 172, 174, 254
hydride migration, 182
hydrogenation, 47, 179, 195, 266
hydrolysis, 23, 35, 47, 67, 68, 71–73, 99, 149, 203–205, 207, 235, 284, 291
hydrophosphination, 180, 183
hydrophosphinylation, 21
hydrophosphonylation, 133, 133f, 138, 140, 140f, 142f
hydroxy-bisphosphonic acids, 200, 201
hydroxymethyl-*H*-phosphinate, 74, 75, 75f

ibandronate, 206, 207, 208f, 210, 211f
imines, 18f, 19, 19f, 108, 109, 131–142, 132f, 133f, 134f, 135f, 136f, 137f, 138f, 139f, 140f, 142f, 152, 235, 256, 257f, 258
iminophosphonate, 287, 288, 288f
iminophosphorane diastereomers, 68
inorganic phosphates, 55, 56, 93
insecticides, 60–62, 61f, 62f, 91, 152
intramolecular cyclization, 4, 15
ionic liquids (ILs), 39, 40, 41f, 94, 99, 110, 119, 120, 120f, 128, 179, 199, 208–210, 210f, 219, 219f
isocyanates, 259
isoxazole moiety, 250, 251f

Kabachnik–Fields condensation, 19, 108, 114, 115, 121, 123, 124f, 126, 128, 148, 248–254
Kabachnik–Fields, 19, 19f, 21, 108–142, 109f, 111t, 113f, 114f, 116t–118t, 119f, 120f, 122f, 123f, 125f, 125t, 126f, 127f, 127t, 129f, 130f, 148, 152, 154f, 218–219, 248–254, 249t
Kabachnik–Fields reaction, 19f, 21, 108–142, 109f, 111t, 113f, 114f, 116t–118t, 119f, 120f, 122f, 123f, 124f, 125f, 125t, 126f, 127f, 127t, 129f, 130f, 152, 154f, 218–219, 249, 249t, 251, 252–254
Kabachnik–Fields reaction of ketones, 251

leaving group, 36, 165, 203, 207
Lewis acid, 113, 181, 235, 295
ligand exchange, 159
lithiation, 21f, 22–27, 22f, 23f, 24f

macrocycle(s), 284, 285, 287–289, 290, 290f, 291, 292f, 293, 293f, 294, 294f, 295, 296, 298, 299, 300–302
macrocyclization, 290, 294, 295
magnetic nanoparticles, 94, 120
McCormack cycloaddition, 46, 47
mechanism, 17, 60, 63, 109, 152, 153f, 154f, 159, 159f, 169, 182, 182f, 187, 193, 194, 199, 204–206, 207f, 208, 209f, 217, 218f, 238, 242
mechanochemistry, 222–223
menthyl chloroformate, 68
menthyl phenyl-*H*-phosphinate, 73, 74, 74f, 141, 142f
menthyl phoshinates, 72
metal hydrides, 181–182
metal triflates, 110
methanesulfonic acid (MSA), 110, 115, 205, 205f, 206, 207, 207f, 208, 209f, 210f, 240
methylpolysiloxane, 189
MgO-nanocatalyst, 220
Michael addition, 15–19, 15f, 16f, 17f, 218
Michaelis–Arbuzov reaction, 5, 44–45, 44f, 69, 71f
Michaelis–Arbuzov type, 217
microwave-assisted addition, 18
Mitsunobu reaction, 38, 38f, 179
monoalkylation, 43
MW, 16f, 35, 37, 38, 40–41, 40f, 42, 45, 94, 97, 110, 115, 119–124, 119f, 122f, 126–130, 130f, 131f, 137, 137f, 150, 156, 168, 168f, 169, 169f, 171, 171f, 172, 173f, 174f, 187, 190–192
MW-assisted, 16f, 37, 40–41, 40f, 45, 115, 121, 122, 122f, 123, 124, 126, 127, 129, 130, 131f, 137, 137f, 168, 169f, 171, 171f, 172, 173f, 190–192
MW-assisted Kabachnik–Fields reaction, 122
MW irradiation, 40, 42, 94, 97, 119, 119f, 128, 150, 156, 168, 171, 172, 187

nano-TiO2, 120, 219
nerve agents, 53, 63–64, 64t, 64f
Ni catalysts, 11, 163–168, 171
NICS values, 268, 270, 274
N-ligand, 163, 165, 167f
non-aromatic character, 270, 273

O,O'-dibenzoyl tartaric acid, 83, 84f, 85f
optically active phosphorus acid derivatives, 47
optically active α-hydroxyphosphonates, 100–104
optically pure crown ether, 299
optimal amounts, 209
organophosphorus pesticides, 60–63, 64t
osteoclasts, 200, 201, 202f
osteoporosis, 56, 200
oxadiazaphosphininyl-acetonitrile, 241
oxalyl chloride, 68, 71, 195, 196
oxaphosphetes, 272
oxaphosphinines, 30f, 31, 171, 240, 240f
oxaphosphocine, 244, 245f
oxaphosphopine, 244, 245f
oxazaphosphinane derivatives, 252
oxazepinephosphonates, 258, 258f
oxidation, 1, 22, 48, 99, 103, 104f, 241, 275, 285, 294, 298, 299, 301
oxidative addition, 159
oxidative P–C coupling, 14, 14f, 15f
oxo compound, 93, 97, 108, 248

Paget-disease, 200
palladium pincer complex, 27
pamidronate, 206
P-arylation, 217
P/B FLPs, 236f
P–C coupling, 7, 9, 10f, 11f, 12f, 14, 45f, 158–175, 165f, 166f, 167f, 169f, 170f, 171f, 172f, 173t, 174f, 175f
P-chemistry, 214–224
P-chiral phosphine oxides, 32, 69f, 76f, 79f, 186f
Pd/C catalyst, 171
phase transfer (PT) catalyst, 42, 43f, 158, 168–171, 170f, 170t, 179
phase transfer catalytic techniques, 168
phenyl-3-phospholene oxide, 186
phenyl-dibenzophosphole oxide, 67, 67f
phenylsilanes, 183–184, 183f, 184f, 193, 193f
phospha-Brook rearrangement, 215
phospha-Mannich, 19, 108, 218
phospha-Michael reactions, 15, 16f, 17
phosphaphenalenes, 237, 238f
phosphate, 1, 19, 37, 53, 55, 56, 56t, 59, 61, 62, 62f, 91, 93, 94, 165, 200, 202, 203, 261, 270f, 284–287, 285f, 286f
phosphate monoesters, 37
phosphate salts, 56, 56t
phosphaylide, 270
phosphetes, 268–273

phosphinates, 35–49, 36f, 44f, 46f, 47f, 48f, 49f, 66, 72, 75, 79f, 80f, 81f, 86, 149, 150, 161, 162f, 217–218, 222, 224
phosphine, 1–32, 8f, 12f, 17f, 23f, 31f, 47, 59, 63, 66–86, 69f, 70f, 71f, 73f, 76f, 79f, 80f, 81f, 84f, 85f, 93, 99, 108, 123, 124, 130, 131, 158, 159, 162f, 163, 174, 175, 179–196, 181f, 182f, 183f, 184f, 185t, 186f, 187f, 188f, 189f, 196f, 217–218, 217f, 221, 221f, 222, 224, 239, 242, 242f, 288, 293–294, 293f, 299–304, 300f, 301f, 302f, 303f
phosphine oxide boronates, 221, 221f, 224
phosphine oxides, 1–32, 26f, 30f, 31f, 38, 66–86, 70f, 72f, 80f, 81f, 93, 123, 131f, 162f, 179–197, 185t, 186, 187f, 189f, 190f, 192f, 195f, 196f, 217, 217f, 222, 239, 288, 289, 290f, 293–294, 293f, 299–302
phosphine–boranes, 181, 182
phosphines, 22, 27, 66, 69, 71, 72f, 179–181, 183, 186, 187, 189, 238, 271, 299–304
phosphinic chlorides, 35–36, 72, 73, 75
phosphinic esters, 75
phosphininofurans, 238, 239f
phosphinite, 69, 187, 296–299, 297f
phosphite, 7, 38–40, 44–45, 91, 93–98, 95t, 98, 98t, 108–110, 112–115, 114f, 119–129, 122f, 129f, 131–140, 132f, 134f, 135f, 139f, 140f, 141f, 152, 154, 154f, 155, 158–161, 163, 164, 168, 171, 172, 173t, 174f, 203, 215, 216, 218, 219, 220, 235, 241, 249–256, 253t, 256f, 257f, 258–261, 296–299, 297f, 299f
phosphole oxide dimers, 276
phosphole ring, 273
Phosphonates, 7, 28f, 35–49, 38f, 44f, 47f, 48f, 155, 156f, 158, 183, 184f, 217, 220, 221f, 222
phosphonic ester-chloride, 35
phosphonite(s), 45, 287, 291, 296–299, 297f, 298f
phosphonium-iodonium, 238
phosphonium salts, 4, 11, 44, 46–47, 46f, 66, 68, 69, 70f, 72, 72f, 76, 100, 101f, 154, 179, 195–196, 205, 270, 274, 295–296, 301
phosphoramidite, 296, 297f, 299, 300f
phosphorane, 66, 272, 273, 295–296
phosphorotropic migration, 3
phosphorous acid, 38, 39, 39f, 204, 205, 208, 210
phosphorus in the human body, 54
phosphorus trichloride, 4, 204–206, 207, 207f, 208, 210, 236
phosphorus trihalides, 46
pillar [5]arene-based phosphine oxides, 7
platinum complexes, 131, 186
P-ligand-free, 158, 171, 172, 174
P-ligand-free Hirao reaction, 12, 171–175, 173f
P-ligands, 160, 161, 163, 168, 171, 174, 174f, 179
polymethylhydrosiloxane, 189
precursor, 12, 19, 27, 45, 48, 77, 130, 131f, 158, 160–163, 160f, 217, 254, 271, 299–301
propylphosphonic anhydride (T3P®), 37, 128, 148, 149
proton-ionizable, 284
PS-BEMP, 215, 216f
P-stereogenic, 66
Pudovik, 248
Pudovik reaction, 18f, 19–21, 19f, 91, 93, 95, 97, 98, 100–102, 101f, 102f, 108–142, 132f, 133f, 134f, 135f, 136f, 137f, 138f, 139f, 140f, 141f, 215, 216, 254–258
pyrazolyloxadiazaphosphorine, 241, 241f

quaternary ammonium salt, 111

reactive intermediates, 37f, 151, 267, 276
reduction, 7, 165, 174, 179–192, 181f, 186f, 187f, 188f, 189f, 190f, 191f, 192f, 193–197, 193f, 194f, 195f, 216, 299–301
reductive amination, 42
reductive elimination, 159
resolution, 27, 66–86, 67f, 72f, 76f, 77f, 78f–79f, 80f–81f, 82f, 83f–84f, 103f
resonance structures, 270, 270f, 278f, 279
retention of configuration, 186–188, 193, 194
ring-opening polymerization, 291, 292f, 298, 298f
ring phosphine oxides, 181–183, 181f, 184f
ring size, 184, 184f, 264, 288, 302
ring system(s), 232, 240, 243, 277
risedronic acid, 205, 205f
(*R*)-menthyl-phenyl-*H*-phosphinate, 141, 142f
rodenticide, 60, 63, 63f

Schiff base, 19, 102, 109, 131–136, 138–140, 139f
secondary phosphine oxide(s), 7, 8, 8f, 10–12, 12f, 17–19, 23, 23f, 24, 47, 71f, 74, 81, 82, 108, 124, 125, 130, 131, 158, 159, 163, 174, 175, 182, 182f, 183, 183f, 218
Shiff-base, 256

solvent, 35, 37, 80, 91, 94, 96, 97, 110, 111t, 115, 122f, 123–125, 131, 135, 138f, 148, 149, 158, 160, 161, 172, 183, 184, 190, 191, 195, 204–210, 207f, 217, 222, 251, 252, 256, 259, 290, 294
solvent-free, 42, 93, 94, 96, 97, 114–121, 116t, 122f, 123, 124, 126, 126f, 127, 129, 130f, 134, 137, 138, 158, 171, 179, 191, 192, 205, 215, 216, 216f, 219, 219f, 222, 223f, 250, 252, 255, 256, 258, 259, 260
stereogenic phosphorus atoms, 180, 302
sterically hindered phospholes, 275, 276
story of phosphorus, 53
substituted salicylic aldehydes, 250
substitution reactions, 81, 180, 203, 244
sulfamic acid, 255
sulfolane, 205, 207, 208f, 209f

T3P, 37, 37f, 38, 128, 129f, 148–156, 149f, 150f, 151, 151f, 151t, 152–156, 152f, 153f, 154f, 155f, 156, 156f
T3P® reagent, 37, 37f, 38, 148–156, 149f, 151, 151f, 151t, 152f, 154f, 155f, 156
TADDOL, 73, 80, 80f
tautomeric equilibrium, 23, 174
tautomerization, 161
terminal alkynes, 44
tert-butylphenylphosphine oxide, 81, 82, 82f, 83, 83f
tetrakis(triphenylphosphine)palladium, 45, 158, 168
tetraphospha-crown ethers, 302
thiophosphetes, 272, 273f
thiophosphonate, 287, 288, 288f, 289, 289f, 290f, 291, 292f, 296
three-component condensation, 108, 110, 113, 119, 128, 129, 218, 220, 248, 252
three-component reactions, 115, 125, 126, 152, 249, 250, 252, 259, 260
tiludronic, 56, 199
titanium tetrachloride, 219
TMSCl-mediated protocol, 258
torsion angle, 278
transition metal(s), 5, 44, 45, 163, 171, 179, 195, 214, 221, 232, 267, 299
transition states, 193, 266
trialkyl phosphite(s), 38–40, 44–45, 93, 97–98, 98t, 114, 114f, 203, 252–254
trialkylphenylphospholes, 276, 279
triarylphosphine, 27, 28f, 29f, 31, 68, 164f, 167, 169, 187, 294, 294f, 300, 301f, 302, 304f
triarylphosphine oxides, 27, 28f, 29f, 31, 68, 164f, 167, 169, 187, 294, 294f
trichlorosilane, 184–187, 185t, 186f, 238
triethylbenzylammonium chloride, 42, 169
trifluoroacetic acid, 112
triphenyl-dichlorophosphorane, 195
tritelluratriphosphorinanes, 243, 244f
trivalent species, 48

ultrasonic irradiation, 97, 110, 248–252, 254, 255, 256, 258, 259, 261
ultrasound-mediated, 134, 248
urea, 123, 252

vinylphosphine oxide, 18, 70
Woolins' reagent, 244
work-up, 91, 94, 96, 97, 98, 216, 222

ylides, 222, 223f, 224, 238

zinc oxide nano-sheets, 254
zoledronic acid, 199, 205

Kolibri 360 GmbH
Ferdinand-Jühlke-Straße 7
99095 Erfurt, Deutschland
produktsicherheit@kolibri360.de